BACTERIAL
ENERGY
TRANSDUCTION

BACTERIAL ENERGY TRANSDUCTION

Edited by

CHRISTOPHER ANTHONY

Department of Biochemistry
University of Southampton
Southampton SO9 3TU
England

1988

Academic Press

Harcourt Brace Jovanovich, Publishers

London   San Diego   New York   Berkeley   Boston
Sydney   Tokyo   Toronto

ACADEMIC PRESS LIMITED
24/28 Oval Road
LONDON NW1 7DX

United States Edition published by
ACADEMIC PRESS INC.
San Diego, CA 92101

British Library Cataloguing in Publication Data

Bacterial energy transduction
1. Microorganisms. Bioenergetics
I. Anthony, C.
576′.119121

ISBN 0-12-058815-3

Typeset by Bath Typesetting Ltd, Bath
Printed in Great Britain by the St Edmundsbury Press, Bury St Edmunds

Contributors

Christopher Anthony   Department of Biochemistry, University of Southampton, Southampton SO9 3TU, UK

Ian R. Booth   Department of Genetics and Microbiology, Marischal College, University of Aberdeen, Aberdeen AB9 1AS, UK

Stuart J. Ferguson   Department of Biochemistry, University of Oxford, South Parks Road, Oxford OX1 3QU, UK

W. Alan Hamilton   Department of Genetics and Microbiology, Marischal College, University of Aberdeen, Aberdeen AB9 1AS, UK

J. Baz Jackson   Department of Biochemistry, University of Birmingham, PO Box 363, Birmingham B15 2TT, UK

Colin W. Jones   Department of Biochemistry, University of Leicester, Leicester LE1 7RH, UK

Douglas B. Kell   Department of Botany and Microbiology, University College of Wales, Aberystwyth, Dyfed SY23 3DA, UK

Robert K. Poole   Department of Microbiology, King's College London Kensington Campus, Campden Hill Road, London W8 7AH, UK

Paul M. Wood   Department of Biochemistry, University of Bristol Medical School, University Walk, Bristol BS8 1TD, UK

To Peter Mitchell

Preface

For more than a decade the starting point for any prospector in the field of bacterial bioenergetics has been the SGM Symposium (No. 27) on 'Microbial Energetics' edited by Allan Hamilton and Bruce Haddock. The meeting of the Society for General Microbiology whose discussions are published in the Symposium volume was memorable and stimulating. I should like to quote from the Editors' preface:

'For many years the controversy over the mechanism of ATP synthesis in mammalian mitochondria has held the centre of the energetic stage. Increasingly, however, the great range of energy conservation mechanisms found among micro-organisms is capturing the imagination and holding the attention of experimenters. These studies both extend our knowledge of energy transduction in the living cell, and strengthen our understanding of the intricacies of microbial metabolism'. The Editors then expressed their belief that the Symposium volume would be 'timely and stimulating to a wide spectrum of teachers and researchers'.

The range and extent of this book on Bacterial Energy Transduction should provide them with a satisfying confirmation that their belief was justified.

The book was conceived with a view to providing a link between the available elementary introductions to the subject (e.g. by Dawes, by Jones, and by Nicholls) and the complexities of the research literature. The authorship was determined by the need to have internationally-acknowledged experts in each field. The fact that they are all British was determined by the need to coordinate their efforts without geographical hindrance and by the encouragement given me when thinking about the project by the other members of the Bioenergetics committee of the Biochemical Society in the U.K.

The book aims to provide complete coverage of all aspects of bacterial energy transduction, starting at the 2nd year undergraduate level all the way up to the frontiers of each developing aspect of the subject.

The first Chapter provides an extensive introduction to all aspects of bacterial energy transduction; it provides a complete overview while being sufficiently detailed to act as a self-contained introductory text on the subject. It is generously littered with cross references to more advanced treatments later in the text; so, in effect each part of the Introduction can also be used as an introduction to each of the separate later Chapters.

The second Chapter covers energy transduction in anaerobes, emphasizing the metabolic and ecological cross-relationships between the varied anaerobic microbes. The Chapter includes the long-established fermentation pathways, anaerobic electron transport, and the rapidly expanding field of energy transduction in methanogens and acetogens.

Chapter three is a brief survey of periplasmic electron transport systems, the appreciation of the importance of which has markedly increased in the last ten years.

Chapter four is a comprehensive account of the diverse range of bacteria that constitutes the chemolithotrophs – bacteria obtaining energy from the chemical oxidation of inorganic molecules. This deals extensively with the proteins involved in energy transduction, and the ways in which they are arranged in order to establish a protonmotive force and hence ATP and NADH synthesis.

Chapter five starts with a survey of what is known about the mitochondrial cytochrome oxidase and then goes on to describe in detail the structure and function (redox and chemiosmotic) of all the bacterial cytochrome oxidases. Because the complexity of spectroscopic techniques can cause problems to anyone starting work in this field, considerable attention is paid to the use of these techniques.

Chapter six is a small Chapter dealing with the quinoproteins – dehydrogenases that have the prosthetic group pyrrolo-quinoline quinone (PQQ). The prosthetic group is described and this is followed by an account of the dehydrogenases and the novel ways in which these are coupled to the electron transport chains and the establishment of a protonmotive force.

Chapter seven is a comprehensive account of bacterial photosynthesis, with special emphasis on the significance of the electrogenic reactions catalysed by the membrane protein complexes. A major aim in this Chapter has been to ensure that those whose knowledge of energy transduction has been obtained during studies of respiration should not have any developing interest in photosynthesis quenched by unexplained esoteric language or experimental design.

Chapter eight is the only Chapter that deals exclusively with the utilization of energy, as it deals with all aspects of energetics and mechanisms in bacterial solute transport.

Chapter nine concludes the book and is designed to be, amongst other things, a final provocation. Whether it provokes agreement, wrath or amusement will depend to some extent on the initial viewpoint of the reader; in all cases, however, it should provoke questioning and wariness in designing and interpreting experiments. Its title is 'Protonmotive energy transducing systems: some physical principles and experimental approaches'. It includes introductions to all those aspects of bioenergetics that may sometimes be relevant but are often avoided; it covers elementary thermodynamics, non-equilibrium thermodynamics, transition state theory, metabolic control theory and Hill diagrams. All previous Chapters of the book, when discussing function and mechanism, have been based (implicitly or explicitly) to a large extent on Mitchell's chemiosmotic hypothesis. The final part of the final Chapter considers the important question – "Is the protonmotive force an energetically-significant intermediate in electron transport-driven phosphorylation?" Needless to say, the importance of the discussion in this context is not the author's conclusions but the challenge constantly to keep in mind the importance of distinguishing intitial assumptions and final conclusions.

Acknowledgements

As Editor I must first acknowledge my gratitude to my friend C. Rajama-nickam, at whose Symposium in Madurai I decided to attempt to produce this book. This must be followed by acknowledging the encouragement of my friends on the bioenergetics group committee of the Biochemical Society (U.K.) and also their willingness to exchange the role of friend for that of author. I trust that the process is reversible.

Acknowledgements for each chapter are as follows;

Chapter 1: We are indebted to Alex Cornish for critically reading the manuscript.

Chapter 2: The following Figures are reproduced with permission: Fig. 2.11 is from Thauer, R. K. and Morris, J. G. (1984) SGM Symp. 36, 123–168. Fig. 2.14 is from Blaut and Gottschalk, G. (1984). Eur. J. Biochem. 141, 217–222. Fig. 2.15 is from Cord-Ruwisch, R., Kleinitz, W. and Widdel, F. (1987). J. Petro. Techno. Jan., 97–106. Fig. 2.16 is from Odom, J. M. and Peck, H. D. (1981). FEMS Microbiol. Lett. 12, 47–50. Fig. 2.17 is from Stams, A. J. M., Kremer, D. R., Nicolay, K., Weenk, G. M. and Hansen, T. A. (1984). Arch. Microbiol. 139, 163–167. Fig. 2.18 is from Kroger, A. (1987). SGM Symp. 27, 61–93. Fig. 2.19 is from Cole, S. T., Condon, C., Lemire, B. D. and Weiner, J. M. (1985). Biochim. Biophys. Acta 811, 381–403. Fig. 2.20 is from Macy, J. M., Schroder, I., Thauer, R. K. and Kroger, A. (1986). Arch. Microbiol. 144, 147–150.

Chapter 3: S. J. Ferguson thanks the SERC for supporting the work reported from his laboratory.

Chapter 5: R. K. Poole thanks J. A. M. Hubbard, M. N. Hughes and especially H. D. Williams for their contributions to the work from his laboratory, and he thanks J. A. Fee and B. Ludwig for access to unpublished typescripts. The following Figures were reproduced with permission (refer-ences at end of Chapter): Fig. 5.3 from Poole *et al.* (1982a). Fig. 5.4 from Poole *et al.* (1979b). Fig. 5.7 from Poole *et al.* (1983a). Fig. 5.8 (with modification) from Poole *et al.* (1986).

Chapter 8: The following Figures were reproduced with permission (refer-ences at end of chapter): Fig. 8.1 from Postma and Lengeler (1985). Fig. 8.3 (modified) from Higgins *et al.* (1986). Fig. 8.5 from Higgins *et al.* (1988). Fig. 8.6 (modified) from Ames, G. F-L. 'Bacterial Periplasmic Transport Systems:

structure, mechanism and evolution' in Annual Reviews of Biochemistry, 1986. 55, 397–425. Fig. 8.8 from Padan *et al.* (1983). Fig. 8.12 from Kashket and Wilson (1973). Fig. 8.13 (data for) from Hirata *et al.* (1974). Fig. 8.15 from Overath and Wright (1983). Fig. 8.16 (data for) from Ghazi and Schechter (1981). Fig. 8.17 (data for) from Ahmed and Booth (1981). Fig. 8.18 from Kaback (1987).

Finally, I should like to thank Gina Fullerlove and Nicki Dennis of Academic Press for their encouragement and hard work.

Christopher Anthony

Contents

Contributors v
Dedication vii
Preface ix
Acknowledgements xiii

1   Membrane-Associated Energy Conservation in Bacteria:
    a General Introduction
    C. W. JONES 1
2   Energy Transduction in Anaerobic Bacteria
    W. A. HAMILTON 83
3   Periplasmic Electron Transport Reactions
    S. J. FERGUSON 151
4   Chemolithotrophy
    P. M. WOOD 183
5   Bacterial Cytochrome Oxidases
    R. K. POOLE 231
6   Quinoproteins and Energy Transduction
    C. ANTHONY 293
7   Bacterial Photosynthesis
    J. B. JACKSON 317
8   Bacterial Transport: Energetics and Mechanisms
    I. R. BOOTH 377
9   Protonmotive Energy-Transducing Systems: Some Physical
    Principles and Experimental Approaches
    D. B. KELL 429

Index 491

1 Membrane-associated energy conservation in bacteria: a general introduction

C. W. JONES

1.1 Introduction 2
1.1.1 Bacterial energy metabolism 2
1.1.2 Thermodynamics and redox potentials; basic concepts 3
1.1.3 Respiration 4
1.1.4 Photosynthetic electron transfer 9
1.1.5 Membrane-associated energy conservation 11
1.1.6 Generation of the protonmotive force 15
1.1.7 Manipulation of the protonmotive force 19
1.1.8 Utilization of the protonmotive force 20
    1.1.8.1 ATP synthesis 21
    1.1.8.2 Secondary solute transport 22
    1.1.8.3 Reversed electron transfer 24
1.1.9 The energy-coupling membrane 26
1.2 Redox components of respiratory chains and
photosynthetic electron transfer systems 27
1.2.1 Nicotinamide nucleotide transhydrogenase 27
1.2.2 Flavoproteins 28
    1.2.2.1 Flavoprotein dehydrogenases 28
    1.2.2.2 Flavodoxin 30
1.2.3 Iron–sulphur proteins 30
    1.2.3.1 Hydrogenases 31
    1.2.3.2 Ferredoxins, HIPIP proteins and molybdo-
             ferredoxins 32
    1.2.3.3 Rubredoxin 32
1.2.4 Iron–sulphur flavoproteins 32
    1.2.4.1 NADH dehydrogenase 32
    1.2.4.2 Succinate dehydrogenase 33
    1.2.4.3 Fumarate reductase 34
1.2.5 Molybdoproteins 35
    1.2.5.1 Carbon monoxide oxidoreductase 36

        1.2.5.2   Formate dehydrogenase 37
        1.2.5.3   Nitrite oxidoreductase 38
        1.2.5.4   Nitrate reductase 38
        1.2.5.5   DMSO/TMAO reductase 39
    1.2.6   Quinoproteins 40
    1.2.7   Quinones 42
    1.2.8   Cuproproteins 45
    1.2.9   Cytochromes 46
        1.2.9.1   Non-autoxidizable cytochromes 48
        1.2.9.2   Cytochrome oxidases 50
        1.2.9.3   Nitrite reductases 52
    1.2.10 Bacteriochlorophylls and bacteriopheophytins 53
    1.3   Respiratory chains and photosynthetic electron transfer
        systems 57
    1.3.1   Sequential organization 57
    1.3.2   Spatial organization 62
    1.3.3   Proton translocation 64
    1.3.4   ATP/$2e^-$ quotients and growth yields 71
    1.4   The ATPase/ATPsynthase complex 73
    1.5   ATP synthesis 76
    References 78

1.1   Introduction

It is a particularly happy coincidence that the contents of this book have mainly been written in 1987, the centenary of the birth of David Keilin (1887–1963), whose classical work on cytochrome spectra and the nature of cellular respiration (Keilin, 1925, 1929; see also Keilin, 1966) laid the foundations of our current understanding of membrane-associated energy conservation in bacteria.

Unfortunately, there is insufficient space within the confines of this chapter to describe the successive developments which have occurred in the last 60 years or so, and which have thus shaped that current understanding. A brief outline only is possible: a resumé of underlying principles and basic facts; a flavour of the impressive ingenuity of bacterial redox systems and a glimpse of their massive diversity. More detailed reading can be found in several recent monographs (Prebble, 1981; Jones, 1981, 1982; Nicholls, 1982; Dawes, 1986; Harold, 1986), in the predominantly review references given at the end of this chapter and in the other contributions to this book.

1.1.1   Bacterial energy metabolism

Bacteria can be broadly divided into four major groups, according to their sources of carbon, energy and reducing power for growth: chemoorganotrophs (chemoheterotrophs), chemolithotrophs, photoorganotrophs (photoheterotrophs) and photolithotrophs (Table 1.1).

Table 1.1 Classification of bacteria according to their sources of carbon, energy and reductant for growth

Type of organism	Carbon source	Energy source	Reductant
Chemoorganotroph	Organic compounds		
Chemolithotroph	CO_2	Inorganic compounds	
Photoorganotroph	Organic compounds	Light	Organic compounds
Photolithotroph	CO_2	Light	Inorganic compounds

Bacteria using CO_2 as their sole source of carbon are also called autotrophs. Bacteria using organic carbon sources are heterotrophs; those heterotrophs able to use reduced C_1-compounds are methylotrophs.

Chemoorganotrophs grow on a wide range of reduced carbon substrates, including carbohydrates, alcohols and organic acids, whereas photoorganotrophs can use only a very restricted range of these substrates. Chemolithotrophs and photolithotrophs assimilate carbon dioxide at the expense of inorganic reductants and, hence, are also classified as autotrophs, but chemolithotrophs use a much wider range of inorganic compounds than photolithotrophs. Chemoorganotrophs conserve energy by either fermentation (substrate-level phosphorylation) or respiration (oxidative phosphorylation); photoorganotrophs predominantly conserve energy by photosynthetic electron transfer (photophosphorylation), although a few species can also carry out fermentation and/or respiration. Chemolithotrophs and photolithotrophs are generally restricted to oxidative or photophosphorylation. This chapter deals primarily with these processes. They are dealt with in more detail in later chapters. The process of fermentation (and anaerobic respiration) is covered in detail in Chapter 2.

1.1.2 Thermodynamics and redox potentials: basic concepts

Respiration and photosynthetic electron transfer are the thermodynamically spontaneous transfer of reducing equivalents (H, H^- or e^-) from a reductant to an oxidant via a series of predominantly membrane-associated redox components. This process is accompanied by the release of free energy, the amount of which is determined by the difference between the redox potentials of an exogenous or endogenous donor couple (D/DH_2) and acceptor couple (A/AH_2). The free energy released is initially conserved as an electrochemical potential difference of H^+, the protonmotive force, which is subsequently used to drive various energy-dependent membrane reactions, including ATP synthesis, reversed electron transfer and certain types of solute transport. In some organisms it is also responsible for motility, sensory perception and the synthesis of inorganic pyrophosphate.

Redox potentials of donors and acceptors are most commonly expressed as standard or midpoint values (E_o' or E_m; mV) measured at 25°C and pH 7.0 on a scale determined by the redox potential of the hydrogen electrode. The difference in redox potential ($\Delta E_o'$ or ΔE_m) between the donor and acceptor couples determines the standard Gibbs free energy change of the overall redox reaction ($\Delta G^{0'}$):

$$-n.\Delta E_m = \Delta G^{0'}/F$$

where n is the number of electrons transferred, and F is the Faraday constant which is used to convert free energy change (kJ.mol^{-1}) into midpoint potential units (mV). However, these standard values are at best crude, and at worst misleading, approximations of the actual values (E_h, ΔE_h and ΔG) since the latter can be significantly altered by the non-standard conditions of concentration, temperature and pH that can pertain experimentally and *in vivo*. Nevertheless, despite these and other limitations, classical equilibrium thermodynamics of this type can be used to predict the direction, but not the rate, of a redox reaction (or of any other bioenergetic reaction including ion transport and ATP synthesis) when that reaction is disturbed from equilibrium. Very importantly, however, information on both the direction and rate of a reaction can be obtained via the use of non-equilibrium (irreversible) thermodynamics since, under near-equilibrium conditions, the rate or flux of a reaction (J) is linearly related to the affinity of the reaction (X; equivalent to $-\Delta G$):

$$J = L.X$$

where L is the coefficient (or proportionality constant) of the reaction. It is beyond the scope of this chapter to discuss either type of thermodynamics in more detail, and the reader is referred to more extensive treatments elsewhere (see Westerhoff and Van Dam, 1979; Nicholls, 1982; Stucki, 1983; Kell, Chapter 9, this volume).

1.1.3   Respiration

In its simplest form, a bacterial respiratory chain consists of a primary dehydrogenase (or very occasionally a 'de-electronase') and a terminal reductase (which is called an oxidase when the oxidant is molecular oxygen). The two enzymes are linked by a single non-enzymic redox carrier such as a quinone, cytochrome or cuproprotein. The composition of the primary dehydrogenase (flavoprotein, iron–sulphur protein, molybdoprotein and/or quinoprotein) and the terminal oxidase or reductase (flavoprotein, iron–sulphur protein, molybdoprotein, cytochrome or cuproprotein) reflects both the chemical structures and the redox potentials of the donor and acceptor couples.

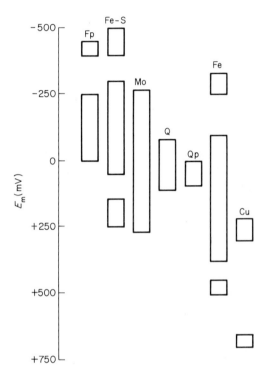

Fig. 1.1 The midpoint redox potentials (E_m) of organic and metal redox centres present in bacterial redox systems. Abbreviations: Fp, flavoproteins; Fe–S, iron-sulphur proteins; Mo, molybdoproteins; Q, quinones; Qp, quinoproteins; Fe, cytochromes; Cu, cuproproteins ('cupredoxins').

Organic substrates react directly with organic redox centres (flavoproteins and quinoproteins), since the latter are generally concerned only with the transfer of H or H^-. As the redox potentials of the organic centres are relatively low and fixed, they are found principally in primary dehydrogenases and appear only rarely in terminal oxidases and reductases. In contrast, inorganic substrates react directly with metal centres (iron–sulphur proteins, cytochromes, cuproproteins and molybdoproteins) which predominantly transfer electrons. Furthermore, the transition metal centres (Fe, Cu,

Mo) exhibit extremely variable redox potentials since they can exist in several different oxidation states, depending on the exact molecular environment in which they are placed. Iron–sulphur proteins and molybdoproteins are thus commonly found in both primary dehydrogenases and terminal reductases, and cytochromes can be found in most regions of respiratory chains, but cuproproteins are largely restricted to the high redox potential end (Fig. 1.1).

The presence of both organic and metal centres in bacterial redox systems is probably extremely important for membrane-associated energy conservation. First, it enables the rapid transfer of reducing equivalents, since the redox potentials of adjacent carriers are fairly well matched. Secondly, it allows the release or binding of protons at organic centre/metal centre junctions; a phenomenon that is crucial to energy conservation.

Aerobic respiration in chemoorganotrophic bacteria is characterized by the transfer of reducing equivalents from an organic donor, principally NADH (E_m $NAD^+/NADH$ -320 mV) to molecular oxygen (E_m $\frac{1}{2}O_2/H_2O$ $+820$ mV). This process is characterized by a large change in redox potential (ΔE_m $+1140$ mV) and, hence, by the release of a large amount of free energy ($\Delta G^{0'}$ -219 kJ.mol^{-1}), and encompasses a large number of sequential redox reactions involving several different types of redox components (Fig. 1.2a). In contrast, the oxidation of weaker reductants such as succinate (E_m fumarate/succinate $+30$ mV) or methanol (E_m formaldehyde/methanol -192 mV) liberates much less free energy ($\Delta G^{0'}$ -152 and -194 kJ.mol^{-1} respectively) and occurs via truncated respiratory chains, in which the succinate and methanol dehydrogenases differ from NADH dehydrogenase in terms of their redox components and/or their spatial interaction with the membrane. The range of reductants that can be oxidized by the respiratory chain of an aerobic chemoorganotroph is thus closely related to the genetic capacity of the organism to synthesize the required types of redox centres and apoproteins, and to insert the finished products into the membrane in the correct position.

Similar considerations also apply to anaerobic respiration in chemoorganotrophs and to aerobic respiration in chemolithotrophs. In the former case, molecular oxygen is replaced by alternative electron acceptors (e.g. fumarate, oxyanions of nitrogen and sulphur, Fe^{3+}) of widely differing redox potential (e.g. E_m sulphite/sulphide -116 mV c.f. N_2O/N_2 $+1355$ mV) (Fig. 1.2b). In the latter case, organic donors are replaced by inorganic reductants (e.g. H_2, oxyanions of nitrogen and sulphur, Fe^{2+}) of equally varied redox potential (e.g. E_m $2H^+/H_2$ -420 mV c.f. nitrate/nitrite $+420$ mV) (Fig. 1.2c). A few species of chemolithotrophic bacteria can replace oxygen with an alternative electron acceptor such as nitrate. The varied redox potentials of these alternative donors and acceptors means that their use is also characterized by wide variations in the amount of free energy released. Furthermore, with the exception of hydrogen, most of the inorganic reductants used by chemolithotrophs are too weak to reduce NAD^+ directly (and in many organisms hydrogen itself does not do so); the NADH required for the assimilation of carbon dioxide is therefore formed by reversed electron transfer from the inorganic reductant to NAD^+, at the expense of the free energy generated by forward respiration from the inorganic reductant to molecular oxygen.

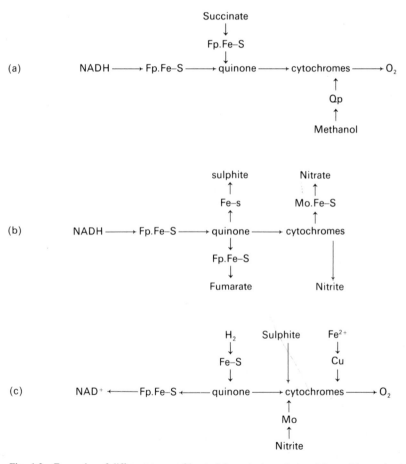

Fig. 1.2 Examples of different types of bacterial respiratory chains. (*a*) aerobic respiration in chemoorganotrophs or facultative photo-organotrophs; (*b*) anaerobic respiration; and (*c*) aerobic respiration in chemolithotrophs Abbreviations: Fp, flavoproteins; Fe–S, iron–sulphur proteins; Qp, quinoproteins; Mo, molybdoproteins; Cu, cuproproteins.

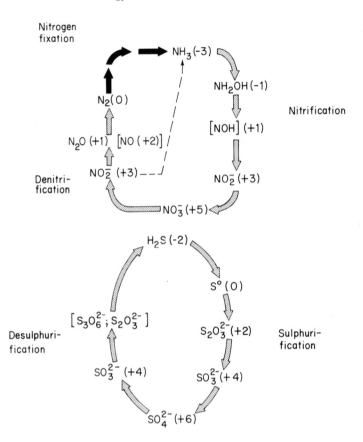

Fig. 1.3   The role of bacterial respiration in the nitrogen and sulphur cycles. The various intermediates are mainly oxidized by molecular oxygen during nitrification (ammonia → nitrate) and sulphurification (sulphide → sulphate), but are reduced by NADH and other high redox potential donors during denitrification (nitrate → N$_2$) and desulphurication (sulphate → sulphide). N$_2$ produced during denitrification is subsequently reduced to ammonia by the process of nitrogen fixation, but some non-denitrifying bacteria form ammonia by the direct six-electron reduction of nitrite. The numbers in brackets refer to the oxidation states of the intermediates.

The majority of these chemolithotrophic and anaerobic respiratory systems utilize various nitrogen and sulphur compounds as reductants or oxidants and are thus major contributors via these dissimilatory reactions to the geochemical cycles of these elements (Ehrlich, 1981; Kelly, 1982, 1985; Payne, 1985; Cole and Ferguson, 1988; Wood, Chapter 4, this volume; Hamilton, Chapter 2, this volume) (Fig. 1.3). Nitrification and sulphurification consist of multistep, 8e^- oxidations, respectively, of ammonia to nitrate and of sulphide to sulphate, using principally molecular oxygen as the acceptor:

$$(-3) \text{ ammonia} \rightarrow \text{hydroxylamine} \rightarrow \text{nitrite} \rightarrow \text{nitrate } (+5)$$

$$(-2) \text{ sulphide} \rightarrow \text{sulphur} \rightarrow \text{thiosulphate} \rightarrow \text{sulphite} \rightarrow \text{sulphate } (+6)$$

whereas denitrification and desulphurification consist overall of essentially the reverse reactions; albeit via several different intermediates and only as far as molecular nitrogen in the case of denitrification, using low redox potential donors such as NADH, hydrogen or lactate:

$$(+5) \text{ nitrate} \rightarrow \text{nitrite} \rightarrow \text{nitric oxide} \rightarrow \text{nitrous oxide} \rightarrow N_2 \text{ (0)}$$

$$(+6) \text{sulphate/APS} \rightarrow \text{sulphite} \rightarrow (\text{trithionate/thiosulphate}) \rightarrow \text{sulphide} (-2)$$

The N_2 produced during denitrification is finally reduced to ammonia by specialized bacteria capable of carrying out nitrogen fixation. It should be noted, however, that some non-denitrifying organisms reduce nitrite directly to ammonia via a concerted $6e^-$ reduction.

Other chemolithotrophs are known which oxidize Fe^{2+}, Mn^{2+} and Sn^{2+}, and organisms may also possibly exist that can conserve energy from the oxidation of selenium and various oxyanions of phosphorous (Ehrlich, 1981; Wood, Chapter 4, this volume). However, compared with various non-enzymic reactions, the contribution of the redox reactions catalysed by chemolithotrophs to the overall geochemical cycles of these elements is probably fairly small.

1.1.4   Photosynthetic electron transfer

Photosynthesis is the reductive assimilation of carbon dioxide (and sometimes other carbon substrates), at the expense of light energy:

$$2DH_2 + CO_2 \rightarrow (CH_2O) + H_2O + 2D$$

It consists of a light phase, in which light-initiated redox reactions conserve energy and generate reducing power (NAD(P)H) at the expense of an exogenous donor (DH_2), and a subsequent dark phase, in which the energy and reducing power so generated are used to drive carbon assimilation.

The nature of the exogenous donor varies between different species of phototrophic bacteria. In the photolithotrophic purple and green sulphur bacteria (the families *Chromatiaceae*, *Chlorobiaceae* and *Chloroflexaceae*, and the recently discovered *Heliobacterium chlorum*) the donor is an inorganic compound such as sulphide, thiosulphate or hydrogen; in the photo-organotrophic purple non-sulphur bacteria (the family *Rhodospirillaceae*) it is an organic donor such as succinate or malate, which also serves as a carbon source. In neither case is oxygen released (anoxygenic photosynthesis). In contrast, blue-green bacteria (*Cyanobiaceae*) use water as a donor and, like algae and green plants, release oxygen (oxygenic photosynthesis).

Phototrophs exhibit two types of photosynthetic electron transfer: cyclic and non-cyclic (Fig. 1.4). During cyclic electron transfer, electromagnetic radiation is absorbed by specialized photopigments (bacteriochlorophylls, bacteriopheophytins and various accessory pigments), which are organized into light-harvesting and reaction centre complexes. The light initiates a series of complex photochemical reactions which lead to the generation of a low redox potential reductant (a reduced quinone or iron–sulphur protein; $E_m < -160 \, \text{mV}$) and a high redox potential oxidant (an oxidized cytochrome; $E_m + 250 \, \text{mV}$). The re-reduction of the latter by the former occurs via a quinone–cytochrome system similar to that found in many respiratory chains and is accompanied by the release of a moderate amount of free energy ($\Delta E_m > +410 \, \text{mV}$; $\Delta G^{0\prime} \geq -79 \, \text{kJ.mol}^{-1}$). Cyclic electron transfer is therefore independent of exogenous donors or acceptors, and its sole function is to conserve energy.

The main function of non-cyclic electron transfer is to generate NAD(P)H at the expense of an exogenous reductant, such as sulphide, thiosulphate, succinate or water, which has a higher redox potential than the NAD(P)$^+$/NAD(P)H couple (e.g. $E_m \, S_o/S^{2-} \, -99 \, \text{mV}$). This energy-dependent process is achieved either by light-dependent electron transfer through the photosystem (sulphide, thiosulphate $\rightarrow$ NAD$^+$) or by reversed electron transfer using energy conserved during cyclic electron transfer (NAD$^+$ $\leftarrow$ succinate).

Blue-green bacteria synthesize an additional photosystem (designated 'photosystem II' to distinguish it from the system which is present in all photosynthetic bacteria, and which is called 'photosystem I' in blue-green bacteria) in order to catalyse the oxidation of water, a very high redox potential donor. Photosystem II is basically similar to photosystem I, but contains a manganese protein that is involved in the oxidation of water to molecular oxygen. The function of photosystem II is therefore to generate a weak reductant (quinol) which donates reducing equivalents to the quinone–cytochrome system. As the redox potential spans of the two photosystems overlap, a small amount of energy conservation is possible during non-cyclic electron transfer. The photosystems present in blue-green bacteria contain photopigments of the type present in higher organisms (chlorophylls and pheophytins) rather than the bacterio- forms found in the purple and green bacteria.

It should be noted that a few species of halobacteria are able to conserve light energy without the mediation of bacteriochlorophylls or other redox components. These organisms use a novel photopigment, bacteriorhodopsin, which is similar in structure to vertebrate rhodopsin (visual purple).

Unfortunately, it is beyond the scope of this chapter to discuss bacteriorhodopsin further (but see Eisenbach and Caplan, 1979; Stoeckenius *et al.*, 1979; Stoeckenius and Bogomolni, 1982; Harold, 1986).

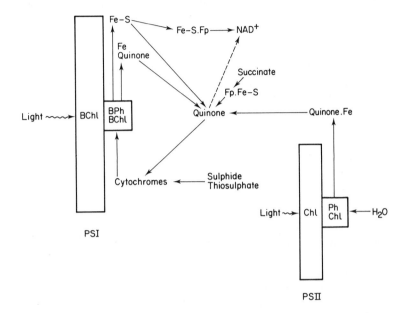

Fig. 1.4 Photosynthetic electron transfer in purple, green and blue-green bacteria. Photosystem I contains (bacterio)chlorophylls, (bacterio)pheophytins and various accessory pigments and is present in all three groups of phototrophs. Photosystem II contains clorophylls, pheophytins and specialized accessory pigments, and is restricted to blue-green bacteria. The large boxes contain the light-harvesting complexes, and the smaller boxes contain the reaction centre pigments. Cyclic electron transfer involves only a single photosystem (PSI) in all organisms. Non-cyclic electron transfer from sulphide or thiosulphate to NAD$^+$ also involves only PSI in purple and green bacteria, but electron transfer from water to NAD$^+$ in blue-green bacteria additionally requires PSII. The reduction of NAD$^+$ by succinate in some purple bacteria occurs by reversed electron transfer ($- - \rightarrow$) at the expense of energy conserved during cyclic electron transfer. (See Chapter 7 for extensive discussion of photosynthesis and Fig. 1.25 for extended versions of this diagram.) Abbreviations: BChl, bacteriochlorophyll; Chl, chlorophyll; BPh, bacteriopheophytin; Ph, pheophytin; Fe, iron associated with bound quinone.

1.1.5 Membrane-associated energy conservation

It is now generally accepted that the major function of membrane-bound redox systems is to translocate protons and thus conserve the free energy released by the constituent redox reactions as an electrochemical potential difference of H$^+$ ($\Delta\bar{\mu}$H$^+$; kJ.mol^{-1}) or protonmotive force (Δp; mV; see

Table 1.2 Formulae relating to ion gradients and the protonmotive force

The Gibbs free energy change associated with the transfer of a neutral chemical species A from the outside of a cell, where its concentration is $[A_o]$ to the inside, where its concentration is $[A_i]$ is given by:

$$\Delta G = -RT \ln \frac{[A_o]}{[A_i]} \quad Jmol^{-1}$$

The Gibbs free energy change associated with the transfer of a charged chemical species B^{n+} from the outside of a cell, where its concentration is $[B_o^{n+}]$ to the inside, where its concentration is $[B_i^{n+}]$ is given by:

$$\Delta G = -RT \ln \frac{[B_o^{n+}]}{[B_i^{n+}]} + nF \quad Jmol^{-1}$$

where $\Delta\psi$ is the membrane potential, defined as the electric potential of the bulk aqueous phase of the cell cytoplasm relative to the bulk aqueous phase of the surrounding medium.

This relationship can also be written in terms of electrochemical potential ($\bar{\mu}$), the difference between the electrochemical potential of the ions in the cytoplasm and the bulk external phase being $\Delta\bar{\mu}_{B^{n+}}$.

For protons: the value of $n = 1$ and the electrochemical gradient of protons is $\Delta\bar{\mu}_{H^+}$

$$\Delta\bar{\mu}_{H^+} = -RT \ln \frac{[H_o^+]}{[H_i^+]} + F\Delta\psi \quad Jmol^{-1}$$

Dividing by F to express the electrochemical gradient of protons as a voltage gives us the protonmotive force, the pmf or Δp:

$$\Delta p = \frac{\Delta\bar{\mu}_{H^+}}{F} = \Delta\psi - \frac{RT}{F} \ln \frac{[H_o^+]}{[H_i^+]} \quad Volts$$

Since $pH_{in-out} = -\log_{10} \dfrac{[H_i^+]}{[H_o^+]}$ $pmf = \Delta p = \Delta\psi - 2.303 \dfrac{RT}{F} \Delta pH \quad Volts$

Thus: $pmf = \Delta p = \Delta\psi - 59 \Delta pH \quad Millivolts$

NB: The ratio $2.303 \dfrac{RT}{F} = 59$ (for 25°C) is sometimes abbreviated to the symbol Z.

$F = $ Faraday's constant $= 96487 \, JV^{-1}$; $R = $ gas constant $= 8.314 \, JK^{-1} \, mol^{-1}$; $T = $ temperature (K).

These concepts are discussed more extensively in Chapter 9 (pages 433–439), and the relationship between solute gradients and pmf is discussed in Chapter 8 (page 414 and Table 8.2, page 402).

Table 1.2). The cytoplasmic membrane also contains an ATPase/ATP synthase complex that similarly conserves the free energy released by the hydrolysis of ATP. Indeed, this is the principal method by which the protonmotive force is generated in fermentative organisms. Both the respiratory chain and the ATPase/ATP synthase complex therefore act as primary solute (H^+) transport systems (primary because H^+ transport is associated with a chemical transformation, viz. oxidation–reduction or ATP hydrolysis). However, if the protonmotive force generated by the redox system is large enough, ATP hydrolysis (the ATPase reaction) is disequilibrated in favour of ATP synthesis (the ATP synthase reaction)

This chemiosmotic mechanism of energy conservation (see, for example, Mitchell, 1961, 1966, 1979; also Nicholls, 1982; Harold, 1986), so-named because it entails both the transfer of chemical groups (H, H^-, e^-) within the membrane and the transport of a solute (H^+) across the membrane, thus requires:

(a) a proton-translocating redox system;
(b) a proton-translocating ATPase/ATP synthase complex; and
(c) a passive topologically-closed membrane which is impermeable to ions, including H^+ and OH^-, except via specific exchange–diffusion systems (protein-based transport systems).

It is envisaged that the free energy released by the redox system is conserved as a transmembrane protonmotive force, i.e. as the electrochemical potential difference of protons between the bulk aqueous phases on either side of the membrane. Energy transduction therefore occurs by way of a delocalized proton current that circulates through an essentially inert insulating membrane and the adjacent bulk aqueous phases (Fig. 1.5). This proton circuit is analogous to an electrical circuit in that it has a current (jH^+; ng-ion $H^+.min^{-1}$), a potential difference (Δp; mV) and a conductance ($C_M H^+$ equivalent to $jH^+/\Delta p$; ng-ion $H^+.min^{-1}. mV^{-1}$).

Under equilibrium conditions, for the transfer of two electrons by the redox system:

$$-2\Delta E_h = \Delta G_{ox}/F = n.\Delta p$$

where ΔG_{ox} is the rather jargonistic term frequently used to describe the free energy change of the overall redox reaction, and n is the number of protons translocated per electron pair transferred (the $\rightarrow H^+/2e^-$ or $\rightarrow H^+/O$ quotient).

The protonmotive force is variably composed of a chemical potential difference (ΔpH, $pH_{out} - pH_{in}$) and an electrical potential difference or membrane potential ($\Delta \psi$; mV):

$$\Delta p = \Delta \psi - Z.\Delta pH$$

where $Z = 2.303RT/F$ and converts the dimensionless ΔpH into electrical units (mV). The respiratory chain translocates H^+ outwards in whole cells, thus generating a protonmotive force that is electrically positive and/or acidic with respect to the cytoplasm.

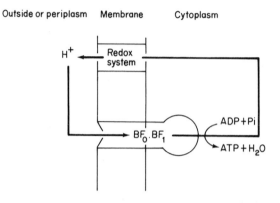

Fig. 1.5   The chemiosmotic mechanism of energy transduction. The membrane contains a proton-translocating redox system and a proton-translocating ATPase/ATP synthase complex ($BF_0.BF_1$). The proton current circulates through the insulating membrane and the adjacent bulk aqueous phases, such that, under steady-state conditions, the rates of H^+ ejection and re-entry are equal ($jH^+_{out} = jH^+_{in}$).   Energy is conserved in the form of a delocalized protonmotive force (Δp) which is additionally able to drive other energy-dependent membrane functions such as certain types of solute transport, reversed electron transfer and motility (see Fig.1.32).

During the subsequent synthesis of ATP at the expense of the proton-motive force:

$$m.\Delta p = \Delta Gp/F$$

where m is the number of protons retranslocated by the ATPase/ATP synthase complex per ATP synthesized (the $\rightarrow H^+$/ATP quotient), and ΔGp is the so-called phosphorylation potential, i.e. the free energy change for the synthesis of ATP from ADP and phosphate:

$$\Delta Gp = \Delta G^{0'} + 2.303RT \log[\text{ATP}]/[\text{ADP}][\text{Pi}]$$

Thus, the overall result of respiratory chain phosphorylation or photophosphorylation is that the free energy released by the redox system (ΔG_{ox}) is conserved as a phosphorylation potential (ΔG_p).

There is substantial experimental support for the basic tenets of the chemiosmotic hypothesis. This includes the observations that:

(a) whole cells or inside-out membrane vesicles generate a protonmotive force of the expected polarity following electron transfer, bacteriorhodopsin activity or ATP hydrolysis;

(b) protonophores dissipate the protonmotive force and hence uncouple respiration from ATP synthesis; and

(c) ATP can be synthesized at the expense of an artificially generated protonmotive force in the complete absence of a redox system or bacteriorhodopsin.

It is very important to note, however, that the concept of a delocalized protonmotive force of the type envisaged by the chemiosmotic mechanism is by no means universally accepted (see Williams, 1978; Kell, 1979, 1987; Kell *et al.*, 1981; Westerhof *et al.*, 1984; Ferguson, 1985; Slater *et al.*, 1985). These doubts are fostered by several recent experimental observations which suggest that Δp may be kinetically and/or thermodynamically incompetent to support ATP synthesis. These observations include:

(a) the absence of a unique relationship *in vitro* between the rate of electron transfer and the protonmotive force, or between the proton-motive force and the rate of ATP synthesis; and

(b) the ability *in vivo* of alkaliphiles, halophiles and mutants of some mesophiles to synthesize ATP when the protonmotive force is either very low or absent.

Furthermore, there is increasing evidence that such a delocalized driving force may exist only under certain essentially artefactual experimental conditions *in vitro* (Kell, 1987), and that the interaction between the redox system and the ATPase/ATP synthase complex *in vivo* may be much more direct than is possible through a purely chemiosmotic mechanism, e.g. via proton channels along the surface of the membrane (Kell, 1979; Kell *et al.*, 1981) or even via collisional interactions that do not necessarily directly involve H^+ (Slater *et al.*, 1985; Slater, 1987). It is beyond the scope of this chapter to discuss this controversy in more detail, and it is extensively considered by Kell in Chapter 9 of this volume. However, since discussion of most of the work carried out on bacterial respiration and photosynthesis during the last two decades has been couched in terms of the chemiosmotic mechanism, this convention will be retained in this introduction.

1.1.6   Generation of the protonmotive force

Several different mechanisms have been proposed to explain the generation of a protonmotive force as a result of redox reactions at different sites within respiratory chains and photosynthetic electron transfer systems (see Mitchell, 1966, 1976; Garland, 1977; Jones, 1977; Hooper and DiSpirito, 1985). They include direct group-translocation mechanisms of fixed $\rightarrow H^+/2e^-$ stoichiometry (redox loop, arm and cycle) and indirect conformational proton pumps of indeterminate, and possibly variable, stoichiometry (Fig. 1.6).

H^+ extrusion by the redox loop mechanism (Fig. 1.6a) is envisaged as being a direct consequence of outwardly directed 2H-transfer from a

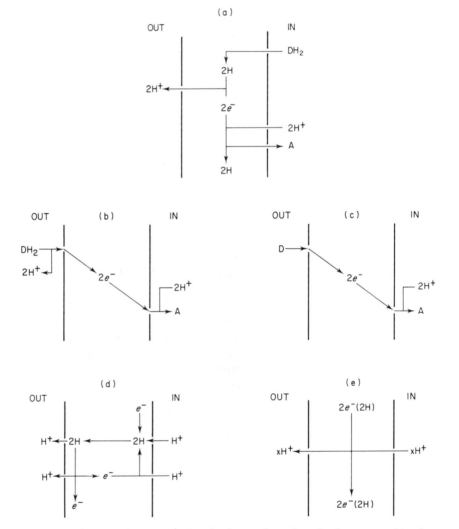

Fig. 1.6 Mechanisms of proton translocation by membrane-bound redox systems. (*a*) redox loop; (*b*) redox arm type I; (*c*) redox arm type II; (*d*) redox cycle; and (*e*) conformational proton pump. Note that each of these protonmotive mechanisms can start or finish either with a substrate or with a membrane-associated redox component.

hydrogen atom donor using an appropriate hydrogen carrier (flavin, quinone); followed by inwardly directed $2e^-$-transfer to a hydrogen atom acceptor using an appropriate electron carrier (Fe–S, Mo, Cu, cytochrome). The redox carriers are spatially arranged within the membrane in such a way that protons are released from the external surface and taken up from the internal surface (but note that this does not necessarily imply that the organic centre/metal centre junctions have to be located on opposite sides of the membrane, merely that they must have access to the aqueous compartment on the appropriate side). As the redox loop starts and finishes on the cytoplasmic side of the membrane, and both the donor and the acceptor carry two hydrogen atoms, proton translocation (H^+) and charge translocation ($H^+ + e^-$) stoichiometries are fixed and equal ($\rightarrow H^+/2e^-$ 2, $\rightarrow$charge$/2e^-$ -2).

The redox arm or semi-loop mechanism involves only the inwardly directed electron transfer half of the loop mechanism. Two types of redox arm have been identified. In the type I mechanism (Fig. 1.6b), the absence of a hydrogen carrier is overcome by the hydrogen atom donor interacting with the redox system from the external surface to release $2H^+$. The proton translocation and charge translocation stoichiometries are again fixed and equal, and are identical to those exhibited by the redox loop ($\rightarrow H^+/2e^-$ 2, $\rightarrow$charge$/2e^-$ -2). In the type II mechanism (Fig. 1.6c), the hydrogen atom donor is replaced by an electron donor; $2H^+$ are no longer released at the external surface, and the proton translocation and charge translocation stoichiometries, although fixed, are no longer equal ($\rightarrow H^+/2e^-$ 0, $\rightarrow$charge$/2e^-$ -2).

At some stage, both the redox loop and redox arm mechanisms involve a hydrogen atom donor or acceptor undergoing a protolytic reaction to help generate the protonmotive force. The release of H^+ on the external surface (redox loop, redox arm type I) can arise either from a membrane-bound redox centre, such as reduced flavin, or directly from a donor, such as methanol, formate, hydroxylamine or H_2, e.g.:

$$\text{flavin.}H_2 \rightarrow \text{flavin} + 2H^+ + 2e^-$$

$$CH_3OH \rightarrow HCHO + 2H^+ + 2e^-$$

It can also arise directly from water, as during the oxidation of nitrite, sulphite or thiosulphate, e.g.:

$$NO_2{}^- + H_2O \rightarrow NO_3{}^- + 2H^+ + 2e^-$$

$$SO_3{}^{2-} + H_2O \rightarrow SO_4{}^{2-} + 2H^+ + 2e^-$$

$$S_2O_3{}^{2-} + 5H_2O \rightarrow 2SO_4{}^{2-} + 10H^+ + 2e^-$$

or from a reaction between the donor and water, as during the oxidation of methylamine or glucose, e.g.:

$$CH_3NH_2 + H_2O \rightarrow HCHO + NH_3 + 2H^+ + 2e^-$$

The non-release of H^+ from the external surface (redox arm type II) results from the oxidation of a membrane-associated redox centre such as cytochrome c, or from a donor such as ferrous iron or thiosulphate, e.g.:

$$2Fe^{2+} \rightarrow 2Fe^{3+} + 2e^-$$

$$2S_2O_3^{2-} \rightarrow S_4O_6^{2-} + 2e^-$$

The consumption of H^+ on the internal surface (redox loop, redox arm types I and II) can arise either via the direct reduction of the terminal acceptor, as during the reduction of oxygen or fumarate, e.g.:

$$\tfrac{1}{2}O_2 + 2e^- + 2H^+ \rightarrow H_2O$$

$$\text{fumarate} + 2e^- + 2H^+ \rightarrow \text{succinate}$$

or by the removal of an atom of oxygen from the acceptor with the concomitant formation of water, as during the reduction of nitrate:

$$NO_3^- + 2e^- + 2H^+ \rightarrow NO_2^- + H_2O$$

The redox cycle mechanism (Fig. 1.6d) was postulated largely because there are insufficient hydrogen carriers within bacterial respiratory chains and photosynthetic electron transfer systems to allow energy conservation to occur via sequential redox loops. The best documented cycle is that involving a quinone, the Q-cycle, which is effectively a fusion of two loops using a single hydrogen carrier (Mitchell, 1975). The Q-cycle involves only membrane-associated redox components, and catalyses the net outward movement of two electrons as well as four protons:

$$2e^-(\text{in}) + 4H^+(\text{in}) \rightarrow 2e^-(\text{out}) + 4H^+(\text{out})$$

The proton translocation and charge translocation stoichiometries are therefore fixed but unequal ($\rightarrow H^+/2e^-$ 4, $\rightarrow$charge$/2e^-$ -2) (see Figs. 1.29 and 1.31). A somewhat analogous cycle, the O_2-cycle, has recently been proposed to explain the proton translocation properties of some cytochrome oxidases, but it remains to be confirmed experimentally. It is envisaged that the cytochrome oxidase catalyses the net inward movement of two electrons concomitant with the outward movement of four protons:

$$2e^-(\text{out}) + 4H^+(\text{in}) + \tfrac{1}{2}O_2 \rightarrow H_2O + 2H^+(\text{out})$$

the proton translocation and charge translocation stoichiometries are again fixed and unequal ($\rightarrow H^+/O$ 2, $\rightarrow$charge$/O$ -4), but are reversed relative to the Q-cycle.

In contrast to these direct mechanisms, the conformational proton pump (Fig. 1.6e) represents an indirect mechanism of proton translocation, i.e. the translocated protons need not bind to the redox centre and, hence, do not exhibit a predictable or fixed stoichiometry. The redox energy is conserved as a conformational change in the redox protein, or possibly even in an adjacent protein. This leads to changes in the pKa values of appropriately located carboxyl or amino groups within the protein and, hence, to the asymmetric uptake and release of protons across the membrane. Although this mechanism envisages no restriction on the stoichiometry of proton translocation, since this is dictated by the number of ionizable groups, the $\rightarrow H^+/2e^-$ and $\rightarrow charge/2e^-$ quotients are expected to remain equal, unless modified by additional protolytic reactions of the type described above. Indeed, it should be noted that these various mechanisms of proton translocation are not mutually exclusive and that there is some evidence that they can operate simultaneously within different regions of a single respiratory chain or photosynthetic electron transfer system.

1.1.7 Manipulation of the protonmotive force

The composition of the protonmotive force (Δp) generated by respiration principally reflects the pH of the external environment (since the internal pH is maintained fairly constant by various homeostatic mechanisms) and the activities of various Δp-dependent solute transport systems. It can easily and reproducibly be manipulated experimentally using specific ΔpH- or $\Delta\psi$-collapsing agents. In the presence of a high concentration of K^+, the ionophorous antibiotic nigericin catalyses H^+ entry in exchange for K^+ exit (electroneutral H^+, K^+ antiport), and thus collapses ΔpH; the respiratory chain responds to the lower Δp by increasing the rate of electron transfer, and hence of H^+ extrusion, thus enhancing $\Delta\psi$ and maintaining the original value of the protonmotive force. Conversely, valinomycin catalyses K^+ entry (electrogenic K^+ uniport), thus collapsing $\Delta\psi$ and enhancing ΔpH to compensate. These effects can also be brought about by the use of permeant weak acids/bases and lipophilic cations/anions respectively, although strictly speaking none of these is an ionophore since they are accumulated in response to ΔpH and $\Delta\psi$ and, hence, do not act catalytically.

Both components of the protonmotive force are collapsed in the presence of a mixture of $\Delta\psi$- and ΔpH-collapsing agents (e.g. nigericin + valinomycin + K^+), which thus mimic the action of conventional protonophores such as carbonylcyanide-*p*-trifluoromethoxyphenylhydrazone (FCCP) or 2,4-dinitrophenol (DNP). Under these circumstances, ATP synthesis and other energy-dependent membrane functions are completely uncoupled from respiration. See Chapter 8 (page 398) for further discussion of uncouplers and ionophores.

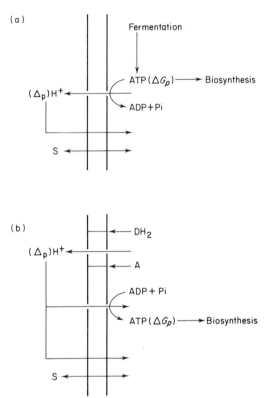

Fig. 1.7   The role of the ATPase/ATP synthase complex in energy conservation during growth under different conditions. The complex acts as a reversible proton pump, interconverting the protonmotive force (Δp) and the phosphorylation potential (ΔGp). During fermentative growth (*a*) the complex acts as a proton-ejecting ATPase, hydrolysing some of the ATP synthesized by substrate-level phosphorylation to generate the protonmotive force which is required for the transport of solutes (S) or for other energy-dependent membrane reactions. Under growth conditions where the protonmotive force is generated directly by respiration or photosynthetic electron transfer (*b*), the complex acts as a proton-injecting ATP synthase and thus catalyses ATP synthesis.

1.1.8   Utilization of the protonmotive force

The protonmotive force generated by respiration and photosynthetic electron transfer is used to drive a wide variety of energy-dependent membrane functions, including ATP synthesis, solute transport, reversed electron transfer, pyrophosphate synthesis, sensory perception and motility. ATP synthesis and solute transport are quantitatively by far the greatest drain on the protonmotive force in most organisms, and reversed electron transfer provides an additional major burden in most chemolithotrophs and some photoorganotrophs. It is beyond the scope of this chapter to discuss the other, relatively minor, uses of the protonmotive force (but see reviews by McNab, 1978, 1985; Koshland, 1981; Taylor, 1983; Mitchell, 1984; Ordal, 1985).

1.1.8.1 ATP synthesis

The ATPase/ATP synthase complex is composed of two oligomeric proteins, BF_1 and BF_0 (where BF stands for bacterial coupling factor). BF_1 is a hydrophilic protein which is attached to BF_0 on the cytoplasmic side of the membrane and is responsible for binding ATP, ADP, inorganic phosphate and divalent metal ions. In contrast, BF_0 is a hydrophobic transmembrane protein which catalyses the reversible transfer of H^+ between the external surface of the membrane and BF_1 (Kagawa, 1978; Downie et al., 1979; Fillingame, 1980, 1981; Walker et al., 1984; Cox et al., 1984; Mitchell, 1985).

During the growth of fermentative bacteria, ATP is synthesized by substrate level phosphorylation and the function of the $BF_0.BF_1$ complex is to act as a proton-ejecting ATPase, i.e. to generate a protonmotive force at the expense of ATP hydrolysis. In contrast, during non-fermentative growth, its function is to act as an ATP synthase, i.e. to catalyse the synthesis of ATP at the expense of the protonmotive force generated by the respiratory chain (Fig. 1.7).

The overall stoichiometry of respiratory chain phosphorylation or photo-phosphorylation is expressed as the ATP/O or $ATP/2e^-$ quotient, and is most commonly determined from simultaneous measurements of the rates or extents of ATP synthesis (or phosphate esterification) and electron transfer in whole cells or inside-out membrane vesicles, i.e.:

$$ATP/O = jATP/jO = \Delta ATP/\Delta O$$

When whole cells· are used correction has to be made for the subsequent utilization of some of the nascent ATP by adenylate kinase (ATP + AMP → 2ADP) and by various anabolic reactions.

The ATP/O quotient also quantitatively reflects the $\rightarrow H^+/O$ quotient (n) and the $\rightarrow H^+/ATP$ quotient (m):

$$ATP/O = n/m$$

The ATP/O quotient determined from kinetic measurements will therefore be significantly diminished if there is significant H^+ leakage through the membrane (as may be the case in some extremophiles) or if secondary solute transport systems compete effectively against the ATP synthase for the available protons.

When the Δp-generating redox system and the Δp-utilizing ATP synthase complex are in equilibrium (static-head conditions):

$$\Delta G_{ox}/n.F = \Delta p = \Delta Gp/m.F$$

Simultaneous measurements of Δp and ΔG_p will therefore allow the determination of the $\rightarrow H^+/ATP$ quotient, a parameter which it is particularly difficult to measure directly. It should be noted, however, that such conditions rarely pertain in bacterial systems, at least in whole cells where competing metabolic reactions diminish Δp and/or ΔGp and thus ensure that respiration and ATP synthesis are under kinetic rather than thermodynamic control (level flow conditions). More usually, therefore, during ATP synthesis:

$$\Delta G_{ox}/n.F > \Delta p > \Delta Gp/m.F$$

whereas during ATP hydrolysis:

$$\Delta p < \Delta Gp/m.F$$

Nevertheless, it is clear that although $\rightarrow H^+/O$ and $\rightarrow H^+/2e^-$ quotients vary dramatically between different species of bacteria, depending on their redox carrier composition and the nature of the donor and acceptor, the $\rightarrow H^+/$ ATP quotient is probably much less variable. Values of approximately 2 to 4 have been reported, but 3 seems increasingly to fit most of the theoretical expectations and experimental data.

There is convincing evidence that it is the magnitude rather than the composition of the protonmotive force that determines the extent of ATP synthesis. ΔpH and $\Delta \psi$ are therefore equally effective driving forces, and either of them will effect a net synthesis of ATP provided it exceeds a threshold value of approximately 150 mV (inside negative). This is reflected in the dramatically increased rate of H^+ retranslocation through the respiratory membrane, when the protonmotive force exceeds this threshold value.

1.1.8.2 Secondary solute transport

The protonmotive force can be used to drive the transport of a wide variety of solutes through the cytoplasmic membrane, leading either to their net accumulation or ejection (West and Mitchell, 1972; Simoni and Postma, 1975; Harold, 1977; Konings, 1977; Rosen and Kashket, 1978; Booth, 1985; Hellingwerf and Konings, 1985; Maloney and Wilson, 1985; see also Booth, Chapter 8, this volume). These solute transport systems are said to be secondary systems, since they are not accompanied by a chemical transformation and are driven by the protonmotive force generated as a result of the primary transport of H^+ by the redox system, bacteriorhodopsin or the ATPase. They comprise tightly membrane-bound carriers which are functionally symmetrical and thus catalyse solute transport in either direction depending on the value of the protonmotive force relative to that of the solute-motive force ($\Delta\bar{\mu}S$). The latter is analogous to the former since:

$$\Delta\bar{\mu}S = \Delta\psi - \Delta\mu S$$

where $\Delta\mu S$ is the chemical potential of the solute and is equivalent to $Z.\log[S_{in}]/[S_{out}]$ (mV). When the coupled transport system reaches equilibrium, or more probably achieves a kinetic steady state by virtue of the activity of accompanying leakage systems, there is no further net uptake or ejection of the solute.

The direction of the protonmotive force in whole cells (electrically negative and/or alkaline inside) dictates that the transported species is either electrically neutral or positively charged. Thus, anions and neutral molecules enter in cotransport with H^+ (H^+.solute symport), whereas cations enter either in cotransport with H^+ or unaccompanied (solute uniport) and exit in exchange for H^+ (H^+.solute antiport) (Fig. 1.8). Unlike ATP synthesis, which is independent of the composition of the protonmotive force and occurs only when Δp exceeds a threshold value, active transport is driven uniquely by $\Delta\mu S$ plus either ΔpH, $\Delta\psi$ or Δp depending on the exact composition of the transported species and it exhibits no threshold value.

In contrast to metabolite entry driven by $\Delta\psi$ and/or ΔpH, there is considerable evidence that H^+ exit can occur at the expense of a solute gradient. Some bacteria appear to use fermentation product gradients to conserve energy in this way (e.g. a lactate gradient to generate a ΔpH) and thus spare ATP produced via substrate level phosphorylation.

The action of these secondary transport systems can obviously generate additional ion gradients. The quantitatively most important of these reflect the accumulation of K^+ and Mg^{2+} for the regulation of enzyme activity, the accumulation of proline and glutamate for osmoregulation, and the ejection of unwanted ions such as Na^+. The latter is effected by an H^+/Na^+ antiport system which ensures not only that the concentration of Na^+ in the cytoplasm is maintained at a desirably low level, but also that the organism has a substantial sodium-motive force ($\Delta\bar{\mu}Na^+$) (West and Mitchell, 1972; Krulwich and Guffanti, 1983; Booth, 1985; Krulwich, 1985). The sodium-motive force acts both as a convenient means of storing energy and as the driving force for Na^+-linked solute transport systems. Increasing numbers of this latter type of transport system have been reported over the last few years, particularly in marine bacteria and also in alkaliphilic bacteria where the reversed ΔpH characteristic of these organisms precludes H^+/anion symport. The H^+/Na^+ antiport system is also intimately involved in homoeostasis of the cytoplasmic pH (Krulwich and Guffanti, 1983; Booth, 1985; Krulwich, 1985). There is also some evidence that the $\Delta\bar{\mu}H_2PO_4^-$

generated by $H^+/H_2PO_4^-$ symport can drive the subsequent uptake of various organic phosphates.

It is increasingly clear that the complete integration of solute transport and metabolism in bacteria is an extremely complicated process which is probably mediated by the action of interlinked cation (H^+, Na^+) and anion (phosphate) circuits (see Chapter 8).

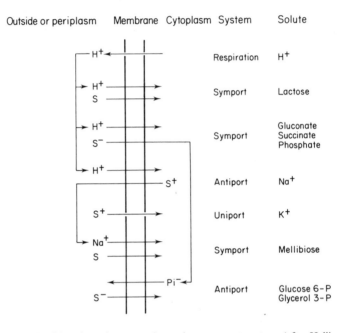

Fig. 1.8   Protonmotive force-dependent secondary solute transport systems (after Hellingwerf and Konings, 1985). A primary H^+ transport system (respiration or photosynthetic electron transfer) generates a protonmotive force (Δp). The secondary transport of other solutes (S, S^+ and S^-), including Na^+ and $H_2PO_4^-$, is driven by Δp, $\Delta\psi$ or ΔpH. The resultant $\Delta\bar{\mu}Na^+$ and $\Delta\bar{\mu}H_2PO_4^-$ can in turn drive the transport of certain sugars and organic phosphates respectively (see Chapter 8 for further discussion of solute transport).

1.1.8.3   Reversed electron transfer

In contrast to forward electron transfer, where thermodynamically spontaneous redox reactions eject H^+ and thus generate a protonmotive force, during reversed electron transfer the protonmotive force drives H^+ influx and thus causes the redox system to run backwards.

For this to happen, the protonmotive force must be generated as a result either of ATP hydrolysis by the ATPase/ATP synthase complex or of forward electron transfer over a different region of the redox system (Fig. 1.9). Reversed electron transfer can occur over the entire respiratory chain or photosynthetic electron transfer system except for the terminal oxidase and light reactions, which are essentially irreversible. ATP-dependent reversed electron transfer occurs principally during fermentative growth using ATP synthesized by substrate level phosphorylation, and is mainly restricted to reversal of the transhydrogenase reaction ($NADP^+ \leftarrow NADH$). In contrast, redox-linked reversed electron transfer is essential to the growth of many chemolithotrophic and phototrophic bacteria and is exemplified, respectively, by the aerobic reduction of NAD^+ by nitrite, sulphite or Fe^{2+} and by the anaerobic reduction of NAD^+ by succinate (Kelly, 1982, 1985; Ingledew, 1982; Wood, Chapter 4, this volume; Jackson, Chapter 7, this volume).

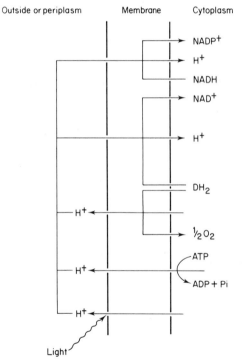

Fig. 1.9 Reversed electron transfer. Energy-dependent reversed electron transfer ($DH_2 \rightarrow NAD^+$, $NADH \rightarrow NADP^+$) occurs at the expense of the protonmotive force generated by ATP hydrolysis, by forward electron transfer over a different region of the respiratory chain ($DH_2 \rightarrow \frac{1}{2}O_2$), or by cyclic electron transfer in some purple bacteria. DH_2 is shown as a relatively high redox potential H-atom donor for the sake of clarity, but in practice the most common donors used by chemolithotrophic bacteria are electron donors such as nitrite, sulphite and Fe^{2+} (see Chapter 4). The exact number of protons translocated at each stage is not shown.

1.1.9 The energy-coupling membrane

Bacterial redox systems and their associated ATPase/ATP synthase complex are located in the cytoplasmic membrane. The latter is the sole boundary membrane present in Gram-positive bacteria, but is accompanied in Gram-negative bacteria by an outer membrane from which the inner membrane is separated by a gel-like periplasm. Cell rigidity and resistance to osmotic damage are conferred by the presence of a peptidoglycan (murein) layer which lies outside the cytoplasmic membrane of Gram-positive organisms and between the outer membrane and the periplasm of Gram-negative organisms (Costerton *et al.*, 1974; Beacham, 1979; Nikaido and Nakai, 1979; Nikaido and Vaara, 1985). In some phototrophic bacteria the inner membrane is differentiated into small invaginations or layered stacks (purple bacteria) (Chapter 7), or remains undifferentiated but has discrete vesicles attached to its cytoplasmic surface (the chlorosomes present in some green bacteria, and the phycobilisomes in blue-green bacteria).

In most bacteria the cytoplasmic membrane conforms to the fluid-mosaic model of membrane structure, i.e. it consists of a tail-to-tail phospholipid bilayer densely interspersed with protein (Singer and Nicholson, 1972). The lipid bilayer acts as a hydrophobic barrier; it is essentially impermeable to the passage of most ions, including H^+, but allows the rapid movement of uncharged species such as water, molecular oxygen, carbon dioxide and ammonia. As the membrane needs to exist in a fairly fluid, liquid-crystalline state in order to exhibit these properties and, hence, to function optimally during energy conservation, most organisms have evolved the capacity to carry out homeoviscous adaptation, i.e. they attempt to maintain this optimum state by varying the phospholipid composition of the cytoplasmic membrane in response to changes in the physical properties of the growth environment (e.g. temperature, pH, salinity). Indeed, archaebacteria and some of the so-called extremophiles have replaced conventional phospholipids with more complex lipids, including novel glycolipids and sulpholipids, or have even abandoned the bilayer structure in favour of a complex-lipid monolayer (Kushner, 1978; Kogut, 1980).

Exposure of whole cells to lysozyme under strictly controlled conditions causes degradation of the peptidoglycan and leads to the formation of osmotically fragile photoplasts and sphaeroplasts from Gram-positive and Gram-negative bacteria respectively, in the latter case with the concomitant release of the periplasm. The subsequent lysis of these structures by exposure to hypotonic conditions generates predominantly right-side-out membrane vesicles in which the orientation of the cytoplasmic membrane is the same as in whole cells. In contrast, the breakage of whole cells using more brutal sonication or shearing procedures tends to produce much smaller, inside-out membrane vesicles (often called chromatophores when produced from phototrophs) in which the orientation of the membrane has been reversed. As a result, BF_1 and the substrate-binding sites of some of the more common dehydrogenases and reductases (e.g. NADH and succinate dehydrogenases, fumarate reductase) are exposed on the outer surface, thus making this type of vesicle particularly suitable for studying electron transfer and associated energy conservation.

By monitoring the interaction between membranes of different orientation (whole cells, protoplasts, sphaeroplasts, and right-side-out membrane vesicles, c.f. inside-out membrane vesicles) and various membrane-impermeant reagents (reductants, oxidants, activators, antibodies, inhibitors, surface-labelling agents and digestive enzymes) it has also been possible to investigate the location of individual redox proteins within the respiratory membrane. This type of approach has been usefully supplemented by measuring the release of redox components from periplasm of Gram-negative organisms following exposure to mild osmotic shock, and by sophisticated spectroscopic analysis of membrane preparations.

The results show that both integral (intrinsic) and peripheral (extrinsic) redox components are present. Integral components are predominantly hydrophobic proteins that are deeply embedded within the lipid bilayer of the membrane and may even span it completely (e.g. bacteriochlorophyll–protein complexes, most iron–sulphur flavoproteins and cytochromes, molybdoproteins, quinones and BF_0), whereas peripheral components are predominantly hydrophilic proteins that are associated only with the surface of the membrane (e.g. quinoproteins, cuproproteins, cytochrome *c* and BF_1). Periplasmic proteins are discussed in more detail in Chapter 3.

1.2 Redox components of respiratory chains and photosynthetic electron transfer systems

Bacterial respiratory chains and photosynthetic electron transfer systems catalyse an extremely wide variety of redox reactions and contain an almost bewildering range of redox centres. The biochemical properties of the major types of redox centres are outlined below, together with a brief description of some of the most interesting and extensively investigated redox proteins.

1.2.1 Nicotinamide nucleotide transhydrogenases

Nicotinamide nucleotide transhydrogenase catalyses the reversible transfer of a hydride ion between NADPH ($NADP^+$) and NAD^+ (NADH):

$$NADPH + NAD^+ \leftrightarrow NADH + NADP^+$$

Since the redox potentials of the two nicotinamide nucleotide couples are very similar (E_m NADP$^+$/NADPH -324 mV; NAD$^+$/NADH -320 mV), the equilibrium constant of this reaction is close to unity.

Membrane-bound and soluble transhydrogenases have both been detected in bacteria. The membrane-bound enzyme is present in a wide range of organisms including *Escherichia coli, Paracoccus denitrificans* and many phototrophs such as *Rhodospirillum rubrum*. It contains no detectable redox centres, but is usually energy-linked when present in the membrane, i.e. the forward reaction generates a protonmotive force, the magnitude of which is proportional to the [NADPH][NAD$^+$]/[NADP$^+$][NADH] ratio, and the reverse reaction is disequilibrated in favour of NADPH formation by the protonmotive force generated as a result of respiration, photosynthetic electron transfer or ATP hydrolysis (see Bragg, 1980; Ingledew and Poole, 1984). The physiological function of the membrane-bound enzyme is almost certainly to catalyse the energy-dependent reduction of NADP$^+$ by NADH in order to generate the high [NADPH]/[NADP$^+$] required by many anabolic reactions.

In contrast to most membrane-bound transhydrogenases, the soluble enzyme contains a flavin prosthetic group. Its activity is unaffected by the protonmotive force, as would be expected of a soluble enzyme, but tends to be allosterically regulated by various signals which reflect intracellular energization. These properties suggest that the physiological function of the enzyme is probably to catalyse the oxidation of NADPH by NAD$^+$, thus ensuring that the excess reducing power generated by the highly active NADP$^+$-linked dehydrogenases present in obligately aerobic bacteria such as *Azotobacter vinelandii* and *Pseudomonas aeruginosa* is rapidly transferred to NAD$^+$, and subsequently oxidized via the NADH oxidase system to conserve energy.

1.2.2   Flavoproteins

Flavoproteins contain either flavin mononucleotide (FMN) or, more usually, flavin adenine dinucleotide (FAD) as a non-covalently-bound prosthetic group (Fig. 1.10). Both FMN and FAD carry a maximum of two hydrogen atoms; in their oxidized forms they are bright yellow in colour but become bleached on reduction (Fig. 1.11). The free flavins exhibit relatively low redox potentials (E_m FMN/FMNH$_2$ -205 mV, FAD/FADH$_2$ -219 mV), but these are often substantially increased when bound to the apoenzyme.

1.2.2.1   Flavoprotein dehydrogenases
The most common flavoprotein dehydrogenases are those which oxidize

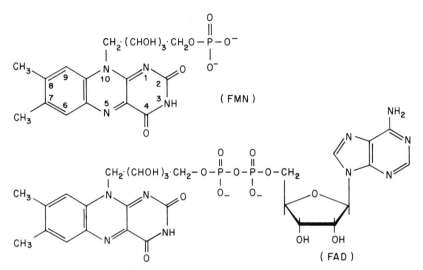

Fig. 1.10 The structures of flavin mononucleotide (FMN) and flavin adenine dinucleotide (FAD). FMN and FAD carry two H-atoms; on reduction they form $FMNH_2$ and $FADH_2$ (possibly via the semiquinones FMNH and FADH). In the reduced form the hydrogen atoms are on N-1 and N-5.

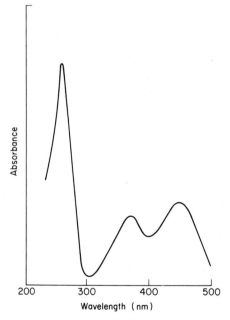

Fig. 1.11 The absorption spectrum of FMN. The band at approximately 450 nm completely disappears on reduction to $FMNH_2$ and the flavin changes from yellow to colourless.

L-lactate, D-lactate, L-malate, L-glycerol 3-phosphate and L-dihydroorotate. These have been isolated from respiratory membranes of *E. coli* and other organisms, and all except L-lactate dehydrogenase contain FAD. They are generally composed of only one subunit and are located on the cytoplasmic side of the membrane to which they are probably anchored by non-specific hydrophobic interactions with phospholipid (Ingledew and Poole, 1984; Anraku and Gennis, 1987). The oxidation of these exogenous charged substrates by whole cells therefore requires the simultaneous presence of appropriate active transport systems.

1.2.2.2 Flavodoxin

A low redox potential flavoprotein, flavodoxin, is present in some anaerobic redox systems (e.g. those catalysing sulphate respiration and electron transfer to nitrogenase during nitrogen fixation). Flavodoxin contains FMN and oscillates between either the oxidized and semiquinone form (E_m Fvd/FvdH -150 mV) or the semiquinone and fully-reduced form (E_m FvdH/FvdH$_2$ -440 mV). Unlike the flavoprotein dehydrogenases it is not membrane-bound and acts only as a non-enzymic redox carrier.

1.2.3 Iron–sulphur proteins

Iron–sulphur proteins contain 2, 4 or 8 atoms of iron per molecule, plus the same amount of acid-labile sulphur. Only the first two types are present in bacterial respiratory chains and photosynthetic electron transfer systems, and these consist of 2Fe–2S and 4Fe–4S clusters (Fig. 1.12), each of which carries only a single electron. The mechanism of electron carriage is extremely complicated and probably involves extensive electron delocalization between the iron atoms that constitute the centre. E_m values range from approximately -500 to $+250$ mV, the very low redox potential centres being found in some photosystems.

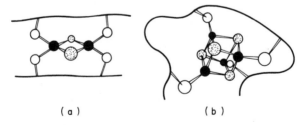

(a)   (b)

Fig. 1.12   The redox centres of iron–sulphur proteins. (*a*) [2Fe–2S], (*b*) [4Fe–4S]. Iron (●), acid-labile sulphur (⊛), cysteine sulphur (○). Both types of iron–sulphur centre carry one electron. Iron–sulphur proteins have rather diffuse absorption spectra, and their redox changes are best monitored using electron paramagnetic resonance (EPR) spectroscopy.

These proteins exhibit rather weak and diffuse absorption spectra, particularly when reduced; their redox states are therefore more easily monitored using electron paramagnetic resonance (EPR) spectroscopy which detects the presence of an unpaired electron. Reduction of most iron–sulphur proteins causes the charge at the redox centre to decrease from $(2+)$ to $(1+)$, although reduction of some types of 4Fe–4S proteins (called high potential iron proteins, HIPIP) is accompanied by a decrease from $(3+)$ to $(2+)$. Since the $(3+)$ and $(1+)$ states contain an unpaired electron, whereas the $(2+)$ state does not (paramagnetic and diamagnetic states respectively), redox changes can be monitored from the appearance or disappearance of the paramagnetism.

1.2.3.1 Hydrogenases

Hydrogenases catalyse the reversible reaction:

$$H_2 \rightleftharpoons 2H^+ + 2e^-$$

They are synthesized by an extremely wide range of chemoorganotrophic and chemolithotrophic bacteria, and catalyse either hydrogen uptake or evolution. The low redox potential of the proton/dihydrogen couple $(E_0 \; -420\,\text{mV})$ dictates that these enzymes contain low redox potential centres. Some hydrogenases, principally those responsible for hydrogen evolution, contain only iron–sulphur centres, whereas those catalysing hydrogen uptake additionally contain a nickel centre and, very occasionally, a flavin. The nickel centre, like the iron–sulphur centres, acts as a low redox potential electron carrier and exists either as Ni^+, Ni^{2+} or Ni^{3+} (Adams *et al.*, 1981; Mayhew and O'Connor, 1982; Keltjens and van der Drift, 1986).

Uptake hydrogenases all contain at least one [2Fe–2S] or [4Fe–4S] centre, plus nickel and, in one case, FMN. However, they exhibit very wide variations in their size and subunit composition, their cellular location and the nature of their physiological electron acceptors. Soluble hydrogenases are found in the cytoplasm of some aerobes (e.g. *Alcaligenes eutrophus*) and in the periplasm of some obligate anaerobes (e.g. *Desulfovibrio*) where they donate electrons respectively to NAD$^+$ and to a low redox potential cytochrome *c*. In contrast, membrane-bound hydrogenases transfer electrons to a quinone and tend to occupy a transmembrane location. *E. coli* contains two such hydrogenases, both of which bind hydrogen on the cytoplasmic side of the membrane, whereas the *Wolinella (Vibrio) succinogenes* enzyme reacts with hydrogen on the periplasmic side. Most membrane-bound hydrogenases appear able to translocate protons $(\rightarrow H^+/2e^- \; 2)$, probably using either a type I redox arm or a redox loop/conformational proton pump (see Fig. 1.6).

1.2.3.2 Ferredoxins, HIPIP proteins and molybdoferredoxins

Ferredoxins and HIPIP proteins are [2Fe–2S] and [4Fe–4S] iron–sulphur proteins which have no enzyme activity and act solely as electron carriers. Ferredoxins are usually soluble, have a low redox potential and function mainly in anaerobic respiratory chains (e.g. sulphate respiration), whereas HIPIP proteins are generally membrane-bound, have a moderately high redox potential and are widely involved in respiration and photosynthetic electron transfer at the level of quinone/cytochrome *b*.

Molybdoferredoxins are iron–sulphur proteins in which at least one of the iron atoms in a [4Fe–4S] centre is replaced by molybdenum. These electron carriers are restricted to a few types of anaerobic respiratory chains.

1.2.3.3 Rubredoxin

Rubredoxin is a small, low-redox-potential (E_m − 60 mV) protein which carries one electron. It is not a true iron–sulphur protein since it contains only one atom of iron and no labile sulphur. Rubredoxin is found in obligate anaerobes such as *Desulfovibrio* and *Desulfotomaculum*, where its main function is to transfer electrons during sulphate respiration.

1.2.4 Iron–sulphur flavoproteins

Iron–sulphur flavoproteins contain both flavin (FAD or FMN) and iron–sulphur centres, comprise several subunits and are firmly membrane-bound. The flavins are usually attached to the protein by an 8-[N(3)histidyl] linkage. Iron–sulphur flavoproteins are well represented in bacterial respiratory chains, most commonly as NADH and succinate dehydrogenases (more correctly, but less commonly, called NADH- and succinate-quinone reductases), fumarate reductase, adenosine phosphosulphate (APS) reductase and trimethylamine dehydrogenase. NADH and succinate dehydrogenases are present in virtually all species of bacteria, whereas the reductases are restricted to those organisms which catalyse anaerobic respiration, and trimethylamine dehydrogenase is mainly restricted to methylotrophs. All five enzymes have been extensively characterized, but considerations of space allow only the first three enzymes to be discussed here.

1.2.4.1 NADH dehydrogenase

NADH dehydrogenase catalyses the transfer of reducing equivalents from NADH to either ubiquinone or menaquinone and, in concert with the soluble transhydrogenase, is the major route through which the terminal quinone–cytochrome system receives the reducing power generated as NAD(P)H by central metabolic pathways. This enzyme complex, unlike succinate dehydrogenase, is usually capable of proton translocation in most chemoheterotrophic bacteria. By contrast, in most chemolithotrophic bacteria it principally catalyses energy-dependent reversed electron transfer from quinol to NAD^+ at the expense of the protonmotive force, thus acting as an NAD^+ reductase and generating NADH for the assimilation of carbon dioxide.

NADH dehydrogenase contains a non-covalently bound flavin (either FMN or FAD) and up to four iron–sulphur centres, and is variably sensitive to rotenone and other inhibitors of the mitochondrial enzyme. It has a transmembrane location and oxidizes NADH on the cytoplasmic side of the membrane (Stouthamer, 1980; Ingledew and Poole, 1984). Unfortunately, the extraction and subsequent purification of the enzyme complex from different sources yields preparations with extremely varied properties that often only marginally resemble those of the membrane-bound dehydrogenase. For example, the NADH dehydrogenase from *Paracoccus denitrificans* comprises up to ten subunits and retains many of its *in vivo* properties, whereas the single subunit enzyme purified from *E. coli* lacks iron–sulphur centres and is probably only the flavoprotein component of the original dehydrogenase (although it is still able to reconstitute NADH oxidase activity when added to membranes prepared from a dehydrogenase-negative mutant). This latter property, coupled with the loss of rotenone sensitivity and proton translocation at the level of NADH dehydrogenase following growth under iron- or sulphate-limited conditions, suggests that one or more iron–sulphur centres may be essential for energy coupling but not respiration.

The marine bacterium *Vibrio alginolyticus* contains a novel three-subunit NADH dehydrogenase, which contains both FAD and FMN but apparently no iron–sulphur centres. The enzyme appears to act as a sodium pump, coupling electron transfer to the indirect translocation of Na^+ and thus conserving free energy in the form of a sodium-motive force which can subsequently be used to drive ATP synthesis and solute transport via an Na^+-translocating ATP synthase and sodium-motive force-dependent transport systems. This arrangement is of considerable benefit to the organism, which grows under alkaline conditions, since it effectively bypasses the energetic problems associated with the synthesis of ATP in the presence of a reversed ΔpH (Krulwich and Guffanti, 1983; Krulwich, 1985).

1.2.4.2 Succinate dehydrogenase

Succinate dehydrogenase catalyses the oxidation of succinate to fumarate and transfers the reducing equivalents to either ubiquinone or menaquinone. The relatively high redox potential of the fumarate/succinate couple (E_m $+30\ mV$) precludes any significant energy conservation by this reaction, and the enzyme does not catalyse proton translocation.

Succinate dehydrogenase has been purified from the respiratory membranes of several species of bacteria including *E. coli, Bacillus subtilis* and *R. rubrum*. Analyses of the purified enzyme, and of the DNA sequence of the cloned *E. coli* structural genes coding for the enzyme, show that it comprises four subunits (Ingledew and Poole, 1984; Cole *et al.*, 1985). Each consists of a flavoprotein which contains covalently bound FAD, an iron–sulphur protein which contains two [2Fe–2S] centres and possibly a [4Fe–4S] centre, and two small hydrophobic proteins which serve to anchor the enzyme into the membrane (one of which is possibly a *b*-type cytochrome).

The genes coding for the four subunits are arranged into an operon which is transcribed in the order *sdhCDAB*. Synthesis of the enzyme by a facultative anaerobe such as *E. coli* is apparently induced by aerobic growth on succinate, but the molecular basis of this regulation has not yet been determined.

1.2.4.3   Fumarate reductase

Many species of bacteria are able to use fumarate as a terminal electron acceptor and thus conserve energy anaerobically via fumarate respiration, most commonly at the expense of NADH:

$$\text{NADH} + \text{fumarate} \rightarrow \text{succinate} + \text{NAD}^+$$

this route therefore involves NADH dehydrogenase, quinone and fumarate reductase (see page 137).

The relatively high redox potential of the fumarate/succinate couple again precludes any significant energy conservation during the reduction of fumarate by quinol, and it also dictates that the donor is menaquinol (MKH_2) rather than the less strongly reducing ubiquinol. Fumarate respiration is thus restricted to organisms such as *E. coli* and the obligate anaerobe *Wolinella (Vibrio) succinogenes*, which can synthesize this quinone (Kroger, 1978; Ingledew and Poole, 1984).

The location, redox carrier composition, subunit structure and genetic organization of the *E. coli* fumarate reductase are remarkably similar to those of succinate dehydrogenase (Cole *et al.*, 1985). The two enzymes have almost identical structures, except that there is some doubt whether one of the anchor subunits of the reductase binds haem *b*. Electron transfer occurs in the order $MKH_2 \rightarrow \text{Fe–S} \rightarrow \text{FAD} \rightarrow \text{fumarate}$. The two catalytic subunits form overlapping knob-like protrusions on the cytoplasmic side of the membrane, which can be seen particularly easily in electron micrographs of *E. coli* strains that have been engineered to overproduce the enzyme.

Both enzymes exhibit dehydrogenase and reductase activities, albeit with greatly different K_m and V_{max} values; i.e. fumarate reductase has a lower K_m for fumarate than for succinate, and a higher V_{max} for fumarate reduction than for succinate oxidation, whereas the converse is true for succinate dehydrogenase.

The structural genes coding for fumarate reductase are organized into an operon, which is not located close to the *sdh* operon on the *E. coli* chromosome and is transcribed in the order *frdABCD*. There is increasing evidence that transcription is subject to both negative and positive control, the enzyme being repressed by nitrate and induced by anaerobic conditions or fumarate. The negative control is specific to fumarate reductase, and probably occurs by way of a nitrate-binding repressor protein (Nir). By contrast, the positive control is probably common to other anaerobic reductases and reflects the presence of an (*fnr*) gene coding for a positively regulating protein (Fnr). This protein, which enhances the binding of RNA polymerase, therefore acts in an analogous manner to the cAMP receptor protein, thus suggesting that the *fnr* gene takes over the function of the *crp* gene under anaerobic conditions. The fumarate reductase operon is therefore regulated in response to the availability of alternative electron acceptors (oxygen, nitrate and fumarate, but not TMAO), such that the enzyme is synthesized maximally only when fumarate is present and when energetically more effective electron acceptors such as oxygen and nitrate are absent.

It has been convincingly speculated that the membrane-bound fumarate reductase evolved from a soluble form of the enzyme that was originally located in the cytoplasm and, in concert with a soluble NADH dehydrogenase, catalysed the oxidation of excess NADH produced by fermentation. It is envisaged that this system later became membrane-associated and acquired the ability to translocate H^+, thus enabling it initially to supplement existing methods of pH regulation and subsequently to conserve energy (Gest, 1980, 1981; Jones, 1985). Succinate dehydrogenase almost certainly evolved much later, probably by duplication of the fumarate reductase structural genes; this was presumably followed by subtle changes in the DNA sequence and regulatory properties of the *sdh* operon in order to optimize the biochemical function of the enzyme.

1.2.5   Molybdoproteins

Several dehydrogenases and reductases are known which contain molybdenum plus at least one other redox centre (e.g. Fe–S, FAD). These include formate dehydrogenase, carbon monoxide dehydrogenase, nitrite dehydrogenase, nitrate reductase and TMAO/DMSO reductase. It is likely that in all of these multi-subunit enzymes the molybdenum is bound to a novel pterin (molybdopterin or bactopterin) to form a molybdenum cofactor (Fig. 1.13) (Johnson, 1980; Rajagopalan, 1985). Bactopterin differs from molybdopterin in that it contains an additional phosphate moiety which is probably bound via an aromatic residue. In both cases the molybdenum is non-covalently attached to the two sulphur atoms and contains additional oxo and/or sulphido groups. It carries one electron and oscillates between either Mo^{4+} and Mo^{5+} or Mo^{5+} and Mo^{6+}, the E_m values of the two couples varying between -270 mV and $+220$ mV depending on the exact environment of the molybdenum cofactor.

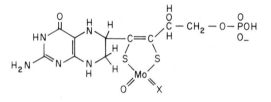

Fig. 1.13   The structure of the molybdenum co-factor. The pterin shown is molybdopterin; bactoperin contains an additional phosphate plus an aromatic residue. X is an oxo- or sulphido-group. The molybdenum atom carries one electron and exists as Mo^{4+}/Mo^{5+} or Mo^{5+}/Mo^{6+}. Redox changes are usually measured using EPR spectroscopy.

1.2.5.1   Carbon monoxide oxidoreductase

This enzyme is found mainly in aerobic carboxydotrophic bacteria (e.g. *Pseudomonas carboxydovorans*), where it catalyses the oxidation of carbon monoxide to carbon dioxide:

$$CO + H_2O \rightarrow CO_2 + 2H^+ + 2e^-$$

(Meyer and Schlegel, 1983; Meyer *et al.*, 1986). Since the formation of the carbon dioxide is at the expense of water, rather than of molecular oxygen plus a reductant, the enzyme is not a mono-oxygenase (hydroxylase). It is properly described as carbon monoxide : acceptor oxidoreductase, but is often incorrectly called carbon monoxide dehydrogenase or even carbon monoxide oxidase. Although the redox potential of the CO_2/CO couple is extremely low ($E_m - 540$ mV), carbon monoxide oxidoreductase is not linked to $NAD(P)^+$ but transfers reducing equivalents directly to cytochrome b/ubiquinone, thus exhibiting a similar efficiency of energy conservation to that of succinate oxidation. The subsequent assimilation of carbon dioxide requires NAD(P)H, which is formed by energy-dependent reversed electron transfer from cytochrome b/ubiquinol to NAD^+ via NADH dehydrogenase.

Carbon monoxide oxidoreductase is a molybdo iron–sulphur flavoprotein comprising three different subunits which bind a molybdenum cofactor, two 2Fe–2S centres and a non-covalently bound flavin. The functional enzyme exists as a dimer and is loosely attached to the cytoplasmic side of the membrane. It is strongly activated by selenite, maximum activity being attained when the selenium and molybdenum contents of the enzyme are equimolar. The function of the selenium, and the order of electron transfer within the enzyme, are unknown.

As carbon monoxide is a potent inhibitor of bacterial cytochrome oxidases, the respiratory chain of *Ps. carboxydovorans* is of neccessity modified to allow CO-insensitive respiration. During growth on heterotrophic substrates the organism exhibits a largely conventional respiratory chain but, during growth on carbon monoxide, it synthesizes a branch that is terminated by an unusually CO-resistant cytochrome oxidase *o* (Meyer and Schlegel, 1983; Meyer *et al.*, 1986; Poole, Chapter 5, this volume). A separate type of carbon monoxide dehydrogenase is involved in the acetyl-CoA pathway in anaerobes (page 108; Fig. 2.10).

1.2.5.2   Formate dehydrogenase

A large number of organisms acquire formate by anaerobic metabolism of pyruvate or from various exogenous sources. Methylotrophs oxidize formate to carbon dioxide using a soluble, NAD^+-linked formate dehydrogenase but, in other organisms, the enzyme is NAD^+-independent and forms part of the respiratory chain. As the redox potential of the CO_2/formate couple is low (E_m $-432\,mV$), formate is a particularly favourable electron donor during anaerobic respiration to relatively weak oxidants such as fumarate and nitrate. Two types of formate dehydrogenase have been reported, namely Fdh_N and Fdh_H. The Fdh_N transfers reducing equivalents to a quinone and is involved in energy conservation during fumarate or nitrate respiration, whereas the Fdh_H transfers reducing equivalents to a hydrogenase and is part of the fermentative formate–hydrogen lyase system which releases carbon dioxide and hydrogen (Ingledew and Poole, 1984):

$$\text{Formate} \rightarrow CO_2 + H_2$$

Fdh_N has been purified from *E. coli* and *W. succinogenes*. It comprises three different subunits, two of which span the membrane, and contains one each of a molybdenum cofactor, protohaem (cytochrome b^{fdh}), iron–sulphur centre and selenium (as a selenocysteine). Electron transfer probably occurs in the order formate $\rightarrow$ Mo $\rightarrow$ b^{fdh}; the role of the iron–sulphur centre and of the selenium remains to be determined. Formate is oxidized on the cytoplasmic side of the membrane by *E. coli*, but on the periplasmic side by *W. succinogenes*, the difference possibly reflecting the physiology of the two organisms and the different sources of formate available to them.

Fdh_H has been purified from *E. coli* and is structurally very similar to Fdh_N, but contains an apparently smaller selenoprotein.

The different physiological roles of the two enzymes is reflected in their regulation: Fdh_N is induced by nitrate and repressed during aerobiosis; Fdh_H is induced by formate and repressed by nitrate. Four structural genes have been identified which code for both enzymes (*fdhA,B,C,D*), and two more (*fdhE,F*) which probably code for the selenoprotein subunits of Fdh_N and Fdh_H respectively, but little is known about their organization and regulation. Several other genes have also been identified that are responsible for controlling the assembly and integration of the molybdenum cofactor into the formate dehydrogenases and nitrate reductase.

1.2.5.3   Nitrite oxidoreductase

The aerobic oxidation of nitrite to nitrate is restricted to a small group of organisms, principally the genus *Nitrobacter*, and is the final step in the process of nitrification. As the redox potential of the nitrate/nitrite couple is high (E_m +420 mV) energy conservation during nitrite respiration is relatively small.

The oxidation of nitrite:

$$NO_2^- + H_2O \rightarrow NO_3^- + 2H^+ + 2e^-$$

is catalysed by nitrite oxidoreductase which appears to contain a non-autoxidisable cytochrome a_1, molybdenum and cytochrome c. Electron transfer is probably in the order nitrite $\rightarrow a_1 \rightarrow$ (Mo) $\rightarrow$ cytochrome c. Nitrite oxidoreductase is almost certainly a transmembrane protein which oxidizes nitrite on the cytoplasmic side of the membrane (with the concomitant release of $2H^+$) and reduces cytochrome c on the periplasmic side.

The reduction of cytochrome c by cytochrome a_1 occurs against a negative ΔE_h and is therefore dependent on the protonmotive force, a conclusion that is confirmed by the observation that nitrite oxidation by whole cells is inhibited by protonophores. Furthermore, as the initial oxidation of nitrite releases $2H^+$ into the cytoplasm (i.e. on the 'wrong' side of the membrane, the terminal reduction of molecular oxygen must be catalysed by a proton-translocating cytochrome oxidase to ensure that energy is conserved during the overall process (Cobley, 1976; Wood, 1987; Wood, Chapter 4, this volume).

1.2.5.4   Nitrate reductase

Nitrate respiration is extremely widely distributed taxonomically and is the first step in the process of denitrification. It has been extensively investigated both in partial denitrifiers such as *E. coli* and in complete denitrifiers such as *Pc. denitrificans* (Stouthamer, 1976; Ingledew and Poole, 1984).

As the redox potential of the nitrate/nitrite couple (E_m +420 mV) is significantly higher than that of the fumarate/succinate couple, the energy yield during nitrate respiration is approximately twice that of fumarate respiration.

The reduction of nitrate:

$$NO_3^- + 2H^+ + 2e^- \rightarrow NO_2^- + H_2O$$

is catalysed by nitrate reductase, which in *E. coli* comprises three subunits and contains iron–sulphur centres, molybdenum and cytochrome b^{nr}. Electron transfer occurs in the order $b^{nr} \rightarrow$ Fe–S $\rightarrow$ Mo, and the enzyme is organized in the membrane such that cytochrome b^{nr} accepts electrons from cytochrome *b* or ubiquinone on the periplasmic side and the molybdoprotein reduces nitrate on the cytoplasmic side.

The genes responsible for the synthesis and regulation of nitrate reductase are organized into the *nar* operon (originally called *chl*), and consist of three structural genes (*narG,H,I*) plus two regulatory genes (*narK,L*). Regulation of nitrate reductase expression is extremely complicated and involves both transcriptional control (induction by nitrate, repression by oxygen) and post-translational modification of the nascent proteins prior to insertion in the cytoplasmic membrane.

Significantly different nitrate reductases to that described above have recently been shown to be present in *Pc. denitrificans* and in several phototrophs within the family *Rhodospirillaceae* (Ferguson *et al.*, 1987; Ferguson, Chapter 3, this volume; Jackson, Chapter 7, this volume). The enzyme in the phototrophs is located in the periplasm and probably consists only of a molybdoprotein in association with a *c*-type cytochrome. Its role, in association with other periplasmic reductases present in these photosynthetic organisms, is to allow anaerobic respiration during phototrophic growth. This has the effect of maintaining the photosynthetic electron transfer system at an optimal redox poise and/or of facilitating the removal of excess electrons during growth on highly reduced carbon substrates.

1.2.5.5 Dimethylsulphoxide reductase and trimethylamine-N-oxide
       reductase
The ability to use dimethylsulphoxide (DMSO) and trimethylamine-N-oxide (TMAO) as terminal electron acceptors for respiration has recently been reported for several organisms, including *E. coli*, *Proteus vulgaris* and *Rb. capsulatus* (Ingledew and Poole, 1984; Ferguson *et al.*, 1987). During this process, DMSO is reduced to dimethylsulphide (DMS):

$$(CH_3)_2SO + 2H^+ + 2e^- \rightarrow (CH_3)_2S + H_2O$$

and TMAO is reduced to trimethylamine (TM):

$$(CH_3)_3NO + 2H^+ + 2e^- \rightarrow (CH_3)_3N + H_2O$$

in reactions that are analogous to the reduction of nitrate to nitrite. However, although the redox potentials of the DMSO/DMS and TMAO/ TM couples ($E_m \simeq +160\,\text{mV}$ and $+130\,\text{mV}$ respectively) are significantly lower than that of the nitrate/nitrite couple, DMSO/TMAO respiration exhibits similar energy yields to that of nitrate respiration (Wood, 1981; Bilous and Weiner, 1985).

There is increasing evidence that a single molybdenum-containing enzyme (DMSO/TMAO reductase) catalyses both reactions. The enzyme in *E. coli* and *P. vulgaris* is membrane-bound and is linked to cytochrome *c*, whereas the enzyme present in the phototroph *Rb. capsulatus* also receives electrons from cytochrome *c* but is located in the periplasm and probably has a similar redox-balance/electron-sink function to the periplasmic nitrate and nitrite reductases (Ferguson *et al.*, 1987).

Superficial considerations might suggest that dimethylsulphone [$(CH_3)_2SO_2$] is also capable of supporting anaerobic respiration. However, the very low redox potential of the dimethylsulphone/DMSO couple ($E_m \simeq -240\,\text{mV}$) probably precludes such a function and no organisms have yet been isolated which can use it as a terminal electron acceptor. On the other hand, the low E_m value suggests that DMSO could act as a useful electron donor for aerobic respiration, and at least one species of a *Thiobacillus*-like organism has been isolated that exhibits this property.

1.2.6   Quinoproteins

Quinoprotein dehydrogenases contain a pyrrolo-quinoline quinone prosthetic group (PQQ; also called methoxatin) (Duine *et al.*, 1986, 1987; Anthony, 1986, and Chapter 6, this volume) (Fig. 1.14). PQQ is a complex *ortho*-quinone which has a characteristic absorption spectrum (Fig. 1.15). It carries a maximum of two hydrogen atoms and thus exists as PQQ, PQQH or PQQH$_2$. Although the redox potential of the free coenzyme is fairly high (E_m PQQ/PQQH$_2$ $+90\,\text{mV}$), it is likely that this value is variably decreased when the PQQ is bound to the apoenzyme, thus allowing different quinoprotein dehydrogenases to interact with the respiratory chain at the level of NAD$^+$, quinone, cytochrome or cuproprotein.

Quinoprotein dehydrogenases are structurally simple enzymes often composed of only one or two identical subunits. They include the methanol and methylamine dehydrogenases of methylotrophic bacteria such as *Methylophilus methylotrophus* and *Methylobacterium (Pseudomonas) AM1*; the ethanol dehydrogenases of the acetic acid bacteria and some species of *Pseudomonas*; the methylamine oxidase of *Arthrobacter P1* and glucose dehydrogenase. PQQ is probably also the prosthetic group of some glycerol, lactate, aldehyde and polyethylene glycol dehydrogenases. Although PQQ is the sole redox centre in most quinoprotein enzymes, it is accompanied by haem *c* in at least one ethanol dehydrogenase, by copper in methylamine oxidase and, possibly, by NAD^+ in a few methanol dehydrogenases.

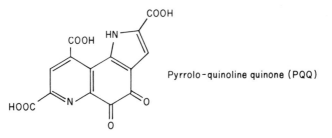

Pyrrolo-quinoline quinone (PQQ)

Fig. 1.14   The structure of pyrroloquinoline quinone (PQQ). PQQ carries two H-atoms and on reduction forms $PQQH_2$ via a semiquinone intermediate (PQQH) (see also Chapter 6, page 295).

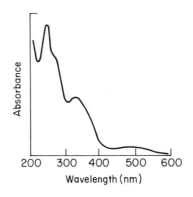

Fig. 1.15   The absorption spectrum of PQQ.

All of the quinoprotein dehydrogenases so far investigated with respect to their location within the cell have been isolated from periplasm fractions, or are membrane-bound with their active sites facing the periplasm. PQQ is bound tightly to the various apoproteins via ionic or covalent interactions, thus preventing significant diffusion of the prosthetic group into the external environment. Interestingly, some species of bacteria produce only inactive forms of glucose dehydrogenase and ethanol dehydrogenase, since they are unable to synthesize PQQ. These organisms appear to rely upon scavenging sufficient PQQ from the natural environment to synthesize functional holoenzymes.

Glucose dehydrogenase has a particularly low affinity for glucose compared with the various active transport systems for this substrate. External oxidation of glucose to gluconolactone, followed by the spontaneous formation and subsequent uptake of gluconate, is therefore used by many organisms as an alternative route for glucose oxidation during growth under excess-glucose conditions.

In contrast to iron–sulphur flavoproteins and molybdoproteins, which can act both as primary dehydrogenases and terminal reductases, no quinoprotein reductases have yet been reported.

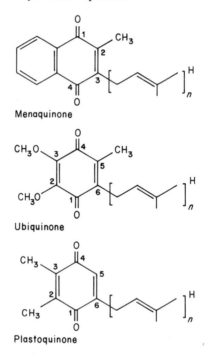

Menaquinone

Ubiquinone

Plastoquinone

Fig. 1.16   The structures of menaquinone (MK), ubiquinone (Q) and plastoquinone (PQ). All three quinones carry two H-atoms and on reduction form MKH_2, QH_2 and PQH_2 via semiquinone intermediates. At least one of these can probably exist in the anionic form (QH^-) which binds to a protein and participates in a protonmotive quinone cycle. The number of isoprene units present in the side chains varies from approximately 6 to 12.

1.2.7   Quinones

The most common quinones present in bacterial redox systems are ubiquinone (Q) and menaquinone (MK). These consist respectively of a substituted 1,4-benzoquinone or naphthoquinone nucleus attached to a long polyisoprenoid side chain. A few organisms contain modified menaquinones which lack the methyl group on the nucleus (demethylmenaquinone; DMK), contain a partially reduced side chain (dihydromenaquinone) or lack the first methylene group on the side chain (chlorobium quinone; CQ). Blue-green bacteria contain a modified ubiquinone in which the two methoxy groups are replaced by methyl groups and the original methyl group is absent (plastoquinone; PQ) (Fig. 1.16).

Novel sulphur-containing polyisoprenoid quinones have recently been isolated from several extremophiles. These include caldariellaquinone from the acidophilic archaebacterium *Caldariella acidophila*, and methionaquinone from the thermophilic hydrogen-oxidizer *Hydrogenobacter thermophilus*, both of which presumably have similar functions to the other quinones.

In contrast to PQQ, which is hydrophilic, all these quinones are lipophilic and are embedded deep within the membrane. They are usually present at high concentration relative to other redox components. It is likely that, in at least some bacteria, these bulk quinones may be attached to specific proteins, but they are easily extracted from the membrane using organic solvents. Some of the quinone present in the photosynthetic electron transfer systems of purple bacteria is strongly bound to the photosystem (Chapter 7).

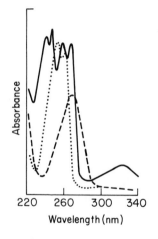

Fig. 1.17 The absorption spectra of MK (—), Q (– – –) and PQ (· · · · ·). The respective quinols exhibit weaker absorption bands which are shifted to slightly longer wavelengths.

Menaquinone, ubiquinone and plastoquinone exhibit characteristic absorption spectra in the ultraviolet region of the spectrum (Fig. 1.17). On reduction, they accept 2H to form the respective quinols, probably via semiquinone intermediates. In at least one case, the semiquinone ionizes to form the anion (QH^-), which is stabilized by association with protein and probably plays an important role in a protonmotive quinone cycle. The oxidized and reduced forms exhibit slightly different absorption spectra but, as these differences lie in the ultraviolet region of the spectrum, their measurement *in vivo* against a background rich in protein absorbance is technically difficult. Nevertheless, such measurements have been carried out successfully and indicate slightly different redox potentials for the major bulk quinones (E_m Q/QH$_2$ +70 mV, PQ/PQH$_2$ +100 mV, MK/MKH$_2$ −74 mV). The quinones that are bound to photosystems of photosynthetic bacteria exhibit much lower redox potentials than their bulk counterparts (E_m < −160 mV).

The roles of these bulk quinones in respiration and photosynthetic electron transfer have been determined using several different approaches, including:

(a) measuring their oxidation-reduction kinetics *in situ*;

(b) investigating the effect on electron transfer of extracting the quinone from the membrane and subsequently adding it back to the depleted membrane; and

(c) examining the respiratory properties of mutants deficient in either ubiquinone *ubi⁻* or menaquinone *men⁻* mutants.

The results of these experiments indicate that the bound quinones are primary electron acceptors in photosynthesis (i.e. they are uniquely involved in the transfer of reducing equivalents from the photosystem to the bulk quinone), whereas the bulk quinones transfer reducing equivalents from various low redox potential donors or bound quinones to the terminal cytochrome system during respiration and photosynthetic electron transfer respectively (see pages 61, 68 and 70).

Ubiquinone and menaquinone usually act as alternative redox components and, with the exception of photosynthetic bacteria, are generally restricted to Gram-negative and Gram-positive organisms respectively (Bishop *et al.*, 1962; Collins and Jones, 1981). It should be noted, however, that facultative anaerobes of the *Enterobacteriaceae* family (e.g. *E. coli*) contain both ubiquinone and menaquinone. Ubiquinone is the major quinone during respiration to oxygen or nitrate, whereas the lower-redox-potential menaquinone is only involved in fumarate respiration (Ingledew and Poole, 1984). Plastoquinone replaces ubiquinone and menaquinone in the photosynthetic electron transfer systems of the blue-green bacteria.

1.2.8 Cuproproteins

The cuproproteins present in bacterial redox systems (type I or blue cuproproteins; sometimes called 'cupredoxins' by analogy with ferredoxins) contain one atom of copper which is tightly bound in a co-ordinate manner to a small monomeric protein, usually via histidine, cysteine and methionine residues (Fig. 1.18) (Gray and Solomon, 1981; Ingledew, 1982; Ryden, 1984). The copper atom carries a single electron and exhibits a high redox potential (E_m Cu^{2+}/Cu^+ $\geqslant 230$ mV). In their oxidized forms, these cuproproteins absorb weakly in the near-red region of the spectrum and thus appear blue, but are completely bleached on reduction (Fig. 1.19).

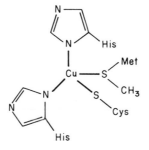

Fig. 1.18   The redox centre of a blue copper protein. The copper atom carries one electron and exists as Cu^+/Cu^{2+}.

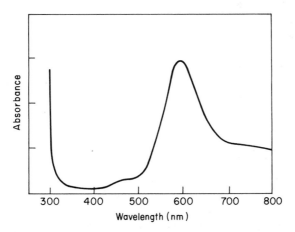

Fig. 1.19   The absorption spectrum of a blue copper protein (Cu^{2+}). The absorption band at approximately 600 nm disappears on reduction, and the Cu^+ form is colourless

Three major species of soluble, blue copper proteins occur in bacterial respiratory chains, viz. azurin, amicyanin and rusticyanin (Ingledew, 1982; Anthony, Chapter 6, this volume; Ferguson, Chapter 3, this volume), and plastocyanin is found widely in the photosynthetic electron transfer systems of blue-green bacteria. All are located in the periplasm, have fairly similar structural properties, but generally have different redox functions. During aerobic electron transfer in methylotrophs, amicyanin transfers electrons from methylamine dehydrogenase to cytochrome c or azurin which subsequently transfer them to the oxidase (Anthony, 1986, and Chapter 6, this volume); azurin also transfers electrons from cytochrome c to nitrite reductase during nitrite respiration in various species of denitrifying bacteria. Rusticyanin is involved in electron transfer from Fe^{2+} to cytochrome c in the acidophilic chemolithotroph *Thiobacillus ferrooxidans* and has an extremely high redox potential (E_m $+680$ mV) which is commensurate with this function (Chapter 4). Plastocyanin transfers electrons from a c-type cytochrome to the photosystem I reaction centre in blue-green bacteria.

Soluble cuproteins act as nitrite reductases and nitrous oxide reductases in some denitrifying bacteria, and are therefore intimately involved in the reduction of nitrite to dinitrogen. All these enzymes are located in the periplasm. By contrast, membrane-bound cuproteins are present in some cytochrome oxidases. All these enzymes exhibit some of the structural and redox properties of the simple electron transfer proteins described above, but they have in general been much less extensively investigated (Ferguson, Chapter 3, this volume; Poole, Chapter 5, this volume).

1.2.9   Cytochromes

Cytochromes are a specialized group of haemoproteins, i.e. they consist of a haem prosthetic group attached to a protein. Haem is composed of porphyrin (four pyrrole rings joined by $-CH=$ bridges) plus a single iron atom located at the centre of the planar molecule (Fig. 1.20). The iron carries one electron and oscillates between the Fe^{3+} and Fe^{2+} forms. It is usually, but not always, in a low-spin configuration such that it forms an octahedral coordination complex which binds six ligands. The four equatorial coordination positions are filled by the pyrrole nitrogens and, in most cytochromes, the fifth and sixth places are occupied by N and/or S atoms from the side chains of histidine or methionine residues in the protein. However, in those cytochromes that catalyse the terminal transfer of electrons to molecular oxygen during aerobic respiration (cytochrome oxidases), one of the axial coordination positions is occupied by water or oxygen, both of which can be replaced by stronger field ligands such as carbon monoxide or nitric oxide, which therefore act as competitive inhibitors. The groups at the C-1, C-3,

C-6 and C-7 positions of the haem are invariant, but the other four positions are available for substitution. It is the varied nature of these substituents that is responsible for the four types of haem: haem *a*, haem *b*, (proto-haem), haem *c* (mesohaem) and haem *d* (chlorins); and hence for cytochromes *a*, *b*, *c* and *d* with their characteristic absorbance properties and E_m values. In the *c*-type cytochromes, haem *c* is additionally attached to the protein via thioether linkages between the substituents at positions C-2 and C-4 on the haem and the side chains of appropriate cysteine residues in the protein. It is likely that this covalent attachment represents an evolutionary modification of the structure of cytochrome *c* to prevent significant dissociation of the mesohaem from the protein and thus allow the cytochrome to function in the periplasm, or on the periplasmic surface of the membrane, without significant loss of function (Wood, 1984).

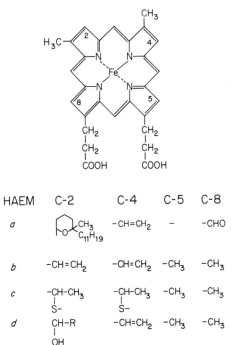

Fig. 1.20 The structures of the various haems found in bacterial cytochromes. Note that haem *d* is a dihydroporphyrin in which the 8,9 double bond is saturated (after Jones and Poole, 1985).

In addition to these simple cytochromes, a number of more complex forms have been reported in which an additional type of redox centre is present. For example, copper in some cytochrome oxidases, and flavin in flavocytochromes such as those catalysing the oxidation of reduced sulphur compounds by some phototrophs.

Cytochrome absorption spectra are characteristic of both the structure and the redox state of the haem prosthetic group, and are most conveniently analysed by measuring difference spectra (Fig. 1.21). Reduced *minus* oxidized difference spectra of bacterial cytochromes generally show the presence of three major absorption bands in the visible region of the spectrum: the α band (within the approximate range 550–640 nm when measured at room temperature); the β band (520–530 nm); and the γ or Soret band (410–450 nm). Exceptionally, *a*-type cytochromes do not exhibit a significant β band and *d*-type cytochromes show only a very weak γ band. The wavelength maximum of the α band is often used to describe the cytochrome (e.g. b_{558}, c_{550}), particularly if more historically based nomenclature (e.g. c_1, c_2, c_3) is unavailable.

Spectral analysis of whole cells and respiratory membranes is significantly improved by freezing them in liquid nitrogen (77 K), since both the sensitivity and resolution of difference spectra are enhanced when measured at lower temperatures. It should be noted, however, that this procedure often significantly shifts the wavelength maxima to a lower wavelength.

The ability of some cytochromes to bind carbon monoxide and other ligands also lends itself to spectral analysis, most commonly by measurement of reduced +CO *minus* reduced difference spectra. This property is indicative, but not by itself confirmatory, of an oxidase function.

E_m values of bacterial cytochromes generally lie between −100 and +500 mV, with the non-autoxidizable cytochromes usually exhibiting lower values than the cytochrome oxidases. A few specialized cytochromes c exhibit E_m values more negative than −100 mV.

1.2.9.1   Non-autoxidizable cytochromes

Non-autoxidizable *b*- and *c*-type cytochromes transfer electrons between the quinones and either the cytochrome oxidases, the terminal reductases or the reaction centre bacteriochlorophyll. The *c*-type cytochromes can be divided into three subgroups: type I, which has a single haem c attached to the protein close to the N-terminal end (e.g. cytochromes c_{551}, c_{555}, c_2 and c_4); type II, which has a single haem c attached to the protein close to the C-terminal end and is a high-spin compound (e.g. cytochrome c'); and type III, which contains several haems c (e.g. the tetrahaem cytochrome c_3 involved in sulphate respiration) (see Pettigrew and Moore, 1987, for a comprehensive review of *c*-type cytochrome).

Aerobic respiratory chains classically contain two cytochromes b and a cytochrome c_1 which, together with a high redox potential iron–sulphur protein, are organized into a tightly membrane-bound cytochrome b–c_1

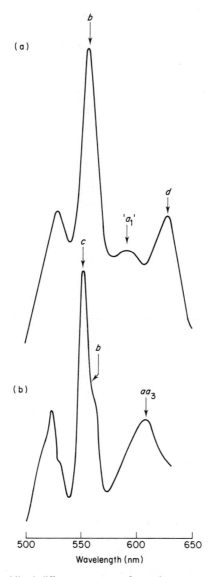

500   550   600   650

Wavelength (nm)

Fig. 1.21 Reduced *minus* oxidized difference spectra of cytochromes present in membranes prepared from aerobic bacteria. (*a*) *Escherichia coli* grown under oxygen-limitation; (*b*) *Methylophilus methylotrophus* grown under carbon-limitation. Spectra were measured at room temperature over the visible region (500–650 nm), and show β bands (λ_{max} 520–530 nm) and α bands (λ_{max} > 550 nm) only. Note that the absorption band of the *b*-type cytochrome (protohaem) at approximately 560 nm has contributions in (*a*) from cytochrome *b*, cytochrome *o* and one of the *b* components of cytochrome oxidase *d*, and in (*b*) from cytochrome *b* and the *o* component of cytochrome oxidase *co*. The absorption band of cytochrome 'a_1' is predominantly due to a long wavelength *b*-type cytochrome which is the other *b* component of cytochrome oxidase *d* (pages 51 and 265).

complex. These chains also contain a second, more soluble cytochrome c which subsequently transfers electrons to the cytochrome oxidase. It should be noted, however, that a few organisms (including *E. coli* and other members of the *Enterobacteriaceae* family) do not synthesize significant concentrations of these high redox potential c-type cytochromes during aerobic growth, and thus lack a $b-c_1$ complex, although they have the capacity to synthesize similar types of cytochromes anaerobically.

The cytochrome $b-c_1$ complex is situated close to the quinone and is part of the protonmotive quinone cycle. The latter involves Q, QH and QH_2, and transfers electrons from the iron–sulphur centres of the primary dehydrogenases to the cytochrome c_1 with concomitant proton translocation. Electrons are thence transferred via the second cytochrome c to one or more cytochrome oxidases. Particularly high concentrations of cytochromes c are present in some aerobic respiratory chains, especially those oxidizing relatively high redox potential substrates such as methanol, methylamine, ammonia and some oxyanions of sulphur, all of which feed reducing equivalents by way of periplasmic dehydrogenases or oxidoreductases direct to cytochrome c. In aerobic respiratory chains that lack cytochrome c_1, a protonmotive quinone cycle is probably absent and electrons are transferred from quinone to cytochrome b and thence to one or more cytochrome oxidases.

Anaerobic respiratory chains invariably contain at least one cytochrome b, and sometimes c-type cytochromes depending on the nature of the terminal acceptor. The former transfer electrons from the quinone to the appropriate terminal reductase during fumarate or nitrate respiration, whereas cytochrome c_{551} and the low redox potential c_3 respectively transfer electrons from cytochrome b to nitrite reductase during denitrification and from hydrogenase to menaquinone or cytochrome b during sulphate respiration (see Chapter 2).

1.2.9.2   Cytochrome oxidases

Three of the four types of haem can transfer electrons to molecular oxygen: haem a (cytochromes a_3 and possibly a_1), protohaem (cytochrome o) and some chlorins (cytochrome d). Each of these cytochromes combines with one or more non-autoxidizable cytochromes or copper centres to form a cytochrome oxidase (aa_3, caa_3, a_1, o, co and d) (see Chapter 5).

All these cytochrome oxidases bind carbon monoxide and exhibit characteristic reduced $+CO$ *minus* reduced difference spectra. However, this property is only suggestive of an oxidase function and should be confirmed by:

(a) comparing difference spectra and photochemical action spectra (the release of CO-inhibited respiration);

(b) measuring a direct reaction between CO and the reduced cytochrome; and

(c) showing that the putative oxidase is kinetically competent to support known electron transfer rates. Various combinations of these sophisticated spectrophotometric procedures have confirmed an oxidase role for most of these cytochromes oxidases.

Cytochrome oxidase aa_3 is widely distributed and closely resembles its mitochondrial counterpart in that it contains the same four redox centres (two haem a and two copper atoms) and probably functions by a similar catalytic mechanism (Ludwig, 1980, 1987; Poole 1983; Poole, Chapter 5, this volume). In contrast, it contains only the core two or three subunits of the mitochondrial enzyme and has an apparently variable capacity to act as a proton pump in different organisms, which may be related to the presence or absence of the third subunit. Cytochrome oxidase caa_3 is similar to cytochrome oxidase aa_3, but contains an additional haem c and is commonly found in thermophilic bacteria.

Cytochrome a_1 acts as an oxidase in the acidophile *Thiobacillus ferrooxidans*, and possibly also in some species of *Acetobacter*, but little is known about its redox centre composition or structure. The cytochrome 'a_1' reported on many occasions to accompany cytochrome oxidase d in the respiratory chains of *E. coli* and related organisms, has recently been shown to contain protohaem rather than haem a. Although it has a high-spin configuration and is able to bind carbon monoxide, it has no oxidase activity and is now classified as a cytochrome b; it forms part of the cytochrome oxidase d complex. It is possible that the cytochrome 'a_1' reported to be present in nitrite oxidoreductase is also a cytochrome b of this type.

Cytochrome oxidases o and co are widely distributed in bacterial respiratory chains, but cytochrome oxidase d is much more restricted and appears to function principally in respiratory chains, or respiratory chain branches, from which cytochrome c is absent. The biochemistry and genetics of cytochrome oxidases o and d from *E. coli* have been extensively investigated (see Poole, 1983; Ingledew and Poole, 1984; Anraku and Gennis, 1987; Gennis, 1987). Both act as ubiquinol oxidases, contain four redox centres and most commonly comprise two subunits. Cytochrome oxidase o contains two haem b and two copper atoms, whereas cytochrome oxidase d contains two haem b and two haem d. Open reading frames coding for these two cytochrome oxidases (*cyo* and *cyd* respectively) have been cloned and sequenced, but as yet no regulatory genes have been identified. Cytochrome oxidase co has been purified from several organisms including *Az. vinelandii* and *M. methylotrophus*. It accepts electrons from cytochrome c, contains two haem b and two haem c, and probably exists as a tetramer (see Chapters 5 and 6).

In spite of their obvious structural diversity, it is increasingly clear that these various cytochrome oxidases are functionally conservative. They all appear to contain four redox centres and thus have the capacity to carry the four electrons required to reduce a molecule of oxygen to water:

$$O_2 + 4e^- + 4H^+ \rightarrow 2H_2O$$

There is mounting evidence that several of these oxidases catalyse this reaction via tightly bound oxy- and peroxy-intermediates. It is noteworthy, however, that the affinity of the various cytochrome oxidases for molecular oxygen increases in the order $aa_3 < o < d$. In order to take maximum advantage of this property, synthesis of the oxidases is under close physiological control: co and d are usually repressed by high concentrations of oxygen in the growth medium, whereas o, aa_3 and caa_3 are either constitutive or are repressed by high concentrations of the carbon source. The result of this regulation is that during oxygen-limited growth $(c)aa_3$ is largely replaced by co, and o is replaced by d. In addition to its role in respiration, cytochrome o is also the receptor for positive aerotaxis which is mediated via the resultant changes in the protonmotive force (Taylor, 1983).

Topographical analyses of respiratory membranes or of cytochrome oxidase structures derived from DNA sequences of cloned genes, indicate that all the enzymes so far examined are transmembrane proteins. The binding site for oxygen is not known for certain, but is likely to be on the cytoplasmic side of the membrane.

All the cytochrome oxidases have a direct role in energy conservation. In the case of cytochrome oxidases o and d in organisms such as *E. coli* this consists of the oxidase, either alone or in combination with other cytochromes, catalysing inwardly directed electron transfer across the membrane from quinol to oxygen, thus acting as a protonmotive redox arm (type I) (Fig. 1.6). Cytochrome oxidases aa_3 and co in organisms such as *M. methylotrophus* and *Pseudomonas* AM1 probably have a similar function, except that they accept electrons from cytochrome c and thus act as protonmotive redox arms (type II) (Chapter 6). However, there is strong evidence that cytochrome oxidase aa_3 from *P. denitrificans*, and some cytochrome oxidases caa_3 from thermophilic bacteria, are additionally able to catalyse H^+ extrusion and thus act as conformational proton pumps in a similar manner to the mitochondrial enzyme. The oxidases briefly described here are discussed extensively in Chapter 5.

1.2.9.3 Nitrite reductases

During nitrite respiration, nitrite is reduced to either nitric oxide or ammonia. In denitrifying bacteria such as *Pc. denitrificans*, the product is nitric oxide:

$$NO_2^- + 2H^+ + e^- \rightarrow NO + H_2O$$

whereas non-denitrifying bacteria, such as *Desulfovibrio desulfuricans* and *E. coli*, produce ammonia:

$$NO_2^- + 8H^+ + 6e^- \rightarrow NH_4^+ + 2H_2O$$

The nitrite reductases responsible for these two reactions are both cytochromes: cytochrome cd_1 and a specialized cytochrome c, respectively. The former enzyme is structurally similar to some cytochrome oxidases, comprising two haem c and two haem b attached to two subunits; it is also able to reduce molecular oxygen to water, albeit slowly and with a low affinity. The latter enzyme contains six haem c per molecule, each of which exhibits a very low redox potential (E_m -240 mV); the haems occupy essentially identical environments and catalyse the concerted $6e^-$-reduction via a series of enzyme-bound intermediates. Both enzymes are located in the periplasm. Since their reactions therefore consume protons on the 'wrong' side of the membrane, the stoichiometry of proton translocation during nitrite respiration is less than expected (see Chapters 3 and 4).

1.2.10   Bacteriochlorophylls and bacteriopheophytins

Bacteriochlorophyll is remarkably similar to haem except that the central iron atom is replaced by magnesium, and pyrrole ring III is fused with an adjacent $-CH=$ to form a cyclopentanone ring (Fig. 1.22). Six types of bacteriochlorophyll have been identified (BChl*a,b,c,d,e* and *g*), and a seventh (BChl*f*) has been predicted to exist. They differ from each other in the nature of the substituents around the ring system, and hence in their absorbance properties and redox potentials. The wavelength maxima of the α-bands vary from approximately 715 to 1035 nm, and the E_m values (BChl$^+$/BChl) from approximately $+250$ to $+450$ mV. For each bacteriochlorophyll there is a corresponding bacteriopheophytin in which the Mg^{2+} is replaced by $2H^+$.

The bacteriochlorophylls, bacteriopheophytins and accessory pigments, mainly carotenoids, present in purple and green bacteria are closely associated with protein and are organized into photosynthetic units comprising light-harvesting complexes plus a reaction centre (Nugent, 1984; Drews, 1985; Siefermann-Harms, 1985; Zuber, 1986; Jackson, Chapter 7, this volume). The light-harvesting complexes contain bulk amounts of bacteriochlorophylls which act as antennae for the absorption of radiant energy and are classified on the basis of their absorbance maxima *in vivo*: B750, B788, B800–850, B870–890, B1015 (Table 1.3). The types of bacteriochlorophylls present are characteristic of the organism and, together with the carotenoids, determine the colour of the cells and hence the niche within the electromagnetic spectrum that the organism can occupy. In contrast, the reaction centre contains very much less bacteriochlorophyll, plus tightly bound bacteriopheophytin, quinone, iron and *c*-type cytochrome. The reaction centre bacteriochlorophyll is in a specialized environment and absorbs light of slightly longer wavelength than the bulk bacteriochlorophyll (P798, P840, P870, P960). The B/P ratio varies with the organisms and the light intensity during growth, but is always extremely high (40–1600). In most phototrophs the light-harvesting complex is an integral part of the membrane, but in some green bacteria (e.g. *Chlorobium limicola*) it is located in discrete 'chlorobium vesicles' which are loosely attached to the cytoplasmic surface.

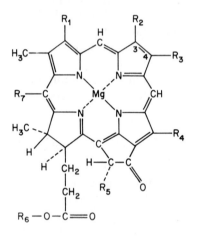

BChl	R_1	R_2	R_3	R_4	R_5	R_6	R_7	max(nm)
a	Ac	Me	Et	Me	CMe	Ph or Gg		850–910
b	Ac	Me	=CH.CH$_3$	Me	CMe	Ph		1020–1035
c	CHOH.CH$_3$	Me	Et or Pr	Et		Fa	Me	745–760
d	CHOH.CH$_3$	Me	Et or Pr	Et		Fa		725–745
e	CHOH.CH$_3$	Fo	Et or Pr	Et		Fa	Me	715–725
f	CHOH.CH$_3$	Fo	Et or Pr	Et		Fa		
g	CH=CH$_2$	Me	=CH.CH$_3$	Me	CMe	Gg		788

Fig. 1.22   The structures of the various bacteriochlorophylls found in photosynthetic bacteria. Abbreviations: Ac, acetyl; CMe, carboxymethyl; Et, ethyl; Fa, farnesyl; Fo, formyl; Gg, geranylgeranyl; Me, methyl; Ph, phytyl and Pr, propyl. The double bond between C3 and C4 is absent from bacteriochlorophylls *a* and *b*. BChl *f* is only hypothetical. The wavelength maxima refer to the long wavelength peak of each bacteriochlorphyll *in vivo*; solvent-extracted bacteriochlorophylls exhibit significantly shorter wavelength maxima due to the absence of BChl-protein interactions. The arrangement of bacteriochlorophyll in the membrane is shown in Figs 7.5 and 7.6 (pages 328 and 331).

Table 1.3 The bacteriochlorophylls of anoxygenic phototrophs

Family	Organism	Bacteriochlorophylls	
Rhodospirillaceae	*Rhodospirillum rubrum*	BChl*a* (B870)	P870
	Rhodobacter capsulatus	BChl*a* (B800, B850)	P870
	Rhodopseudomonas viridis	BChl*b* (B1015)	P960
Chromatiaceae	*Chromatium vinosum*	BChl*a* (B800, B820, B850, B890)	P870
Chlorobiaceae	*Chlorobium limicola*	BChl*c* (B750); BChl*a* (B870)	P840
Chloroflexaceae	*Chloroflexus aurantiacus*	BChl*c* (B750); BChl*a* (B870)	P865
—	*Heliobacterium chlorum*	BChl*g* (B788)	P798

Abbreviations: BChl, bacteriochlorophyll; B, bulk BChl; P, reaction centre BChl.
Numbers represent the approximate wavelengths of major absorption bands *in vivo*. See Fig. 1.22 for structures of these chlorophylls.

The absorption of a photon of light energy by a molecule of the bulk bacteriochlorophyll in the light-harvesting complex raises an electron to a higher energy level (singlet excitation), and the energy is then transferred through other bacteriochlorophyll molecules in the complex, until it eventually reaches the reaction centre.

In purple bacteria such as *Rhodopseudomonas viridis* and *Rhodobacter capsulatus* the reaction centre generally contains four molecules of bacteriochlorophyll *b*, two of which (the so-called special pair, $BChl_2$ or P870) become excited by accepting energy from the bulk bacteriochlorophyll. A single electron is then transferred very rapidly from P870 to one of the two molecules of bacteriopheophytin, thence to a quinone (Q_A) which lies close to the iron; and, finally, to a second quinone (Q_B). The transfer of a single electron from the tetrahaem *c*-type cytochrome to the oxidized P870 completes the function of the reaction centre. The role of the iron has not yet been determined. In green bacteria such as *C. limicola*, *Chloroflexus aurantiacus* and the recently discovered *Heliobacterium chlorum*, the quinone is replaced by a low redox potential iron–sulphur protein.

Elegant structural analyses of reaction centre crystals prepared from *R. viridis* have shown that the reaction centre comprises four subunits (H, M, L and a *c*-type cytochrome). The H subunit and the *c*-type cytochrome are on the cytoplasmic and periplasmic sides of the membrane respectively, and are linked by the hydrophobic L and M subunits which span the membrane and carry the bacteriochlorophylls, bacteriopheophytins and iron (Deisenhofer *et al.*, 1985a,b; see Chapter 7).

The overall result of these photochemical reactions is that an electron is transferred inwards across the membrane from a bound cytochrome *c* on the periplasmic side to a bound quinone or iron–sulphur protein on the cytoplasmic side, thus producing a moderately strong oxidant and reductant. The resultant electron transfer between these two redox components via the bulk quinone/cytochrome system completes the process of cyclic electron transfer which is predominantly responsible for the generation of the protonmotive force in photosynthetic bacteria.

A rather similar photosystem to that present in green bacteria is found in blue-green bacteria, except that chlorophylls (B680, P700) and pheophytins replace their bacterio- counterparts. These organisms also contain a second, higher-redox potential photosystem (photosystem II), which is also based on chlorophylls and pheophytins (B620, B670, P680), and contains a manganese protein which is probably responsible for oxygen evolution. Phycobiliproteins replace carotenoids as the main accessory pigments. (A complete description of bacterial photosynthesis is given in Chapter 7.)

1.3 Respiratory chains and photosynthetic electron transfer systems

1.3.1 Sequential organization

Bacteria exhibit considerable diversity in the complexity of their respiratory chains. This diversity reflects several different properties including:

(a) the redox potentials of the donor and acceptor couples;
(b) the genetic information inherent within the organism;
(c) the ability of the organism to regulate the synthesis of redox components according to the needs of the cell under different growth conditions; and
(d) the extent to which the respiratory chain has to interact simultaneously or sequentially with more than one donor and/or acceptor (e.g. during denitrification or desulphurification).

Most of these properties also apply to photosynthetic electron transfer systems.

The simplest types of bacterial respiratory chains are found where the difference in redox potential between the donor and acceptor couples (ΔE_h) is relatively small, thus restricting the range of redox components present. Such systems are exemplified by the aerobic oxidation of relatively high-redox potential donors such as iron, nitrite or methanol, by nitrate respiration at the expense of succinate, and by the anaerobic reduction of relatively low redox potential acceptors such as fumarate or adenosine phosphosulphate (APS) (Fig. 1.23a). These three groups of respiratory chains contain some quite dissimilar redox components, since the latter are primarily determined by the chemical structures and E_h values of the donor and acceptor couples rather than by the ΔE_h *per se*. In contrast, the most complex types of bacterial respiratory chains are found where the overall ΔE_h is very large and where the range of redox components present is therefore correspondingly broad. Such systems are exemplified by the oxidation of NADH or H_2 by oxygen or nitrite (Fig. 1.23b). These respiratory chains contain many common types of redox components since the E_h values of the donor (and acceptor) couples are not dramatically different.

It is important to note, however, that not all respiratory chains catalysing the same reaction (e.g. the aerobic oxidation of NADH) necessarily contain identical types of redox components. Significant differences occur between species, and also within a single species following growth under different conditions.

Interspecies differences usually reflect either the replacement of one redox

(a)

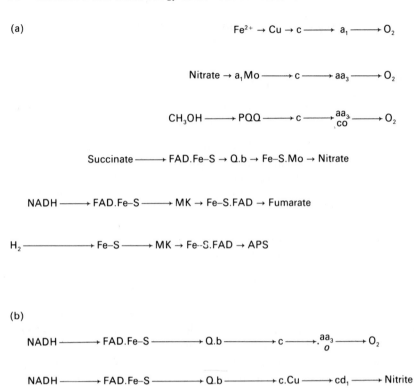

(b)

Fig. 1.23 Bacterial respiratory chains. (*a*) Simple systems (small ΔG); (*b*) complex systems (large ΔG). Abbreviations: APS, adenosine phosphosulphate (the activated, higher redox potential form of sulphate). Note that cytochromes *b* and *c* are each shown only as single species but in reality may exist as two or three different species.

carrier by another with basically similar properties, or the deletion of a carrier. Replacement is fairly common and is exemplified by the alternative use of ubiquinone and menaquinone (predominantly by Gram-positive and Gram-negative organisms, respectively), and of several cytochrome oxidases (aa_3, caa_3, o, co, or d). Deletion is rather less common and is largely restricted to the absence of an energy-linked transhydrogenase from many organisms, and of a high-redox-potential cytochrome *c* from the *Entero-bacteriaceae*.

Intraspecies differences usually reflect changes in the growth environment; most commonly, the availability of a particular nutrient. Versatile organisms such as *E. coli* or *Pc. denitrificans* contain the genetic information to synthesize at least two cytochrome oxidases, and several dehydrogenases and reductases, but not all of these enzymes are maximally expressed at the same time. For example, synthesis of the cytochrome oxidases is regulated by the availability of oxygen and/or the carbon source such that the oxidase with the highest affinity for oxygen is dominant during oxygen-limited growth (e.g. *d* largely replaces *o* or *co*, and *o* or *co* largely replaces aa_3 under these conditions). Similarly, the dehydrogenases and reductases are subject to induction or repression by various signals in response to the presence of various respiratory chain donors and acceptors, or of a catabolite repressor such as glucose. The regulation of reductase synthesis can be extremely complex, particularly where the acceptor can be sequentially reduced by the action of several different enzymes (e.g. the reduction of nitrate to N_2 during denitrification, or of APS to sulphide during desulphurification).

As a result of these control mechanisms the synthesis of a particular dehydrogenase or reductase is maximized under conditions where that enzyme is specifically required for the organism to respire rapidly, and hence grow as fast as possible, in a given growth medium. The nascent enzyme is inserted into its required position in the membrane, usually at the expense of a redundant dehydrogenase or reductase, by mechanisms which are as yet unclear. Bacterial respiratory chains are therefore particularly dynamic, and essentially modular, entities which respond rapidly to changes in the growth environment.

The varying ability of different organisms to synthesize alternative dehydrogenases, oxidases and reductases leads to considerable variation in the extent of respiratory chain branching (Fig. 1.24). All chains exhibit branching at the level of the dehydrogenases, albeit to a variable extent in different organisms, since this allows reducing equivalents from several different substrates to be chanelled simultaneously into a common, terminal respiratory chain and thus enables the effective integration of oxidative metabolism. In contrast, the terminal respiratory chain can be either branched or linear. The latter type of pathway is relatively rare and is limited to those systems which contain only one functional oxidase or reductase. The simplest form of terminal branching is usually associated with the presence of two oxidases or reductases, both of which accept electrons from the same penultimate redox carrier (usually a *b*- or *c*-type cytochrome). More complex branching occurs when the oxidases or reductases receive electrons from different *b*- or *c*-type cytochromes; in these systems one branch usually contains cytochrome *c* and is terminated by cytochrome oxidases aa_3, caa_3, *o*, *co* or nitrite reductase, whilst in the other branch cytochrome *b* donates electrons directly to cytochrome oxidases *d* or *o* or to nitrate reductase (Jones, 1977; Stouthamer, 1980).

The main interspecies variations in photosynthetic electron transfer systems involve replacements rather than deletions. Within the photosystems this includes the use of different (bacterio)chlorophylls, (bacterio)pheophy-tins and accessory pigments, and the replacement of bound quinone by a low redox potential iron–sulphur protein. The bulk quinone is ubiquinone or menaquinone in purple bacteria, menaquinone or chlorobium quinone in green bacteria; and plastoquinone in blue-green bacteria.

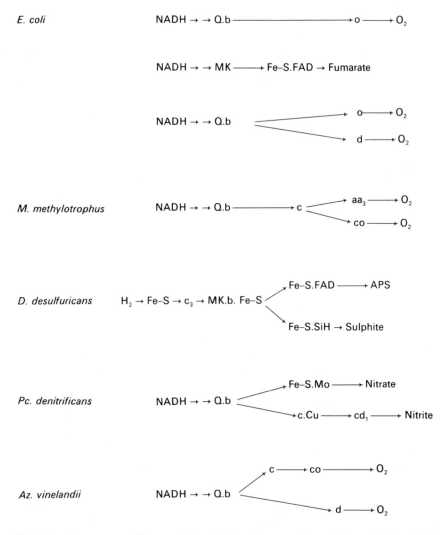

Fig. 1.24 The presence of linear and branched respiratory pathways in bacteria. Note that cytochromes *b* and *c* are each shown only as a single species but in reality may exist as two or three distinct species.

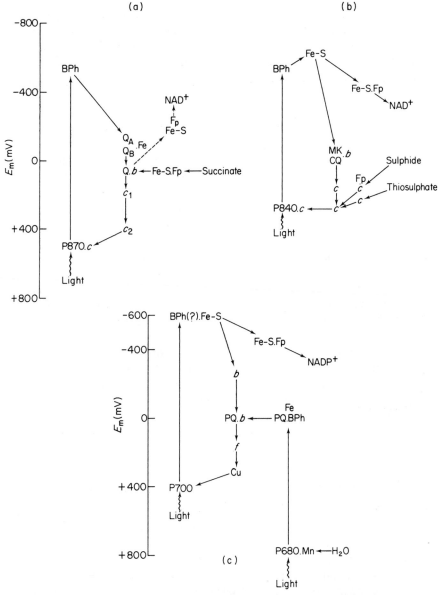

Fig. 1.25 Electron transfer systems in photosynthetic bacteria. (*a*) Purple bacteria; (*b*) green bacteria and some purple bacteria; and (*c*) blue-green bacteria. The bulk (bacterio)-chlorophylls and various accessory pigments responsible for light harvesting have been omitted. Q_a and Q_B are bound quinones that are present in the reaction centre in close association with iron. Δp-dependent reversed electron transfer is shown as a dashed line. Abbreviations: CQ, chlorobium quinone; *f*, cytochrome *f* (a *c*-type cytochrome); Mn, manganese protein; Cu, plastocyanin.

These differences have little obvious effect on cyclic electron transfer (Fig. 1.25), but the presence of a bound quinone rather than an iron–sulphur protein has a profound effect on non-cyclic electron transfer, i.e. the route by which NAD(P)H is generated from exogenous reductants. As the redox potential of the bound quinone is too high to allow the direct reduction of NAD^+, most purple bacteria catalyse reversed electron transfer from succinate to NAD^+ at the expense of the protonmotive force generated by cyclic electron transfer. In contrast, the redox potential of the bound iron–sulphur protein present in green bacteria is substantially lower than that of the quinone and, therefore, these organisms catalyse the direct, light-dependent transfer of electrons from sulphide or thiosulphate to NAD^+ via the photosystem without the mediation of the protonmotive force. This is also the case with blue-green bacteria, but in these organisms the use of water as the reductant necessitates the involvement of a second photosystem, since the larger redox span requires an additional input of radiant energy.

Intraspecies variations in the electron transfer systems of obligate phototrophs are less marked than in respiratory chains, and are largely restricted to the light-harvesting complexes, which increase dramatically in concentration during growth at low light intensities in order to maintain a high rate of photon capture. However, these variations are much more substantial in facultative phototrophs. Thus, in organisms such as *R. rubrum* and *Rb. capsulatus*, oxygen strongly represses synthesis of the photosystem and of cytochrome c_2, and increases the cytochrome oxidase content, thus allowing the organism to grow aerobically in the dark by switching from a phototrophic to a chemotrophic mode of growth. Indeed, *Rb. capsulatus* is particularly flexible in this respect since it can grow in the dark using aerobic respiration, anaerobic respiration or fermentation as means of energy conservation (see Fig. 7.1, page 321).

1.3.2 Spatial organization

The spatial organization of bacterial respiratory chains and photosynthetic electron transfer systems has been extensively investigated. The substrate-binding sites of transhydrogenase, flavoprotein and iron–sulphur flavoprotein dehydrogenases and reductases, most molybdoprotein dehydrogenases and reductases, and probably all cytochrome oxidases, are on the cytoplasmic side of the membrane. These enzymes are therefore either peripheral or transmembrane proteins. However, it should be noted that, in the latter case, the exact location of the individual redox centres within the membrane is rarely known. In contrast, c-type cytochromes, quinoprotein dehydrogenases, and most cuproproteins are attached to the external side of the membrane or are present in the periplasm; the redox centres are therefore tightly bound to the apoproteins to avoid significant loss into the environment, and their locations are obviously known with some precision (Fig. 1.26).

The light-harvesting complexes of phototrophic bacteria are embedded within the membrane, or are housed in membranes of accessory vesicles (chlorosomes or phycobilisomes) which are attached to the cytoplasmic side of the inner membrane. The reaction centres span the membrane, the cytochrome *c* and the bound quinone/iron–sulphur protein being located on the periplasmic and cytoplasmic sides of the membrane, respectively, and linked by the bacteriochlorophylls and bacteriopheophytins.

Relatively little is known about the exact position within the membrane of ubiquinone, menaquinone, the *b*-type cytochromes and, where present, the high-redox-potential iron–sulphur protein. However, there is some evidence that some semiquinone anions are bound to proteins located towards the sides of the membrane, the quinone and quinol forms being free to diffuse across the membrane; whereas the iron–sulphur protein and the lower redox potential cytochrome *b* are probably close to the periplasmic surface.

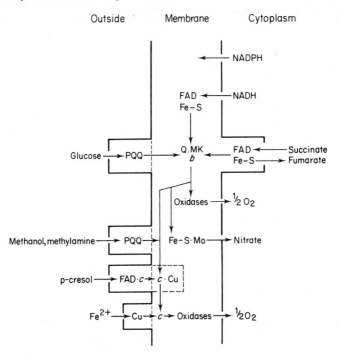

Fig. 1.26 The spatial organization of bacterial respiratory chains. The dashed lines indicate that the redox component is only loosely attached to the membrane or is even present in the periplasm.

1.3.3   Proton translocation

There is now considerable evidence that under conditions where the $\Delta\psi$ component of the protonmotive force is artificially collapsed, bacterial respiration is accompanied by the rapid translocation of H^+ across the respiratory membrane (although there is some doubt as to the extent to which this occurs *in vivo*; see Kell, Chapter 9, this volume).

This phenomenon has been extensively investigated by adding a small aliquot of dissolved oxygen, or of an alternative oxidant, to lightly buffered anaerobic cell suspensions (oxygen-pulse experiments), or by adding a small aliquot of reductant to aerobic cell suspensions (reductant-pulse experiments). Under these conditions, the resultant rate of H^+ extrusion by the cells is usually commensurate with the respiration rate, and the extent of acidification is stoichiometric with the amount of oxidant or reductant added (i.e. $\rightarrow H^+/O = \Delta H^+/\Delta O$) (Fig. 1.27a). The subsequent decay of the pH gradient is generally first order with respect to the concentration of H^+ and is relatively slow ($t_{\frac{1}{2}} > 45$ s), although a few organisms exhibit strikingly faster rates. The slow decay reflects the inherently low permeability of the respiratory membrane to H^+, and is dramatically accelerated by protonophores such as FCCP. Respiration-linked H^+ translocation has also been investigated, albeit less extensively, by adding excess reductant to aerobic cell suspensions (initial rate experiments) and then simultaneously measuring the rates of H^+ ejection, K^+ uptake and oxygen consumption ($\rightarrow H^+/O = jH^+/jO$; $\rightarrow$charge$/O = -jK^+/jO$) (Fig. 1.27b).

The number and location of the Δp-generating sites within respiratory chains has been further investigated by functionally dissecting the chains into discrete proton- and/or charge-translocating regions using exogenous donors and acceptors, both physiological and artificial, in the presence of appropriate inhibitors to block electron transfer elsewhere.

These methods have been used successfully with many species of bacteria, particularly aerobes. Whole cell $\rightarrow H^+/O$ and/or $\rightarrow$charge$/O$ quotients of between 2 and 10 have been reported for the aerobic oxidation of endogenous substrates (predominantly NADP(H)) by chemoorganotrophs. This reflects the presence of up to four Δp-generating sites, each with its characteristic $\rightarrow H^+/2e^-$ and $\rightarrow$charge$/2e^-$ quotient, viz. site 0 (NADPH $\rightarrow$ NAD$^+$), site 1 (NADH $\rightarrow$ quinone), site 2A or 2B (quinol $\rightarrow$ cytochrome c or quinol $\rightarrow O_2$ respectively) and site 3A or 3B (cytochrome $c \rightarrow O_2$ via either a non H^+-pumping cytochrome oxidase or via an H^+-pumping cytochrome oxidase respectively) (Table 1.4) (Fig. 1.28).

The molecular mechanisms via which the transfer of reducing equivalents through these sites leads to proton and/or charge translocation have not yet been fully resolved (see Fig. 1.6 for summary of types of mechanism).

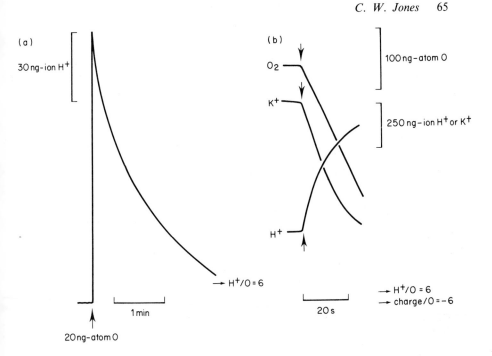

Fig. 1.27 Measurement of respiration-linked proton translocation in bacteria. (*a*) Oxygen-pulse experiment; a small aliquot of dissolved oxygen (20 ng-atom) is added to a suspension of anaerobic cells ($\rightarrow H^+/O = \Delta H^+/\Delta O = 6$). (*b*) Initial-rate experiment; excess reductant is added to aerobic cells ($\rightarrow H^+/O = jH^+/jO = 6$, $\rightarrow$charge/O = $-jK^+/jO = -6$). In both experiments the cells are suspended in lightly buffered medium and pre-incubated with valinomycin + K^+ to collapse $\Delta\psi$.

However, the lack of detectable redox carriers in transhydrogenase indicates a conformational proton pump mechanism at site 0 which, in a rather different form and in association with a type II redox arm, probably also operates at site 3B. Site 1 involves either a redox loop or a conformational proton pump, whereas site 2A is a Q-cycle involving quinone plus the cytochrome b-c_1 complex, and sites 2B and 3A are probably a redox loop and a type II redox arm, respectively (Jones, 1977; Stouthamer, 1980; Ingledew and Poole, 1984). It should be noted that site 3A is readily converted from a type II to a type I redox arm during the oxidation of methanol or methylamine, since the dehydrogenases for these H-atom donors interact with cytochrome c on the periplasmic side of the membrane.

These large interspecies variations in $\rightarrow H^+/O$ and $\rightarrow$charge/O quotients, and the rather smaller ones which occur within a single organism following growth under different conditions, clearly reflect the presence or absence of

Table 1.4 H^+ translocation and respiratory chain composition

Site	Respiratory chain region	$\rightarrow H^+/2e^-$	$\rightarrow$charge/$2e^-$	Mechanism
0	Transhydrogenase	2	-2	Indirect pump
1	NADH $\rightarrow$ quinone	2	-2	Indirect pump/redox loop
2A	Quinol $\rightarrow$ cytochrome c	4	-2	Q-cycle
B	Quinol $\rightarrow O_2$	2	-2	Redox loop
3A	Cytochrome $c \rightarrow O_2$	0	-2	Redox arm (II)
B	Cytochrome $c \rightarrow O_2$ (H^+ pumping aa_3 or caa_3)	2	-4	Indirect pump + Redox arm (II)

A summary of the mechanisms is given in Fig. 1.6 (page 16).

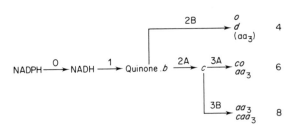

Fig. 1.28 The effect of the composition and sequential organisation of aerobic respiratory chains on the number and type of Δp-generating sites used by a chemoorganotroph. The $\to H^+/O$ quotient shown is for NADH as substrate; for the oxidation of NADPH by organisms which contain an energy-linked transhydrogenase (site 0), the $\to H^+/O$ quotient should be increased by 2. It should be noted that some organisms contain branched respiratory chains and will thus exhibit variable $\to H^+/O$ quotients depending on the exact route of electron transfer, and that the growth of some organisms (e.g. *Az. vinelandii*) under extremely high oxygen tension leads to the loss of proton translocation at the level of NADH dehydrogenase (site 1). $\to H^+/O$ and $\to$charge/O quotients for the oxidation of endogenous substrates thus range from a minimum of approximately 2 to a maximum of approximately 10.

certain respiratory chain components (transhydrogenase, cytochrome c or proton-pumping cytochrome oxidases), and also the sequential organization of the respiratory chain (linear or branched), all of which affect the number of actual or effective Δp-generating sites used (Fig. 1.29).

Wide variations in the stoichiometry of respiration-linked proton and/or charge translocation also occur during anaerobic respiration in chemoorganotrophs and during the oxidation of inorganic donors by chemolithotrophs. They principally reflect the difference in the redox potentials of the various donors and acceptors used by these organisms. It should be noted, however, that the spatial location of some reductases is very important in this respect. For example, the sequential reduction of nitrite to dinitrogen by denitrifying bacteria ($3e^-$ transfer) and the concerted reduction of nitrite to ammonia by some non-denitrifying bacteria ($6e^-$ transfer) respectively consume $4H^+$ and $8H^+$ on the periplasmic side of the membrane and thus significantly diminish the $\to H^+/2e^-$ quotient. Interestingly, many of the dehydrogenases and oxidoreductases used by chemolithotrophs are also periplasmic, but with no detriment to proton translocation. Type I redox arms are therefore commonly found both at site 1 ($H_2 \to$ quinone) and, by analogy with methanol oxidation, at site 3 (sulphite $\to c \to O_2$) (Fig. 1.30).

Proton translocation during photosynthetic electron transfer is measured under essentially the same conditions as for respiration except that it is initiated by light instead of by the addition of an oxidant or reductant.

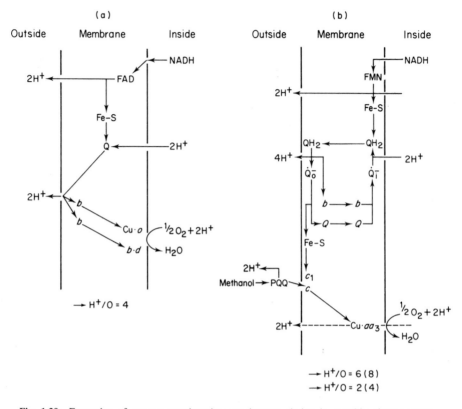

Fig. 1.29   Examples of proton-translocating respiratory chains in aerobic chemoorganotrophs. (*a*) High redox potential cytochrome *c* absent (e.g. *E. coli*); (*b*) high redox
potential cytochrome *c* present (e.g. *P. denitrificans, M. methylotrophus*). Site 1
could be a redox loop or a conformational pump in either respiratory chain,
although there have been claims that site 1 in *P. denitrificans* translocates more than
$2H^+$ and would therefore more probably involve a conformational pump mechanism. Site 3A is present in *M. methylotrophus* (non proton-pumping cytochrome
oxidase), whereas *P. denitrificans* contains site 3B (a proton-pumping cytochrome
oxidase). Note that site 2A, the protonmotive quinone cycle, translocates $2H^+$ for
the net transfer of a single electron ($4H^+/2e^-$).

Following illumination, H^+ ejection occurs until a steady state is achieved,
i.e. until the rates of ejection and backflow are equal ($jH^+_{out} = jH^+_{in}$) and
there is no further net movement. Since cyclic electron transfer involves no
net oxidation or reduction, the determination of $\rightarrow H^+/2e^-$ quotients is
extremely difficult and requires the measurement of H^+ movement following
illumination by flashes of light which are sufficiently brief ($<20\,\mu s$) as to
allow only one complete turnover of the photosystem. These complicated

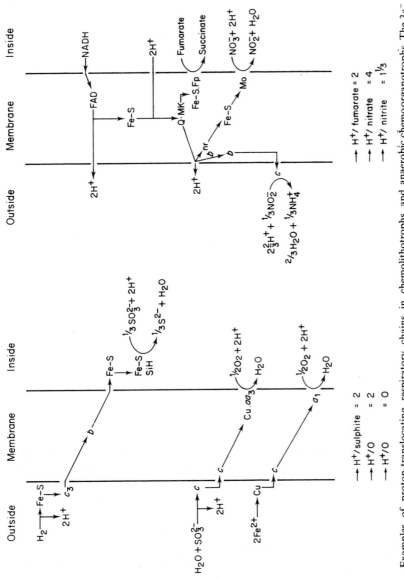

Fig. 1.30 Examples of proton-translocating respiratory chains in chemolithotrophs and anaerobic chemoorganotrophs. The $3e^-$ reduction of nitrite to ammonia is normalized to a $2e^-$ reaction.

and difficult experiments have been carried out using chromatophores from several species of purple bacteria and indicate an $\rightarrow H^+/2e^-$ quotient of approximately 4 during cyclic electron transfer. This reflects the presence of inwardly directed electron transfer from cytochrome c_2 to bound quinone through the reaction centre (type II redox arm), followed by a protonmotive quinone cycle involving ubiquinone and the cytochrome $b–c_1$ complex (Fig. 1.31). It is beyond the scope of this chapter to discuss these experiments in more detail (but see Jackson, Chapter 7, this volume).

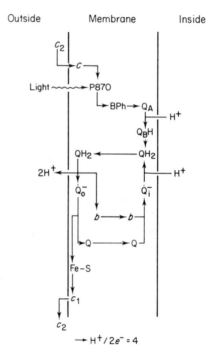

Fig. 1.31   Proton translocation during cyclic electron transfer in some purple photosynthetic bacteria (e.g. *Rb. capsulatus*). The tightly bound quinones (Q_o and Q_i) are also called Q_z and Q_c, as shown in the expanded version of this diagram in Fig 7.20 (page 355).

The protonmotive quinone cycle which operates during cyclic electron transfer is therefore closely related to site 2A in the respiratory chains of chemotrophic bacteria. Indeed, it takes on this latter function when facultative phototrophs are grown aerobically in the dark, under which conditions it is accompanied by sites 1 (NADH $\rightarrow$ quinone) and 3A ($c \rightarrow O_2$ via a non-proton-pumping cytochrome oxidase).

1.3.4 ATP/2e⁻ quotients and growth yields

The capacity of a redox system to translocate H^+ and thus generate a protonmotive force should, of course, be reflected in the subsequent ability of the organism to carry out various energy-dependent membrane functions, principally ATP synthesis and solute transport, and ultimately to synthesize cell material. There is now convincing evidence from the results of *in vitro* experiments with whole cells or membrane vesicles that the various H^+- and/or charge-translocating sites in these redox systems:

(a) allow the stimulation of electron transfer by protonophores;
(b) drive ATP synthesis and solute transport; and
(c) catalyse reversed electron transfer at the expense of the protonmotive force generated by ATP hydrolysis or by forward electron transfer through a different energy-coupling site.

ATP/O (ATP/2e^-) quotients have been widely determined using whole cells and inside-out membrane vesicles by measuring, respectively, the endogenous and exogenous changes in the concentrations of adenine nucleotides and inorganic phosphate as a function of electron transfer. In many cases the values obtained are probably underestimates due either to failure to correct for ATP turnover or adenylate kinase activity in whole cells, or to the presence of broken vesicles. The most reliable determinations generally indicate values of up to 2 for respiration (depending on the nature of the donor and acceptor, and on the redox carrier composition of the respiratory chain); and of less than 2 for cyclic electron transfer in phototrophs (although this is again particularly difficult to measure since there is no net oxidation–reduction during this process).

These studies have been supplemented by *in vivo* determinations of the molar growth yields (Y; g cells. mol substrate^{-1}) of many chemotrophs. During growth in continuous culture:

$$q = D/Y$$

where q is the rate of substrate utilization (mol. g cells^{-1}. h^{-1}) and D is the dilution rate (equivalent to the specific growth rate, μ; h^{-1}). In terms of non-equilibrium thermodynamics, q and D are regarded as input and output flows respectively, and hence Y is a flow ratio (D/q). By measuring the rate of substrate utilization of an energy-limited culture growing at a fixed dilution rate, the growth yields of chemoorganotrophs can be determined with respect both to the utilization of the carbon substrate (e.g. $Y_{glucose}$) and, more appropriately for analysis of respiratory chain energy conservation, the terminal oxidant (e.g. Y_{O_2}, $Y_{nitrate}$). For the growth of chemolitho-trophs, the yield with respect to the utilization of the inorganic donor (e.g. $Y_{thiosulphate}$) is also a good index, although it additionally reflects energy consumption for reversed electron transfer to $NAD(P)^+$.

Comparative studies with various species of aerobic chemoorganotrophs indicate that organisms that contain a high redox potential cytochrome c, and hence conserve energy at site 3, generally exhibit substantially higher growth yields than organisms that lack this cytochrome and hence terminate respiration after site 2A; this has been confirmed using cytochrome c-deficient mutants. In contrast, the presence or absence of an energy-linked transhydrogenase (site 0) does not significantly affect the growth yields, thus supporting the view that it has no direct function in energy conservation; as yet no data are available that confirm the expectation of enhanced energy conservation through the use of a proton-pumping cytochrome oxidase (site 3B). Growth yield determinations for chemolithotrophs and anaerobic chemoheterotrophs have also largely confirmed parallel measurements of respiratory chain energy conservation *in vitro*.

The quantitative relationship between growth yield and the overall ATP/O ($ATP/2e^-$) quotient is an apparently simple one, viz.

$$Y_{O_2} = Y_{ATP} \cdot 2[ATP/O]$$

where Y_{ATP} is the yield of cells with respect to ATP utilization (g cells. mol ATP^{-1}). However, the quantitative determination of the ATP/O quotient from the growth yield is fraught with problems since the value of Y_{ATP} varies with the nature of the carbon substrate, and is probably only known with any degree of confidence for growth on glucose; even in the latter case there remains some doubt, since it is calculated from anaerobic growth yields using the known ATP yields of the fermentation pathways employed and is then used on the assumption that $Y_{ATP(fermentative)}$ and $Y_{ATP(aerobic)}$ are identical.

ATP/O quotients calculated in this way range from approximately 1 to possibly as high as 3 for different organisms, and generally reflect the ATP/O quotients and $\rightarrow H^+/O$ quotients measured *in vitro*. In some cases, however, this relationship is complicated by additional factors including the presence of branched respiratory chains, enhanced H^+ leakage through the membrane, increased ATP turnover via futile cycles, further oxidation/reduction of the initial donor/acceptor, and even by the possibility of slippage in proton pumps (see Chapter 9, this volume). It also seems increasingly clear that energy conservation in bacteria is geared to maximizing growth rate rather than growth efficiency. Unfortunately, it is beyond the scope of this chapter to discuss these problems in more detail (but see Stouthamer and Bettenhaussen, 1973; Jones, 1977; Stouthamer, 1977; Tempest, 1978; Westerhoff *et al.*, 1982; Tempest and Neijssel, 1984; Kell, 1987; Stouthamer and van Verseveld, 1987).

1.4   The ATPase/ATP synthase complex

The ATPase/ATP synthase complex is a membrane-bound assembly of two oligomeric proteins, BF_1 and BF_0, which has been found in all organisms so far investigated. Its function is to catalyse the Mg^{2+}- or Ca^{2+}-dependent hydrolysis or synthesis of ATP with the concomitant generation or utilization of the protonmotive force. ATP hydrolysis occurs predominantly in fermentative organisms and uses ATP produced by substrate-level phosphorylation to generate the protonmotive force that subsequently drives various energy-dependent membrane functions. ATP synthesis occurs in respiratory and photosynthetic organisms at the expense of the protonmotive force generated by the redox reactions or bacteriorhodopsin.

The structure and function of the ATPase/ATP synthase complex has been extensively investigated in several organisms, particularly *E. coli* and the thermophile *Bacillus stearothermophilus* PS3. Early work with the thermophile was directed very much towards the reconstitution of an active complex from purified subunits (Kagawa, 1978), whereas studies with *E. coli* have mainly concentrated on the analysis of uncoupled (*unc*) mutants and, latterly, on the use of recombinant DNA technology to derive detailed structural information via DNA sequencing and site-directed mutagenesis (Downie *et al.*, 1979; Walker *et al.*, 1984; Cox *et al.*, 1984).

BF_1 is a hydrophilic complex which is attached to BF_0 on the cytoplasmic side of the membrane, whence it can easily be released by sonication, by washing in low ionic strength buffers, or by exposure to detergents and chaotropic reagents. Purified BF_1 binds ATP, ADP and inorganic phosphate, and catalyses the divalent metal ion-dependent hydrolysis of ATP (but not ATP synthesis, since a protonmotive force cannot be brought to bear upon the enzyme in the absence of BF_0 and a vesicular membrane structure). It generally comprises five subunits in the ratio $\alpha_3\beta_3\gamma\delta\epsilon$, although structurally simpler forms are present in some fermentative organisms and archaebacteria. In contrast, BF_0 is a hydrophobic complex which facilitates the transport of H^+ across the membrane, and is composed of three subunits in the probable ratio ab_2c_{8-10}. Membranes stripped of BF_1, and thus containing only the BF_0 part of the overall complex, have no ability to catalyse either ATP hydrolysis or ATP synthesis, but they exhibit high rates of H^+ leakage. When BF_0 and BF_1 are recombined by adding BF_1 back to stripped membranes under the appropriate conditions, proton-translocating ATPase/ATP synthase activity is regained. It is clear, therefore, that the co-ordinated action of approximately 20 proteins, eight of them different, is required to catalyse this process.

Synthesis of the $BF_0.BF_1$ complex is coded for by the *unc* operon (sometimes also called the *atp* operon). The operon has been completely sequenced and shown to consist of nine genes; eight of these code for the subunits of the $BF_0.BF_1$ complex, and the remaining gene (gene 1) codes for a membrane protein of unknown function. The operon is transcribed in the order $1acb\delta a\gamma\beta\epsilon$, indicating that BF_0 is synthesized prior to BF_1. The potentially deleterious effects on the organism of inserting BF_0 into the membrane prior to capping with BF_1 (enhanced H^+ leakage) or of synthesizing soluble BF_1 (increased ATP hydrolysis) are self-evident, and several different mechanisms by which the complex may be assembled to avoid such problems have been proposed. These include direction of the assembly process by the gene 1 product, and mutual dependence for assembly on the simultaneous presence of BF_0 and BF_1. The mechanism by which the cell synthesizes different amounts of each subunit from a polycistronic message has not yet been determined, but may involve mRNA loops or differential codon usage.

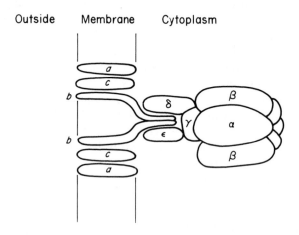

Fig. 1.32   A diagrammatic representation of the structure of the *E. coli* ATPase/ATP synthase complex ($BF_0.BF_1$). The complex contains eight different subunits in the ratio $\alpha_3\beta_3\gamma\delta\epsilon ab_2c_{8-10}$. As the *c* subunits are arranged in the form of a circular picket fence, only two are shown. The exact location of the *a* subunit is not known, but it possibly surrounds the *c* subunits; the 'two' *a* subunits shown are therefore two of the transmembrane helices that comprise this subunit. (After Walker *et al.*, 1984; Cox *et al.*, 1984) (see Fig. 1.5).

The two larger subunits in BF_1 exhibit significant sequence homology to each other, and much of the β subunit in particular is very highly conserved in both prokaryotes and eukaryotes. The α and β subunits are arranged alternately to form a hollow, planar hexagon ($\alpha_3\beta_3$), the centre hole of which is partly filled by the γ subunit (Fig. 1.32). DNA sequences indicative of adenine nucleotide binding sites have been identified in the α and β subunits; of the six such binding sites present in BF_1, the three on the α subunits bind nucleotides with high affinity and probably have a regulatory function, whereas the three on the β subunits (or possibly at the αβ interfaces) bind nucleotides with low affinity and almost certainly have a catalytic function. The available evidence therefore indicates that the $\alpha_3\beta_3$ hexagon is responsible for the catalytic activity of the complex, whereas the γ subunit probably binds the hexagon headpiece to the stalk.

The two *b* subunits each contain a hydrophilic and hydrophobic region which respectively protrude from the surface of the membrane into BF_1 and are embedded deep within the membrane. The hydrophilic region of each subunit, together with the δ and ε subunits, constitutes a stalk that connects the catalytic/regulatory headpiece of BF_1 to the H^+ channel portion of BF_0. In contrast, the hydrophobic region of each *b* subunit contains an essential lysine residue (lys-23) which is in contact with, and surrounded by, a picket fence of eight to ten *c* subunits. The amino acid sequence of the *c* subunit, like that of the β subunit, is very highly conserved between species and produces a hairpin structure comprising two transmembrane α-helical regions. It is characterized by the presence of an essential aspartate residue (asp-61), or in some species a glutamate residue, which is probably involved in the binding and/or translocation of H^+. Replacement of the acidic residue with a neutral one using site-directed mutagenesis, or exposure of the $BF_0.BF_1$ complex to carboxyl-group reagents such as DCCD (dicyclohexyl carbodiimide), leads to the complete inhibition of ATP hydrolysis/synthesis, even when only a small fraction of the acidic residues are modified, thus indicating that the *c* subunit oligomer is required for activity. BF_0 therefore resembles its eukaryote counterpart (F_0) in terms of its sensitivity to DCCD, although it is largely insensitive to the other classical inhibitor of F_0, oligomycin.

Remarkably little is known about either the physiology of $BF_0.BF_1$ synthesis or the regulation of ATPase/ATP synthase activity. The complex is clearly synthesized by a facultative anaerobe such as *E. coli* during both oxidative or fermentative growth, but whether the mode or rate of growth has any significant effect on the amount of enzyme produced or on its catalytic or structural properties is not known. It is noteworthy, however, that the enzymes from obligately aerobic and obligately anaerobic eubacteria exhibit significantly different kinetic and structural properties. At the molecular level, DNA sequencing of the *unc* operon has revealed a noncoding region adjacent to the promotor which may serve to concentrate the RNA polymerase; as the latter can apparently switch between at least two different conformational states depending on the ambient concentration of various nucleotides (e.g. ATP, ADP, GTP, ppApp, ppGpp), it is possible that this offers a mechanism by which the expression of the operon can be modulated by the metabolic status of the cell. There is some evidence that this transcriptional regulation may be accompanied, at least in some aerobic organisms, by kinetic regulation of the enzyme to prevent ATP hydrolysis (mediated in some way by the ε subunit).

1.5   ATP synthesis

It is now clear that net synthesis of ATP by whole cells, membrane vesicles or artificial liposomes containing $BF_0.BF_1$ generally occurs when the protonmotive force is of the desired direction and exceeds a threshold value of approximately 150 mV (although it should be noted that a much lower value is apparently sufficient in alkaliphilic bacteria). ATP synthesis can therefore be inhibited for either thermodynamic or kinetic reasons. These include conditions which dissipate the protonmotive force, such as the presence of protonophoric uncoupling agents (although some interesting mutants have been isolated which are resistant to uncouplers) or the absence of a vesicular membrane; and the presence of reagents, which block H^+ transport through BF_0 (e.g. DCCD) or prevent substrate binding by BF_1 (e.g. aurovertin).

The synthesis of ATP is largely unaffected by whether the protonmotive force is composed of a ΔpH and/or a $\Delta \psi$ (see Kagawa, 1978; Fillingame, 1980, 1981). The latter observation implies that the role of the BF_0 includes the ability not only to act as a proton 'channel' through the membrane, but also to convert the $\Delta \psi$ into an additional ΔpH by acting as a 'well' within the membrane into which protons from the aqueous phase can be attracted by the $\Delta \psi$ and thus accumulated.

Early ideas envisaged that ATP synthesis occurs by a direct chemiosmotic mechanism in which protons move through BF_0 all the way to the catalytic site in BF_1 headpiece, where they react chemically with appropriately ionized forms of ADP and inorganic phosphate to effect dehydration and consequently ATP synthesis (Mitchell, 1974, 1977).

More recently, however, it has been proposed that energy is required for physical rather than chemical events during ATP synthesis, i.e. to facilitate the binding of the reactants to BF_1 and/or the release of bound ATP from BF_1 rather than to drive the condensation of ADP and inorganic phosphate *per se*. This mechanism would require the protonmotive force to catalyse protonation/deprotonation reactions at various sites in the $BF_0.BF_1$ complex and thus eventually cause conformational changes at the three substrate-binding sites present in the headpiece. Such changes would significantly alter the binding constants for the adenine nucleotides and inorganic phosphate, thus allowing ATP to be bound much more tightly than ADP.Pi; this would minimize the amount of free energy required for their interconversion, but increase the amount required to release the bound ATP.

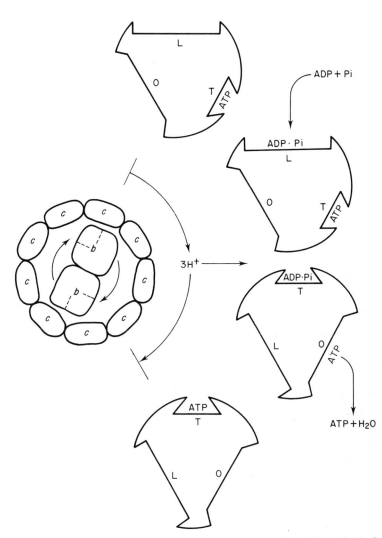

Fig. 1.33 Conformational changes in the $BF_0.BF_1$ complex during ATP synthesis. In the presence of an appropriate protonmotive force, the b-subunit dimer revolves within the ring of nine c subunits, covering one-third of a full revolution for every three H^+ translocated inwards across the membrane. Only the portion of the b subunit enclosed by the dashed line traverses the membrane and thus interacts with the c subunits; the remainder extends into the stalk where it interacts with the γ, δ and ε subunits. The latter interaction produces a conformational change in the $\alpha_3\beta_3$ hexagon, and hence a change in the nature of the three substrate-binding sites ($T \rightarrow O$, $O \rightarrow L$, $L \rightarrow T$), which alters the binding constants for the substrates and leads to the synthesis and release of ATP ($\rightarrow H^+/ATP = 3$). One complete revolution of the b-subunit dimer is accompanied by the translocation of nine protons and drives the synthesis of three molecules of ATP. It is envisaged that during ATP hydrolysis the cycle operates in reverse to eject H^+ and thus generates a protonmotive force. (After Cross, 1981; Cox *et al.*, 1984).

Substantial experimental evidence has accumulated in support of this mechanism, mainly as a result of kinetic and substrate-affinity studies using the mitochondrial enzyme (Boyer, 1975; Cross, 1981; Mitchell, 1985), but the further elucidation of the exact mechanism requires detailed structural information.

The recent analysis of the molecular architecture of the ATPase/ATP synthase complex from *E. coli* has therefore proved particularly useful in this respect (Cox *et al.*, 1984). The results indicate that the lys-23 residues in the two *b* subunits interact with the asp-61 residues in two of the encircling *c* subunits, and suggest that the passage of a proton through BF_0, by an as yet undetermined route, interferes with this interaction and causes a partial rotation of the *b*-subunit dimer. This is transmitted to the headpiece via further interactions between the dimer and the other stalk subunits, and causes a significant conformational change in the $\alpha_3\beta_3$ hexagon (Fig. 1.33).

At any given moment, the three substrate-binding sites in the headpiece exist in different forms, i.e. a tight-binding/active form (T site), a loose-binding/inactive form (L site) and an open/low affinity/inactive form (O site). The conformational changes in the headpiece brought about by the successive passage of three protons through BF_0 cause interconversion of the three binding sites (T → O, O → L, L → T), thus allowing the release of tightly bound ATP from the original T-site and the formation of tightly bound ATP from ADP + Pi at the original L-site. These events are thus completely compatible with an alternating catalytic site mechanism based upon co-operative interactions between three substrate-binding sites, and represent one-third of the entire catalytic cycle. During the complete cycle, nine protons pass through BF_0 causing one complete revolution of the *b*-subunit dimer within the *c*-subunit ring and, hence, one complete series of conformational changes in the binding sites; thus causing the ultimate synthesis of three molecules of ATP. The $\rightarrow H^+/ATP$ quotient of 3 inherent in this mechanism is within the range of approximately 2 to 4 obtained experimentally by comparing values of $\Delta Gp/F$ and Δp, and $\rightarrow H^+/2e^-$ and $ATP/2e^-$.

References

Adams, M. W. W., Mortenson, L. E. and Chen, J. S. (1981). *Biochimica et Biophysica Acta* **594**, 105–176.
Anraku, Y. and Gennis, R. B. (1987). *Trends in Biochemical Sciences* **12**, 262–266.
Anthony, C. (1986). *Advances in Microbial Physiology* **27**, 113–210.
Beacham, I. R. (1979). *International Journal of Biochemistry* **10**, 877–881.
Bilous, P. T. and Weiner, J. H. (1985). *Journal of General Microbiology* **163**, 369–375.
Bishop, D. H. L., Pandya, K. P. and King, H. K. (1962). *Biochemical Journal* **83**, 550–554.

Booth, I. R. (1985). *Microbiological Reviews* **49**, 359–378.

Boyer, P. D. (1975). *FEBS Letters* **58**, 1–6.

Bragg, P. D. (1980). *In* 'Diversity of Bacterial Respiratory Systems' (Ed. C. J. Knowles), Vol. 1, pp. 115–136. CRC Press, Boca Raton.

Cobley, J. G. (1976). *Biochemical Journal* **156**, 493–498.

Cole, J. A. and Ferguson, S. J. (Eds.) (1988). 'The Nitrogen and Sulphur Cycles'. SGM Symposium 42. Cambridge University Press, Cambridge.

Cole, S. T., Condon, C., Lemire, B. D. and Weiner, J. H. (1985). *Biochimica et Biophysica Acta* **811**, 381–403.

Collins, M. D. and Jones, D. (1981). *Microbiological Reviews* **45**, 316–354.

Costerton, J. W., Ingram, J. M. and Cheng, K.-J. (1974). *Bacteriological Reviews* **38**, 87–110.

Cox, G. B., Fimmel, J. A. L., Gibson, F. and Hatch, L. (1984). *Biochimica et Biophysica Acta* **768**, 201–208.

Cross, R. L. (1981). *Annual Review of Biochemistry* **50**, 681–714.

Dawes, E. A. (1986). 'Microbial Energetics', 187 pp. Blackie, London.

Deisenhofer, J., Epp, O., Miki, K., Huber, R. and Michel, H. (1985a). *Nature* **318**, 618–624.

Deisenhofer, J., Michel, H. and Huber, R. (1985b). *Trends in Biochemical Sciences* **10**, 243–248.

Downie, J. A., Gibson, F. and Cox, G. B. (1979). *Annual Review of Biochemistry* **48**, 103–131.

Drews, G. (1985). *Microbiological Reviews* **49**, 59–70.

Duine, J. A., Frank, Jzn., J. and Jongejan, J. A. (1986). *FEMS Microbiological Reviews* **32**, 165–178.

Duine, J. A., Frank, Jzn., J. and Jongejan, J. A. (1987). *Advances in Enzymology* **59**, 169–212.

Eisenbach, M. and Caplan, S. R. (1979). *Current Topics in Membranes and Transport* **12**, 166–240.

Erlich, H. L. (1981). 'Geomicrobiology'. Marcel Dekker, New York.

Ferguson, S. J. (1985). *Biochimica et Biophysica Acta* **811**, 47–95.

Ferguson, S. J., Jackson, J. B. and McEwan, A. G. (1987). *FEMS Microbiology Reviews* **46**, 117–143.

Fillingame, R. H. (1980). *Annual Review of Biochemistry* **49**, 1079–1113.

Fillingame, R. H. (1981). *Current Topics in Bioenergetics* **11**, 35–106.

Garland, P. (1977). *In* 'Microbial Energetics' (Eds. B. A. Haddock and W. A. Hamilton). Society for General Microbiology Symposium **27**, 1–21. Cambridge University Press, Cambridge.

Gest, H. (1980). *FEMS Microbiology Letters* **7**, 73–77.

Gest, H. (1981). *FEMS Microbiology Letters* **12**, 209–215.

Gennis, R. B. (1987). *FEMS Microbiology Reviews* **46**, 387–399.

Gray, H. B. and Solomon, E. I. (1981). *In* 'Copper Proteins' (Ed. T. G. Spiro), pp. 1–39. John Wiley, New York.

Harold, F. M. (1977). *Current Topics in Bioenergetics* **6**, 84–151.

Harold, F. M. (1986). *In* 'The Vital Force: a Study of Bioenergetics'. 577 pp. W. H. Freeman, New York.

Hellingwerf, K. J. and Konings, W. N. (1985). *Advances in Microbial Physiology* **26**, 125–154.

Hooper, A. B. and DiSpirito, A. A. (1985). *Microbiological Reviews* **49**, 140–157.

Ingledew, W. J. (1982). *Biochimica et Biophysica Acta* **683**, 89–117.

Ingledew, W. J. and Poole, R. K. (1984). *Microbiological Reviews* **48**, 222–271.

Johnson, J. L. (1980). *In* 'Molybdenum and Molybdenum-containing Enzymes' (Ed. P. Coughlan), pp. 347–383. Pergamon, Oxford.

Jones, C. W. (1977). *In* 'Microbial Energetics' (Eds. B. A. Haddock and W. A. Hamilton). Society for General Microbiology Symposium 27, 23–59. Cambridge University Press, Cambridge.

Jones, C. W. (1981). 'Biological Energy Conservation: Oxidative Phosphorylation', 80 pp. Chapman and Hall, London.

Jones, C. W. (1982). 'Bacterial Respiration and Photosynthesis', 106 pp. Van Nostrand Reinhold, London.

Jones, C. W. (1985). *In* 'Evolution of Prokaryotes' (Eds. K. H. Schleifer and E. Stackebrandt), pp. 175–204. Academic Press, London.

Jones, C. W. and Poole, R. K. (1985). *Methods in Microbiology* 18, 285–328.

Kagawa, Y. (1978). *Biochimica et Biophysica Acta* 505, 45–93.

Keilin, D. (1925). *Proceedings of the Royal Society (London)* B98, 312–339.

Keilin, D. (1929). *Proceedings of the Royal Society (London)* B104, 206–252.

Keilin, D. (1966). 'The History of Cell Respiration and Cytochrome', 416 pp. Cambridge University Press, Cambridge.

Kell, D. B. (1979). *Biochimica et Biophysica Acta* 549, 55–99.

Kell, D. B. (1987). *Journal of General Microbiology* 133, 1651–1665.

Kell, D. B., Clarke, D. J. and Morris, J. G. (1981). *FEMS Microbiology Letters* 11, 1–11.

Kelly, D. P. (1982). *Philosophical Transactions of the Royal Society (London)* B298, 499–528.

Kelly, D. P. (1985). *Microbiological Sciences* 2, 105–109.

Keltjens, K. T. and van der Drift, C. (1986). *FEMS Microbiol. Revs.* 39, 259–303.

Kogut, M. (1980). *Trends in Biochemical Sciences* 5, 47–50.

Konings, W. N. (1977). *Advances in Microbial Physiology* 15, 175–253.

Koshland, D. E. (1981). *Annual Review of Biochemistry* 50, 765–782.

Kroger, A. (1978). *Biochimica et Biophysica Acta* 505, 129–145.

Krulwich, T. A. (1985). *Biochimica et Biophysica Acta* 726, 245–264.

Krulwich, T. A. and Guffanti, A. A. (1983). *Advances in Microbial Physiology* 24, 173–213.

Kushner, D. J. (1978). 'Microbial Life in Extreme Environments'. Academic Press, London.

Ludwig, B. (1980). *Biochimica et Biophysica Acta* 594, 177–189.

Ludwig, B. (1987). *FEMS Microbiology Reviews* 46, 41–56.

McNab, R. M. (1978). *Critical Reviews in Biochemistry* 5, 291–341.

McNab, R. M. (1985). *Methods in Enzymology* 125, 563–581.

Malony, P. C. and Wilson, T. H. (1985). *BioScience* 35, 43–48.

Mayhew, S. G. and O'Connor, M. E. (1982). *Trends in Biochemical Sciences* 7, 18–21.

Meyer, O. and Schlegel, H. G. (1983). *Annual Review of Microbiology* 37, 277–310.

Meyer, O., Jacobitz, S. and Kruger, B. (1986). *FEMS Microbiology Reviews* 39, 161–179.

Mitchell, P. (1961). *Nature (London)* 191, 144–148.

Mitchell, P. (1966). *Biological Reviews* 41, 445–502.

Mitchell, P. (1974). *FEBS Letters* 43, 189–194.

Mitchell, P. (1975). *FEBS Letters* 59, 137–139.

Mitchell, P. (1976). *Journal of Theoretical Biology* 62, 327–367.

Mitchell, P. (1977). *FEBS Letters* 78, 1–20.

Mitchell, P. (1979). *In* 'Les Prix Nobel en 1978', pp. 135–172. Nobel Foundation, Stockholm.

Mitchell, P. (1984). *FEBS Letters* **176**, 287–294.

Mitchell, P. (1985). *FEBS Letters* **182**, 1–7.

Nicholls, D. G. (1982). 'Bioenergetics: An Introduction to the Chemiosmotic Theory', 190 pp. Academic Press, London.

Nikaido, H. and Nakai, T. (1979). *Advances in Microbial Physiology* **20**, 163–250.

Nikaido, H. and Vaara, M. (1985). *Microbiological Reviews* **49**, 1–32.

Nugent, J. H. A. (1984). *Trends in Biochemical Sciences* **9**, 354–357.

Ordal, G. W. (1985). *Critical Reviews in Biochemistry* **12**, 95–130.

Pettigrew, G. W. and Moore, G. R. (1987). 'Cytochromes *c*: Biological Aspects'. Springer-Verlag, Berlin, Heidelberg.

Poole, R. K. (1983). *Biochimica et Biophysica Acta* **726**, 205–243.

Prebble, J. (1981). *In* 'Mitochondria, Chloroplasts and Bacterial Membranes', 378 pp. Longman, New York.

Rajagopalan, K. V. (1985). *Biochemical Society Transactions* **13**, 401–403.

Rosen, B. P. and Kashket, E. R. (1978). *In* 'Bacterial Transport' (Ed. B. P. Rosen), pp. 559–620. Marcel Dekker, New York.

Ryden, L. (1984). *In* 'Copper Proteins and Copper Enzymes' (Ed. R. Lontie), Vol. 1, pp. 157–182. CRC Press, Boca Raton.

Siefermann-Harms, D. (1985). *Biochimica et Biophysica Acta* **811**, 325–355.

Simoni, R. D. and Postma, P. W. (1975). *Annual Review of Biochemistry* **44**, 523–554.

Singer, S. J. and Nicholson, G. L. (1972). *Science* **175**, 720–731.

Slater, E. C. (1987). *European Journal of Biochemistry* **166**, 489–504.

Slater, E. C., Berden, J. A. and Herweijer, M. A. (1985). *Biochimica et Biophysica Acta* **811**, 217–231.

Stoeckenius, W. and Bogomolni, R. A. (1982). *Annual Review of Biochemistry* **52**, 587–616.

Stoeckenius, W., Lozier, R. H. and Bogomolni, R. A. (1979). *Biochimica et Biophysica Acta* **505**, 215–278.

Stouthamer, A. H. (1976). *Advances in Microbial Physiology* **14**, 315–375.

Stouthamer, A. H. (1980). *Trends in Biochemical Sciences* **5**, 164–166.

Stouthamer, A. H. and Bettenhausen, C. W. (1973). *Biochimica et Biophysica Acta* **301**, 53–70.

Stouthamer, A. H. and van Verseveld, H. W.. (1987). *Trends in Biotechnology* **5**, 149–155.

Stucki, J. W. (1983). *Biophysical Chemistry* **18**, 111–115.

Taylor, B. L. (1983). *Annual Review of Microbiology* **37**, 551–573.

Tempest, D. W. (1978). *Trends in Biochemical Sciences* **3**, 180–184.

Tempest, D. W. and Neijssel, O. M. (1984). *Annual Review of Microbiology* **38**, 459–486.

Walker, J. E., Saraste, M. and Gay, N. J. (1984). *Biochimica et Biophysica Acta* **768**, 164–200.

West, I. C. and Mitchell, P. (1972). *Journal of Bioenergetics* **3**, 445–462.

Westerhof, H. V. and van Dam, K. (1979). *Current Topics in Bioenergetics* **9**, 1–62.

Westerhof, H. V., Lolkema, J. S., Otto, R. and Hellingwerf, K. J. (1982). *Biochimica et Biophysica Acta* **683**, 181–220.

Westerhof, H. V., Melandri, B. A., Venturoli, G., Azzone, G. F. and Kell, D. B. (1984). *FEBS Letters* **165**, 1–6.

Williams, R. J. P. (1978). *Biochimica et Biophysica Acta* **505**, 1–44.
Wood, P. M. (1981). *FEBS Letters* **124**, 11–14.
Wood, P. M. (1983). *FEBS Letters* **164**, 223–226.
Wood, P. M. (1984). *Biochimica et Biophysica Acta* **768**, 293–317.
Wood, P. M. (1987). *In* 'Nitrification' (Ed. J. Prosser), Society of General Micro-
    biology Special Publication. Cambridge University Press, Cambridge (in press).
Zuber, H. (1986). *Trends in Biochemical Sciences* **11**, 414–419.

2 Energy transduction in anaerobic bacteria

W. A. HAMILTON

2.1 Introduction 84
2.2 Mechanisms of anaerobic energy coupling 86
2.3 Fermentation 89
2.3.1 Introduction 89
2.3.2 Ethanol fermentations 90
2.3.3 Lactate fermentations 92
2.3.4 Butyrate and butanol/acetone fermentations 96
2.3.5 Mixed-acid fermentations 100
2.3.6 Succinate and propionate fermentations 102
2.3.7 Formation of a protonmotive force in fermenters 104
    2.3.7.1 Anaerobic citrate degradation 104
    2.3.7.2 Acidic fermentation products 105
2.3.8 Acetate fermentations 105
    2.3.8.1 Introduction 105
    2.3.8.2 The homoacetogens: the acetyl-CoA pathway 106
    2.3.8.3 The hydrogen-producing acetogens 109
2.3.9 Fermentation: general comments 111
2.4 Anaerobic respiration: methanogenesis and sulphate
   reduction 112
2.4.1 Methanogenic bacteria 112
2.4.2 Chemolithotrophic and methylotrophic metabolism 114
2.4.3 Carbon assimilation in methanogens 115
2.4.4 Energy coupling in methanogens 116
    2.4.4.1 Introduction 116
    2.4.4.2 Unusual coenzymes in methanogens 117
    2.4.4.3 Energy transduction and methanogenesis from
            carbon dioxide 119
    2.4.4.4 Energy coupling during methanogenesis from
            formaldehyde and acetate 120
    2.4.4.5 The role of Na in the reactions of methanogenesis 122
2.4.5 Sulphate-reducing bacteria 123
    2.4.5.1 Introduction 123
    2.4.5.2 Components of redox systems 126
    2.4.5.3 Hydrogen metabolism 128
    2.4.5.4 Nutrition and carbon flux 133

	2.4.5.5	Sulphite dismutation 135
2.5	Anaerobic microbial consortia 135
2.6	Anaerobic respiration: reduction of fumarate and nitrate 136
2.6.1	Introduction 136
2.6.2	Fumarate as terminal electron acceptor 137
	2.6.2.1	Introduction 137
	2.6.2.2	Energy coupling 138
	2.6.2.3	Fumarate reductase 139
	2.6.2.4	Sulphur reduction and sulphide oxidation coupled to fumarate reductase 141
2.6.3	Oxides of nitrogen as terminal electron acceptor 141
	2.6.3.1	Introduction 141
	2.6.3.2	Nitrate reductase in *Pa. denitrificans* and *E. coli* ... 142
	2.6.3.3	Nitrate reductase in *Rhodobacter, Rhodopseudomonas* and *Rhodospirillum* 144
	2.6.3.4	Nitrite reductase and nitrous oxide reductase 144
	2.6.3.5	Nitric oxide reductase 145
	2.6.3.6	TMAO reductase 145
2.7	Concluding remarks 146
References 146

2.1   Introduction

Since the discovery of oxygen in the eighteenth century by Priestley and his recognition of its production by plants and utilization by animals, oxygen and life have been inextricably linked in our minds. Even at the level of molecular mechanisms, most of us have our first experience of bioenergetics in the study of mitochondrial oxidative phosphorylation. Yet it is now more than a hundred years since Pasteur discovered the existence of microbial life in the total absence of oxygen.

While studies since that time have confirmed that life in anaerobic environments is confined to lower forms (bacteria, fungi, protozoa, helminths) it is evident that the environments themselves are in no way confined to fermenting wine vats. In fact, a very large proportion of the biosphere is likely to be anaerobic and there is certainly a considerable diversity in the nature of such environments: the animal intestinal tract; anaerobic digesters and compost heaps; aquatic sediments and, on occasion, deep water columns; and biofilms, including marine fouling deposits. Many of these environments owe their anaerobic character to the presence of aerobic and facultative organisms, again most often microbial, which remove oxygen at a rate faster than it can diffuse into the system. This type of situation arises from the high metabolic activity of microorganisms, while a second characteristic property, their small size, is responsible for another intriguing feature of many anaerobic environments. The distances separating aerobic and anaerobic zones can be very short, with the interface between the two phases being extremely sharp. For example, a biofilm only 25 μm thick may be anaerobic at its base, even where the system as a whole is aerobic. At least in terms of microbial ecology and interaction between species, microenvironments constituted in this way are of major significance.

Anaerobic heterotrophic microorganisms are not very different from their aerobic cousins, however; at least not as far as their chemical composition and modes of anabolic metabolism are concerned. Nor are their mechanisms of energy generation fundamentally different: chemical oxidation of nutrient materials coupled to the synthesis of the universal cellular energy currency, ATP. What is different is the redox span between the electron donors and acceptors available within the anaerobic environments. This is directly responsible for both the lower energy yields available to the anaerobes and the various mechanisms they must adopt in order to gain sufficient energy to drive their anabolic and maintenance functions. In considering these mechanisms, however, it is not possible to divorce the study of energetics and ATP production from questions of redox balance and carbon flux. Within the restraints, primarily redox in character, imposed by the anaerobic environments, the interdependencies of these three metabolic strands become very much more marked. For example, in the process of fermentation with no added external electron acceptor, the maintenance of redox balance with the reduction of pyrimidine nucleotides stoichiometrically coupled to their reoxidation is the determinant influence on the pathways of carbon flux and, therefore, on the consequent energy yield. Further, in processes of anaerobic respiration with a terminal electron acceptor other than oxygen, the redox potential of that electron acceptor not only determines the potential energy yield but often also restricts the organic compounds that can serve as carbon and energy sources for the organism in question.

Consequently, in dealing with the mechanisms of bioenergetics as demonstrated by anaerobic microorganisms, this chapter will not confine itself solely to the turnover of ATP but will set that activity within the wider context of anaerobic metabolism as a whole.

Even such a metabolic approach is not of itself sufficient to achieve a wholly satisfactory understanding of anaerobic bioenergetics in terms of either cellular mechanisms or their wider ecological implications. Whereas with aerobic or phototrophic species we are conditioned to think of metabolic processes solely with reference to the organism under study as a pure culture, and to see any involvement in, say, the nitrogen cycle as being a separate and secondary issue, the reality under anaerobic conditions is that such interspecies metabolic links are integral to the energetics and general metabolism of the individual species and, indeed, may be obligatory for their survival. Throughout this chapter, therefore, continual reference will be made to those threads whose interweaving creates the backcloth against which the energetic mechanisms of anaerobic chemoheterotrophs must be set: microenvironments and the aerobic/anaerobic interface, narrow redox span, and interspecies metabolic coupling.

Reference has already been made to the seminal findings of Priestley and of Pasteur in the latter parts of the eighteenth and nineteenth centuries; but perhaps a more directly relevant starting point for our discussion of anaerobic metabolism in the latter part of the twentieth century would be 1977. By that time the chemiosmotic hypothesis had been effectively universally accepted and the central role of the protonmotive force in energy transduction established. This in turn had led to a deeper appreciation of the essential unity of energetic mechanisms. Along with particularly full and well-presented accounts of the then-current state of our detailed knowledge of the systems in question, these facets were the subject of three publications in that year: the Society for General Microbiology's Symposium on 'Microbial Energetics' (Haddock and Hamilton, 1977), and the two articles in *Bacteriological Reviews* by Haddock and Jones (1977) on 'Bacterial Respiration' and by Thauer, Jungermann and Decker (1977) on 'Energy Conservation in Chemotrophic Anaerobic Bacteria'. Building on the foundation of these publications, this chapter will pay particular attention to those areas where there have since been the most noticeable advances. In respect of increased factual knowledge these have included: details of the mechanisms of coupling between electron transport and phosphorylation; ecology, nutrition and physiology of the sulphate-reducing, methanogenic and acetogenic bacteria; and the acetyl-CoA pathway for CO_2 fixation. Major conceptual advances have been made with regard to syntrophism and microbial consortia, and the putative role of energy-driven reversed electron transport in achieving both energy yield and redox balance under conditions of minimal availability of free energy.

2.2   Mechanisms of anaerobic energy coupling

The point has already been made that the chief characteristics of energy-yielding catabolic sequences in the absence of oxygen as the terminal electron acceptor, are the identities of the alternative acceptors and the magnitude of the free energy potentially available from their use. With the exception of materials such as the hydrocarbons (the initial steps of whose biodegradation require molecular oxygen), the same extensive range of organic nutrients is available to both aerobic and anaerobic chemoheterotrophic organisms as electron donors, and as sources of energy as well as of carbon.

On the basis of their electron acceptors, therefore, anaerobic chemoheterotrophs can be divided into two broad classes, using as their modes of energy metabolism either fermentation or anaerobic respiration. In fermentation, an organic substrate is oxidized in a series of anaerobic reactions in which the electron acceptor is a product of the metabolism of the substrate, with the difference in redox potential of the substrate and the electron acceptor derived from it providing the energy for ATP synthesis. In fermentation, all components of the reaction schemes are normally soluble (fumarate reduction is an exception) and ATP is generated by the mechanism of substrate level phosphorylation (SLP) (Table 2.1). The sole reason for accumulating the reduced fermentation product is to achieve redox balance. Common examples of electron acceptors in fermentation are pyruvate (reduced to lactate), acetyl CoA (reduced to ethanol), fumarate (reduced to succinate) and H^+ (reduced to H_2). By definition, therefore, fermentation products are both reduced and only partially degraded, and are consequently potential sources of energy, reducing power and (with the exception of H_2) carbon for other organisms.

Anaerobic respiration is the name given to the membrane-associated processes of electron transport (using the same or similar redox carriers but with a terminal acceptor other than oxygen) directly coupled to the phosphorylation of ADP to ATP. Electron acceptors include: nitrate (reduced to nitrite); nitrite (reduced to N_2); trimethylamine oxide (TMAO) (reduced to TMA); fumarate (reduced to succinate); sulphate (reduced to sulphide); and CO_2 (reduced to methane). As opposed to its role in fermentation, where it is an important product, H_2 is a common energy substrate in anaerobic respiration. As we shall see, this forms an important practical difference between the two anaerobic classes of organism, and one of considerable significance with regard to the interspecies coupling in which hydrogen plays a key role.

As with fermentation, the reduced products of anaerobic respiration are potential sources of energy and reducing power for other organisms. In most cases, however, these are obligatory aerobic chemolithotrophs and the methane, etc. must first diffuse across the aerobic/anaerobic (O_2/AnO_2) interface. As the electron acceptors such as CO_2 or sulphate are themselves generally not products of the anaerobe's own metabolism and are in fact more associated with aerobic environments, we can readily see the marked fluxes necessary across the O_2/AnO_2 interface, and so identify a major driving force for the carbon, nitrogen and sulphur cycles in nature.

Table 2.1   Substrate-level phosphorylation reactions occurring during fermentations

Substrate-level phosphorylation (SLP) is the phosphorylation of ADP to ATP by a process of phosphate transfer from an 'energy-rich' phosphorylated substrate produced during degradation of the fermentation substrate. The initial incorporation of inorganic phosphate is, as seen below, coupled to exergonic oxidation or lyase reactions. Other 'energy-rich' compounds such as thio-esters, phosphoacyl anhydrides and phosphoenol esters are sometimes intermediates. The formation of phosphoenolpyruvate (PEP) catalysed by pyruvate kinase is not, strictly speaking, a site of SLP because no phosphate is consumed in the production of PEP; it is provided by ATP at an earlier step in the glycolytic sequence. Besides acetyl-CoA (below), other thioesters (propionyl-CoA, butyryl-CoA, succinyl-CoA) can be converted to the acyl phosphate derivatives (in a thiokinase reaction) and, hence, to ATP plus the free acid.

The enzymes involved in these reactions are as follows: a, pyruvate dehydrogenase; b, phosphotransacetylase; c, acetate kinase; d, 2-oxoglutarate dehydrogenase; e, succinate thiokinase; f, acetaldehyde dehydrogenase; g, glyceraldehyde phosphate dehydrogenase; h, phosphoglycerate kinase; i, pyruvate kinase; j, pyruvate:formate lyase; k, ketothiolase; l, phosphoketolase; m, ornithine transcarbamoylase (ornithine carbamoyl transferase); n, carbamate kinase; o, formyltetrahydrofolate synthetase.

(a) *Oxidation reactions coupled to phosphorylation reactions*

1. Pyruvate $\xrightarrow{\text{a}}$ acetyl-CoA $\xrightarrow{\text{b}}$ acetyl phosphate $\xrightarrow{\text{c}}$ acetate
   (CoA, CO$_2$, '2H') (Pi, CoA) (ADP, ATP)

2. 2-Oxoglutarate $\xrightarrow{\text{d}}$ succinyl-CoA $\xrightarrow{\text{e}}$ succinate
   (CoA, CO$_2$, '2H') (ADP, ATP, CoA)

3. Acetaldehyde $\xrightarrow{\text{f}}$ acetyl-CoA $\xrightarrow{\text{b}}$ acetyl phosphate $\xrightarrow{\text{c}}$ acetate
   (CoA, '2H') (Pi, CoA) (ADP, ATP)

4. Glyceraldehyde-P $\xrightarrow{\text{g}}$ 1,3-diphosphoglycerate $\xrightarrow{\text{h}}$ 3-phosphoglycerate
   (Pi, NAD, NADH) (ADP, ATP)

5. Phosphoenolpyruvate $\xrightarrow{\text{i}}$ pyruvate
   (ADP, ATP)

(b) *Lyase reactions coupled to phosphorylation reactions*

1. Pyruvate $\xrightarrow{\text{j}}$ acetyl-CoA $\xrightarrow{\text{b}}$ acetyl phosphate $\xrightarrow{\text{c}}$ acetate
   (CoA, formate) (Pi, CoA) (ADP, ATP)

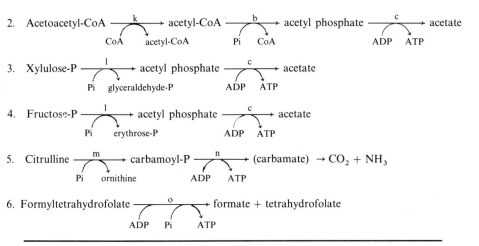

2. Acetoacetyl-CoA $\xrightarrow{\text{k}}$ acetyl-CoA $\xrightarrow{\text{b}}$ acetyl phosphate $\xrightarrow{\text{c}}$ acetate

       CoA acetyl-CoA   Pi CoA   ADP ATP

3. Xylulose-P $\xrightarrow{\text{l}}$ acetyl phosphate $\xrightarrow{\text{c}}$ acetate

     Pi glyceraldehyde-P   ADP ATP

4. Fructose-P $\xrightarrow{\text{l}}$ acetyl phosphate $\xrightarrow{\text{c}}$ acetate

     Pi erythrose-P   ADP ATP

5. Citrulline $\xrightarrow{\text{m}}$ carbamoyl-P $\xrightarrow{\text{n}}$ (carbamate) $\rightarrow CO_2 + NH_3$

     Pi ornithine   ADP ATP

6. Formyltetrahydrofolate $\xrightarrow{\text{o}}$ formate + tetrahydrofolate

           ADP Pi ATP

2.3 Fermentation

2.3.1 Introduction

Fermentation processes are capable of degrading almost the complete range of nutrients available for the chemoheterotrophic growth of microorganisms, including: sugars; organic, fatty and amino acids; purine and pyrimidine bases; heterocyclic compounds; and polysaccharide, protein and lipid polymers. Lignin and saturated hydrocarbons are generally considered to be recalcitrant in anaerobic environments due to the requirement for molecular oxygen in the initial catabolic reactions. There have been recent reports, however, of the slow breakdown of lignin (up to 15% in 300 days), and of unsaturated hydrocarbons by mixed enrichment microbial populations.

Those microorganisms capable of fermentation also span a huge range. Although they include certain protozoa and fungi, with the exception of yeasts we shall be concerned here only with bacterial species. Fermentation reaction sequences are found amongst facultative (e.g. Enterobacteriaceae) and obligate (e.g. *Clostridia*) anaerobes. Whereas the *Clostridia* are obligately fermentative, the sulphate reducers are facultative because, in the presence of sulphate as terminal electron acceptor, their mode of energy-generating metabolism is anaerobic respiration whereas, in the absence of sulphate, they can ferment a somewhat restricted range of substrates. The methanogens generally lack the normal respiratory cofactors. Despite this lack, and its implications for the mechanism of energy transduction in these organisms, it is most convincing to deal with them under the general heading of anaerobic respiration, with CO_2 fulfilling the role of alternative electron acceptor.

In contrast to the extensive array of potential substrates for fermentation, only a limited number of metabolic pathways is required for transformation into what is also a relatively small number of fermentation products. Classification is usually based on these products, rather than on substrates or organisms.

2.3.2   Ethanol fermentations

Ethanol is perhaps the most widely recognized fermentation product; it is certainly the one with the longest history as man's earliest biotechnological product. In fact, ethanol arises from fermentation reactions in both yeasts, principally *Saccharomyces* species, and bacteria; it is instructive to examine the different mechanisms employed.

The major substrates giving rise to ethanol are the sugars, which in yeast are degraded to pyruvate by the Embden–Meyerhof–Parnas (EMP) or glycolytic pathway (Fig. 2.1). There is a net yield of 1 ATP for each pyruvate formed from glucose as a net result of the two kinase reactions to fructose 1,6-bisphosphate, with ATP production by SLP at the reactions 1,3-diphosphoglycerate to 3-phosphoglycerate and phosphoenol pyruvate to pyruvate. There is also the reduction of 1 NAD^+ in the oxidative conversion of glyceraldehyde 3-phosphate to 1,3-diphosphoglycerate. The redox balance is achieved by the formation of ethanol via the intermediary production of acetaldehyde.

A key element in comparing fermentation reactions is the nature of the enzymic conversion of pyruvate, which is itself the key intermediate in the majority of fermentations. In the *Saccharomyces* ethanol fermentation, the enzyme is pyruvate decarboxylase which has thiamine pyrophosphate as a cofactor and which degrades pyruvate to CO_2 and enzyme-bound hydroxyethyl thiamine pyrophosphate which in turn hydrolyses to free acetaldehyde. Alcohol dehydrogenase reduces the acetaldehyde to ethanol, with the stoichiometric reoxidation of the reduced pyrimidine nucleotide.

The identical metabolic route for ethanol formation is found in the bacterial species *Sarcina ventriculi* and *Erwinia amylovora*, although small amounts of other fermentation products are also found (H_2, acetate and lactate). Two other bacteria, *Zymomonas mobilis* and *Zymomonas anaerobica*, also possess the enzymes pyruvate decarboxylase and alcohol dehydrogenase, although they degrade glucose to pyruvate by the Entner–Doudoroff (ED) pathway with a consequently reduced ATP yield of only 0.5 per pyruvate (Fig. 2.2).

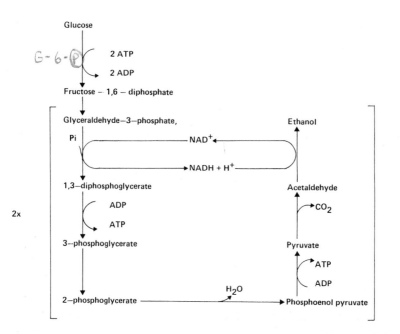

Fig. 2.1 Glycolytic pathway for conversion of glucose to pyruvate, with redox balance maintained by formation of ethanol as the reduced fermentation product formed by way of pyruvate decarboxylase.

Balance: Glucose → $2CO_2$ + 2 ethanol (+ 2ATP)

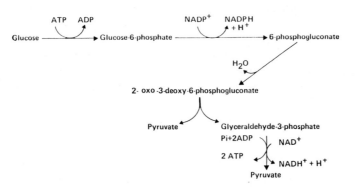

Fig. 2.2 Entner-Doudoroff pathway for fermentation of sugars. In this pathway the pyruvate is decarboxylated to acetaldehyde which is used as the oxidant for NADH (giving ethanol; as in Fig. 2.1).

Balance: Glucose → $2CO_2$ + 2 ethanol (+1ATP)

Ethanol is, in fact, a common fermentation product from a wide range of bacterial species but it is generally a component only of a mixed fermentation in, for example, Enterobacteriaceae, *Clostridia* and lactic acid bacteria. Also, the normal bacterial mechanism does not involve pyruvate decarboxylase. Rather, acetyl-CoA is formed as an intermediate (for mechanisms of its formation in these organisms, see later sections) and then reduced to ethanol by *two* NADH-requiring reactions involving acetaldehyde dehydrogenase and alcohol dehydrogenase:

$$CH_3CO\text{-}SCoA + NADH + H^+ \rightarrow CH_3CHO + CoASH + NAD^+$$

$$CH_3CHO + NADH + H^+ \rightarrow CH_3CH_2OH + NAD^+$$

It is of considerable importance that this bacterial mechanism of ethanol production reoxidizes twice as much NADH as is reduced in the formation of pyruvate by either glycolysis or the ED pathway. Redox balance is therefore only attained if ethanol formation is accompanied by an equal conversion of pyruvate to another non-reduced fermentation product. This redox-based consideration has direct implications for both carbon flux and energy yield. For example, where pyruvate does not require to be reduced to a fermentation product it is then free to serve as a carbon intermediate in biosynthesis. Alternatively, the most common non-reductive fermentation route from pyruvate is via acetyl-CoA and acetyl phosphate to acetate as product, with an extra ATP yield by SLP in the final reaction (see Section 2.3.4).

From the biotechnological point of view, the bacterial fermentations to ethanol have a number of advantages over the more traditional use of *Saccharomyces cerevisiae*. Amongst bacteria it is possible to find strains capable of directly fermenting not only the glucose, fructose, sucrose and maltose characteristic of *S. cerevisiae*, but starch and cellulose polymers and also the pentose xylose which arises from the hydrolysis of hemicellulose. These processes are further favoured by the existence of thermophilic bacteria capable of these reactions at elevated temperatures. On the other hand, the bacterial fermentations are more prone to inhibition by high substrate concentrations, produce a lower yield of ethanol (of the order of 1% rather than 10%), and do so along with contaminating acetate and lactate as additional fermentation products.

2.3.3 Lactate fermentations

The term 'lactic acid bacteria' is used to cover the genera *Lactobacillus, Sporolactobacillus, Streptococcus, Leuconostoc, Pediococcus* and *Bifidobacterium*, which have in common the property of producing lactic acid as a major fermentation product.

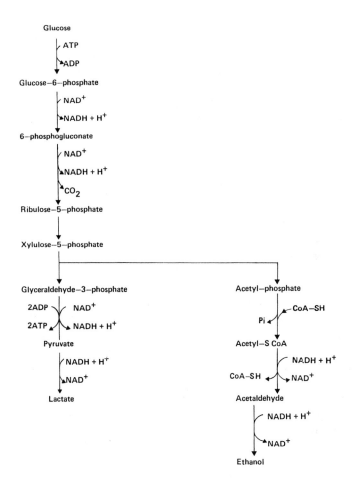

Fig. 2.3 The heterolactic fermentation pathway.
Balance: Glucose $\rightarrow CO_2$ + lactate + ethanol (+1ATP)

Where lactate is the sole product, as with the *Streptococci, Pediococci, Sporolactobacilli* and the majority of species of *Lactobacilli*, the mechanism is described as 'homofermentative'. Glucose is degraded to pyruvate by glycolysis, and NADH is reoxidized by the action of lactate dehydrogenase reducing pyruvate to lactate. The product (2 lactate/glucose) and energy (1 ATP/lactate) yields are identical with those obtained in the yeast alcohol fermentation.

The heterofermentative metabolic sequence found in *Leuconostoc* and some species of *Lactobacilli* ferments glucose according to the equation:

$$glucose \rightarrow lactate + ethanol + CO_2$$

The reactions are shown in Figure 2.3, where the unique enzyme is the phosphoketolase which hydrolyses xylulose 5-phosphate to glyceraldehyde 3-phosphate and acetyl phosphate (Table 2.1). The net energy yield (1 ATP/glucose) is only half that of the homofermentative pathway. In this connection it is interesting to note that, with ribose as substrate the heterofermentative equation alters to

$$\text{ribose} \rightarrow \text{lactate} + \text{acetate}$$

and the energy yield rises to 2 ATP/substrate. This results from the absence of the two oxidation reactions from glucose to ribulose-5-phosphate, with the consequent freedom to convert acetyl phosphate directly to acetate via acetate kinase with an extra ATP being formed by SLP (see Fig. 2.4 and Section 2.3.4).

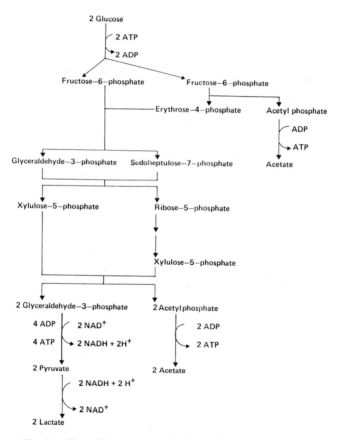

Fig. 2.4   The *Bifidum* pathway of glucose fermentation to lactate. Balance: 2 glucose → 2 lactate + 3 acetate (+ 5ATP)

A third mechanism of lactate formation is that known as the bifidum pathway, and found in *Bifidobacterium bifidum*. Here a second phosphoketolase is involved, cleaving fructose 6-phosphate to acetyl phosphate and erythrose 4-phosphate (Table 2.1). The complete pathway (Fig. 2.4) also employs enzymes of the glycolytic and pentose phosphate pathways and ferments glucose according to the equation:

$$2 \text{ glucose} \rightarrow 2 \text{ lactate} + 3 \text{ acetate}$$

Each of the acetyl phosphate to acetate reactions is coupled to ATP synthesis with a consequent overall energy yield of 2.5 ATP/glucose, i.e. higher than either the homofermentative or heterofermentative pathways.

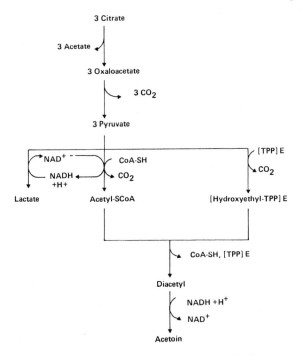

Fig. 2.5 Fermentation of citrate to diacetyl and/or acetoin, e.g. in *Streptococcus lactis* var. *diacetylactis* or *Leuconostoc cremoria*. [TPP]-Enz is enzyme-bound thiamine pyrophosphate.
Balance: 3 citrate → 5CO$_2$ + lactate + 3 acetate + diacetyl
    or: 2 citrate → 4CO$_2$ + 2 acetate + acetoin

Although not of direct bioenergetic consequence, it should be noted that the lactic acid bacteria have the capacity to ferment a wide range of saccharides, including fructose, galactose, mannose, saccharose, lactose, maltose and various pentoses. Malate can also act as a fermentative substrate being oxidized directly to lactate plus CO_2 by a special NAD^+-dependent malic enzyme in the so-called malo-lactate fermentation. Among the other products of fermentations within this group of organisms are diacetyl and acetoin, most commonly produced from citrate in, for example, *Streptococcus lactis* var. *diacetylactis* and *Leuconostoc cremoria* (Fig. 2.5).

A point of major energetic significance, however, arises from the fact the lactic acid bacteria are most correctly described as being aerotolerant or microaerophilic. Because of the action of superoxide dismutase and peroxidases, NADH can be reoxidized by reaction with oxygen. Pyruvate need not then be reduced to lactate, but may be converted to acetyl-CoA *en route* to acetate and an increased ATP yield.

There are three separate enzymic mechanisms for the synthesis of acetyl-CoA or acetyl phosphate from pyruvate in lactic acid bacteria. Streptococci have the pyruvate dehydrogenase multienzyme complex normally associated with aerobic metabolism and the operation of the TCA cycle. The pyruvate-formate lyase found in the Enterobacteriaceae is also present in, for example, *S. faecalis*, *B. bifidum* and *L. casei*. A third mechanism in *L. plantarum* and *L. delbruckii* is a pyruvate dismutation, where both pyruvate oxidase and lactate oxidase enzymes are flavoproteins:

$$pyruvate + P_i + FAD \rightarrow acetyl\ phosphate + CO_2 + FADH_2$$

$$pyruvate + FADH_2 \rightarrow lactate + FAD$$

2.3.4   Butyrate and butanol/acetone fermentations

Reference has already been made to the production of ethanol by the obligately fermentative *Clostridia*. This group also produce the short-chain fatty acids, acetate and butyrate, and the solvents, acetone and butanol, with hydrogen being a further important fermentation product (Figs 2.6, 2.7, 2.8).

Central to this fermentation pattern is the enzyme pyruvate-ferredoxin oxidoreductase. As with pyruvate decarboxylase (ethanol fermentation with acetaldehyde as intermediate) and pyruvate dehydrogenase (aerobic production of acetyl-CoA for entry to the TCA cycle), pyruvate-ferredoxin oxidoreductase is a thiamine pyrophosphate (TPP)-containing enzyme. In the breakdown of pyruvate, enzyme-bound hydroxyethyl-TPP is an intermediate:

$$pyruvate + Enz\text{-}TPP \rightleftharpoons Enz\text{-}TPP\text{-}hydroxyethyl + CO_2$$

$$Enz\text{-}TPP\text{-}hydroxyethyl + ferredoxin_{ox} + CoASH \rightleftharpoons$$
$$Enz\text{-}TPP + ferredoxin_{red} + acetyl\text{-}CoA$$

These reactions are reversible and can also serve as a mechanism for CO_2 fixation; for example, in *Clostridium kluyveri* and the sulphate-reducing and green sulphur phototrophic bacteria.

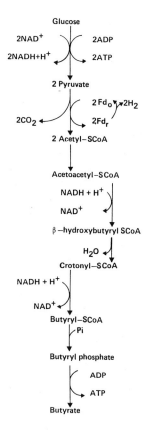

Fig. 2.6 Fermentation of glucose to butyrate by way of glycolysis and pyruvate:ferredoxin oxidoreductase, e.g. in *Clostridium butyricum, Cl. kluyveri* and *Cl. pasteurianum*.
Balance: glucose → $2CO_2$ + butyrate + $2 H_2$ ($+ 1ATP$)
Note: A proportion of the acetyl-CoA can give rise to acetate plus ATP (by way of acetyl phosphate—catalysed by phosphotransacetylase and acetate kinase). When this is the case the 'surplus' NADH is oxidized by way of hydrogenase (giving more hydrogen).

The ferredoxins are a group of iron–sulphur proteins, widespread in nature, which function as low-redox-potential electron carriers (page 32). In *Clostridia*, their $E_{m,7}$ values are around -400 mV and, through the enzyme hydrogenase, the reduced ferredoxin is thus close to equilibrium with molecular hydrogen (E_{m7} -420 mV):

$$\text{ferredoxin}_{red} + 2H^+ \rightleftharpoons \text{ferredoxin}_{ox} + H_2$$

The identity of the remaining fermentation products depends upon the reactions involving acetyl-CoA. As detailed earlier, this may be reduced to ethanol, or give rise to acetate by the action of phosphotransacetylase and acetate kinase:

$$\text{acetyl-CoA} + P_i \rightleftharpoons \text{acetyl phosphate} + \text{CoASH}$$

$$\text{acetyl phosphate} + \text{ADP} \rightarrow \text{acetate} + \text{ATP}$$

Alternatively, two molecules of acetyl-CoA can condense to give acetoacetyl-CoA and from there butyrate (Fig. 2.6); or butanol and acetone (Fig. 2.7) are the final fermentation products. Noteworthy is the association of butyrate production with ATP formation by SLP, in parallel with the extra energy yield also from acetate; while butanol parallels ethanol as a highly reduced fermentation product in that four molecules of reduced pyrimidine nucleotide are reoxidized during its formation. This reaction thus plays an important role, referred to as 'electron sink', in achieving overall redox balance in Clostridial fermentations. The same end is served by the generation of hydrogen directly from the pyruvate-ferredoxin oxidoreductase; this does not occur in organisms with pyruvate decarboxylase for which NAD^+ is the electron carrier coenzyme.

However, these pathways serve to illustrate the possible fermentation products rather than defining in quantitative terms what is observed in any given organism under a particular set of conditions. For example, with *Cl. acetobutylicum*, butanol and acetone are formed in the molar ratio of 2 : 1, and there is generally some isopropanol also present. Whereas *Cl. butyricum* yields butyrate and acetate in the approximate proportions 2 : 1, with *Cl. perfringens* the ratio of the acids produced is closer to 1 : 2, with ethanol and lactate also being significant products in this latter organism. *Cl. pasteurianum* and *Cl. sporogenes* produce only small amounts of butanol and no acetone or isopropanol, while *Cl. acetobutylicum* is a major solvent-producing strain.

Since *Cl. acetobutylicum* will also produce butyrate and acetate, it is instructive to consider the possible mechanism for the switch to solvent production. Butanol and acetone can be major products in continuous culture under sulphate- or phosphate-limitation at pH values below 5, although smaller amounts of butyrate and acetate are also produced. In batch culture, initial acid formation is succeeded, after the pH has dropped below 5, by a phase of solvent production: butanol, acetone and ethanol. Although the exact mechanism triggering the shift to solvent production has still to be elucidated, it is likely to involve intracellular pH, levels of free CoA–SH and of CoA esters, and the induction of the enzymes of the butanol and acetone pathways.

A striking feature of these Clostridial fermentation patterns of very great significance in energetic terms, is their branching at the level of acetyl-CoA,

with ethanol, acetate, butanol, acetone and butyrate as alternative fermentation products. To reiterate an earlier point, some of these are highly
reduced (ethanol and butanol); others are associated with additional ATP
synthesis (acetate and butyrate). As a result of variations in the flux through
particular branches, it is therefore possible to maintain redox balance while
showing variable fermentation products and energy yield. The determining
factor is the amount of the NADH produced at glyceraldehyde 3-phosphate
dehydrogenase which can be reoxidized by NADH: ferredoxin oxidoreductase and hydrogenase, with the consequent production of hydrogen. Since
the E_{m7} for the $NAD^+/NADH$ couple is $-320\,mV$ (ferredoxin $-410\,mV$,
hydrogen $-420\,mV$) this is an endergonic reaction, with hydrogen production largely dependent on a low hydrogen partial pressure.

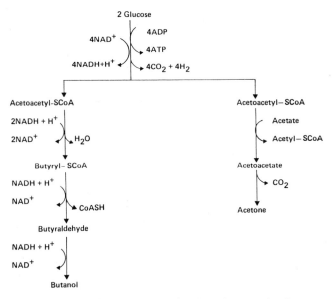

Fig. 2.7   Butanol and acetone fermentation: e.g. in *Clostridium acetobutylicum*.
Balance: (a) Glucose $\rightarrow$ 2CO$_2$ + butanol + 2 H$_2$ (+ 2ATP)
    or: (b) Glucose $\rightarrow$ 3CO$_2$ + acetone + 4 H$_2$ (+ 3ATP)
Note: The acetyl-CoA produced during acetone formation yields the extra ATP (by
way of acetyl phosphate): the extra hydrogens are produced from NADH by way of
hydrogenase. The first part of the pathway is as described in Fig. 2.6. When some of
the pyruvate produced during the initial glycolysis is converted to acetate, then more
hydrogen is produced (see legend to Fig. 2.6).

With the hydrogen partial pressure at 1 atm, the observed products of
glucose fermentation by *Cl. pasteurianum* are (moles per mole glucose
fermented):

0.6 acetate + 0.7 butyrate + 2.0 CO$_2$ + 2.6 H$_2$ + 3.3 ATP (Fig. 2.8).

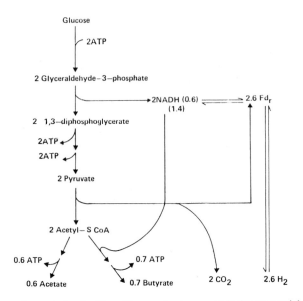

Fig. 2.8 Glucose fermentation by *Clostridium pasteurianum* at a hydrogen partial pressure of 1 atmosphere.

If the hydrogen partial pressure were to be kept low, through continuous removal of the gas, it would be theoretically possible to modify the fermentation products (per mole of glucose) to:

$$2.0 \text{ acetate} + 2\,CO_2 + 4\,H_2 + 4\,ATP$$

with all the NADH now being reoxidized via NADH-ferredoxin oxido-reductase and hydrogenase. Exactly this situation has, in fact, been demonstrated in the analogous acetate/ethanol fermentation with *Ruminococcus albus* when it is grown in co-culture with the hydrogen-oxidizing *Wolinella succinogenes*. This is an example of interspecies hydrogen transfer which succinctly demonstrates the energetic advantage gained by both hydrogen-producing (increased ATP yield) and hydrogen-oxidizing (provision of energy source) organisms, and in so doing underlines the great importance of such interactions in microbial ecosystems.

2.3.5   Mixed-acid fermentations

Like the *Clostridia*, the enterobacteria display a complex fermentation pattern, involving an extensive array of pathways and end-products. Like the *Clostridia*, they also employ the EMP glycolytic pathway for hexose breakdown but, although acetyl-CoA, CO_2 and hydrogen are again key products of pyruvate catabolism, the mechanism of their formation is quite different in the enterobacteria.

Unlike the *Clostridia*, the enterobacteria are facultative anaerobes and, under aerobic conditions, acetyl-CoA is formed from pyruvate by the pyruvate dehydrogenase multienzyme complex, with NAD^+ as cofactor. This enzyme system is repressed under anaerobic conditions while any pre-existing activity is inhibited by NADH. Instead, pyruvate-formate lyase is synthesized, giving rise to acetyl-CoA plus formate:

$$pyruvate + Enz \rightarrow acetyl\text{-}Enz + formate$$

$$acetyl\text{-}Enz + CoA\text{-}SH \rightarrow Enz + acetyl\text{-}CoA$$

Acetyl-CoA may then either be reduced to ethanol or transformed to acetate with concomitant ATP synthesis. Formate is excreted as a fermentation product by *Shigella* and *Erwinia*, but in *Escherichia coli* and *Enterobacter aerogenes* it is largely converted to CO_2 and hydrogen by the formate hydrogen lyase system. This consists of a formate dehydrogenase (FDH_{II}) and a hydrogenase, both of which are membrane-bound.

$$formate \rightarrow CO_2 + 2\,H^+ + 2e$$

$$2\,H^+ + 2e \rightarrow H_2$$

In the presence of nitrate or fumarate, however, a second formate dehydrogenase (FDH_I) channels the electrons to nitrate or fumarate reductase (see later sections). Both formate dehydrogenases are iron–sulphur proteins containing selenium, molybdenum and cytochrome b as cofactors (see page 37).

Lactate is another significant product of the mixed-acid fermentation and it is formed directly from pyruvate by the NADH-dependent lactate dehydrogenase.

These products, along with succinate, are characteristic of the genera *Escherichia*, *Salmonella* and *Shigella*. *Enterobacter*, *Serratia* and *Erwinia* on the other hand produce less acid, more CO_2 and ethanol and, in particular, acetoin and 2,3-butanediol. The key enzyme involved in the formation of these last two is α-acetolactate synthase, which again possesses enzyme-bound TPP:

$$pyruvate + Enz\text{-}TPP \rightarrow Enz\text{-}TPP\text{-}hydroxyethyl + CO_2$$

$$Enz\text{-}TPP\text{-}hydroxyethyl + pyruvate \rightarrow \alpha\text{-}acetolactate + Enz\text{-}Tpp$$

The α-acetolactate is then decarboxylated to acetoin which is dehydrogenated to give 2,3-butanediol. Since α-acetolactate synthase is activated at pH 6 this ensures a less acid fermentation pattern in those genera possessing the butanediol pathway.

2.3.6   Succinate and propionate fermentations

Although succinate is a fermentation product in the enterobacterial mixed-acid fermentation, it is instructive to consider it separately, along with its production and that of propionate, in other genera.

In the enterobacteria, the pathway leading to succinate begins with PEP carboxylase to give oxalacetate, which is then converted to succinate by the reductive reactions of a reversed TCA cycle. Since there are two reduction steps at malate dehydrogenase and fumarate reductase, succinate is a highly reduced bacterial fermentation product in the sense already discussed with reference to ethanol and butanol. The formation of succinate therefore fulfils an electron sink role and (indirectly) gives the potential for a higher energy yield from the fermentation by SLP in the formation of acetate.

The reduction of fumarate to succinate can, however, be coupled directly to the synthesis of ATP. The clue to this unique mechanism in anaerobic fermentations is the enzyme, fumarate reductase; the reaction does not proceed by the simple reversal of succinate dehydrogenase. Fumarate reductase is a membrane-bound enzyme associated with the electron carrier menaquinone and a *b*-type cytochrome (page 139). Both where fumarate is an intermediate metabolite in a fermentation pathway and where it is added to the medium as an alternative electron acceptor in anaerobic respiration, its reduction to succinate can be directly coupled to the formation of a protonmotive force (pmf) and consequently to the synthesis of ATP. The details of the mechanism will be considered more fully in Section 2.6.2 under the heading of the use of fumarate as an externally added alternative electron acceptor; but it is generally assumed that the ATP yield is 1 per mole of fumarate reduced. NADH may serve as electron donor for fumarate reductase, and NAD^+-independent donors include H_2, formate, glycerol 3-phosphate and, for the majority of the propionic acid bacteria, lactate. Amongst fermentative organisms, electron transport-dependent phosphorylation associated with fumarate reductase is found in several enterobacteria (including *E. coli*) and in the propionibacteria, in which succinate is an intermediate and also sometimes an end-product in its own right.

Propionate is produced as a fermentation product from glucose according to the equation:

$$1.5 \text{ glucose} \rightarrow 2 \text{ propionate} + \text{acetate} + CO_2$$

There are three mechanisms found amongst the propionibacteria for the necessary carboxylation step to oxaloacetate. One is the PEP carboxylase found also in the enterobacteria; a second, PEP carboxytransphosphorylase, conserves the phosphate bond energy with the formation of inorganic pyrophosphate (PP_i).

$$PEP + CO_2 + P_i \rightarrow oxaloacetate + PP_i$$

These enzymes must be involved where succinate appears as a fermentation product along with propionate.

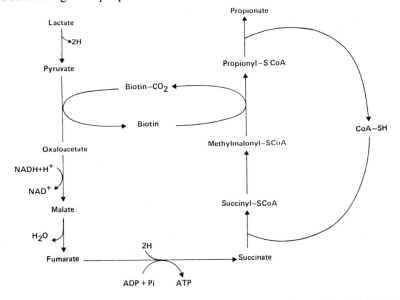

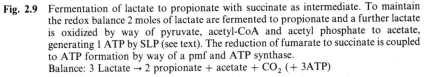

Fig. 2.9   Fermentation of lactate to propionate with succinate as intermediate. To maintain the redox balance 2 moles of lactate are fermented to propionate and a further lactate is oxidized by way of pyruvate, acetyl-CoA and acetyl phosphate to acetate, generating 1 ATP by SLP (see text). The reduction of fumarate to succinate is coupled to ATP formation by way of a pmf and ATP synthase.
Balance: 3 Lactate → 2 propionate + acetate + CO_2 (+ 3ATP)

The third carboxylation reaction uses pyruvate as acceptor and does not involve free CO_2. Lactate is a common substrate for the propionibacteria according to the equation:

$$3 \text{ lactate} \rightarrow 2 \text{ propionate} + \text{acetate} + CO_2$$

The fermentation pathway, including methylmalonyl-CoA-pyruvate trans-carboxylase, is depicted in Figure 2.9. Biotin acts as an enzyme-bound CO_2 carrier linking the carboxylation of pyruvate to the decarboxylation of methylmalonyl-CoA to give propionyl-CoA. The formation of propionate is also linked to the conversion of succinate to succinyl-CoA by a CoA transferase. The lactate dehydrogenase is a flavoprotein which can donate the hydrogens through menaquinone and cytochrome *b* to fumarate reductase. Overall, redox balance is maintained by the oxidation of one lactate to acetate which supplies the necessary reducing equivalents for the oxaloacetate to malate step in the reduction of two lactates to propionate.

This pathway is particularly efficient in that the energy yield is 3 ATP per 3 lactate fermented to propionate and acetate; 1 from SLP, and 2 from electron transport-linked phosphorylation. Notably, this result is achieved with a substrate that is, in fact, a discarded product of a number of other bacterial fermentations.

In a very few of the propionibacteria, e.g. *Cl. propionicum* and *Megasphaera elsdenii*, propionate is formed by another route in which succinate is not an intermediate but acrylyl-CoA is. Although the fermentation balance from lactate is identical to that obtained with the succinate pathway, the energy yield is less by two-thirds, due to the absence of the fumarate reductase electron transport-linked phosphorylation.

2.3.7   Formation of a protonmotive force in fermenters

In addition to fumarate reductase there are two other mechanisms known to give rise to a protonmotive force during fermentations and thus increase the overall energy yield.

2.3.7.1   Anaerobic citrate degradation

Anaerobic citrate degradation is found among the lactic acid bacteria, some *Clostridia* and certain of the enterobacteria. Citrate is first cleaved to acetate and oxaloacetate, which may then be decarboxylated to pyruvate plus CO_2. Whereas the enzyme is generally soluble, in *Enterobacter aerogenes* oxaloacetate decarboxylase is membrane-bound and contains biotin as a cofactor. Also, and most significantly, the enzyme is sodium-requiring and it has been shown that decarboxylase activity results in the electrogenic translocation of sodium ions across the membrane with the establishment of an electrochemical sodium gradient.

Other examples of sodium-translocating decarboxylases are the methylmalonyl-CoA decarboxylase of *Veillonella alcalescens* and *Propiogenium modestum* (these propionibacteria do not possess the methylmalonyl-CoA-pyruvate transcarboxylase discussed above), and the glutaconyl-CoA decarboxylase involved in glutamate fermentation by *Cl. symbosium*. The operation of an electroneutral Na^+/H^+ antiport would allow for the transformation of the electrochemical sodium gradient into a protonmotive force and, assuming a H^+ translocating stoichiometry of 3 for the ATPase, it can be estimated on thermodynamic grounds that the energy yield should be 1/3 ATP per substrate molecule decarboxylated.

In the case of *Pr. modestum* growing on succinate (the products being propionate $+CO_2$), this is the only energy-yielding reaction available to the cells.

2.3.7.2   Acidic fermentation products

The formation and excretion of acidic fermentation products can affect membrane energization, either detrimentally or beneficially. The undissociated forms of acetic and butyric acids (pKa values around 4.8) freely permeate the bacterial membrane and allow the accumulation of these fermentation products in the medium, where the relative amounts of the dissociated and undissociated forms will reflect the prevailing pH. As increasing amounts are excreted by the cells, the pH of the medium will be lowered until, at values around 4, the great excess of the acids will be in the undissociated membrane-permeable form which will thus re-enter the cell where they will tend to dissociate at the higher intracellular pH values around 6. The effect of this process will be to translocate protons into the cell and so partially de-energize the membrane by collapsing the pH gradient. (Note that this is an electroneutral phenomenon which does not affect the membrane potential and thus should not be referred to as 'uncoupling'.)

Lactic acid, however, has its pKa at 3.86 and its translocation across the membrane is dependent upon a carrier mechanism, which appears to be a proton symport with a proton/lactate stoichiometry of 2/1 (see Chapter 8). This means, in effect, that the membrane translocation of lactate is electrogenic, with its intracellular accumulation being driven by the full protonmotive force, while its efflux from the cell can at least potentially generate such a protonmotive force. Evidence that this is indeed the case has been obtained from studies with the lactic acid bacterium *Streptococcus cremoris* and *E. coli*. Particularly significant is the observation that the growth yield of *Strep. cremoris* is increased when it is grown in co-culture with the lactate-utilizing organism.

In one respect, this situation parallels that referred to already where, in the co-culture of hydrogen-producing and hydrogen-consuming species, not only does the second organism gain from the provision of an energy source but there is also a direct energy gain to the first organism resulting from the removal of its fermentation products. The molecular mechanisms of the two effects are, however, quite different.

2.3.8   Acetate fermentations

2.3.8.1   Introduction

We have already noted that acetate is both a common product of several fermentation patterns in a range of organisms, and of particular importance in as much as its production is associated with an increased energy yield by SLP. When considering the sulphate-reducing and methanogenic bacteria, acetate will again be seen to play a key role, although this time predominantly as a substrate. In fact, the production and consumption of acetate and of hydrogen, by different species of microorganisms, forms the focal point in the complex series of interacting metabolic activities that characterize the anaerobic biodegradative microbial food chain. This chain leads from hydrocarbon, carbohydrate, protein and lipid polymers to the final products of mineralization—CO_2 and either sulphide or methane, depending on environmental conditions.

These processes are greatly facilitated by the so-called acetogenic bacteria, knowledge of whose existence and physiological activities has increased greatly in recent years. Operationally they can be considered as two subgroups: the homoacetogens which ferment hexoses, $H_2 + CO_2$, and occasionally carbon monoxide or methanol $+ CO_2$, to acetate as sole fermentation product; and the hydrogen-producing acetogens which ferment alcohols and organic acids to acetate plus H_2. In concert, therefore, with the other fermentative bacterial genera already described, the acetogens ensure that the full rage of nutrient materials available to anaerobic microbial ecosystems can ultimately be converted to acetate and hydrogen.

With regard to the acetogens themselves, two metabolic processes are of particular interest: the synthesis of acetate from CO_2; and the wider implications of the production of H_2 in the fermentation of short-chain fatty acids such as propionate and butyrate.

2.3.8.2 The homoacetogens: the acetyl-CoA pathway

The route whereby acetogenic bacteria such as *Cl. aceticum*, *Cl. thermoaceticum* and *Aceobacterium woodii* ferment hexoses to three molecules of acetate, involves glycolysis and the formation of acetate from both molecules of pyruvate by the action of pyruvate-ferredoxin oxidoreductase; the third acetate comes from the reduction of the two molecules of CO_2 arising from this last reaction. These organisms can also be grown on CO_2 as sole carbon source according to the equation:

$$2 CO_2 + 4 H_2 \rightarrow \text{acetate} + 2 H_2O$$

These mechanisms of heterotrophic and autotrophic CO_2 fixation are, in fact, identical and are referred to as the acetyl-CoA pathway (Fig. 2.10).

However, the acetyl-CoA pathway is confined neither to the acetogens nor to CO_2 fixation. It appears that in both eubacterial sulphate-reducing and archaebacterial methanogenic bacteria the route of carbon fixation from CO_2 is via the acetyl-CoA pathway (with the likely exception of the genera

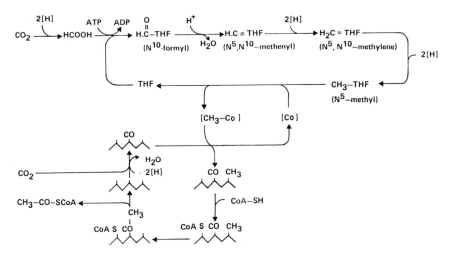

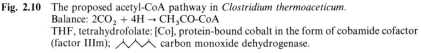

Fig. 2.10 The proposed acetyl-CoA pathway in *Clostridium thermoaceticum*.
Balance: $2CO_2 + 4H \rightarrow CH_3CO\text{-}CoA$
THF, tetrahydrofolate: [Co], protein-bound cobalt in the form of cobamide cofactor
(factor IIIm); ⋀⋀⋀ carbon monoxide dehydrogenase.

Desulfobacter and *Desulfuromonas*). At the same time, the acetyl-CoA
pathway, or a slight variant of it, has also been found to be the basis for the
reduction of acetate to methane, and for the oxidation of acetate to CO_2,
again in a range of sulphate-reducing and methanogenic bacteria. Clearly,
these reactions are central to all considerations of carbon and energy
metabolism in these anaerobic chemotrophs and it would be seriously to
miss the point, and to underestimate its crucial importance, to refer to the
acetyl-CoA pathway solely as a route of CO_2 fixation. Further, when it is
appreciated that, even in the CO_2 fixation mode the pathway is associated
with energy generation and ATP synthesis, it would clearly be wise to
consider the reaction sequence rather as a mechanism for the oxidation of
hydrogen with CO_2 as electron acceptor. In this respect, therefore, the
acetogens directly parallel the methanogens.

Most information on the details of the acetyl-CoA pathway has come
from studies with *Cl. thermoaceticum*, and to a lesser extent with *Cl.
formicoaceticum* and *A. woodii*. While the data are as yet incomplete and
subject to some minor variations of interpretation, the general picture now
appears to be clearly established. The complete reaction sequence can
perhaps best be considered in terms of two partial reactions: the formation
of methyl-tetrahydropteridine; and its carboxylation to acetyl-CoA through
the mediation of a corrinoid methyl-transferring protein and the enzyme
carbon monoxide dehydrogenase.

The general scheme proposed for the acetyl-CoA pathway in *Cl. thermo-aceticum* is given in Figure 2.10. The pteridine derivative has been identified as tetrahydrofolate (THF); in methanogens it is tetrahydromethanopterin, while it remains unidentified in those sulphate-reducers which employ the pathway. Formate is a free intermediate and the formate dehydrogenase catalysing its synthesis from CO_2 is an NADPH-dependent flavoprotein containing tungsten, selenium and iron–sulphur centres, with reduced ferredoxin being the likely physiological electron donor. The formation of N^{10}-formyl-THF requires ATP activation. The transformation through N^5-methenyl-THF to N^5,N^{10}-methylene-THF is carried out by an NADPH-dependent bifunctional enzyme; in *Cl. formicoaceticum* there are two mono-functional enzymes, the reduction being NADH-dependent. The N^5,N^{10}-methylene-THF reductase catalysing the reduction to N^5-methyl-THF is a flavoprotein containing zinc and iron–sulphur centres; it is NAD(P)H-independent with reduced ferredoxin again being the likely physiological electron donor.

The next stage of the reaction sequence involves a methyltransferase and a non-enzymic corrinoid protein. The methyl group is transferred to the reduced cobalt within the protein-bound corrinoid nucleus; Factor 111_m in *Cl. thermoaceticum*, but vitamin B_{12} in *A. woodii*, with the cofactor identity not yet established for the sulphate-reducing bacteria and methanogens.

The heart of the acetyl-CoA pathway lies with the enzyme carbon monoxide dehydrogenase. This is a tetrameric protein, with nickel, zinc and iron–sulphur centres, which catalyses the addition of a carboxyl group to methyl-THF with the formation of acetyl-CoA, where the carboxyl donor may be CO_2 (plus H_2, hydrogenase and ferredoxin), pyruvate (plus pyruvate-ferredoxin oxidoreductase and ferredoxin) or carbon monoxide. Carbon monoxide dehydrogenase has three sites which can bind a C_1 group from carbon monoxide (possibly a nickel carbonyl), a methyl group and coenzyme A. Intramolecular transfer of these groups leads to the synthesis and release of acetyl-CoA.

Cells capable of the reactions of the acetyl-CoA pathway possess very large amounts of the pteridine and corrinoid cofactors. *Cl. thermoaceticum* additionally has the standard respiratory hydrogen and electron carriers, ferredoxin, flavodoxin (under iron-limitation), NAD^+, rubredoxin, mena-quinone and cytochrome *b*. *A. woodii*, however, lacks the quinone and cytochrome. The exact nature of the electron transport chains leading from H_2, reduced ferredoxin or pyrimidine nucleotides to the reactions of acetyl-CoA synthesis has still to be established. Equally, the site and mechanism of any electron transport-linked phosphorylation remains a matter for specu-lation; the yield of ATP by SLP from the formation of acetate from acetyl-CoA is offset by the ATP used in the activation of formate. The largest drop in free energy is associated with the reduction of methylene-THF to methyl-

THF ($E_{m,7}$ -117 mV, being the most positive in the sequence) and this reaction is therefore the one most likely to be coupled to the development of a protonmotive force and the synthesis of ATP. The electron and energy donors for the reaction are either ferredoxin (from pyruvate–ferredoxin oxidoreductase) or NADH (from glyceraldehyde phosphate dehydrogenase) and there is a sufficient ΔE for the synthesis of at least 1 ATP, although details of the mechanism remain unknown.

With reference to the methanogens, it has been suggested that membrane-bound cobamides might have a role to play in electron transport and energy generation. If substantiated, such a mechanism might also be applicable to the acetogens.

The formation of acetyl-CoA from $H_2 + CO_2 +$ methyl-THF + coenzyme A arises from a reaction sequence close to equilibrium, but the initial formation of enzyme-bound CO is likely to require the input of significant activation energy. This could account for an ATP yield of less than 1 for the pathway as a whole, and it has even been suggested that the mechanism of activation is by energy-driven reversed electron transport with a direct coupling between two proton-translocating enzymes, methylene-THF reductase and carbon monoxide dehydrogenase.

A second route of acetate synthesis is found in organisms characterized by the fermentation of purines and amino acids. It is known as the glycine synthase pathway and is present in, for example, *Cl. acidi-urici* and *Peptococcus glycinophilus*. The reactions are identical to those of the acetyl-CoA pathway leading to the formation of methylene-THF, which is then transformed to glycine by glycine synthase in a reaction involving the addition of a second CO_2, reducing equivalents and ammonium ions, and the release of free THF. The glycine reductase then forms acetate in a reaction sequence consuming two further reducing equivalents and giving rise to ATP by SLP from an as-yet unidentified phosphoryl intermediate. The glycine synthase and reductase system is a complex of proteins, at least one of which is a selenium metallo-enzyme.

Amongst the several laboratories engaged in elucidating the pathways of acetate synthesis in recent years, those of Wood and Fuchs have been particularly prominent. It is indeed fortunate that both these authors have published comprehensive reviews of the subject area (Wood *et al*, 1986b; Fuchs, 1986).

2.3.8.3 The hydrogen-producing acetogens

The most celebrated example of an H_2-producing acetogen is to be found in *Methanobacillus omelianskii*. This 'organism', first isolated in 1940, grows on ethanol with the formation of methane, but was shown in 1967 to be in fact an obligately syntrophic association of two bacteria; the so-called 'S' organism, an acetogen converting ethanol to acetate plus H_2; and the methogen *Methanobacterium bryantii*, which oxidizes H_2 with the production of methane:

$$2 CH_3CH_2OH + 2 H_2O \rightarrow 2 CH_3COOH + 4 H_2 \quad \Delta G^{0\prime} +9.6 \, kJ/reaction$$

$$4 H_2 + CO_2 \rightarrow CH_4 + 2 H_2O \quad \Delta G^{0\prime} -136 \, kJ/reaction$$

The ability of the S organism to obtain energy from the oxidation of ethanol is dependent upon the removal of H_2 by the methanogen, with the consequent alteration of the reaction from being endergonic under standard conditions to being exergonic under the conditions of the two-organism consortium. The term 'interspecies hydrogen transfer' is used to describe such an association, which has a structural as well as a metabolic manifestation in that the bacterial partners are found to be in close physical contact. In the cases cited earlier where co-culture with a H_2-oxidizing species was shown to confer an energetic advantage on the branched fermentation pattern found with *Cl. pasteurianum* and *Ruminococcus albus*, the association was seen to be optional. With the S organism and methanogen in *Methanobacillus omelianskii*, however, the association is obligate, and the S organism is unable to grow on ethanol in pure culture.

Other examples of H_2-producing acetogenic bacteria that can only exist in nature as components of syntrophic associations demonstrating interspecies hydrogen transfer are *Syntrophobacter wolinii* (propionate) and *Syntrophomonas wolfei* (butyrate):

$$CH_3CH_2COOH + 3 H_2O \rightarrow CH_3COOH + 3 H_2 + H_2CO_3$$
$$\Delta G^{0\prime} + 76.1 \, kJ/reaction$$

$$CH_3CH_2CH_2COOH + 2 H_2O \rightarrow 2CH_3COOH + 2 H_2$$
$$\Delta G^{0\prime} + 48.1 \, kJ/reaction$$

Both in natural ecosystems and in the laboratory either a sulphate-reducing bacterium or a methanogen may serve as the H_2-oxidizing partner and by this means reduce the H_2 concentration to about $0.1 \, \mu M$. A similar high affinity for acetate might also result in its concentration being maintained at around $10 \, \mu M$. Even under such ideal conditions, however, the fermentation of, for example, butyrate by *Syntrophomonas wolfei* would still only give a low net yield ($\Delta G_0' = -15 \, kJ/reaction$).

In a thought-provoking article Thauer and Morris (1984) introduced the concept of ATP yields less than 1 per substrate molecule catabolized, and discussed the part that might be played by energy-driven reversed electron transport in integrating redox balance and net energy yield in anaerobic heterotrophic metabolism. These authors identified $-44 \, kJ$ as being the minimum free energy requirement for the synthesis of 1 ATP under quasi-reversible conditions and suggested that, where $-\Delta G$ was numerically less than this figure, reversed electron transport reactions might feature in pathways leading to fractional ATP yields.

The catabolic route from butyrate most probably proceeds through butyryl-CoA, crotonyl-CoA, β-hydroxybutyryl-CoA and acetoacetyl-CoA to two acetyl-CoA, with a net yield of 1 ATP by SLP from acetyl-CoA (the second acetyl-CoA being required for butyrate activation by a Co-ASH transferase). However, H_2 production from the butyryl-CoA to crotonyl-CoA step ($E_{m,7}$ − 15 mV) is energy-requiring as, even at 0.1 μM H_2 the H^+/H_2 couple has an E_h of − 200 mV. Assuming a H^+-translocation to ATP stoichiometry from the ATPase of 3, the hypothesis is that 2/3 ATP is hydrolysed with the translocation of 2 H^+ which then re-enter the cell accepting two electrons from the butyryl-CoA to crotonyl-CoA reaction. By this means, the conversion of butyrate to two acetates is facilitated with an ATP yield of 1/3 which agrees with the ΔG for the overall reaction of − 15 kJ (Fig. 2.11).

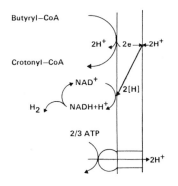

Fig. 2.11 Proposed mechanism of energy-driven reversed electron transport during production of hydrogen by oxidation of butyryl-CoA to crotonyl-CoA in *Syntrophomonas wolfei*.

2.3.9 Fermentation: general comments

In this summary of microbial fermentation reaction sequences, the energetic consequences of the particular routes of carbon flux and mechanisms of redox balance have been stressed. An especially rich overall presentation of fermentation processes, including those concerned with amino acids, purines and pyrimidines, can be found in the second edition of Gottschalk's book, 'Bacterial Metabolism' (Gottschalk, 1986).

One particularly striking feature of fermentation as a source of energy for anaerobic microbial growth stems from the relative inefficiency of SLP and the limited occurrence of pmf-linked ATP synthesis (in terms of energy yield per mole of substrate used). For a given amount of cellular growth, large amounts of substrate are required with the consequent formation of similar quantities of a variety of reduced fermentation products. Where, in environmental or biotechnological processes, either the maximum turnover of substrate or the maximum formation of product along with the minimum wasteful production of microbial biomass is being sought, then clearly fermentative schemes have much to recommend them.

In terms of the thesis being developed here, however, the most significant feature is the excretion by organism A of reduced fermentation products which are potential sources of carbon, energy and reducing equivalents for organisms B, C and D, even within the same anaerobic environments. Ethanol, lactate, propionate, butyrate, for example, are all capable of acting as substrates for other fermentative species. Along with such food-chain considerations, there are the optional or obligate energetic benefits to be gained from lactate efflux and interspecies hydrogen transfer. It can be concluded that fermentative microorganisms are not likely to function in nature as single species in metabolic isolation, but that the natural habitat of anaerobic heterotrophic bacteria is within a closely functionally-integrated microbial consortium. In such a scenario, the acetogens take on a particularly important role in that they determine the potential for the anaerobic biodegradation of almost all organic primary substrates to the products acetate, CO_2 and H_2.

These last transformations, as already seen, are often only possible in the presence of H_2-oxidizing bacteria, which therefore ultimately control the fermentative and biodegradative activities of the consortium as a whole. These key H_2-oxidizing bacteria, which incidentally further stimulate the overall process by their utilization of acetate, are the methanogenic and sulphate-reducing bacteria.

2.4 Anaerobic respiration: methanogenesis and sulphate reduction

2.4.1 Methanogenic bacteria

The methanogenic bacteria are usually considered to belong to the group of bacteria that grow anaerobically using the process of anaerobic respiration for energy generation. The reducing substrate is usually hydrogen gas, with CO_2 acting as terminal electron acceptor. As noted later, in the absence of oxygen, a fairly wide range of facultative genera can use nitrate, nitrite, trimethylamine oxide or fumarate as alternative terminal electron acceptors in a redox mechanism that requires a standard, if slightly modified, electron transport chain coupled to ATP synthesis through a protonmotive force. This also holds true for the sulphate-reducing bacteria, save only that they are obligate rather than facultative anaerobes. In their reduction of CO_2 to methane, the methanogenic bacteria appear to be employing the same basic energetic mechanism and, certainly in ecological terms the sulphate-reducers and methanogens can be seen to function, and therefore must be considered, in a parallel fashion. Both groups oxidize H_2 and metabolize acetate, with the methanogens predominating in more reducing ($E_h = -300\,mV$) and sulphate-free environments.

Unlike the more conventional anaerobic respiratory bacteria, however, the methanogens are largely without cytochromes and other usual respiratory carriers. Both in respect of their oxidation of H_2 with CO_2 as electron acceptor and their use of the acetyl-CoA pathway, the methanogens bear a close physiological relationship to the fermentative acetogens; yet they clearly do not accord with our normal perception of fermentation since all their ATP arises from a chemiosmotic mechanism with no evidence for substrate level phosphorylation.

The most striking feature of the methanogens, however, lies not in their points of similarity but rather in the dramatic dissimilarity they show to other bacteria, whether aerobic or anaerobic. The methanogens are Archaebacteria, a group of organisms probably of early evolutionary origin, distinct from both eukaryotes and eubacteria and yet sharing certain characteristics with each group. The Archaebacteria were recognized as a separate Kingdom on the basis of studies of their 16S ribosomal RNA, and features of their uniqueness have now extended to various other ribosomal, transfer and messenger RNA properties, along with histones, ribosome architecture and the mechanisms of transcription and translation. From the point of view of cell structure and metabolism a characteristic feature identifying the Archaebacteria is the chemistry of their cell walls and membranes. Here there is an absence of peptidoglycan in the wall, and the membrane comprises glycerolipids in which L-glycerol is ether-linked to long-chain phytyl groups. Along with the methanogens, the Archaebacterial Kingdom contains the extreme halophiles and the sulphur-dependent thermoacidophiles.

Although methanogenesis is only found amongst the Archaebacteria, the methanogen group in fact displays a wide range of bacterial morphologies. There are four distinct types: rods, cocci, sarcinae and spirilla; demonstrating both positive and negative Gram reactions. The rods and spirillum-type cells may grow as individual cells or as long filaments.

2.4.2   Chemolithotrophic and methylotrophic metabolism

With respect to their metabolism, the methanogens can be subdivided into two groups. The obligate chemolithotrophs obtain their energy and carbon from the reduction of CO_2 with H_2 as electron and energy donor. Most species can also grow on formate. On the other hand, the methylotrophic methanogens can grow on H_2 plus CO_2, but they are also able to grow on methanol, on mono-, di- or tri-methylamine, or on acetate as energy and carbon source. *Methanobrevibacter ruminantium* is unusual in that it can grow on H_2 plus CO_2, but additionally requires acetate as a source of carbon. Whereas there is no evidence for the presence of quinones or cytochromes among the chemolithotrophic methanogens, they are found in those methylotrophic methanogens able to oxidize methyl groups to CO_2. For example, *Methanosarcina barkeri* has two *b*-type and one *c*-type cytochromes.

Recently, two intriguing lithotrophic methanogens have come to light. A freshwater *Methanospirillium* sp. and a marine *Methanogenium* sp. have been isolated in pure culture using propan-2-ol as hydrogen donor for methanogenic growth on CO_2 according to the reactions:

$$4CH_3.CHOH.CH_3 \rightarrow 4CH_3.CO.CH_3 + 8[H]$$

$$CO_2 + 8[H] \rightarrow CH_4 + 2H_2O$$

H_2, formate and butan-2-ol may also serve as hydrogen donor for both species, and ethanol and *n*-propanol additionally for the marine strain. Propan-2-ol has also been noted, however, to be a substrate for the growth of the syntrophic consortium designated *Methanobacterium omelianski*, and this raises a most interesting point regarding metabolic mechanisms and energy coupling at the cellular level. Where '*M. omelianski*' grows on ethanol or propan-2-ol it does so by the combined fermentative reactions of the acetogenic S organisms and the H_2-oxidizing capacity of its partner *Methanobacterium bryantii*, with energy coupling at the cellular level being achieved by interspecies H_2 transfer. However, in the newly isolated *Methanogenium* sp., the same two substrates can support growth by reactions that can be represented mechanistically and energetically by the same equations, but within a single organism and with no evidence for intermediary H_2 production.

An exactly analogous comparison will be drawn later between the growth on propionate by *Syntrophobacter wolinii* along with an H_2-oxidizing methanogen or sulphate-reducer, and by *Desulfobulus propionicus* in pure culture.

Detailed knowledge of metabolism and energetics of the methanogenic bacteria, is largely dependent on extensive studies with two particular species. *Methanobacterium thermoautotrophicum* is an obligate chemolithotroph which grows at temperatures in excess of 40°, with an optimum between 65° and 70°. *M. barkeri* on the other hand is the most versatile methanogen known, although it is worth pointing out that, whereas growth on H_2 plus CO_2, methanol or methylamines may take from three to seven days, growth on acetate may take as long as three weeks. Despite this, *M. barkeri* is generally isolated from such environments as anaerobic digesters which are high in acetate, and it is not found in the rumen where acetate is at a very low concentration. Epithelial cells rapidly absorb volatile fatty acids, and the dilution rate in the rumen exceeds the maximum growth rate (μ_{max}) of the chemolithotrophic methanogens.

2.4.3 Carbon assimilation in methanogens

No methanogen uses the usual methylotrophic pathways for the assimilation of reduced C-1 compounds (ribulose monophosphate pathway, serine pathway, dihydroxyacetone pathway); none uses the Calvin (ribulose bisphosphate) pathway or reductive TCA cycles of CO_2 fixation. The route for assimilation of CO_2 in methanogens is the acetyl-CoA pathway, in which a major intermediate is methyl tetrahydromethanopterin ($H_4MPT-CH_3$) (Fig. 2.12). As with acetogens (Fig. 2.10), the methyl group is transferred (by a methyltransferase) to a corrinoid protein. It is then carbonylated to acetyl-CoA by way of the nickel-containing CO dehydrogenase which produces the bound CO by reduction of carbon dioxide. An F_{420}-dependent reductive carboxylation then gives pyruvate. The overall reaction is thus:

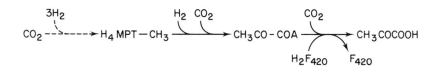

The pyruvate is then further carboxylated to oxaloacetate by the action of PEP synthetase and PEP carboxylase. In *M. barkeri*, 2-oxoglutarate is then produced by the condensation of oxaloacetate and acetyl-CoA to give citrate followed by the oxidative reactions of a conventional TCA cycle. These reactions are absent from *M. thermoautotrophicum*, however, and 2-oxoglutarate is synthesized from oxaloacetate by the reductive reactions of the TCA cycle, with F_{420} again being the electron donor in the reductive carboxylation of succinyl-CoA to 2-oxoglutarate. In neither organism is there a complete cycle operating in an oxidative energy-generating mode.

Methanogenesis from acetate in *M. barkeri* proceeds by the reversal of the pathway described above: carbon monoxide dehydrogenase, corrinoid protein, methyl-tetrahydromethananopterin, methyl-CoM, methane. Growth of *M. barkeri* on acetate induces a five-fold increase in the levels of carbon monoxide dehydrogenase as compared with H_2O/CO_2-grown cells.

2.4.4   Energy coupling in methanogens

2.4.4.1   Introduction

The whole purpose of methanogenesis is to make ATP. All the evidence so far is consistent with the conclusion that methanogens obtain ATP by way of a protonmotive force (pmf) which is generated by electron transport, with hydrogen as the electron donor and CO_2 usually providing the electron acceptor. The overall reaction is:

$$CO_2 + 4H_2 \rightarrow CH_4 + 2H_2O \ \Delta G^{0\prime} = -131 \ kJmol^{-1}$$

In non-standard conditions, particularly with the usual low concentrations of hydrogen, the free energy change will be more positive and it is unlikely that more than one ATP per mole of methane could be produced. In 1956 Barker proposed a scheme for reduction of CO_2 to CH_4 which, in essence, remains a valid summary of this process:

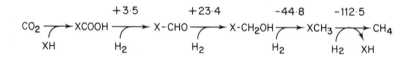

The carrier of C_1 units (X) is not the same for every level of reduction. The numbers over each step are the standard free energy changes ($\Delta G^{0\prime}$ in $kJ \ mol^{-1}$) for each reaction (on the basis of the free intermediates). The hydrogen in methane and its precursors does not come directly from the H_2 energy source but from the protons of water. The process of electron transport has an extra degree of complexity in methanogens because there are, in effect, four different terminal electron acceptors which must be reduced in sequence; and, hence, four electron transport chains having the same donor but different acceptors. Not all of these chains can be coupled to ATP synthesis. As shown above, the first two reductions are endergonic and they must be driven by the second pair of reductions which are exergonic. This may be by way of a pmf established during one or both of the exergonic reactions.

It is remarkable that, in general, the methanogens lack quinones and cytochromes and that, where present, they do not appear to play a direct part in the generation of a pmf for ATP synthesis. None the less, the mechanism for driving the phosphorylation of ADP in methanogens is chemiosmotic, as has been shown by

(a) the measurement of a pmf;
(b) ATP synthesis driven by a valinomycin-induced K^+ efflux potential or an artificially generated proton potential; and
(c) sensitivity to protonophore uncouplers and the ATPase inhibitor DCCD.

2.4.4.2 Unusual coenzymes in methanogens

Some of the special coenzymes of methanogens appear to be functionally analogous to those involved in aerobic electron transport systems: they differ mainly in their low redox potentials. Others have the function of carriers of the reduced carbon precursors of methane; and yet others have a function primarily in biosynthesis. Some of the coenzymes function in more than one role. Their structures are given in Figure 2.12 and their involvement in the reduction of Co_2 to methane is shown in Figure 2.13.

Factor F_{420} is a 5-deazaflavin analogue of FMN. It functions as a soluble hydrogen carrier with a low redox potential ($E_{m7} = -360\,mV$), is the hydrogen donor in the reduction of methyl-Coenzyme M, and may also be the hydrogen donor in the earlier steps in reduction of CO_2. It can be reduced by hydrogen (hydrogenase), by formate (formate dehydrogenase) and by NADPH (NADP$^+$ reductase). It may thus mediate between oxidation of hydrogen or formate and reduction of NADP$^+$. As mentioned above, coenzyme F_{420} is also the hydrogen donor in the reductive carboxylation of pyruvate (and oxoglutarate) during assimilation of cell carbon.

Factor F_{430} is a nickel tetrapyrrole which is also probably involved in the reduction of methyl-CoM to methane.

Methanofuran, previously called the CO_2 reduction factor (CDR), is the C_1-carrier at the level of oxidation of formate: it is thus the first bound intermediate in the reduction of CO_2.

Tetrahydromethanopterin (H_4MPT) is the next carbon carrier, the formyl group being transferred from formyl-methanofuran to yield formyl-H_4MPT which is dehydrated to methenyl-H_4MPT before reduction to methylene-H_4MPT and thence to methyl-H_4MPT.

The best-known of the novel coenzymes, coenzyme M, is the smallest of all known coenzymes. Coenzyme M is 2-mercaptoethane sulphonate ($HSCH_2CH_2SO_3{}^-$). This acts as the methyl carrier ($CH_3 - SCH_2CH_2SO_3{}^-$) in the last stage of methanogenesis catalysed by the methyl-CoM reductase complex.

a) F$_{420}$

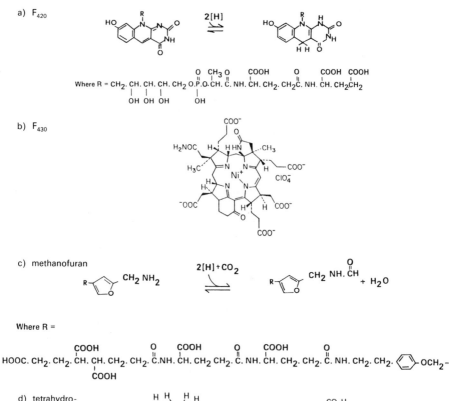

Where R = CH$_2$. CH. CH. CH. CH$_2$ O.$\overset{O}{\overset{\|}{P}}$.O.$\overset{CH_3}{\overset{|}{CH}}$. $\overset{O}{\overset{\|}{C}}$. NH. $\overset{COOH}{\overset{|}{CH}}$. CH$_2$. CH$_2\overset{O}{\overset{\|}{C}}$. NH. $\overset{COOH}{\overset{|}{CH}}$. $\overset{COOH}{\overset{|}{CH_2CH_2}}$
          OH OH OH   OH

b) F$_{430}$

c) methanofuran

Where R =

HOOC. CH$_2$. CH$_2$. $\overset{|}{\underset{COOH}{CH}}$. CH. CH$_2$. CH$_2$. $\overset{O}{\overset{\|}{C}}$.NH. $\overset{COOH}{\overset{|}{CH}}$. CH$_2$ CH$_2$. $\overset{O}{\overset{\|}{C}}$. NH. $\overset{COOH}{\overset{|}{CH}}$. CH$_2$. CH$_2$. $\overset{O}{\overset{\|}{C}}$. NH. CH$_2$. CH$_2$. ⬡-OCH$_2$-

d) tetrahydro-
methanopterin

Fig. 2.12   Structure of cofactors found in methanogens.

Component B of the methyl-CoM reductase complex is also probably a coenzyme: it has recently been shown to be 7-mercaptoheptanoylthreonine phosphate (HTP)

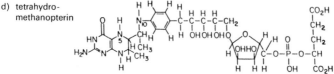

Its role is unknown, but it may act as a methyl carrier, the terminal thiol having the same role as that in coenzyme M.

2.4.4.3   Energy transduction and methanogenesis from carbon dioxide

Figure 2.13 summarizes the route for methanogenesis from CO_2; most of the steps of which have been discussed in the previous section.

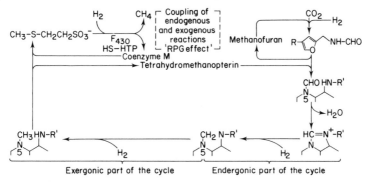

Exergonic part of the cycle   Endergonic part of the cycle

Fig. 2.13   Methanogenesis from CO_2. The 'RPG effect' is the driving of methanogenesis from CO_2 by addition of (for example) methyl-CoM. It is probably an expression of the necessary coupling of the endergonic steps with the exergonic steps of methanogenesis. This may well be achieved by way of a pmf (see text). The hydrogenases involved in each reduction step are not necessarily identical. If they are arranged across the membrane so that only electrons pass across to a hydrogen carrier on the inside then they will contribute to a pmf. For discussions of the involvement of the pmf in coupling these reactions see the reviews by Anthony (1982) and Blaut and Gottschalk (1985); for reviews of the intermediates see Anthony (1982) and Wolfe (1985). For an excellent overview, including the role of hydrogenases and the relevant thermodynamics see Keltjens and van der Drift (1986). MFR, methanofuran; H_4MPT, tetrahydromethanopterin; CoM, coenzyme M.

The final step, which is the most complex, is the formation of methane by the methyl-CoM reductase complex which is located in the cytoplasmic membrane. This complex consists of many proteins: component A-1 is a crude fraction containing hydrogenase; component A-2 is a single protein of unknown function; component A-3 consists of several proteins. Component C is the site of reduction of methyl-CoM to methane; it is a large protein associated with two molecules of the nickel tetrapyrrole F_{430} and two of coenzyme M. Component B is not a protein but another coenzyme of uncertain function (see above). Although a great deal is known about this part of methanogenesis, Ralph Wolfe (who has been responsible for much of what is known) has pointed out that much is yet to be learned about this multienzyme system before any reaction mechanisms can be taken seriously.

In Figure 2.13 the whole process of methanogenesis from CO_2 is drawn as a cycle. This emphasizes the essential fact that the later exergonic part of the process must be coupled somehow to the earlier endergonic part. The coupling of the first and last reactions has long been appreciated, and is illustrated by the RPG effect (named after its discoverer R. P. Gunsalus): the stimulation by methyl-CoM (and some other compounds) of CO_2 reduction to methane.

This coupling is almost certainly not via a direct chemical linkage of the reactions. The obvious way of coupling the last reaction of methanogenesis to the first reaction is by way of the pmf. If the methyl-CoA reductase complex is arranged suitably in the membrane then reduction of methyl-CoM could establish a pmf; this could then drive the endergonic reactions, which would also have to be arranged across the bacterial membrane.

2.4.4.4   Energy coupling during methanogenesis from methanol, formaldehyde and acetate

Methanol can also act as a methanogenic substrate for *M. barkeri* according to the reactions:

$$CH_3OH + H_2O \rightarrow CO_2 + 6[H]$$
$$3CH_3OH + 6[H] \rightarrow 3CH_4 + 3H_2O$$

$$4CH_3OH \rightarrow 3CH_4 + CO_2 + 2H_2O$$

One molecule of methanol is oxidized to supply the reducing equivalents required for the reduction of a further three molecules of methanol to methane (Fig. 2.14). The overall process is found to be Na^+-requiring and uncoupler-sensitive.

If, as seems likely, the electron carrier between the oxidation and reduction reactions is F_{420}, then the transfer of reducing equivalents from the redox level of methanol ($E_{m7} = -182\,mV$) to F_{420} ($E_{m7} = -360\,mV$) will be energy-requiring. Interestingly, however, *M. barkeri* can grow on methanol in the presence of H_2 according to the equation:

$$CH_3OH + H_2 \rightarrow CH_4 + H_2O$$

Clearly this reaction, mediated by hydrogenase through F_{420} and coupled to the reduction of methanol to methane is exergonic and potentially capable of developing a pmf and consequent ATP synthesis. Under these conditions methanogenesis is neither Na^+-requiring nor uncoupler-sensitive. Treatment with the ATPase inhibitor DCCD, however, causes a rapid exhaustion of the ATP pool without the pmf being reduced from a measured value of $-130\,mV$. Methane production is inhibited under these conditions, and this inhibition can be relieved by the addition of an uncoupler. In other words, methanogenesis from methanol plus H_2 is under a form of respiratory control.

The reduction of methanol to methane involves only two enzymes: a methyltransferase catalysing formation of methyl-CoM from methanol and the methyl-CoM methyl reductase. This second reaction is strongly exergonic and it is therefore by way of the membrane-bound methyl-CoM reductase complex that redox energy must be transduced into a transmembrane proton gradient; although the molecular mechanism is not presently identifiable.

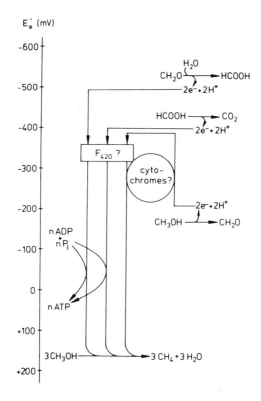

Fig. 2.14 Proposed reaction scheme for methanogenesis in *M. barkeri* from methanol. The redox scale indicates the standard mid-point potentials of the formal redox couples; the intermediates are most likely carrier-bound.

It has been suggested that the cytochromes present in *M. barkeri* are required for an energy-driven reversed electron transport in the transfer of reducing equivalents from methanol to F_{420}, and that this is also the Na^+-requiring and uncoupler-sensitive step under growth on methanol alone. *Methanosphaera stadtmaniae* provides circumstantial support for this idea in that it can only grow on methanol under an H_2 gas phase and is found to lack any cytochrome. Figure 2.14 presents a scheme for the growth of *M. barkeri* on methanol.

Methanogenesis from formaldehyde plus hydrogen mirrors the characteristics noted above for methanol plus hydrogen, i.e. respiratory control, resistance to uncouplers, no requirement for Na^+, and coupled ATP synthesis by a chemiosmotic mechanism, presumably again at the methyl-CoM reductase. However, the reduction of CO_2 to the level of formate is an endergonic reaction. This energy-requiring reductive C_1-transfer (to metha-nofuran) is thus a site of the uncoupler sensitivity of the overall process of methanogenesis from CO_2 plus H_2.

The considerations above have led Blaut and Gottschalk (1985) to propose that Na^+ may have a role in the coupling of the endergonic and exergonic reactions of methanogenesis.

2.4.4.5   The role of Na^+ in the reactions of methanogenesis
The role of Na^+ remains at least partially obscure in this instance. A Na^+ requirement has been identified for methane formation from H_2 plus CO_2, methanol, and acetate in a number of methanogenic species, and for the synthesis of ATP in *M. thermoautotrophicum* driven by a K^+ diffusion potential. However, the fact that methanogenesis and ATP formation from formaldehyde, and from methanol plus H_2 is Na^+-independent, suggests that Na^+ does not play a direct role in either the methyl reductase or the ATPase. The characterization of an Na^+/H^+ antiport in *M. thermoauto-trophicum* has led to the proposal that most probably the regulation of intracellular pH may be the focus of the Na^+ requirement, which, from these data, appears to be closely linked with the various energy-dependent steps in methanogenesis. Inhibition of the antiport also inhibits methanogenesis from H_2 plus CO_2. Interestingly, the antiport is sensitive to amiloride, harmaline and NH_4^+, which are the characteristic inhibitors of the Na^+/H^+ antiport in eukaryotic cells.

The methanogenic pathway from acetate also poses some intriguing energetic questions. The need to expend 1 ATP in the formation of acetyl-CoA determines that methanogenesis from acetyl-CoA via carbon monoxide dehydrogenase must be associated with an ATP yield greater than 1. It is therefore particularly interesting that, where methanogenesis in *M. barkeri* has been inhibited with bromoethanesulfonate, the oxidation of carbon monoxide

$$CO + H_2O \rightarrow CO_2 + H_2$$

gives rise to an increase in the pmf from $-100\,mV$ to $-150\,mV$. Addition of an uncoupler leads to rapid decreases in both the pmf and the ATP level, along with an increase in the rate of oxidation. This is directly paralleled by the demonstration that the oxidation of carbon monoxide to CO_2 in *Acetobacterium woodii* can be energetically coupled to histidine uptake. The reverse has also been demonstrated: the reduction of CO_2 with H_2 to carbon monoxide plus water is dependent upon methanogenesis and energy gener-ation. Studies with uncoupler-insensitive methane formation from methanol plus H_2 in *M. barkeri* have shown that CO_2 reduction by carbon monoxide dehydrogenase is uncoupler-sensitive and DCCD-insensitive, and thus appears to be driven directly by the protonmotive force rather than by ATP. Unlike other energy-requiring reactions in methanogenesis, however, CO_2 reduction to CO is not Na^+-dependent.

The enzyme carbon monoxide dehydrogenase, like the methyl reductase, might therefore be directly responsible for energy transduction and the generation of a pmf, although once again the mechanism involved cannot yet be identified.

2.4.5 Sulphate-reducing bacteria

2.4.5.1 Introduction

Like the methanogens, the sulphate-reducing bacteria constitute a phylogenetically diverse group of organisms which none the less share a common physiological character which, in turn, dictates a common ecology; they form a so-called physiological–ecological group. They play a key role in the natural sulphur cycle, catalysing the process of sulphurication (page 8). The chemistry of the sulphur compounds discussed in this section is covered in detail in Chapter 4, page 208 and Figs. 4.5 and 4.6.

The sulphate-reducing bacteria are strict anaerobes requiring Eh values of −100 mV or less for growth; they obtain their energy by an anaerobic respiratory process, using sulphate as terminal electron acceptor, i.e. dissimilatory sulphate reduction. The product of this reduction is sulphide and, in parallel with the acetogens and methanogens there is now widespread adoption of the nomenclature 'sulphidogens' rather than 'sulphate-reducing bacteria'. Perhaps a more pressing reason for basing identification on the product of respiration is the recent appreciation that there are a number of sulphur-reducing bacteria (again giving rise to sulphide) which clearly belong within the same physiological–ecological group. The vast majority of the organisms within the group, however, use sulphate (and often other oxidized derivatives such as thiosulphate) as terminal electron acceptor, and are completely unreactive to sulphur itself. The term 'sulphate-reducing' bacteria retains by far the wider general usage. For these reasons, but while recognizing the degree of inexactitude involved, the author shall continue to refer to organisms within this group as the sulphate-reducing bacteria. (They are often in fact referred to by the abbreviation SRB, but this does not represent correct usage in the present context.)

As is well documented in the second edition (1984) of Postgate's excellent monograph 'The Sulphate-Reducing Bacteria', before the late 1970s two genera only were recognized: *Desulfovibrio* and the spore-forming *Desulfotomaculum*. The 12 or so species identified were characterized by a restricted metabolic capability, both with respect to the spectrum of nutrients that could serve as carbon and energy sources for heterotrophic growth (typically lactate), and the incomplete oxidation due to the absence of a functional TCA cycle with the consequent production of acetate as a metabolic end-product. The identification of two new species by Pfennig's laboratory in 1976 and 1977 heralded what has turned out to be something of a revolution in the knowledge and understanding of the sulphate-reducing bacteria: *Desulfuromonas acetoxidans* and *Desulfotomaculum acetoxidans* are both capable of oxidizing acetate and using it as a source of carbon and energy to support growth, while the former organism shows the other (at that time new) property of reducing elemental sulphur as terminal electron acceptor. In the intervening ten years there has been a massive increase in both the number and the activity of the research groups working with the sulphate reducers. For example, Pfennig and his co-workers, notably Widdel, have isolated and characterized many more new genera and species (Table 2.2); cell physiology has been studied by many groups, in particular those of Thauer and Fuchs (H_2 oxidation and acetate metabolism) and of Peck and Le Gall (hydrogenase and electron transport carriers); Jørgensen and his colleagues have examined their ecological impact; Stetter's laboratory has extended the range of sulphur- and sulphate-reducers to include certain of the Archaebacteria; and the whole has been leavened by the active interest of the oil-, gas- and chemical-processing industries as a result of their involvement in problems such as corrosion and reservoir souring.

Recent studies of phylogenetic relationships among the sulphate- and sulphur-reducing bacteria, based on comparative oligonucleotide cataloguing of their 16S ribosomal RNA, suggest that the Gram-positive spore-forming *Desulfotomaculum* and *Clostridium* genera are, in fact, closely related. The other genera tested represent a single distinctive, but not very coherent, cluster showing some relatedness to the aerobic myxobacteria and *Bdellovibrio*; i.e. *Desulfuromonas* (sulphur-reducing) and *Desulfovibrio*, *Desulfobacter*, *Desulfosarcina*, *Desulfonema*, *Desulfococcus*, *Desulfobulbus* (all sulphate-reducing). Whereas the majority of sulphate- and sulphur-reducers are mesophilic, there are now examples of thermophilic species being isolated from hot springs, hydrothermal vents and the formation waters associated with petroleum reservoirs. Both eubacterial and archaebacterial species have been recognized, with examples of both sulphur and sulphate reduction. Amongst the extreme thermophilic subdivision of the Archaebacteria, *Thermoproteus*, *Thermodiscus* and *Pyrodictium* are capable of autotrophic growth on H_2, CO_2 and sulphur. Some species of *Thermoproteus* are facultative, and *Desulfurococcus*, *Thermofilum* and *Thermococcus* are other examples of heterotrophic sulphur-reducing thermophilic Archaebacteria.

Table 2.2 Characteristics of representative sulphate-reducing bacteria.

	Cell Form	Approximate Optimum Temperature for Growth (°C)	Compounds Oxidized				
			H_2	Acetate	Fatty Acids	Lactate	Others
Incomplete oxidation							
Desulfovibrio							
desulfuricans	Curved	30	+	–	–	+	Ethanol
vulgaris	Curved	30	+	–	–	+	Ethanol
gigas	Curved	30	+	–	–	+	Ethanol
salexigens	Curved	30	+	–	–	+	Ethanol
sapovorans	Curved	30	–	–	C_4 through C_{16}	+	
thermophilus	Rod-shaped	70	+	–	–	+	
Desulfotomaculum							
orientis	Rod-shaped	30 to 35	+	–	–	+	Methanol
ruminis	Rod-shaped	37	+	–	–	+	
nigrificans	Rod-shaped	55	+	–	–	+	
Desulfobulbus							
propionicus	Oval	30 to 38	+	–	C_3	+	Ethanol
Complete oxidation							
Desulfobacter							
postgatei	Oval	30	–	+	–	–	
Desulfovibrio							
baarsii	Curved	30 to 38	–	(+)	C_3 through C_{18}	–	
Desulfotomaculum							
acetoxidans	Rod-shaped	35	–	+	C_4, C_5	–	Ethanol
Desulfococcus							
multivorans	Spherical	35	–	(+)	C_3 through C_{14}	+	Ethanol, benzoate
niacini	Spherical	30	+	(+)	C_3 through C_{14}	–	Ethanol, nicotinate, glutarate
Desulfosarcina							
variabilis	Cell packets	30	+	(+)	C_3 through C_{14}	+	Ethanol, benzoate
Desulfobacterium							
phenolicum	Oval	30	–	(+)	C_4	–	Phenol, *p*-cresol, benzoate, glutarate
Desulfonema							
limicola	Filamentous	30	+	(+)	C_3 through C_{12}	+	Succinate

Symbols + = utilized (+) = slowly utilized – = not utilized

The other main subdivision of the Archaebacteria comprises the methanogens and the extreme halophiles, and here again, within the orders Methanobacteriales, Methanococcales and Methanomicrobiales, there are species demonstrating sulphur reduction, and indeed doing so in preference to methanogenesis in the presence of elemental sulphur. Sulphate reduction has now also been found in an extreme thermophilic Archaebacterium, with the capability of growing heterotrophically on a wide range of relatively complex organic nutrients including glucose, peptone and bacterial cell homogenates. This latter property differentiates this organism markedly from the eubacterial sulphate reducers.

It is thus likely that the many and varied bacterial genera showing sulphur or sulphate reduction are examples of convergent evolution. For this reason one must guard against treating them as a uniform group and too freely extrapolating metabolic and energetic findings from one species to another. This warning is particularly apposite to the study of this bacterial group at the present time as, not only is information still fragmentary, but there is already clear evidence of divergent mechanisms of energy generation and carbon flux in species that might otherwise be thought of as closely related.

2.4.5.2   Components of redox systems

The sulphate- (and sulphur-) reducing bacteria, unlike perhaps the methanogens, are true anaerobic respirers in that substrate oxidation is coupled to the reduction of a terminal electron acceptor through the alternating reduction and oxidation of a chain of redox carriers, at least some of which are membrane-associated. Although the molecular structure of many of these redox carriers is now known in some detail, knowledge of their structural organization and integrated functioning, particularly with regard to the generation of a pmf, remains circumstantial and inconclusive.

The genus *Desulfotomaculum* has membrane-bound cytochrome *b* but no cytochrome *c*, whereas *Desulfovibrio* and the other Gram-negative sulphate reducers and the sulphur-reducing *Desulfuromonas* have various *c*-type cytochromes, and possibly also cytochrome *b*, depending on species and growth mode. Menaquinone would seem to be the principal quinone found in all species. Also, when present, the enzymes fumarate reductase, nitrate reductase and nitrite reductase are membrane-bound; *Desulfovibrio gigas* and *Desulfuromonas acetoxidans* can grow with fumarate as terminal electron acceptor, while *Desulfobulbus propionicus* and *Desulfovibrio desulfuricans* can reduce nitrate and *D. gigas* reduces nitrite. Although the data are somewhat ambiguous, it is possible also that lactate dehydrogenase may be associated with the inner face of the cytoplasmic membrane.

The redox carriers ferredoxin and, where present, flavodoxin, rubredoxin and an octa-haem cytochrome c_3 are all cytoplasmic, as are the enzymes APS reductase and the bisulphite reductase.

Uniquely amongst terminal electron acceptors, sulphate requires to be activated. This occurs by reaction with ATP and the formation of adenosine phosphosulphate (APS). This reaction, catalysed by ATP sulphurylase, is itself unfavourable thermodynamically and therefore has to be coupled to the hydrolysis of pyrophosphate which is catalysed by pyrophosphatase.

$$ATP + SO_4{}^{2-} \rightarrow APS + PP_i \rightarrow 2Pi$$
$$H_2O$$

The need for this activation is primarily thermodynamic. The standard redox potential for the sulphate–bisulphite couple is $-516\,mV$ and so sulphate itself is quite unable to act as electron acceptor even from $H_2(2H^+/H_2, E_{m,7} = -414\,mV)$ in an energy-generating reaction. However, the redox potential of the APS/bisulphite + AMP couple is $-60\,mV$. APS is thus the true terminal electron acceptor and its reduction to bisulphite plus AMP is the first step in the 8 electron reduction of sulphate to sulphide:

$$APS + 2[H] \rightarrow HSO_3{}^- + AMP$$

The net cost of two ATP (the product of APS reductase is AMP, not ADP) means that the 6 electron reduction of bisulphite to sulphide must yield in excess of 2 ATP, presumably by a chemiosmotic mechanism.

It has been suggested that *Desulfotomaculum* species do not carry out electron transport-linked phosphorylation (but see later). It is proposed that sulphate merely stimulates metabolism and facilitates redox balance by acting as an external electron acceptor with all ATP being generated by substrate level phosphorylation. Further, these cells lack high levels of inorganic pyrophosphatase but are said to possess instead pyrophosphate: acetate phosphotransferase and acetate kinase which trap the potential energy of pyrophosphate with the production of 1 ATP:

$$PP_i + acetate \rightarrow Pi + acetyl\ P \rightarrow acetate + ATP$$
$$ADP$$

Such a reaction sequence would have the effect of reducing the net cost of sulphate activation from 2 to 1 ATP.

There are four bisulphite (sulphite) reductases recognized: desulfoviridin, the principal reductase in the genus *Desulfovibrio*, and also present in *Desulfococcus multivorans* and *Desulfonema limicola*; desulforubidin, found in some *Desulfovibrio* including *D. desulfuricans* Norway 4; desulfofuscidin from *Thermodesulfobacterium commune* and *D. thermophilus*; and P582 from *Desulfotomaculum nigrificans* and *Desulfonema magnum*. All forms of the enzyme have sirohaem (an iron tetrahydroporphyrin) as prosthetic group. It remains unresolved whether the reduction to sulphide is direct or via the trithionite pathway with trithionite and thiosulphate as either free or enzyme-bound intermediates.

Direct evidence for a chemiosmotic mechanism employing a pmf to drive ATP synthesis, reversed electron transport and the uptake of ions and nutrients in the sulphate-reducing bacteria is almost totally lacking, although such data as are available are entirely consistent with such a view:

(a) many species can grow on H_2 as a source of energy or on acetate as a source of energy and carbon, where substrate level phosphorylation is not possible with either substrate;

(b) H_2 oxidation with nitrite as electron acceptor in *D. desulfuricans* and *D. gigas*, or with sulphite in *D. vulgaris* and *D. desulfuricans* has been shown to be accompanied by an uncoupler-sensitive transmembrane proton translocation, in each case with an approximate stoichiometry $\rightarrow H^+/2e$ of 2;

(c) whereas the uptake of sulphate, sulphite and thiosulphate in *D. desulfuricans* appears to be by electroneutral proton symport, sodium is pumped out of *D. vulgaris* by an electrogenic proton antiport ($H^+/Na^+ > 1$) such that the addition of sodium acetate to resting cells creates an uncoupler-sensitive pmf of 60–90 mV (inside negative);

(d) as discussed above, the presence of dehydrogenase and reductase enzymes and standard redox carriers, and their transmembrane organization is fully consistent with the operation of chemiosmotic energy transduction mechanisms.

2.4.5.3   Hydrogen metabolism

A major part of our understanding of bioenergetics in the sulphate-reducing bacteria has come from studies of hydrogen metabolism. The ability to oxidize H_2 as a source of energy is widespread amongst all eubacterial and archaebacterial genera of sulphate- and sulphur-reducing bacteria, growing either autotrophically or using acetate as carbon source (Table 2.2). At the same time, H_2 is also produced by many species either during fermentative growth in the absence of sulphate or even in the presence of sulphate as terminal electron acceptor; this latter is particularly marked during growth on lactate.

Growth on H_2 is taken as incontrovertible evidence for the operation of electron transport-linked phosphorylation and thus negates earlier suggestions that all ATP generation in *Desulfotomaculum* is by substrate level phosphorylation. (This is in accord with the presence of cytochrome *b* in this genus and the capacity of *Dt. acetoxidans* to grow on acetate as source of energy and carbon; it does not necessarily rule out the proposed acetyl phosphate pathway, although high levels of the pyrophosphate:acetate phosphotransferase have yet to be unequivocally demonstrated.) From growth yield studies on *D. vulgaris* with H_2 as energy source and either sulphate or thiosulphate (formally and energetically equivalent to sulphite) as electron acceptor, Thauer and his colleagues have deduced that the ATP yield from reduction of sulphate to sulphide is between 1.0 and 1.3, while for sulphite reduction the figures are between 3.0 and 3.5; the difference is due to the ATP cost in the required activation of sulphate to APS. Similar studies with *Desulfotomaculum orientis*, although of a more preliminary nature, also show a higher yield from thiosulphate or sulphite reduction than from that of sulphate, but it is not yet possible to draw firm conclusions on either the ATP yields from reduction of electron acceptor, or the ATP cost for sulphate activation in this genus.

Reference has already been made to the ability of sulphate reducers to produce H_2 by fermentation in the absence of an external electron acceptor. Species (and substrates) capable of this mode include: *D. vulgaris* and *D. desulfuricans* (lactate and ethanol); *Desulfotomaculum orientis* and *Dt. nigrificans* (lactate and ethanol); *Thermodesulfobacterium commune* (pyruvate); *Desulfobulbus propionicus* (lactate, pyruvate and ethanol); *Desulfococcus multivorans* (lactate and pyruvate); and *Desulfosarcina variabilis* (lactate, pyruvate and fumarate). A number of studies have shown that, when one of these *Desulfovibrio* or *Desulfotomaculum* species are grown in co-culture with a H_2-oxidizing methanogen, the fermentation of lactate becomes characterized by the production of methane and acetate. This effect arises from interspecies H_2 transfer with a shift in the fermentation to acetate and H_2 production (exactly as with *Cl. pasteurianium* or *Ruminococcus albus* under reduced H_2 partial pressure), with the latter serving as the substrate for methanogenesis.

Intriguingly, a co-culture involving interspecies H_2 transfer has also been reported where the methanogen is the H_2 donor and the sulphate-reducer is the H_2-oxidizing partner. *D. vulgaris* cannot grow on either methanol or acetate whereas growth of *Methanosarcina barkeri* on these substrates is associated with a build-up of H_2. Co-culture of these two organisms in sulphate medium gives substrate utilization with the production of sulphide, along with less methane and more CO_2.

These phenomena, and more particularly the observation that the growth of *D. vulgaris* in sulphate medium with lactate (and to a lesser extent with pyruvate) is associated with significant H_2 production, led Odom and Peck (1981) to propose the hydrogen cycling model for energy coupling in the sulphate-reducing bacteria (Fig. 2.15). The essential features of the model are that the reducing equivalents from the two substrate dehydrogenation steps should reduce protons to produce H_2 in a reaction catalysed by a cytoplasmic hydrogenase. This H_2 would then diffuse across the membrane and be oxidized by a second, periplasmic hydrogenase. The specific electron

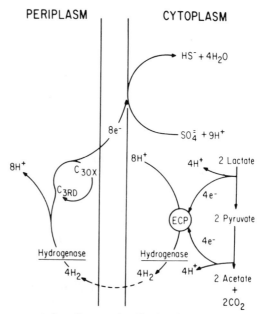

Fig. 2.15 A schematic representation of enzyme localization, hydrogen cycling and vectorial electron transfer by *Desulfovibrio* with lactate and sulfate as substrates. ECP, electron carrier proteins; c_3, cytochrome c_3. For a recent review of sulphate reduction and hydrogen cycling see Peck and Lissolo (1988).

acceptor for this hydrogenase is the low potential tetrahaem cytochrome c_3, which is also found in the periplasm. Transmembrane electron transport would then give both sulphate reduction in the cytoplasm and a transmembrane electrochemical gradient of protons: the protonmotive force. It is envisaged, therefore, that only periplasmic (or extracellular) H_2 oxidation is directly coupled to energy transduction and that organic substrate catabolism is fermentative in character, with ATP synthesis being mediated through an *intra* species H_2 transfer.

The principal points of evidence in favour of the H_2 cycling model are: the widespread occurrence of periplasmic hydrogenases and cytochrome c_3 in *Desulfovibrio*; the loss of lactate oxidation using sulphate in spheroplasts of *D. gigas* which have lost periplasmic hydrogenase and cytochrome c_3; some degree of restoration of lactate oxidation by addition of partially purified hydrogenase and cytochrome c_3; and complete reduction of cytochrome c_3 added to *D. gigas* spheroplasts along with lactate and dehyrogenase, with reduction being only 60% if sulphate is also present.

However, significant amounts of data are not consistent with the H_2 cycling model. A number of species are capable of growth on lactate and related organic substrates but do not oxidize H_2, for example: *D. sapovorans*,

Desulfococcus multivorans and a recently described hydrogen-inhibited mutant of *D. desulfuricans*. Further, the model cannot apply to the Grampositive *Desulfotomaculum* since the cell wall structure of this genus precludes a periplasmic space. It has been demonstrated that the H_2-oxidizing hydrogenase of *Dt. orientis* has an intracellular location. While these discrepancies from the model predictions might be considered as only species variations, the alterations proposed for their mechanism of energy coupling during the catabolism of organic nutrients are so fundamental that they could not be expected to be present in one species but absent from a closely related organism, that might even be from the same genus.

Other more serious problems arise from a consideration of the thermodynamics of the reaction sequence constituting the H_2 cycling model. The standard mid-point redox potentials of the pyruvate/lactate ($-197\,mV$) and $2H^+/H_2$ ($-420\,mV$) couples are such that the production of H_2 from lactate dehydrogenation is a highly endergonic reaction. Even with a lactate: pyruvate ratio of 100:1 the reaction would only be energy-generating at H_2 partial pressures of less than 10^{-5} atm. Under these circumstances, therefore it would be predicted from the model that lactate oxidation, which might proceed in sulphate medium under a N_2 gas phase, would be inhibited by the addition of H_2. A number of studies have now shown this not to be the case. It is perhaps more reasonable, therefore, to suggest that H_2 production from lactate, during either fermentative or sulphate respiration modes, could arise by energy-driven, reversed electron transport between lactate dehydrogenase and hydrogenase, both of these enzymes being associated with the inner face of the cytoplasmic membrane.

The H_2 cycling model suggests some interesting speculation regarding the growth of *Desulfobulbus propionicus* on propionate (or equally the growth of *D. sapovorans* and other species on butyrate). Reference has already been made to the obligately syntrophic acetogens *Syntrophobacter wolinii* and *Syntrophomonas wolfei*, capable of growth on propionate and butyrate, respectively, only when in co-culture with a H_2-oxidizing sulphate reducer (or methanogen) (page 110). The question therefore arises whether *Db. propionicus* adopts what is formally the same mechanism but within a single organism, by virtue of H_2 cycling or intraspecies H_2 transfer. Since the pathway of propionate catabolism to acetate appears to involve a methylmalonyl CoA:pyruvate transcarboxylase and that part of the TCA cycle from succinyl-CoA to oxaloacetate (Fig. 2.17), the alternative mechanism would involve direct transfer of reducing equivalents to sulphate reduction, with energy-driven reversed electron transport from succinate (fum/succ $E_{m,7} = +33\,mV$). Even if molecular H_2 were to be produced from this dehydrogenation (and from malate to oxaloacetate), the thermodynamics make even more pressing the need to invoke energy-driven reversed electron transport

$(2H^+/H_2, E_{m,7} = -420\,\text{mV}; HSO_3^-/APS, E_{m,7} = -60\,\text{mV}; HSO_3^-/HS^-, E_{m,7} = -116\,\text{mV})$.

Whatever the final verdict on the H_2 cycling model, it has undoubtedly stimulated a very significant amount of research into the identification and characterization of hydrogenase enzymes in the sulphate-reducing bacteria. There appear to be three quite distinct periplasmic hydrogenases with high specific activity in the H_2 oxidation assay. *D. vulgaris* has a non-haem iron enzyme with no nickel and a two subunit structure (45.8 KD and 13.5 KD). The *D. gigas* enzyme also has two sub-units (62 KD and 26 KD) and non-haem iron but additionally it has 1 g atom nickel mol^{-1}. On the other hand, the hydrogenase from *D. desulfuricans* Norway and the closely related *D. baculatus* is a nickel selenium non-haem iron enzyme.

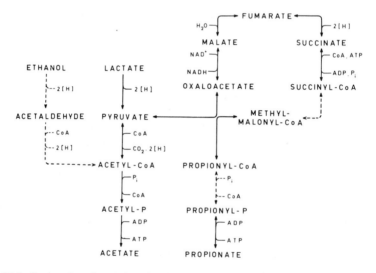

Fig. 2.16 Proposed pathway for the formation and degradation of propionate in *Db. propionicus*.

In line with the predictions of the H_2 cycling mode, the search is on for unequivocal evidence for a cytoplasmic hydrogenase operating in the proton reduction mode. At least two studies have identified membrane enzymes from *D. vulgaris* antigenically different from the periplasmic hydrogenase. In one study three such enzymes were noted, none of which reacted with antibodies raised to the periplasmic enzyme. Two, however, showed a degree of cross-reaction with antibodies to the nickel enzyme from *D. gigas*; while the third reacted with antibodies to the *D. desulfuricans* nickel selenium hydrogenase. It remains to be demonstrated what are the true physiological functions of these enzymes and how, if at all, they relate to the H_2 cycling model.

The non-haem iron periplasmic hydrogenase of *D. vulgaris* has recently been the subject of some intriguing molecular biological analysis. The gene for the large subunit has been cloned, although its expression in *E. coli* gives an enzymically inactive protein. Sequence studies show a considerable degree of homology between the N-terminal end of the molecule and (8Fe–8S) ferredoxin. The gene for the smaller subunit is downstream from that for the larger subunit and it is suggested the two might constitute an operon. Whereas there is no evidence of a signal peptide for the 45.8 KD subunit, the mature 13.5 KD protein subunit lacks the hydrophobic N-terminal amino acid sequence encoded by its gene, which has a sequence with the general characteristics of a signal peptide. Interestingly, there is a hydrophobic region located between residues 99 and 132 in the large subunit which has the general properties of an amino-terminal signal peptide; a situation similar to that with ovalbumin, a protein which is also post-translationally translocated across a membrane.

2.4.5.4   Nutrition and carbon flux

Table 2.2 clearly indicates how far our appreciation of the catabolic potential of the sulphate- and sulphur-reducing bacteria has advanced in the last ten or so years. The list of growth-supporting substrates continues to expand, and recent additions include: sugars (sulphate- and sulphur-reducing Archaebacterial thermophiles and *Dt. nigrificans*); methanol (*Dt. orientis* and a recent Gram-negative non-sporing isolate); catechol, resorcinol and hydroquinone (*Desulfobacterium catecholicum*); indole, anthranilate and hydroxy-substituted benzoate and phenylacetate (*Desulfobacterium phenolicum*).

The sulphate- and sulphur-reducing bacteria can be divided into two groups: those species capable of only incomplete substrate oxidation with acetate as a metabolic product, and those capable of complete oxidation to CO_2 with some species able to grow on acetate as the sole source of energy and carbon. It was long considered impossible, on thermodynamic grounds, that the TCA cycle could operate in an energy-generating mode with sulphate as terminal electron acceptor. The problem focused on the dehydrogenation of succinate to fumarate ($E_{m,7} = +33$ mV). Clearly, the situation is equally adverse with sulphur reduction (S^0/HS^-, $E_{m,7} = -270$ mV). Under these circumstances it should not be possible to oxidize acetate as a source of energy for growth. However, the discovery of *Desulfuromonas acetoxidans* and *Desulfotomaculum acetoxidans*, and the identification of acetate as a major respiratory substrate in sulphidogenic ecosystems has forced a radically altered view to prevail.

Enzymic and ^{14}C-labelling experiments have now clearly shown that an oxidative TCA cycle does operate in *Desulfobacter postgatei* and *Desulfuro-monas acetoxidans*, with some relatively minor variations:

(a) ferredoxin is the cofactor for 2-oxoglutarate dehydrogenase;

(b) acetate activation and conversion of succinyl CoA to succinate are catalysed by a single enzyme, succinyl-CoA; acetate CoA transferase;

(c) in *D. postgatei* the succinate and malate dehydrogenases are membrane-bound and menaquinone-dependent, whereas in *D. acetoxiodans* the malate dehydrogenase is NAD-dependent and located in the cytoplasm;

(d) the glyoxylate cycle is absent from both bacteria and the route of acetate assimilation is by reductive carboxylation to oxaloacetate, with ferredoxin being the electron donor for reductive carboxylation of acetyl-CoA to pyruvate (shown in *D. postgatei*);

(e) it was deduced for *D. postgatei* that the succinate to fumarate reaction is most likely to require energy-driven reversed electron transport, with the menaquinone ($E_{m,7} = -74\,mV$) coupling of malate dehydrogenation rendering that reaction virtually irreversible and thus having the effect of 'pulling' the succinate dehydrogenation. ATP-dependent, and uncoupler- and DCCD-sensitive dehydrogena-tion of succinate to fumarate has now been directly demonstrated in membrane preparations of *D. acetoxidans* with either sulphur or NAD as electron acceptor.

It turns out, however, that these are not the most common routes of acetate oxidation and assimilation amongst the sulphate-reducing bacteria. Other species, including *D. baarsii*, *Dt. acetoxidans*, *Desulfosarcina variabilis*, *Desulfobacterium autotrophicum*, *Desulfococcus niacini* and *Desulfococcus multivorans*, lack 2-oxoglutarate dehydrogenase and a complete TCA cycle, although they do use the reactions to 2-oxoglutarate for glutamate and amino acid synthesis. What they do possess is the enzyme carbon monoxide dehydrogenase, and the acetyl-CoA pathway appears to be used for acetate oxidation and, where operative, CO_2 fixation. It is significant that the standard mid-point redox potential of the most positive step in the acetyl-CoA pathway (methyl-THF to methylene-THF) is $-117\,mV$, which could thus readily be coupled to APS and bisulphite reduction in an exergonic reaction sequence.

Although *D. vulgaris* has also been shown to have an active carbon monoxide dehydrogenase, its physiological role remains unclear as *D. vulgaris* lacks the acetyl-CoA pathway and cannot oxidize acetate as a source of energy.

2.4.5.5   Sulphite dismutation

Recently, a completely novel mechanism of energy generation has been reported in a newly identified isolate; *D. sulfodismutans*. The reaction sequence has been called an inorganic or chemolithotrophic fermentation since it involves the coupled oxidation and reduction of either sulphite or thiosulphate, according to the equations:

$$4SO_3^{2-} + H^+ \rightarrow 3SO_4^{2-} + HS^- \qquad \Delta G^{0\prime} = -58.9 \, \text{kJ mol}^{-1} \text{ sulphite}$$

$$S_2O_3^{2-} + H_2O \rightarrow SO_4^{2-} + HS^- + H^+$$
$$\Delta G^{0\prime} = -21.9 \, \text{kJ mol}^{-1} \text{ thiosulphate}$$

The mechanism of energy conservation remains unclear at present.

2.5   Anaerobic microbial consortia

A constantly recurring theme in this presentation of the bioenergetics of anaerobic microorganisms has been the relatively restricted metabolic potential of individual species, with the consequent interdependencies between species. Our consideration of the acetogens and methanogens has noted in particular the key roles taken by H_2 and acetate in these interdependencies. It can now be seen that, depending on species and conditions, the sulphate-reducing bacteria can act both as H_2-producing acetogens and as terminal oxidizers paralleling the methanogens. Any analysis of anaerobic microbial consortia, therefore, must be constantly aware of the interactions between individual components of the community, and expect the sulphate reducers to be of central importance in sulphate-containing environments. These considerations are not minor or confined to a few laboratory models. They are, in fact, the dominant and controlling features of a host of naturally occurring and economically important microbial ecosystems such as the rumen (methanogenic), anaerobic digesters (methanogenic), polluted estuarine or marine sediments (sulphidogenic) and microbial biofilms associated with metal corrosion (sulphidogenic). A striking feature of such consortia is that, while they can be characterized in terms of the principal terminal oxidant species (methanogen or sulphate reducer), the primary nutrient for the consortium is invariably a substrate or substrates to which the methanogenic or sulphate-reducing bacteria are themselves totally unreactive. They have an absolute requirement for the heterotrophic and fermentative organisms to perform the initial partial decomposition, and often also to create anaerobic conditions within the consortium through reacting with any available oxygen. At the same time, the methanogens or sulphate-reducers stimulate the overall conversion of substrate by oxidative removal of the products of fermentation.

In view of the central metabolic importance of H_2 and acetate, the acetogenic bacteria often act as intermediaries between the initial fermentative reactions and the terminal oxidative steps. As well as stimulating the degradative capacity of microbial ecosystems in this manner, significantly a major proportion of the energy input to the systems remains in the form of the reduced terminal electron acceptor. In a very real sense, methane and H_2S are extracellular energy currencies that are available to other microorganisms, perhaps after diffusion to an aerobic microenvironment. Primary nutrient sources that are biodegraded in this manner include complex biopolymers such as cellulose, hydrocarbons, and more or less recalcitrant molecules such as aromatics, phenols and sugars. Although individual sulphate-reducing species have recently been isolated that are capable of reacting with some of these substrates directly, their quantitative importance in most natural ecosystems remains to be established.

It is generally accepted that in sulphate-containing environments the terminal oxidant species will be sulphidogenic rather than methanogenic. This was previously thought to be due to a sulphide toxicity effect, but this is no longer considered a likely explanation. In fact, the production of sulphide is likely to be beneficial to both groups of organisms because of its effect of making the environment more reducing. It has now been shown that the sulphate reducers have higher affinities than the methanogens for both the major substrates for which they compete. The K_s values for H_2 for the sulphate reducers and methanogens are, respectively, around $2\,\mu M$ and between 6 and $20\,\mu M$. For acetate the figures are approximately $200\,\mu M$ and $3\,mM$ respectively.

In a number of environments evidence has been obtained that methanogenesis and sulphidogenesis proceed simultaneously. Methanogenesis in these habitats is due to the methylotrophic methanogens (oxidizing methanol and the methylamines) that do not compete with the sulphate reducers for H_2. There have been recent reports of methanogenesis and sulphidogenesis in the non-traditional province for them of the open ocean; presumably this involves anoxic microparticulates. As understanding of these intriguing groups of anaerobic bacteria increases, along with improved techniques for enrichment, isolation and growth, one might reasonably expect such reports to be substantiated and extended.

2.6 Anaerobic respiration: reduction of fumarate and nitrate

2.6.1 Introduction

Where fumarate or nitrate operate as terminal electron acceptors rather than

CO_2 or sulphate, the differences in cell physiology are qualitatively major and not simply restricted to the quantitative effects resulting from the more positive mid-point potentials of the fumarate–succinate and nitrate–nitrite redox couples.

(a) The organisms capable of these activities are facultative anaerobes and will always use oxygen preferentially as terminal electron acceptor; fumarate and nitrate are therefore genuine alternatives to oxygen.

(b) A wide spectrum of bacterial genera have the capacity for fumarate and/or nitrate respiration.

(c) At the cellular level, the mechanism of these energy conserving processes requires only minimal modification of the normal oxygen-linked respiratory pathways. Consequently, the major thrust of research has been directed at elucidating the details of such modifications, and in the process much seminal information has been gained on the structural and functional bases of membrane-dependent energy conservation mechanisms in general.

(d) Cell–cell interactions and implications for microbial ecology are much less to the fore as, in terms of energy yield and consequent carbon flux, each organism is capable of being self-sufficient.

2.6.2   Fumarate as terminal electron acceptor

2.6.2.1   Introduction

The ability to use fumarate as electron acceptor is very widespread amongst Gram-negative bacteria, including enterobacteria; it is also found in some Gram-positives, notably strains of *Bacillus* and *Staphylococcus*. In comparison with other electron acceptors, although fumarate may be added as an external acceptor, it is more commonly produced directly by the cell's own metabolic activity. Under no circumstances can the TCA cycle be operational in an oxidative energy-generating mode during fumarate reduction. For example, the fermentative production of propionate by the propionibacteria (or of succinate by *Streptococcus faecalis* or *E. coli*), is by the reductive arm of the TCA cycle with the penultimate (or final) step being the reduction of fumarate to succinate. In cytochrome-deficient organisms such as *Strep. faecalis*, fumarate reduction serves only as an electron sink allowing redox balance within the fermentation along with increased production of the more oxidized acetate. The increase in energy yield from fumarate reduction is therefore gained indirectly from substrate level phosphorylation during conversion of acetyl-CoA to acetate. With *E. coli* on the other hand, and with other cytochrome-containing organisms, there is clear evidence for electron transport-linked phosphorylation coupled to fumarate reduction. The most striking metabolism of fumarate, however, is by *Proteus rettgeri*, which can grow anaerobically on the dicarboxylic acid as the sole source of carbon, energy and reducing equivalents (Fig. 2.17).

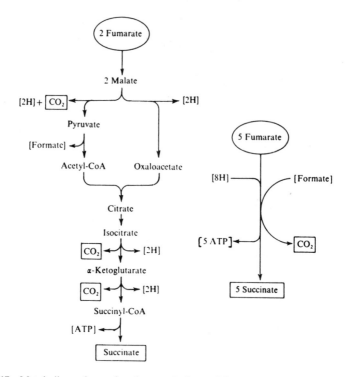

Fig. 2.17 Metabolic pathway for the metabolism of fumarate as the only substrate for anaerobic growth of *Proteus rettgeri*.

2.6.2.2   Energy coupling

The fumarate reductase system and its coupling to cell metabolism has been most extensively studied in *E. coli* and the rumen bacterium *Wolinella succinogenes*. Both organisms can obtain energy from the oxidation of H_2 or formate coupled to fumarate reduction, while *E. coli* can use fumarate as electron acceptor for anaerobic growth on glucose, glycerol or lactate. In glycerol/fumarate-grown cells the main dehydrogenases donating reducing equivalents are those oxidizing formate, NADH, lactate and glycerol 3-phosphate. Electron flow to the fumarate reductase is dependent upon menaquinone ($E_{m,7} = -74\,\text{mV}$) or desmethylquinone ($E_{m,7} = -36\,\text{mV}$), and a *b*-type cytochrome. From *W. succinogenes*, a cytochrome b_{560} co-purifies with fumarate reductase and contains equal amounts of two components with redox potentials of -20 mV and -200 mV, which are considered to act as electron donor to the fumarate reductase and electron acceptor from the formate dehydrogenase, respectively.

The evidence for energy coupling from fumarate reductase by a chemiosmotic mechanism is undeniable but remains incomplete in respect of quantitation. Anaerobic growth of *E. coli* on H_2, formate, lactate or glycerol is not possible in the absence of fumarate reductase, or where the enzyme has been subject to mutation. Similar results were obtained with uncoupled (*unc*) mutants. Diminished growth with haem-deficient mutants, however, and the loss of fumarate reduction-linked uptake of amino acids has been interpreted as demonstrating the need for cytochrome in order to develop a pmf, with fumarate reduction in the absence of haem being capable of acting as an electron sink and increasing ATP yield only indirectly through substrate level phosphorylation.

Growth yield data with H_2 and formate-coupled fumarate reduction in both *E. coli* and *W. succinogenes* are consistent with an ATP yield of 1 per fumarate. This would be in agreement with thermodynamic predictions for the oxidation of H_2, formate or NADH. On the basis of standard redox potentials, oxidation of lactate or glycerol 3-phosphate might be expected to yield less than 1 ATP. While fractional ATP yields are now considered perfectly feasible, it must also be remembered that conditions within the cell will not be 'standard'. If for example, the ratio of fumarate:succinate were 100:1 then the energy yield when coupled to the oxidation of lactate or glycerol 3-phosphate would indeed be sufficient for the synthesis of 1 ATP.

The oxidation of endogenous reserves, NADH or formate by whole cells of *E. coli* has been shown to generate a pmf of the order of -100 mV. The equivalent proton uptake by inverted vesicles was uncoupler-sensitive, and inhibited by the quinone analogue HOQNO. The $\rightarrow H^+/2e$ ratios for these substrates in *E. coli* and for formate in *W. succinogenes* are all around 2.

While clearly these combined data do not allow final decisions to be reached regarding absolute values for the pmf, stoichiometries or ATP yield from the coupling of substrate oxidation to fumarate reduction, there can be no doubt that such coupling does occur by a chemiosmotic mechanism.

2.6.2.3   Fumarate reductase

Fumarate reductase is a membrane-associated enzyme. It comprises four subunits: A (69 KD), B (27 KD), C (15 KD) and D (13 KD). A and B are extrinsic membrane proteins exposed on the cytoplasmic side; they are also the catalytic subunits. A has the substrate-binding site and FAD as cofactor, while B has 3 non-haem iron–sulphur centres. In *W. succinogenes*, A alone is sufficient for catalytic activity, but in *E. coli* both A and B are required. Subunits C and D are needed for energy coupling; these are hydrophobic intrinsic membrane proteins whose function is to anchor the catalytic subunits to the membrane, thus facilitating electron transport and making possible energy coupling by transmembrane charge separation (Fig. 2.18). It is now clear that formate dehydrogenase is on the periplasmic side of the membrane and that the pmf is generated in this case by an electron transfer arm rather than by a conventional Mitchellian loop (see pages 16 and 37).

Fumarate reductase and succinate dehydrogenase are separate enzymes which as a group are highly conserved and individually show structural similarities. In *E. coli*, the two enzymes show marked homology in amino acid sequences but are none the less immunologically distinct. Although the amino acid sequences of the C and D subunits are not greatly conserved, the distribution of hydrophilic and hydrophobic domains is conserved, in line with their proposed transmembrane organization. At the functional level, the fumarate reductases of *E. coli* and *W. succinogenes* and the succinate dehydrogenases of *E. coli*, *Bacillus subtilis* and bovine mitochondria all contain three similar iron–sulphur centres.

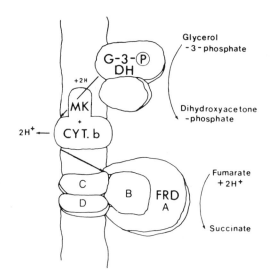

Fig. 2.18   Proposed anaerobic electron-transport chain from glycerol-3-P to fumarate. Reducing equivalents from the primary dehydrogenase (glycerol-3-P dehydrogenase, G-3 PDH) are believed to be transferred to the lipid soluble hydrogen carrier menaquinone (MK). The electrons are transferred to the *b*-type cytochrome and the protons extruded, the electrons are then transferred to fumarate reductase (FRD). The four subunits of the reductase (FRD, A, B, C, D) are shown. It should be noted that the exact route taken by the reducing equivalents from the dehydrogenase to the reductase is unclear, and that the role of the cytochrome is controversial.

2.6.2.4 Sulphur reduction; and sulphide oxidation coupled to fumarate reduction

While certain of the sulphate-reducers can use fumarate as terminal electron acceptor, a further link is established by the ability of *W. succinogenes* to reduce sulphur.

As noted above, *W. succinogenes* can grow by formate oxidation coupled to fumarate reduction with an estimated yield of 1 ATP from the oxidation-reduction. It has now been shown that the organism can also grow by formate oxidation coupled to the reduction of sulphur, and by the oxidation of hydrogen sulphide coupled to reduction of fumarate, according to the scheme suggested in Figure 2.19. The enzymic and redox carrier mechanisms of the two partial reactions remain to be elucidated, and the energy yields that allow each independently to support growth have yet to be established.

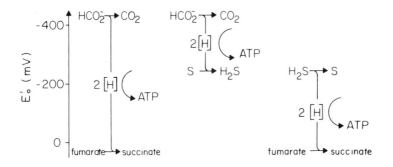

Fig. 2.19   Redox potentials of the electron donors and acceptors which support growth of *W. succinogenes*.

2.6.3   Oxides of nitrogen as terminal electron acceptor

2.6.3.1   Introduction

There are two principal activities associated with the dissimilatory reduction of nitrate by bacteria. Denitrification is the reduction of nitrate to nitrogen through the intermediates nitrite, nitric oxide (possibly) and nitrous oxide, although generally there is no accumulation of the intermediates as such. In nitrate respiration, on the other hand, nitrate is only reduced to nitrite. Both of these processes are coupled to energy generation through the development of a pmf. A third mechanism of nitrate reduction to nitrite is found in *Clostridia*, where the activity is not energy-coupled but fulfils only the role of an electron sink.

Denitrification occurs in a wide range of facultative chemo- and photo-trophic organisms, including *Paracoccus denitrificans, Pseudomonas aeru-ginosa, Alcaligenes faecalis, Bacillus licheniformis, Thiobacillus denitrificans, Achromobacter cyloclastes*, and *Rhodobacter sphaeroides* f.sp. *denitrificans* and strains of *R. capsulata*. There appear to be significant differences between the chemotrophic and phototrophic organisms, most particularly with regard to the cellular site of the nitrate reductase (see Chapters 3 and 4).

Nitrate respiration is found, for example, in most enterobacteria and *Staphylococci*. The nitrite produced can be further reduced, but on this occasion to ammonia rather than to nitrogen. In anaerobic species such as *Propionibacteria, Cl. perfringens* and *Veillonella alcalescens* the nitrite reduc-tase is soluble and not energy-linked. However, in *E. coli* and other organisms there appear to be three pathways for the reduction of nitrite to ammonia. There is a soluble NADPH-dependent sulphite reductase which has activity also for nitrite. The major soluble enzyme, however, is the NADH-dependent nitrite reductase. This nitrite reductase is probably a dimer of two identical subunits (88 KD), with each subunit having one iron–sulphur centre, one FAD and one sirohaem. Sirohaem is also the cofactor in the NADPH-sulphite reductase. The function of these soluble enzymes is to regenerate NAD^+ according to the equation:

$$3NADH + 5H^+ + NO_2^- \rightarrow 3NAD^+ + NH_4^+ + 2H_2O$$

About 20% of the nitrite reduction, however, proceeds via a membrane-dependent, energy-linked system that incorporates a periplasmic cytochrome c_{552}. Although not necessarily exclusive to these two substrates, this activity has been demonstrated with formate and lactate. Formate-driven reduction of nitrite to ammonia in *E. coli* can generate a pmf of the order of $-150 \, mV$.

The evidence suggests that where *D. desulfuricans* and *Db. propionicus* use nitrate as terminal electron acceptor in the absence of sulphate, the reaction sequence is reduction of nitrate to nitrite, followed by nitrite reduction to ammonia, with both reactions being energy coupled. Cytochrome *c* and desulfoviridin (sulphite reductase) are always present.

2.6.3.2   Nitrate reductase in *Pa. denitrificans* and *E. coli*
The initial reaction in both denitrification and nitrate respiration is the reduction of nitrate to nitrite. The nitrate reductases of denitrifying *Pa. denitrificans* and nitrate respiring *E. coli* are essentially similar, while that of the phototrophic *Rb. sphaeroides* and *Rb. capsulata* appears markedly different.

Nitrate reductase in *E. coli* and *Pa. denitrificans* is membrane-bound and comprises three non-identical subunits, in the probable configuration

$\alpha_2\beta_2\gamma_4$. The α subunit (127–150 KD, depending on source and methodology) contains the active site with an iron–sulphur centre and pterin-bound molybdenum as cofactors (page 39). The active site is located on the cytoplasmic side of the membrane and this raises the question, still not satisfactorily answered, of the entry and exit of nitrate and nitrite (see Chapter 3). The β subunit (60–61 KD) is concerned with membrane attachment. It is an intrinsic membrane protein with groups accessible to chemical treatment from the cytoplasmic side. The γ subunit (20–21 KD) contains a special nitrate reductase b-type cytochrome which is required for the transfer of electrons from ubiquinol to the α subunit. Interestingly, the artificial electron donor benzyl viologen can transfer electrons to the active site in the absence of the nitrate reductase cytochrome b. The γ subunit is at least partially exposed on the periplasmic side in terms of the reactivity of certain amino acid groups.

Although nitrate can permeate through the cytoplasmic membrane, it has been calculated that the rate of such permeation at low concentrations would not be sufficient to support the observed rates of nitrate reduction and for this reason a specific transport mechanism has been evoked. Since the entry of negatively charged nitrate itself poses a problem in terms of the prevailing pmf, inside negative, a nitrate/nitrite antiporter seems the most likely mechanism. Not only would this electroneutral system facilitate entry of nitrate, but it would also bring the nitrite to the site of its reductase in the periplasm. Indeed, nitrite has been shown to potentiate the activity of nitrate reductase in *Pa. denitrificans*. While this argument fits with the situation of denitrification in *Pa. denitrificans*, it is less convincing for the cytoplasmic nitrite reductase in *E. coli*.

A further piece of evidence for a nitrate transport system is the high affinity found with whole cells (K_m for NO_3^- < 5 μM) as compared with that measured with the partially purified enzyme using reduced viologen dyes as electron donor (K_m for NO_3^- = 283 μM). This argument has less force now that it has been appreciated that viologen dyes donate electrons to nitrate reductase without the involvement of the b-type cytochrome. Duroquinol is therefore closer to the natural electron donor, electron transfer being mediated by this cytochrome b. With duroquinol as electron donor for the purified enzyme, the K_m for NO_3^- is 13 μM, suggesting that the apparent low affinity of the enzyme is not due to the absence of a high affinity transport system but is simply a result of using an artificial electron donor that does not transfer electrons to the active site by the normal physiological mechanism involving the nitrate reductase cytochrome b.

While on balance the evidence favours the existence of nitrate/nitrite antiport, particularly in *Pa. denitrificans*, the matter will remain unresolved until direct structural evidence for a transport protein is obtained.

Nitrate reduction in *E. coli* is associated with the generation of a pmf of 165 mV in inverted vesicles with formate, NADH, lactate or glycerol 3-phospate as substrate. Proton translocation stoichiometries have also been recorded. With endogenous substrates, malate and formate, $\rightarrow H^+/NO_3^-$ was 4, and for succinate, lactate and glycerol, $\rightarrow H^+/NO_3^-$ values of 2 were found. Similar values were obtained when oxygen was the oxidant, which fits with the known absence of site III phosphorylation in *E. coli* and the conclusion that nitrate reduction is coupled to phosphorylation at sites I and II, i.e. NADH-linked substrate to ubiquinone and ubiquinol to nitrate as terminal oxidant. Although there is no proton translocation directly linked to the nitrate reduction, protons are released from the ubiquinol into the periplasm and the electrons are passed to the nitrate, where its reduction in the cytoplasm involves the scalar consumption of protons which contributes to the generation of the pmf (see Chapter 3):

$$NO_3^- + 2H^+ + 2e \rightarrow NO_2^- + H_2O$$

Thus, in theory at least, nitrate should be as efficient a terminal electron acceptor as oxygen in *E. coli*. That this is not, in fact, the case arises from alterations in metabolic carbon flux. Whereas in *B. licheniformis* it is claimed that substrates such as glucose can be oxidized completely to CO_2 by glycolysis and the TCA cycle during denitrification, with glucose or lactate catabolism in *E. coli* under nitrate respiration, only a proportion of the substrate appears as CO_2 and there is a build-up of formate and acetate. The TCA cycle does not operate under these ciucumstances and there is consequently a greatly diminished ATP production from electron-transport-linked phosphorylation.

2.6.3.3   Nitrate reductase in *Rhodobacter* and *Rhodospirillum*

As mentioned above, the nitrate reductase of denitrifying species of the phototrophic genera *Rhodobacter* and *Rhodospirillum* is periplasmic rather than membrane-bound (see Chapter 3). Since proton translocation arises from the oxidation of the ubiquinol, the location of the electron accepting nitrate reduction in the periplasm need pose no problem for development of the pmf. The $\rightarrow H^+/2e$ stoichiometry in *Rb. sphaeroides* f. sp. *denitrificans* for NO_3^- reduction to NO_2^- with endogenous substrates, and corrected for scalar consumption of protons in the periplasm, has been measured at 4.05. At present, however, little is known about the structure of the reductase other than that it is a molybdoprotein.

2.6.3.4   Nitrite reductase and nitrous oxide reductase

The enzymes responsible for the further reactions in denitrification, nitrite and nitrous oxide reductases, are found located in the periplasm of both chemotrophic *Pseudomonas* and *Paracoccus*, and of phototrophic *Rhodobacter* and *Rhodospirillum*. Both reductions are coupled to energy generation and the measured $\rightarrow H^+/2e$ stoichiometries (corrected for scalar proton consumption in the periplasm) in *Rb. sphaeroides* f. sp. *denitrificans* with endogenous substrates are: 4.95 for NO_2^- reduction to $\frac{1}{2}N_2O$; and 6.01 for N_2O reduction to N_2. In *Pa. denitrificans*, both reductases accept electrons from ubiquinol via a cytochrome bc_1 complex. This is true also of nitrite reductase in *Rb. sphaeroides*, but the bc_1 complex is not involved in electron transfer to the nitrous oxide reductase in the phototroph.

Not a great deal is known about the protein composition and structure of the two enzymes. There are two different nitrite reductases involved in denitrification. The enzyme in *Pa. denitrificans* is a dimer with c- and d- type cytochrome associated with each subunit. The enzyme can also reduce oxygen but this is unlikely to be of physiological significance (see Chapter 5). A second nitrite reductase is found in *Alcaligenes faecalis*. This is a copper-containing tetrameric protein, each subunit having a molecular weight of 30 KD.

The nitrous oxide reductase characterized from *Rhodobacter* species is a dimer of subunits with molecular mass of 74 KD, and containing eight copper atoms per molecule. Such an enzyme has still not been positively identified in *Pa. denitrificans*.

2.6.3.5 Nitric oxide reductase
The evidence that nitric oxide is an intermediate between nitrite and nitrous oxide in the denitrification sequence remains equivocal. The $\rightarrow H^+/NO$ stoichiometries have, however, been measured with several bacterial species: *Pa. denitrificans*, 3.65; *Rb. sphaeroides* f. sp. *denitrificans*, 4.96; *Achromobacter cycloclastes* 1.94; *Rhizobium japonicum*, 1.12. Nitric oxide reduction has also been coupled to amino acid uptake in *Pa. denitrificans*. Under certain conditions, nitric oxide can be a major product of both the cytochrome *cd*-nitrite reductase in *Ps. aeruginosa* and the non-haem, copper nitrite reductase of *A. cycloclastes*. The data from these experiments are consistent with a periplasmic location for the proton-consuming sites. Although the enzyme has yet to be separately identified and characterized, a nitric oxide reductase activity does co-purify with the cytochrome bc_1 complex from *Rb. sphaeroides* f. sp. *denitrificans*, which additionally suggests that it is likely to be a membrane protein.

2.6.3.6 TMAO reductase
The ability to use trimethylamine N-oxide (TMAO) as an anaerobic alterna-

tive terminal electron acceptor, reducing it to trimethylamine (TMA), is very widespread and is found, for example, in enterobacteria, *Rhodobacter*, marine *Alteromonas*, and possibly *Campylobacter*. In terms of supporting active transport and growth, being coupled to proton translocation, and generating a protonmotive force, TMAO reduction has been shown to be energy coupled. Whereas electron transfer to nitrate reductase is mediated by ubiquinone, with TMAO (as with fumarate) the quinone is menaquinone. Cytochromes (*b*- and *c*-type) are also components of the TMAO reduction electron chain, with the cytochrome *b* being non-identical with the specific nitrate reductase cytochrome *b*. The TMAO reductase itself is a molybdoprotein and is membrane-bound in *E. coli* but periplasmic in *Rb. capsulata*. There appear to be two separate enzymes, one inducible and the other constitutive. The TMAO reductase also has dimethylsulfoxide (DMSO) reductase activity (see page 39).

2.7   Concluding remarks

Looking at anaerobic microbial energetics as a whole, their study can be seen to offer much to the microbiologist, whether his or her particular viewpoint be that of the biochemist, cell physiologist, ecologist or biotechnologist. For example, there is the unique cellular and cofactor chemistry found in the methanogens; the carbon monoxide dehydrogenase and the acetyl-CoA pathway; the metabolic interdependence of species within consortia; and the application of such systems in anaerobic digesters and industrial fermentations. However, what is particularly significant in the present context is the centrality of reversible redox systems, determining both energy yield and the supply of reducing equivalents, while at the same time directing and controlling carbon flux. The thermodynamic limitations imposed on both organisms and ecosystems experiencing anaerobiosis throw these relationships into sharp relief, and highlight such ingenious mechanisms as the methyl-CoM methyl reductase, which have evolved to overcome particular energetic barriers.

References

General interest and essential background
Gottschalk, G. (1986). 'Bacterial Metabolism', 2nd Ed. Springer-Verlag, New York.
Haddock, B. A. and Jones, C. W. (1977). *Bacteriol. Revs.* **41**, 47–99.
Haddock, B. A. and Hamilton, W. A. (1977). *Symp. Soc. Gen. Microbiol.* **27**, 1–442.
Thauer, R. K., Jungerman, K. and Decker, K. (1977). *Bacteriol. Revs.* **41**, 100–180.
Thauer, R. K. and Morris, J. G. (1984). *Symp. Soc. Gen. Microbiol.* **36**, 123–168.

Fermentations, including acetogens
Braun, K. and Gottschalk, G. (1981). *Arch. Microbiol.* **128**, 294–298.
Fuchs, G. (1986). *FEMS Microbiol. Revs.* **39**, 181–213.
Gottwald, M. and Gottschalk, G. (1985). *Arch. Microbiol.* **143**, 42–46.
McInerney, M. J., Bryant, M. P., Hespall, R. B. and Costerton, J. W. (1981). *Appl. Environ. Microbiol.* **41**, 1029–1039.
Ragsdale, S. W., Ljundahl, L. G. and DerVartanian, D. V. (1983). *J. Bacteriol.* **185**, 1224–1237.
Tewes, F. J. and Thauer, R. K. (1980). *In* 'Anaerobes and Anaerobic Infections', (Eds. G. Gottschalk, N. Pfennig and H. Werner), pp. 97–104. Fischer, Stuttgart.
Winter, J. and Wolfe, R. S. (1980). *In* 'Anaerobes and Anaerobic Infections' (Eds. G. Gottschalk, N. Pfennig and H. Werner), pp. 105–115. Fischer, Stuttgart.
Wood, H. G., Ragsdale, S. W. and Pezacka, E. (1986). *Trends Biochem. Sci.* **11**, 14–18.
Wood, H. G., Ragsdale, S. W. and Pezacka, E. (1986). *FEMS Microbiol. Revs.* **39**, 345–362.

Methanogens
Anthony, C. (1982). 'The Biochemistry of Methylotrophs'. Academic Press, London.
Blaut, M. and Gottschalk, G. (1984). *FEMS Microbiol. Lett.* **24**, 103–107.
Blaut, M. and Gottschalk, G. (1984). *Eur. J. Biochem.* **141**, 217–222.
Blaut, M. and Gottschalk, G. (1985). *Trends. Biochem. Sci.* **10**, 486–489.
Blaut, M., Müller, V., Fiebig, K. and Gottschalk, G. (1985). *J Bacteriol.* **164**, 95–101.
Bott, M. and Thauer, R. K. (1987). *Eur. J. Biochem.* (in press).
Bott, M., Eikmanns, B. and Thauer, R. K. (1986). *Eur. J. Biochem.* **159**, 393–398.
Butsch, B. M. and Bachofen, R. (1984). *Arch. Microbiol.* **138**, 293–298.
Jones, W. J., Donnelly, M. I. and Wolfe, R. S. (1985). *J. Bacteriol.* **163**, 126–131.
Jones, W. J., Nagle, D. P. Jr. and Whitman, W, B. (1987). *Microbiol. Revs.* **51**, 135–177.
Keltjens, K. T. and van der Drift, C. (1986). *FEMS Microbiol. Revs.* **39**, 259–303.
Krzycki, J. A., Lehman, L. J. and Zeikus, J. G. (1985). *J. Bacteriol.* **163**, 1000–1006.
Schönheit, P. and Beimborn, D. B. (1985). *Arch. Microbiol.* **142**, 354–361.
Schönheit, P. and Beimborn, D. B. (1985). *Eur. J. Biochem.* **148**, 545–550.
Widdel, F. (1986). *Appl. Environ. Microbiol.* **51**, 1056–1062.
Wolfe, R. S. (1985). *Trends Biochem. Sci.* **10**, 396–399.

Sulphate-reducing bacteria
Bak, F. and Cypionka, H. (1987). *Nature* **326**, 891–892.
Belkin, S., Wirsen, C. O. and Jannasch, H. W. (1985). *Appl. Environ. Microbiol.* **49**, 1057–1061.
Brandis, A. and Thauer, R. K. (1981). *J. Gen. Microbiol.* **126**, 249–252.
Brandis-Heep, A., Gebhardt, N. A., Thauer, R. K., Widdel, F. and Pfennig, N. (1983). *Arch. Microbiol.* **136**, 222–229.
Cord-Ruwisch, R., Kleinitz, W. and Widdel, F. (1987). *J. Pet. Technol.* **Jan.**, 97–106.
Cypionka, H. (1987). *Arch. Microbiol.* **148**, 144–149.
Cypionka, H. and Dilling, W. (1986). *FEMS Microbiol. Lett.* **36**, 257–260.
Cypionka, H. and Pfennig, N. (1986). *Arch. Microbiol.* **143**, 396–399.
Fowler, V. J., Widdel, F., Pfennig, N., Woese, C. R. and Stackebrandt, E. (1986). *System. Appl. Microbiol.* **8**, 32–41.
Gebhardt, N. A., Thauer, R. K., Linder, D., Kaulfers, P.-M. and Pfennig, N. (1985). *Arch. Microbiol.* **141**, 392–398.

Gow, L. A., Pankhania, I. P., Ballantine, S. P., Boxer, D. H. and Hamilton, W. A. (1986). *Biochim. Biophys. Acta.* **851**, 57–64.
Hamilton, W. A. (1985). *Ann. Rev. Microbiol.* **39**, 195–217.
Jansen, K., Fuchs, G. and Thauer, R. K. (1985). *FEMS Microbiol. Lett.* **28**, 311–315.
Kröger, A., Schröder, J., Paulsen, J. and Beilmann, A. (1988). *In* 'The Nitrogen and Sulphur Cycles'. *S. G. M. Symp.* **42**, 133–145.
Le Gall, J. and Fauque, G. (1988). *In* 'Biology of Anaerobic Microorganisms'. (Ed. A. Zehnder), Williams & Wilkins, New York.
Lissolo, T., Choi, E. S., Le Gall, J. and Peck, H. D. Jr. (1986). *Biochem. Biophys. Res. Comm.* **139**, 701–708.
Nethe-Jaenchen, R. and Thauer, R. K. (1984). *Arch. Microbiol.* **137**, 236–240.
Odom, J. M. and Peck, H. D. Jr. (1981). *FEMS Microbiol. Lett.* **12**, 47–50.
Odom, J. M. and Peck, H. D. Jr. (1984). *Ann. Rev. Microbiol.* **38**, 551–592.
Paulsen, J., Kröger, A. and Thauer, R. K. (1986). *Arch. Microbiol.* **144**, 78–83.
Peck, H. D. Jr., Le Gall, J., Lespinat, P. A., Berlier, Y. and Fauque, G. (1987). *FEMS Microbiol. Lett.* **40**, 295–299.
Peck, H. D. Jr. and Lissolo, T. (1988). *In* 'The Nitrogen and Sulphur Cycles'. *S. G. M. Symp.* **42**, 99–132.
Pfennig, N. and Biebl, H. (1976). *Arch. Microbiol.* **110**, 3–12.
Phelps, T. J., Conrad, R. and Zeikus, J. G. (1985). *Appl. Environ. Micro.* **50**, 589–594.
Postgate, J. R. (1984). 'The Sulphate-Reducing Bacteria', 2nd edn. Cambridge University Press, Cambridge.
Postgate, J. R., and Kelly, D. P. (1982). *Phil. Trans. R. Soc. Lond.* B **298**, 431–602.
Prickril, B. C., Czechowski, M. H., Przybyla, A. E., Peck, H. D. Jr. and Le Gall, J. (1986). *J. Bacteriol.* **167**, 722–725.
Schauder, R., Eikmanns, B., Thauer, R. K., Widdel, F. and Fuchs, G. (1986). *Arch. Microbiol.* **145**, 162–172.
Stams, A. J. M., Kremer, D. R., Nicolay, K., Weenk, G. M. and Hansen, T. A. (1984). *Arch. Microbiol.* **139**, 167–173.
Stetter, K. O. (1984). *Origins of Life* **14**, 809–815.
Stetter, K. O. and Gaag, G. (1983). *Nature* **305**, 309–311.
Stetter, K. O., Lauerer, G., Thomm, M. and Neuner, A. (1987). *Science* **236**, 822–824.
Traore, A. S., Hatchikian, C. E., Belaich, J.-P. and Le Gall, J. (1981). *J. Bacteriol.* **145**, 191–199.
Voordouw, G. (1988). *In* 'The Nitrogen and Sulphur Cycles'. *S. G. M. Symp.* **42**, 147–160.
Voordouw, G. and Brenner, S. (1985). *Eur. J. Biochem.* **148**, 515–520.
Voordouw, G., Walker, J. E. and Brenner, S. (1985). *Eur. J. Biochem.* **148**, 509–514.
Widdel, F. and Pfennig, N. (1977). *Arch. Microbiol.* **112**, 119–122.

Microbial consortia
Hamilton, W. A. (1984). *In* 'Aspects of Microbial Metabolism and Ecology' (Ed. G. A. Codd), pp. 35–57. Academic Press, London.
Kristjansson, J. K., Schönheit, P. and Thauer, R. K. (1982). *Arch. Microbiol.* **131**, 278–282.
Laanbroek, H. J. and Pfennig, N. (1981). *Arch. Microbiol.* **128**, 330–335.
Laanbroek, H. J., Geerligs, H. J., Peijnenburg, Ad. A. C. M. and Seisling, J. (1983). *Microbial Ecol.* **9**, 341–354.
McInerney, M. J. and Bryant, M. P. (1981). *Appl. Environ. Micro.* **41**, 346–354.
Oremland, R. S., Marsh. L. M. and Polcin, S. (1982). *Nature* **296**, 143–145.

Schönheit, P., Kristjannson, J. K. and Thauer, R. K. (1982). *Arch. Microbiol.* **132**, 285–288.

Sørensen, J., Christensen, D. and Jørgensen, B. B. (1981). *Appl. Environ. Micro.* **42**, 5–11.

Zeikus, J. G. (1983). *Symp. Soc. Gen. Microbiol.* **34**, 423–462.

Fumarate, nitrate and TMAO reduction

Alef, K., Jackson, J. B., McEwan, A. G. and Ferguson, S. J. (1985). *Arch. Microbiol.* **142**, 403–408.

Barret, E. L. and Kwan, S. (1985). *Ann. Rev. Microbiol.* **39**, 131–149.

Cole, J. A. (1982). *Biochem. Soc. Trans.* **10**, 476–477.

Cole, J. A. (1988). *In* 'The Nitrogen and Sulphur Cycles'. *S. G. M. Symp.* **42**, 281–329.

Cole, S. T., Condon, C., Lemicke, B. D. and Weiner, J. M. (1985). *Biochim. Biophys. Acta* **811**, 381–403.

Craske, A. and Ferguson, S. J. (1986). *Eur. J. Biochem.* **158**, 429–436.

Ferguson, S. J. (1987). *Trends Biochem. Sci.* **12**, 354–357.

Graf, M., Bokrang, M., Böcher, R., Frierdl, P. and Kröger, A. (1985). *FEBS Lett.* **184**, 100–103.

Ingledew, W. J. and Poole, R. K. (1984). *Bacteriol. Revs.* **48**, 222–271.

Kröger, A. (1977). *Symp. Soc. Gen. Microbiol.* **27**, 61–93.

McEwan, A. G., Jackson, J. B. and Ferguson, S. J. (1984). *Arch. Microbiol.* **137**, 344–349.

McEwan, A. G., Wetzstein, M. G., Ferguson, S. J. and Jackson, J. B. (1985). *Biochim. Biophys. Acta* **806**, 410–417.

Macy, J. M., Schröder, I., Thauer, R. K. and Kröger, A. (1986). *Arch. Microbiol.* **144**, 147–150.

Mitchell, E. J., Jones, J. G. and Cole, J. A. (1986). *Arch. Microbiol.* **144**, 35–40.

Pope, N. R. and Cole, J. A. (1982). *J. Gen. Microbiol.* **128**, 219–222.

Seitz, H.-J. and Cypionka, H. (1986). *Arch. Microbiol.* **146**, 63–67.

Shapleigh, J. P. and Payne, W. J. (1985). *J. Bacteriol.* **163**, 837–840.

Urata, K., Shimada, K. and Satoh, T. (1983). *Plant & Cell Physiol.* **24**, 501–508.

Zumft, W. G., Viebrock, A. and Körner, H. (1988). *In* 'The Nitrogen and Sulphur Cycles'. *S. G. M. Symp.* **42**, 245–279.

3 Periplasmic electron transport reactions

S. J. Ferguson

3.1 Introduction 151
3.2 The periplasmic electron transfer proteins 160
3.3 The rationales for periplasmic location 162
3.3.1 Periplasmic oxidation reactions 164
3.3.2 Periplasmic reductase reactions 166
3.3.3 Full occupancy of sites on the cytoplasmic surfaces of the
      plasma membrane? 169
3.4 Pathways of electron flow to and from periplasmic enzymes . 170
3.5 Mobility and organization of electron transfer proteins
    within the periplasm: mini-electron transfer chains? 172
3.6 Are there electron transport proteins in Gram-positive
    organisms that have equivalent functions to the periplasmic
    enzymes of Gram-negative bacteria? 176
3.7 A doubt over the accepted role for cytochrome c_2 as a
    periplasmic electron shuttle in *Rhodospirillaceae* 179
3.8 Conclusion 179
References 180

3.1 Introduction

A general review in 1979 of periplasmic enzymes in Gram-negative bacteria listed only three examples of proteins involved in electron transport processes (Beacham, 1979). Although this list may not have been complete in every respect, it nevertheless contrasts sharply with the situation in 1987, when more than 20 types of electron transport protein have been assigned to the periplasm. It is now clearly established that this region of the bacterial cell often has a major function in electron transport. Such a function for the periplasm is additional to its previously recognized roles in accommodating degradative and detoxifying enzymes together with binding proteins for nutrient uptake and chemotactic processes. The purpose of this chapter is to describe and classify the electron transport reactions of the periplasm.

First of all, what is meant by 'periplasm' needs to be defined. This chapter regards as the periplasm the region in Gram-negative bacteria that lies between the cytoplasmic membrane and the outer membrane (Fig. 3.1). The term 'periplasmic space' is avoided because, as will be seen, it creates a false impression of emptiness. The conventional view is that, for Gram-positive organisms, the peptidoglycan layer is closely juxtaposed to the cytoplasmic membrane and that, consequently, there is no region equivalent to the periplasm. This point will be returned to later in the chapter. The possible existence and role of a periplasm in archaebacteria is not discussed because of the paucity of data concerning this topic. A periplasm has been tentatively identified in *Halobacterium halobium* (Blaurock *et al.*, 1976).

Next, what will be termed a 'periplasmic protein' must be defined. Here are included peripheral membrane proteins that are located at the periplasmic surface of the cytoplasmic membrane, as well as proteins that may have no direct interaction with the membrane (Fig. 3.1). This definition is made because, in reality, there may be little or no distinction between these two classes of proteins. It is probable that many peripheral proteins have specific binding sites on one or more integral membrane proteins. There will be an equilibrium between bound and free forms of such peripheral proteins, so it is senseless to argue whether they are membrane or periplasmic proteins. For completeness, and because the bioenergetic implications can be similar, reactions that may be catalysed at the periplasmic surface by integral membrane proteins will be discussed (Fig. 3.1) although such proteins will not be categorized as periplasmic.

The periplasm is not a negligible part of the bacterial cell. There is evidence that between 10% and 20% of cell volume is located in this region (Stock *et al.*, 1977). Recognition of the relatively large number of proteins now being assigned to the periplasm has led to considerations of the extent to which proteins can interact through diffusional processes (e.g. Beardmore-Gray and Anthony, 1984). Indeed, measurements of laser-induced fluorescence bleaching and recovery, made with a fluorescent derivative of a maltose-binding protein incorporated into the periplasm of *Escherichia coli*, have indicated that periplasmic proteins have diffusion coefficients that are only 1–10% of the values determined for cytoplasmic proteins (Brass *et al.*, 1986; Brass, pers. comm.) The exact basis for this restricted mobility is not properly understood but can perhaps be correlated with the recent suggestion that the periplasm is a gel phase (Hobot *et al.*, 1984). The gel is postulated to arise from the presence throughout the periplasm of hydrated peptidoglycan. The peptidoglycan is envisaged as more cross-linked where it underlies the outer membrane than where it approaches the cytoplasmic

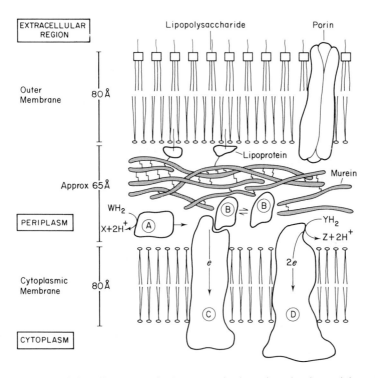

Fig. 3.1 Diagram of the cell envelope of a Gram-negative bacterium showing periplasm. The murein layer is shown as increasing in extent of cross-linking as it approaches the outer membrane, to which it is attached through covalent links to the hydrophilic ends of lipoproteins. The sketch is meant to convey that the periplasm is a filled region of the cell. There is probably not a discrete boundary between the periplasm and the murein layer (c.f. Hobot *et al.*, 1984). Protein A represents a water soluble dehydrogenase which does not directly donate electrons to the membrane-bound electron carriers but could use a protein such as B as electron acceptor. The latter protein represents an electron carrier (e.g. a cytochrome) which has a binding site on the integral membrane protein C which is also an electron carrier. Protein D represents an integral membrane protein that oxidizes substrate at the periplasmic surface. Proteins similar to A and D with respect to location may also occur as reductases. (The distance scale is approximate.)

membrane (Hobot *et al.*, 1984; Fig. 3.1). The concept that peptidoglycan might extend up to the cytoplasmic membrane suggests that in this respect there may not be a great difference between Gram-positive and Gram-negative organisms, a point that will be discussed later.

A central question needing to be addressed at the outset is that of how one might, in the sense defined above, establish a periplasmic location for a protein. There are a number of available methods to choose from according to the circumstances of a particular investigation:

(*1*)   A traditional criterion for a periplasmic protein is that it should be released upon digesting and/or disrupting the cell wall with lysozyme and/or EDTA. The procedure must, of course, be done in a medium of sufficiently high osmolarity to prevent release of cytoplasmic proteins following lysis of the plasma membrane. Indeed, the basis for assigning a periplasmic location actually depends upon the retention by the lysozyme/EDTA-treated cells of enzymes of known cytoplasmic location such as malate dehydrogenase. The activity of such an enzyme is assayed after lysis of the osmotically fragile, treated cells often called 'sphaeroplasts'. In practice, this method has to be used with care. First, there are always some cells in the population that release the cytoplasmic marker protein; secondly, the release of the periplasmic protein is rarely complete. Consequently, a careful balance sheet of proteins of interest plus a cytoplasmic marker must be constructed in terms of:

(a)   Content in a supernatant following collection by centrifugation of the lysozyme/EDTA-treated cells;

(b)   Content in both the membrane and soluble fractions obtained after lysis of the lysozyme/EDTA-treated cells.

The degree of release of a periplasmic protein depends upon the extent of any interactions with other membrane proteins and also on the extent of cell wall degradation. Thus it was found that rendering *Paracoccus denitrificans* cells osmotically fragile by treatment with lysozyme was inadequate to release nitrite reductase. This enzyme appeared in a supernatant only after treatment with lysozyme plus EDTA (Alefounder and Ferguson, 1980). Variation of the buffer and ionic strength used during incubation with lysozyme can be important for release of periplasmic proteins (e.g. Wu, 1986). Thus, an important point is that lack of release of a protein under a restricted set of experimental conditions cannot be used as evidence against a periplasmic location. Tables 3.1–3.3 list a number of examples of periplasmic proteins that have been identified using lysozyme/EDTA treatment of cells.

It should also be kept in mind that a protein such as B in Figure 3.1 could not be expected to be completely released from lysozyme/EDTA-treated cells, unless such materials were to be repeatedly washed or the interaction between proteins B and C were to be weak.

(*2*)   Release of periplasmic proteins by osmotic shock treatment of cells of certain bacteria (Heppel, 1971) is similar in principle to the first method and the same comments about interpretation are appropriate. This procedure and related methods (Hooper and DiSpirito, 1985) have not been widely used for analysing the location of electron transport proteins.

(*3*)   Chemical modification of proteins by reagents, such as isethionyl

Table 3.1 Periplasmic enzyme-catalysed oxidation reactions

Reaction	Organism	Criterion	Reference
$CH_3NH_2 + H_2O \rightarrow HCHO + NH_3 + 2H^+ + 2e$	Organism 4025	1	Lawton and Anthony (1985b)
	Hyphomicrobium X	4	Kasprzak and Steenkamp (1983)
	Organism W3A1	4	Kasprzak and Steenkamp (1983)
	P. denitrificans	1	Husain and Davidson (1985)
$CH_3OH \rightarrow HCHO + 2H^+ + 2e$	Organism 4025	1	Lawton and Anthony (1985b)
	Hyphomicrobium X	4	Kasprzak and Steenkamp (1983)
	Organism W3A1	4	Kasprzak and Steenkamp (1983)
	M. methylotrophus	1	Burton *et al.* (1983)
	M. methylotrophus	4	Quilter and Jones (1984)
	P. denitrificans	1	Alefounder and Ferguson (1981)
p-cresol $+ H_2O \rightarrow p$-hydroxybenzylalcohol $+ 2H^+ + 2e$	*P. putida*	1	Hopper *et al.* (1985)
p-hydroxybenzylalcohol $\rightarrow$ p-hydroxybenzaldehyde $+ 2H^+ + 2e$	*P. putida*	1	Hopper *et al.* (1985)
$Fe^{2+} \rightarrow Fe^{3+} + e$	*T. ferrooxidans*	1	Ingledew *et al.* (1977)
$H_2 \rightarrow 2H^+ + 2e$	*D. gigas*	1	Bell *et al.* (1974)
	D. vulgaris	1	Van der Western *et al.* (1978)
	D. vulgaris	1	Badziong and Thauer (1980)
	D. vulgaris	6[a]	Pickrill *et al.* (1986)
	D. sulfuricans	5[c]	Liu and Peck (1981); Steenkamp and Peck (1981)
$HS^- \rightarrow S + H^+ + 2e$	*R. sulfidophilus*	[b]	Brune and Truper (1986)
	Various	[b]	Van Gemerden (1984)
$NH_2^+OH + H_2O \rightarrow NO_2^- + 5H^+ + 4e$	*N. europaea*	1	Olson and Hooper (1983)
$S_2O_3^{2-} + 5H_2O \rightarrow 2SO_4^{2-} + 10H^+ + 8e$	*T. versutus*	1	Wu (1986)

The numbered criteria refer to the six methods for establishing a periplasmic location that are described in the text. Superscripts have the following meanings: [a] the small subunit has an identifiable signal sequence and site of protease-cleavage but it is inferred that the large subunit must have an internal signal sequence; [b] inferred from comparison of pH dependence and apparent K_m values for sulphide oxidation by whole cells or inside out vesicles (chromatophores) (Brune and Truper, 1986) or from external site of elemental sulphur deposition by cells (Van Gemerden, 1984); [c] accessibility to electron acceptors also used as criteria. As pointed out by Hooper and DiSpirito (1985) it is very probable that a number of other metal ions and also elemental sulphur and its compounds will prove to be oxidized in the periplasm. Electron transport from methylamine and methanol is discussed further in Chapter 6; oxidation of inorganic substances is the subject of Chapter 4. Organisms 4025 and W3A1 are obligate methylotrophs with similarities to *Methylophilus methylotrophus*.

Table 3.2 Periplasmic enzyme-catalysed reduction reactions

Reaction	Organism	Criterion	Reference
$(CH_3)_3NO + 2H^+ + 2e \rightarrow (CH_3)_3N + N_2O$	R. capsulatus	1	McEwan et al. (1985b)
$(CH_3)_2SO + 2H^+ + 2e \rightarrow (CH_3)_2S + H_2O$	R. capsulatus	1	McEwan et al. (1985b)
fumarate $\rightarrow$ succinate	C. formicoaceticum	1	Dorn et al. (1979)
$N_2O + 2H^+ + 2e \rightarrow N_2 + H_2O$	P. denitrificans	5	Boogerd et al. (1981)
	P. denitrificans	1	Alefounder et al. (1983)
	R. capsulatus	1	McEwan et al. (1985a)
	R. sphaeroides f.sp. denitrificans	1	Urata et al. (1982)
$NO_2^- + 2H^+ + e \rightarrow NO + H_2O$	P. denitrificans	5	Meijer et al. (1979)
	P. denitrificans	1	Alefounder and Ferguson (1980)
	P. aeruginosa	1	Wood (1978)
	R. sphaeroides f.sp. denitrificans	1[a]	Sawada and Satoh (1980)
$NO_2^- + 8H^+ + 6e \rightarrow NH_4^+ + 2H_2O$	E. coli	1[b]	Fujita and Satoh (1966); Cole (1968); Kajie and Anakaru (1986)
	D. desulfuricans	5[c]	Liu and Peck (1981); Steenkamp and Peck (1981)
$NO_3^- + 2H^+ + 2e \rightarrow NO_2^- + H_2O$	R. capsulatus	1	McEwan et al. (1984)
	R. sphaeroides f.sp. denitrificans	1	Sawada and Satoh (1980)
$H_2O_2 + 2H^+ + 2e \rightarrow 2H_2O^{[d]}$	P. aeruginosa	1	Pettigrew and Moore (1987)

Details of numbered criteria are given in Table 3.1 and the text. Superscripts have the following meanings: [a] Note that there are two types of enzyme catalysing nitrite reduction to nitric oxide; a cytochrome cd (e.g., P. denitrificans and P. aeruginosa) as well as a copper-containing enzyme (e.g., R. sphaeroides f.sp. denitrificans); [b] Fujita and Satoh (1966) and Cole (1968) showed a cytochrome c_{552} to be periplasmic whilst Kajie and Anakaru (1986) identified this cytochrome as the nitrite reductase; [c] although this nitrite reductase is probably a similar c-type cytochrome to the E. coli enzyme, it is less certainly established as water-soluble and possibly ought to be included in Table 3.5. [d] Cytochrome c peroxidase. It is probable that a number of further reactions will in due course be shown to occur in the periplasm, e.g., $Fe^{3+} \rightarrow Fe^{2+}$. See Chapter 2 for further discussion of these reduction reactions.

Table 3.3 Periplasmic electron transport proteins

Protein	Organism	Criterion	Reference
Amicyanin	Organism 4025	1	Lawton and Anthony (1985b)
Amicyanin	P. denitrificans	1	Husain and Davidson (1985)
Azurin	P. aeruginosa		Wood (1978)
Azurin	P. aeruginosa	6	Canters (1986)
Cytochrome b_{562}	E. coli		Nikkila et al. (unpublished)
Cytochrome c_{550}	P. denitrificans		Scholes et al. (1971)
Cytochrome c_{551}	P. denitrificans	1	Husain and Davidson (1986)
Cytochrome c_{553}	P. denitrificans	1	Husain and Davidson (1986)
Cytochrome c_2	R. sphaeroides	1	Prince et al. (1975)
Cytochrome c_2	R. capsulatus	1	Prince et al. (1975)
Cytochrome c'	R. sphaeroides	1	Prince et al. (1975)
Cytochrome c'	R. capsulatus	1	Prince et al. (1975);
			McEwan et al. (1985a)
Cytochrome c_{552}	N. europaea	1	DiSpirito et al. (1985)
Cytochrome c_{554}	N. europaea	1	DiSpirito et al. (1985)
Cytochrome c_3	D. vulgaris	1	Badziong and Thauer (1980)
	D. vulgaris	6	Voordouw and Brenner (1986)
	D. gigas	1	LeGall and Peck (1987)
Cytochrome c_{553}	D. vulgaris	1	LeGall and Peck (1987)
Cytochrome c_H	Various	1	Anthony (1986)
Cytochrome c_L	methylotrophs	1	Anthony (1986)
Cytochrome c_{552}	T. thermophilus	1	Lorence et al. (1981)
Cytochrome c_{549}	Alteromonas haloplanktis	1	Knowles et al. (1974)
Cytochrome c_{550}	Alcaligenes eutrophus	1	Probst and Schlegel (1976)
Rusticyanin	T. ferrooxidans	1	Ingledew et al. (1977)

Details of numbered criteria are given in Table 3.1 and in the text.

acetamidate, that cross the outer wall but are unable to penetrate the cytoplasmic membrane is also applicable. This method proved particularly useful for locating the major dehydrogenases in two methylotrophs (Kasprzak and Steenkamp, 1983; Table 3.1).

(4) Histochemical studies, especially using antibodies, can in principle be used to define the location of a membrane protein. But this is a method that can give false results and requires some disruption of the cell wall to provide access of the antibody to the protein under study. An example of ambiguity with this approach is a study of the location of nitrite reductase (cytochrome *cd*) in *Pseudomonas aeruginosa*. Antibody staining techniques gave evidence that the protein was located on the cytoplasmic surface (Saraste and Kuronen, 1978), but enzyme release studies with both *P. aeruginosa* (Wood, 1978) and *P. denitrificans* (Alefounder and Ferguson, 1980) indicated a periplasmic position, a view that was also indicated by the experimental approach to be outlined next.

(5) Many oxidations and reductions catalysed by electron transport proteins involve the uptake or release of protons. Such changes in proton concentration that take place in the periplasm are more readily detected, in lightly buffered suspensions of cells, than those which occur in the cytoplasm. Using this rationale Stouthamer and colleagues, for example, were able to show that the protons required for nitrite and nitrous oxide reduction by *P. denitrificans* were taken from the periplasm (Meijer *et al.*, 1979; Boogerd *et al.*, 1981). Such a finding is, of course, consistent with a periplasmic location but does not rule out the possibility that, although the active site of the enzyme is at the periplasmic surface, the protein itself is integral to the membrane. Indeed, unlike methods (1) to (4), this approach has the advantage that it can provide evidence for the function of a protein such as D shown in Figure 3.1.

(6) The proteins of the periplasm are synthesized in the cytoplasm and therefore must be translocated across the cytoplasmic membrane. A general feature of such protein translocation is that the polypeptide as synthesized on the ribosomes possesses an N-terminal extension of approximately 20 amino acids. On arrival in the periplasm, a peptidase cleaves this N-terminal signal sequence. Thus the actual N-terminus of a mature, functional periplasmic protein may be expected to have features in its amino sequence that served as signal peptidase recognition sites in the maturation process. Accordingly, LeGall and Peck (1987) have argued that where N-terminal amino acid sequences are known for water-soluble proteins, they may be used as an indication for a periplasmic or cytoplasmic location. They showed that this method of analysis correlated well with previously estab-

lished periplasmic locations for cytochrome c_3 (a tetrahaem species) from two species of sulphate-reducing bacteria, and predicted the same location in four other cases. A second type of cytochrome, cytochrome c_{553}, also had an N-terminus that correlated with a known periplasmic location, as did the small subunit of the periplasmic hydrogenase from *Desulfovibrio vulgaris* (Prickrill *et al.*, 1986). In the latter case, as well as for the cytochrome c_3 from *D. vulgaris*, the availability of gene sequences has confirmed the presence of N-terminal extensions on the polypeptide that can serve as signal sequences for direction of the polypeptides to the periplasm.

For inspection of N-terminal sequences to be a reliable guide to a periplasmic location, cytoplasmic proteins must have distinct N-terminal sequences. Inspection of a number of such proteins from sulphate-reducing bacteria revealed that there was no instance of the N-terminus having a typical signal peptidase site. However, as LeGall and Peck (1987) pointed out, this criterion is not rigorous because there may yet exist unrecognized peptidase cleavage sites. There is also the possibility that some periplasmic proteins may have internal rather than N-terminal signals that relate to translocation of the protein across the membrane.

Clearly, the analysis of N-terminal amino acid sequences, when available, has something to contribute for determination of cellular localization of proteins, especially if other methods (see above) prove ambiguous. It will be interesting to learn if two predictions made on the basis of sequence (LeGall and Peck, 1987), a cytoplasmic location for a mono-haem cytochrome c_{553} from *Desulfovibrio baculatus* and a periplasmic location for the trihaem cytochrome c_3 from *Desulforomonas acetoxidans*, can be confirmed by other methods.

From the foregoing discussion it will also be apparent that, in cases where the gene sequence is known, it may be possible to identify a typical signal sequence at the amino terminus even when there are no data available from direct protein sequencing of the mature protein. Such identification can be assisted by location of the Shine–Dalgarno sequence that occurs upstream of the translational start signal on an RNA transcript. Clearly, when the N-terminal amino acid sequence of the mature protein is known, comparison with the gene sequence can firmly identify the signal sequence and, hence, indicate a periplasmic location. Again, the absence of such an N-terminal signal sequence does not preclude a periplasmic location because of the possibility of internal signal sequences.

Thus a good selection of methods is available for determining whether a protein is periplasmic. There are few incorrect assignments of proteins to the periplasm, but more ambiguity has been associated with proteins not positively identified as periplasmic. In other words, negative results obtained with any of the techniques outlined above are not necessarily evidence for an electron transport protein being either an integral membrane protein or associated with the cytoplasmic surface of the plasma membrane. Indeed, periplasmic locations have been overlooked in the past. Good illustrative examples are the dimethylsulphoxide (DMSO)/trimethylamine N-oxide (TMAO) reductase and nitrate reductase of *Rhodobacter capsulatus*. Total water-soluble extracts of cells were found to have NADH-DMSO, -TMAO and -NO$_3^-$ reductase activities in the presence of FMN. Hence it was assumed that these two reductases were cytoplasmic which, of course, is where NADH is to be found. Subsequent work showed that the NADH-dependent activity is absent from the periplasmic fraction that contains the two reductases (Ferguson *et al.*, 1987). The NADH-dependent activity was an artefact resulting from the mixing of the periplasmic and cytoplasmic fractions in the total extract, with an unidentified cytoplasmic enzyme catalysing reduction of FMN by NADH.

3.2   The periplasmic electron transfer proteins

In the above section several criteria were given for possible assignment of a protein to the periplasm. Such methods have now demonstrated the existence of a substantial number of periplasmic electron transport proteins. The extensive lists of Tables 3.1–3.3 contrast with the compilation made by Beacham in 1979, but nevertheless it should be kept in mind that many proteins have not been scrutinized for a periplasmic or cytoplasmic location. The peripheral membrane protein dehydrogenases for D-alanine, allohydroxy-D-proline, choline and sarcosine in *E. coli* (Bater and Venables, 1977) are examples of enzymes for which the location is unknown.

Tables 3.1 and 3.4 are compilations of the substrate oxidation reactions that are known to occur in the periplasmic region. The list is dominated by substrates that are small organic molecules or inorganic reductants, but note the appearance of glucose in the list. It must be stressed, however, that many bacteria take up glucose and feed it via glucose 6-phosphate into classic pathways of catabolism. The oxidants handled in the periplasm also tend to be relatively small molecules (Table 3.2).

The electron transfer proteins found in the periplasm are dominated by *c*-type cytochromes (Table 3.3). Indeed, there is little evidence for *c*-type cytochromes occurring on the cytoplasmic side (Ferguson, 1982; Wood, 1983). Probably the most persuasive evidence for a cytoplasmic location for a *c*-type cytochrome can be marshalled for an octa-haem cytochrome c_3 ($M_r = 26\,000$) that is found in all species of *Desulfovibrio* tested (Odom and Peck, 1984). This type of cytochrome, which is quite distinct from the tetra-haem cytochrome c_3 ($M_r = 13\,000$) listed in Table 3.3, is reportedly not

Table 3.4 Integral membrane proteins thought to catalyse oxidation reactions at the periplasmic surface of the cytoplasmic membrane

Reaction	Organism	Criterion	Reference
glucose + $H_2O \rightarrow 2H^+ + 2e$ + gluconic acid	P. aeruginosa	[a]	Midgley and Dawes (1973)
	A. calcoaceticus	[b]	Dokter et al. (1985)
$H_2 \rightarrow 2H^+ + 2e$	Wolinella (Vibrio) succinogenes	[c]	Kröger et al. (1980)
	Campylobacter sputorum	5	De Vries et al. (1984)
$HCOOH \rightarrow 2H^+ + CO_2 + 2e$	W. succinogenes	[c]	Kröger et al. (1980)
	C. sputorum	5	De Vries et al. (1982)
	D. desulfuricans	5[c]	Steenkamp and Peck (1981)

Details of numbered criteria are given in Table 3.1 and the text. Superscripts have the following meanings: [a] established from comparison of kinetics of glucose uptake by wild-type and mutant cells; [b] released from cells by mild detergent treatment; [c] location deduced from studies of substrate and electron acceptor accessibility together with chemical labelling studies. See Chapter 6 for further discussion of glucose oxidation.

found among the periplasmic proteins identified in *Desulfovibrio* (Odom and Peck, 1984). Guerlesquin *et al.* (1982) purified the octa-haem cytochrome c_3 from the water-soluble fraction obtained upon disrupting cells of *Desulfovibrio baculatus* Norway 4 (formerly *Desulfovibrio desulfuricans* Norway strain) and determined the N-terminal amino acid sequence. The latter does not have a recognizable signal peptidase cleavage site and thus according to criterion 6 outlined above a cytoplasmic location must be considered.

Despite the above example, careful consideration must clearly be given to the likelihood of a periplasmic location for any water-soluble *c*-type cytochrome isolated from a bacterium. At least some membrane-bound *c*-type cytochromes also function effectively in the periplasm because the haem group is located on that side of the membrane (Wood, 1983), but these are not listed in this chapter. Indeed, Wood (1983) has suggested that the covalent attachment of the haem to the polypeptide in *c*-type cytochromes is a device to prevent loss by dissociation of haem from cytochromes in the periplasm where, unlike the cytoplasm, replenishment of haem by *de novo* synthesis is presumed to be impossible (Wood, 1983). Whilst this idea is attractive it should not obscure the fact that other types of cytochrome with non-covalently bound haem are found in the periplasm. For example, one type of periplasmic reductase that converts nitrite to nitric oxide contains *d*-type as well as *c*-type haem (Table 3.2). As a second example, consider cytochrome b_{562} from *E. coli*, which is a water-soluble protein and has some structural homology (Weber *et al.*, 1981) with cytochrome *c'* of photosynthetic bacteria, a known periplasmic protein (Table 3.3). On this basis it would be tempting to postulate a periplasmic location for cytochrome b_{562} of *E. coli*, and, indeed, this has recently been confirmed (Nikkila, H., Gennis, R. B. and Sligar, S., unpublished work). This location now needs to be correlated with a function for the protein.

Tables 3.4 and 3.5 list enzymes that are believed to be integral membrane proteins catalysing either oxidation or reduction reactions at the periplasmic surface (i.e. proteins of type D shown in Fig. 3.1).

3.3 Rationales for periplasmic location

The extensive lists of periplasmic electron transfer proteins (Tables 3.1–3.3) suggest that it may be possible to discern some general rules to account for the location of so many activities in the periplasm. Oxidation and reduction reactions will now be discussed in turn.

Table 3.5 Integral membrane proteins thought to catalyse reduction reactions at the periplasmic surface of the cytoplasmic membrane

Reaction	Organism	Criterion	Reference
$8H^+ + 6e + NO_2^- \rightarrow NH_4^+ + 2H_2O$	C. sputorum	5	De Vries et al. (1982)

Detail of the numbered criterion is given in Table 3.1 and the text. The nitrite reductase of *D. desulfuricans* is sometimes regarded as membrane-bound (Steenkamp and Peck, 1981, c.f. Table 3.2). This serves to illustrate the problem in some instances of discriminating between proteins of types A and B (Fig. 3.1). Further work may produce additional examples of this type of reaction. It is possible, for example, that the membrane-bound nitric oxide reductase of denitrifying bacteria will prove to reduce its substrate at the periplasmic surface.

3.3.1   Periplasmic oxidation reactions

If an oxidation reaction releases protons as well as electrons then a periplasmic location for the enzyme can contribute to the generation of a pH gradient, acidic outside the cell, that can form part of the total proton electrochemical gradient. The membrane potential component of this gradient can be formed if the electrons released by the periplasmic oxidation are transferred across the membrane to a site of combination with protons and an oxidant on the cytoplasmic surface of the membrane (Fig. 3.2). Actually, the same result in thermodynamic terms is achieved if the electrons are not transported across the membrane but rather protons are moved outwards (Fig. 3.3).

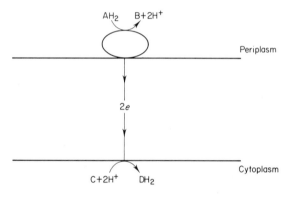

Fig. 3.2 Generation of a proton electrochemical gradient through linkage via an electron transfer system of a periplasmic site of substrate oxidation to a cytoplasmic site of reduction of an oxidant.

It is often very difficult to distinguish by experiment between these two possibilities. The release of protons in the oxidation reaction will very often play a quantitatively subsidiary role to the movement of charge, as shown in Figures 3.2 and 3.3, in the generation of a proton electrochemical gradient. Hence the oxidation of, for example, methanol (which releases protons) and iron II (which does not release protons) in the periplasm rather than in the cytoplasm can be rationalized on an equivalent basis. The schemes shown in Figures 3.2 and 3.3 have been appreciated for some time (e.g. Ingledew *et al.*, 1977; Alefounder and Ferguson, 1981) and have been particularly discussed by Hooper and DiSpirito (1985). The latter authors have suggested that oxidations of simple reductants and the concomitant release of protons, can always be expected to occur in the periplasm because of the consequent contribution to the generation of a proton electrochemical gradient (Figs 3.2

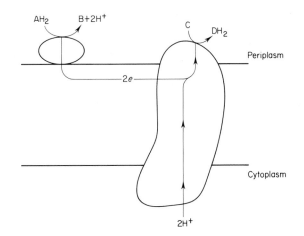

Fig. 3.3 Generation of a proton electrochemical gradient as electrons flow from a periplasmic dehydrogenase to a reductase which is an integral membrane protein that abstracts protons for the reduction specifically from the cytoplasm, but accepts its substrate, C, directly from the periplasm. The generation of the gradient would be enhanced if the reductase were to pump additional protons across the membrane.

and 3.3). With the present state of knowledge this is an uncertain generalization to make. There is, for example, evidence that the generalization does not apply to nitrite oxidation by *Nitrobacter* (Ferguson, 1982, 1987; Ballard and Ferguson, 1987; Wood, Chapter 4, this volume). It should also be appreciated that the location of some reductases in the periplasm can, together with a periplasmic site of oxidation, result in electron flow down a sizeable drop in redox potential not being coupled to generation of a protonmotive force (Fig. 3.4 and Section 3.2.2).

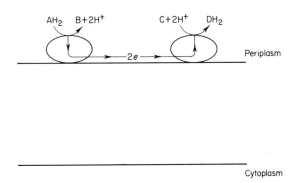

Fig. 3.4 Absence of proton electrochemical gradient generation when a periplasmic dehydrogenase is directly linked to a periplasmic reductase.

A periplasmic location for an enzyme involved in substrate oxidation has to be correlated with a pathway for electrons from the enzyme into the electron transfer chain. As discussed later, one of the few known routes for such transfer is via the *c*-type cytochromes. Employment of this route with relatively reducing substrates such as methanol means that available redox energy is not fully harnessed by the cell because the redox drop between methanol dehydrogenase and cytochrome *c* is not linked to the generation of a protonmotive force. Thus a periplasmic location is of no apparent advantage in this respect.

A periplasmic location for an enzyme means that the bacterium need not possess a transport system for the substrate. Energetic costs of active transport are thus avoided, but perhaps equally important is the exclusion from the cytoplasm of potentially toxic or insoluble substrates such as p-cresol (Hopper *et al.*, 1985), iron II, sulphur, sulphide and thiosulphate. Arguments based on avoidance of production of toxic compounds in the cytoplasm can also be applied as a rationale for periplasmic locations. This has been suggested as a reason why methanol is oxidized to the toxic formaldehyde in the periplasm (Alefounder and Ferguson, 1981). Although formaldehyde must subsequently enter the cytoplasm for oxidation or incorporation into cell carbon, one can suggest that the concentration in the cytoplasm is kept at a minimum level by relying on formaldehyde diffusion into the cytoplasm.

3.3.2   Periplasmic reductase reactions

In terms of generation of a proton electrochemical gradient (pmf), the converse of the argument used with respect to the oxidation enzymes in Section 3.3.1 applies to the periplasmic reductase reactions which are listed in Tables 3.2 and 3.5. A periplasmic location means that the necessary electrons must either originate from the periplasm (Fig. 3.4) or they must be moved from sites of substrate oxidation (e.g. NADH or succinate) at the cytoplasmic side of the membrane (Fig. 3.5). Such movement of negative charge is counter-productive in terms of pmf generation, as is consumption of protons in the periplasm. This factor appears even more acute when considering the case of nitrous oxide reduction which, at least in the case of *P. denitrificans*, accepts electrons from *c*-type cytochromes which have $E_{m,7}$ values in the range of 200–300 mV. The $E_{m,7}$ value for N_2O reduction is +1100 mV, so considerable energy appears to be lost by allowing electrons to flow within the periplasm from cytochrome *c* to nitrous oxide. On first consideration it would appear advantageous for nitrous oxide to be reduced on the inner surface of the plasma membrane so that the pathway of electron flow from cytochrome *c* to nitrous oxide would be inwards (cf. Fig. 3.2) and thus contribute to the generation of a membrane potential. However, such a view overlooks the possibility that the physiological concentration of nitrous oxide is likely to be very low but the concentration of its reduction product (nitrogen) is high. Thus the operating redox potential (E_h) for nitrous oxide reduction is likely to be very much lower than 1100 mV. Hence the thermodynamics of nitrous oxide reduction at the periplasmic surface may not be as wasteful as they appear on first consideration.

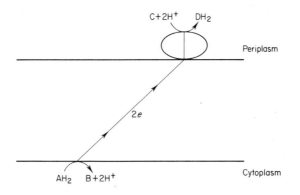

Fig. 3.5 Connection of a cytoplasmic site of dehydrogenase activity to a periplasmic reductase solely via a transmembrane electron carrying system opposes the formation of a normal proton electrochemical gradient across the bacterial plasma membrane.

Inspection of Figures 3.4 and 3.5 does not indicate how electron flow to a water-soluble periplasmic reductase might be linked to the generation of a proton electrochemical gradient. Unlike the integral membrane protein shown in Figure 3.3, a water-soluble reductase cannot abstract protons from the cytoplasm. Figures 3.6 and 3.7 reveal how the gradient can be generated. In the case when enzymes for both the oxidation and reduction reactions are in the periplasm, the gradient can develop if electron flow between the two enzymes is catalysed by proton-translocating integral membrane proteins. An example of such behaviour is the oxidation of hydrogen by nitrite in *Desulfovibrio desulfuricans* (Steenkamp and Peck, 1981). This should be contrasted with the oxidation of methanol by nitrite or nitrous oxide in *P. denitrificans*. Here, periplasmic *c*-type cytochromes are thought to act both as electron acceptors from methanol dehydrogenase and as donors to nitrite and nitrous oxide reductases, with the consequence that the electron flow pathway conforms to the pattern of Figure 3.4 and generation of a gradient cannot occur (Alefounder and Ferguson, 1981). Inspection of the redox potentials for the couples formaldehyde/methanol ($E_{m7} = -0.182$ V) and nitrous oxide/nitrogen ($E_{m7} = 1.1$ V) would lead one to suppose that electron flow from methanol to nitrous oxide would be associated with the generation of a proton electrochemical gradient and thereby the synthesis of ATP. Understanding of the organizational aspects of the system (Fig. 3.4) shows that inspection of redox potentials alone is an inadequate guide to cellular energetics.

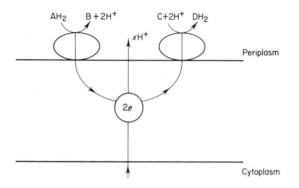

Fig. 3.6   Generation of a proton electrochemical gradient by integral membrane proteins that translocate protons and connect a periplasmic dehydrogenase to a periplasmic reductase.

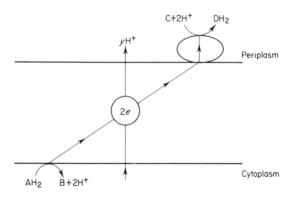

Fig. 3.7   Generation of a proton electrochemical gradient by integral membrane proteins that pump protons as they transfer electrons from the cytoplasmic to the periplasmic sides of the plasma membrane. The value of y must be greater than 2. It is widely considered to be 4 for the cytochrome bc_1 complex that functions according to the scheme shown in the diagram (see also Chapter 1).

Figure 3.7 deals with the situation when the electrons originate on the cytoplasmic side of the membrane and terminate on the periplasmic side. A specific example is electron transport from succinate to nitrous oxide in *P*.

denitrificans. In this instance the proton electrochemical gradient develops because the value of y is 4, and thus the transfer of two electrons from succinate to nitrous oxide is associated with the net movement of two positive charges out of the cell (Ferguson, 1986). The exact mechanism of this transfer is thought to involve the operation of a ubiquinone cycle (e.g. see page 68; Ferguson 1986). It should also be remembered that in the bacterial cell many of the electrons destined for reduction of nitrous oxide will originate from NADH and that net proton translocation out of the cell will also be effected by the NADH-ubiquinone oxidoreductase.

At present, the only example of uncompensated outward electron flow across the bacterial cytoplasmic membrane (Fig. 3.5) appears to be from nitrite to cytochrome c in *Nitrobacter*. As discussed elsewhere (Ferguson, 1982, 1987), this contributes to the driving of an otherwise energetically uphill reaction and so is not energetically wasteful for the cell, as would be the case if succinate to nitrous oxide (an energetically downhill reaction) were also to conform to Figure 3.5 rather than Figure 3.7.

With some substrates (e.g. nitrate), the operation of a periplasmic reductase obviates the requirement for a transport system. A curious feature of bacterial electron transport to emerge recently is that in some organisms the respiratory nitrate reductase is periplasmic, whereas in others it is an integral membrane protein with the active site facing the cytoplasm (Ferguson *et al.*, 1987). No explanation can be offered for this phenomenon but, as with the oxidation reactions, a further advantage of a periplasmic location for a reductase may be to exclude toxic reactants and products (e.g. nitrite and sulphur compounds) from the cytoplasm.

3.3.3   Full occupancy of sites on the cytoplasmic surfaces of the plasma membrane?

The F_1 part of the ATP synthase enzyme is known to project as an approximate 90 Å diameter sphere from the inner surface of the bacterial plasma membrane (Fig. 1.32). As this enzyme is a major component of the plasma membrane a substantial proportion of the inner surface will be taken up by the ATP synthase. When it is also considered that other enzymes, including nitrate reductase, cytochrome bc_1 and cytochrome aa_3, also project beyond the boundary of the bilayer on the cytoplasmic side it can be envisaged that there may be little room available for additional dehydrogenases or reductases. Thus a factor in the development of periplasmic electron transport proteins may have been avoidance of overcrowding of the inner surface of the plasma membrane.

3.4   Pathways of electron flow to and from periplasmic enzymes

If electron transfer to or from a periplasmic redox protein is to be coupled to the generation of a protonmotive force then the overall path of the electron must involve integral membrane proteins (Figs 3.2, 3.3, 3.6, 3.7). Thus those points at which electrons can pass into or out of the periplasm to or from membrane-bound electron carriers need to be identified. Knowledge in this respect is by no means complete. The best characterized system is *P. denitrificans*. Figure 3.8 shows how electrons are thought to enter and leave the periplasm via cytochrome *c*. This is assumed to be the same cytochrome *c* that also transfers electrons between the cytochrome bc_1 complex and cytochrome aa_3. It should nevertheless be stressed that this is an incomplete picture because other electron transport proteins of the periplasm in this organism have not been fully characterized. There are certainly other components that mediate between cytochrome *c* and the methanol and methylamine dehydrogenases (see later). Whilst it has been shown that cytochrome c_{550} can donate electrons to the purified cytochrome *cd* type nitrite reductase, this has not been directly demonstrated for the nitrous oxide reductase. There is also azurin, which is presumably periplasmic, in anaerobically grown *P. denitrificans*, but there is no clear evidence as to its physiological role (Martinkus *et al.*, 1980).

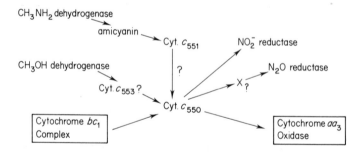

Fig. 3.8   Entry and exit of electrons to/from periplasmic enzymes in *Paracoccus denitrificans*. The boxes indicate integral membrane proteins. As explained in the text, the pathways of electron flow to and from the periplasm have not been determined with certainty; hence the query marks.

A role for *c*-type cytochrome in mediating transfer of electrons to or from the cytoplasmic membrane and into or out of the periplasm is also found in other bacteria, e.g. methylotrophs (Beardmore-Gray and Anthony, 1984; Lawton and Anthony, 1985a,b; Chapter 6, this volume). Here again more than one *c*-type cytochrome as well as a copper protein can be involved in the electron transport processes within the periplasm (see page 174 and Chapter 6). Other examples can be found in Chapter 4 by Wood.

Not all electron transport between the membrane and the periplasm is mediated by c-type cytochromes as shown by the scheme in Figure 3.8. In at least certain strains of the two photosynthetic bacteria, *Rhodobacter capsulatus* and *Rhodobacter sphaeroides*, there is evidence that electron flow to the periplasmic reductases for nitrate, DMSO/TMAO and nitrous oxide does not involve the cytochrome bc_1 complex (Ferguson *et al.*, 1987). In the case of N_2O reduction there is evidence that a periplasmic cytochrome c' plays a role in the electron transport process (McEwan *et al.*, 1985b). But how electrons might pass to this cytochrome from ubiquinol is at present a mystery.

Two other examples of electron flow from periplasmic oxidation reactions to the membrane-bound electron transport system warrant discussion. The first of these is the oxidation of hydroxylamine (see page 199, Chapter 4, this volume). In this instance it has been argued that electrons pass via a periplasmic cytochrome c_{554} from a hydroxylamine dehydrogenase to the ubiquinone of the respiratory chain. However, that this postulated process takes place is not certain although, energetically, there is much to be gained by feeding the electrons into the chain at this point rather than via the cytochrome c that is the donor to the oxidase cytochrome aa_3. Similar thinking suggests that electrons from the active site of glucose dehydrogenase, which faces the periplasm (Table 3.4), would be transferred to the chain nearer to ubiquinone than to cytochrome c. Indeed, there is evidence that electron flow is via ubiquinone in *Acinetobacter calcoaceticus* (Beardmore-Gray and Anthony, 1986; page 311, Chapter 6, this volume). It is possible that in this organism the glucose dehydrogenase can be coupled to a periplasmic water-soluble b-type cytochrome ($M_r = 18\,000$) which would be the electron donor to ubiquinone (Duine *et al.*, 1987). Confirmation of this point is required.

The glucose dehydrogenase enzyme has a puzzling feature; it is synthesized by some bacteria, e.g. *E. coli*, as an inactive apoprotein. Only when PQQ, its prosthetic group, is added can activity be observed. This enzyme is evidently able to connect to the electron transport chain of *E. coli* because Van Schie *et al.* (1985) have been able to link the oxidation of glucose by oxygen to the generation of a proton electrochemical gradient in vesicles. In itself this work does not indicate the pathway of electron flow but more information has recently been obtained by Matsushita *et al.* (1987), who have reconstituted glucose dehydrogenase and *E. coli* o-type cytochrome oxidase into a functional system. The latter results suggest that electrons can be fed from the glucose dehydrogenase to ubiquinone with the resulting ubiquinol acting as a donor to cytochrome o.

Many bacteria do not contain either the *c*-type cytochrome or the antimycin-sensitive cytochrome bc_1 complex that provide the pathway for transferring electrons from ubiquinol to cytochrome *c* (Chapter 1). Thus in *E. coli*, for instance, there is a route from ubiquinol to the periplasmic nitrite reductase that is independent of this pathway (Motteram *et al.*, 1981). Nothing is known about the nature of such pathways but it seems improbable that quinols directly reduce periplasmic proteins, because it is generally thought that the head groups of quinones are confined to the interior of the membrane phospholipid bilayer. One must postulate additional membrane-bound electron carriers, with the redox active group near the surface of the membrane, that transfer the electrons to the periplasm.

3.5 Mobility and organization of electron transfer proteins within the periplasm: mini-electron transfer chains?

Knowledge of electron transfer processes within the periplasm is incomplete. At the simplest level it is not known if the water-soluble periplasmic proteins undergo independent diffusion within the periplasm. Two gross features of the periplasm need to be considered here.

The first is the contention of Cook *et al.* (1986) that the periplasm is not a continuous aqueous compartment. The effect of exposing cells to hypertonic sucrose solutions is the formation of bays of swollen periplasm rather than a continuum. These observations, it is argued, mean that between the cytoplasmic membrane and the outer layer there are points of attachment that constitute a barrier to the free flow of molecules throughout the periplasm. These attachment points are suggested to be related to the cell division processes. Even if such points of attachment are common features of Gram-negative bacteria throughout their various stages of growth, the consequences for electron transport may not be very significant because the number of discrete zones within the periplasm would be quite small.

The second relevant aspect of the periplasm is the contention mentioned earlier that its proteins do not diffuse as if in dilute aqueous medium. Brass *et al.* (1986) have developed a technique for introducing fluorescently labelled proteins into the periplasm of *E. coli*. One such protein was mitochondrial cytochrome *c* and its mobility in the periplasm was estimated from the rate at which fluorescence returned to a small region of the periplasm of a tethered cell, after this region had been bleached with a laser. In the original publication (Brass *et al.*, 1986) measurements of this kind indicated a diffusion coefficient of $5 \times 10^{-10} \, \text{cm}^2 \, \text{s}^{-1}$. However, further refinement of the method now suggests a value of approximately $5 \times 10^{-9} \, \text{cm}^2 \, \text{s}^{-1}$ (J. Brass, pers. comm.). This value is about 100-fold lower than would be expected for diffusion in aqueous solution and tenfold less than estimates of diffusion in the bacterial cytoplasm. An important point of comparison is with estimates that have been made with similar methods of diffusion kinetics for mitochondrial electron transfer proteins (Hackenbrock *et al.*, 1985). In the latter system, a diffusion coefficient for cytochrome *c* of $1.9 \times 10^{-9}\,\text{cm}^2\,\text{s}^{-1}$ is estimated at a reasonable ionic strength, whereas the much larger integral membrane protein complexes, NADH-ubiquinone oxidoreductase, ubiquinol-cytochrome *c* oxidoreductase and cytochrome oxidase have coefficients of $4 \times 10^{-10}\,\text{cm}^2\,\text{s}^{-1}$ (Hackenbrock *et al.*, 1985). These data are consistent with a model in which electron transfer from ubiquinol to oxygen occurs as a result of random collisions between ubiquinol-cytochrome *c* oxido-reductase and cytochrome *c* together with similar collisions between the latter and cytochrome oxidase. Thus, a widely held contemporary view is that mitochondrial electron transport is catalysed by a set of independent protein systems. An additional mobile component is the ubiquinone/ubiquinol redox pair which acts between NADH-ubiquinone oxidoreductase and ubiquinol cytochrome *c* oxidoreductase. There is no compelling evidence in favour of an alternative model in which a set of electron transport components would be permanently associated so as to provide a 'wire-like' structure for the flow of electrons from ubiquinol to oxygen.

Appreciation of the current model for mitochondrial electron transport is important for two reasons in the consideration of the properties of periplasmic electron transfer systems. First, these latter systems will usually, if not always, interact with membrane-bound electron transfer components, as discussed earlier. Such components may have similar diffusional characteristics as their mitochondrial counterparts. Secondly, how electron transport proteins are organized within the periplasm needs to be considered. Specifically, are all such proteins independently and randomly dispersed throughout the periplasm or do, for example, a pair of cytochromes and a terminal reductase form a permanently associated complex that acts as a 'wire-like' mini-electron transport chain? Information on this subject is sparse, but we can suggest by analogy with the mitochondrial system that the estimated diffusion coefficient for a protein of similar size to cytochrome *c* in the periplasm is likely to be adequate to allow electron transfer by collisions between the diffusing electron transfer proteins. This question is best discussed with reference to proteins involved in oxidation of methanol or methylamine, for which more relevant measurements have been made than for other periplasmic systems.

An organism known only as 4025 (similar to *Methylophilus methylotrophus*) grows aerobically on methanol or methylamine. Growth on the latter compound, but not the former, resulted in synthesis of a periplasmic copper protein, amicyanin, that was first described for *Pseudomonas* AM1 (Lawton and Anthony, 1985a,b; Tobari and Harada, 1981). In organism 4025, amicyanin was not the only periplasmic electron transfer protein; there was also a second copper protein of the azurin type together with two *c*-type cytochromes known according to their respective high and low pI values as cytochrome c_H and cytochrome c_L (Table 3.3) (Anthony, 1986; pages 303–305, Chapter 6, this volume). All four redox proteins were tested for their ability to act as electron acceptors from methylamine dehydrogenase. Only amicyanin functioned as an electron acceptor. Michaelis-Menten kinetics were followed, with a K_m value of 20 µM for amicyanin. Very similar results had been obtained earlier for *Pseudomonas* AM1. Lawton and Anthony (1985a) went on to consider whether treating the methylamine dehydrogenase–amicyanin interaction as an enzyme–substrate interaction would be consistent with reasonable rates of electron transport in the periplasm. From knowledge that methylamine dehydrogenase and amicyanin constitute 5 and 1.6% respectively of the total cell protein, together with an estimate of the periplasmic volume, they calculated the periplasmic concentrations of methylamine dehydrogenase and amicyanin to be 180 µM and 580 µM, respectively. As the K_m for amicyanin had been estimated to be 20 µM, Lawton and Anthony (1985a) concluded that methylamine dehydrogenase would be able to operate at its maximum rate. Such a conclusion is based on the supposition that binding of amicyanin in its oxidized form to the dehydrogenase is followed by its dissociation in the reduced form, and that diffusion in the periplasm does not restrict the rate of productive collisions between the dehydrogenase and amicyanin.

It can be argued that these suppositions may be justified because Lawton and Anthony (1985a) converted their measurement of the specific rate of reduction of amicyanin by the dehydrogenase, obtained from *in vitro* experiments, into an estimated rate of methylamine oxidation by the cells. To do this they made a reasonable assumption about the fraction of the total cell mass that is water-soluble protein and combined this with the estimates of the cell content of the dehydrogenase and amicyanin (see above). The estimated rate so obtained was similar (within a factor of 3) to the measured rate of respiration by cells supplied with methylamine. The success of this comparison suggests not only that the reaction between amicyanin and methylamine dehydrogenase is of physiological importance but also that amicyanin could diffuse within the periplasm independently of the dehydrogenase.

Although amicyanin was the only protein that would accept electrons from methylamine dehydrogenase, no such specificity was observed in tests of proteins that could oxidize amicyanin; both *c*-type cytochromes as well as azurin were equally effective in this respect within the kinetic resolution available from the methods used. There was also rapid electron transfer between the cytochromes and azurin. These observations therefore do not provide any information about the onward route of electrons from amicyanin to the membrane-bound cytochrome oxidase, which in this organism is of the *o*-type.

Studies on the methanol dehydrogenase system from *M. methylotrophus* led to a slightly different conclusion from that for the study of methylamine oxidation (see pages 299–301, Chapter 6, this volume). Of the cytochromes tested, cytochrome c_L appeared to be specific as an electron acceptor from methanol dehydrogenase (Anthony, 1986). The K_m value for cytochrome c_L was 1.2 μM. In contrast to the methylamine dehydrogenase system in organism 4025, the rate of electron transfer from methanol dehydrogenase to cytochrome c_L *in vitro* was estimated to be only 1% of the rate in whole bacteria. Consequently, it was argued that the low rate *in vitro* arose because the dehydrogenase and cytochrome c_L are each present in the assay at low concentration. In the periplasm, it is estimated that both these proteins are present at 0.5 mM. This would mean that methanol dehydrogenase could always be associated with cytochrome c_L, possibly as a complex covering most of the outer surface of the bacterial cytoplasmic membrane (Anthony, 1986; Beardmore-Gray and Anthony, 1984). Thus the change in environment on going from the periplasm to the *in vitro* conditions might account for the large decrease in activity. If this explanation is valid it is difficult to understand why a similar decrease was not observed for the methylamine system. Possibly, there is an additional protein factor associated with methanol dehydrogenase in the periplasm which is absent from the *in vitro* experiments. At present, the contrast between the results with methanol and those with methylamine can also be taken as indicative of uncertainty over the physical behaviour of proteins in the periplasm.

Investigations of *P. denitrificans* have led to a similar conclusion about the pathway of electron flow from methylamine dehydrogenase to amicyanin, but with the difference that study of several water-soluble *c*-type cytochromes from this organism has indicated that a cytochrome c_{551} (probably analogous to cytochrome c_H mentioned earlier) is the kinetically most competent electron acceptor from amicyanin (Husain and Davidson, 1986). There is, however, one inconsistency in this scheme: cytochrome c_{551} does not oxidize reduced amicyanin when these two electron carriers are mixed, in agreement with the relative redox potentials of the two species (Gray *et al.*, 1986). The latter authors therefore suggest that the redox potential of amicyanin may decrease when the protein forms a complex with methylamine dehydrogenase. This explanation implies, in turn, that electrons would be passed to cytochrome c_{551} from the dehydrogenase–amicyanin complex, contrary to the independent collisional model discussed earlier for organism

4025. Further studies are evidently needed in order to distinguish between the collisional and permanently associated models for electron transfer between periplasmic proteins.

Returning to *P. denitrificans*, it is tempting to suggest that cytochrome c_{553} (analogous to cytochrome c_L mentioned earlier) could be the electron acceptor from methanol dehydrogenase, and that cytochromes c_{551} and c_{553} might both donate electrons to cytochrome c_{550}, which in turn acts as the reductant for the oxidase cytochrome aa_3. A difficulty in seeking to confirm or deny this scheme is that methanol dehydrogenase exhibits only sluggish activity when assayed *in vitro* (see earlier and Chapter 6, this volume, by Anthony).

3.6   Are there electron transport proteins in Gram-positive organisms that have equivalent functions to the periplasmic enzymes of Gram-negative organisms?

There is, by definition, no periplasm in Gram-positive bacteria. It is often assumed that proteins translocated across the plasma membranes of such organisms must also be excreted, as the cell wall would not retain them. Thus, as *c*-type cytochromes are predominantly, if not exclusively, located in the periplasms of Gram-negative organisms, it might be anticipated that water-soluble *c*-type cytochromes would be absent from Gram-positive bacteria. Yet there are a number of instances of *c*-type cytochromes having been described for Gram-positive organisms. For example, Miki and Oku-nuki (1969) purified two water-soluble *c*-type cytochromes, cyt *c*-550 and cyt *c*-554, from a strain of *Bacillus subtilis*. The former was suggested to have similarities with mitochondrial cytochrome *c* and, therefore, presumably by analogy also with cytochrome c_{550} of *P. denitrificans* and cytochrome c_2 of the *Rhodospirillacae*. Accordingly, it is suggested to be the electron donor to the aa_3-type cytochrome oxidase now identified in *B. subtilis* (de Vrij *et al.*, 1987). However, a feature of cytochrome c_{550} from *B. subtilis* is that it is not very readily released from cytoplasmic membranes (de Vrij *et al.*, 1987). This observation, along with an earlier similar finding by Vernon and Magnum (1960), suggests that this type of cytochrome *c* could be firmly attached to the external surface of the cytoplasmic membrane so as to guard against its loss by diffusion through the relatively permeable cell wall. In this context one should note the recently demonstrated occurrence of fatty acid covalently attached by a thioether bond to the *N*-terminus of a *c*-type cytochrome in the Gram-negative organism *Rhodopseudomonas viridis* (Weyer *et al.*, 1987). A post-translational modification of this type could conceivably contribute to the binding of *c*-type cytochromes to the outer surface of the cytoplasmic membrane in Gram-positive organisms. A location for c-type cytochrome on the external surface of the cytoplasmic membrane of Gram-positive organisms is indicated by the findings of Jacobs *et al.*, (1979) who were able, with protoplasts of *Mycobacterium phlei*, both to remove a cytochrome c with 0.15 M KCl and to label it with a membrane-impermeant chemical label.

Three water-soluble c-type cytochromes can be released from *Bacillus licheniformis*, c_{551}, c_{552} and c_{554} (Woolley, 1987). The role of these is unknown and it was not possible with confidence to group either c_{551} or c_{552} with mitochondrial cytochrome c, in contrast to the suggestion for *B. subtilis*. Two points of comparison between c-type cytochromes from *B. subtilis* and those from the Gram-negative *Rhodospirillacae* are worth noting. First, the amount of c-type cytochrome produced by the *Bacillus* was considerably less than that typically produced by the latter group of organisms. Whilst this may relate to the different metabolic activities of the two groups of organisms, it could conceivably also relate to a requirement in the Gram-positives that such cytochromes make good contact with the cytoplasmic membrane and, therefore, are limited in amount. In Gram-negative organisms the cytochromes can be dispersed throughout the periplasm and need not be immediately juxtaposed to the membrane. The second point of comparison is that the cytochromes of *B. licheniformis*, unlike those of *Rhodospirillacae*, were released from the membrane only by lysozyme/protease digestion, suggesting perhaps that the *Bacillus* cytochromes were not free to diffuse in an aqueous zone of the cell.

The above discussion suggests that proteins with not only electron transport, but also other, functions may be associated with the outer surface of the plasma membrane of Gram-positive organisms, and retained for the cell by the peptidoglycan layer and/or non-covalent protein–protein or protein–lipid interactions. In the case of larger molecules, the peptidoglycan layer may present a sufficient retaining layer. This factor may be suggested to account for the location of a nucleoside diphosphate sugar hydrolase ($M_r = 137\,000$) outside the plasma membrane but inside the cell wall of *B. subtilis* (Mauk and Glaser, 1970).

Earlier in this chapter two important questions concerning the organization and function of the periplasm were discussed:

(a) are there more or less permanently associated sets of electron transfer proteins that form mini electron transfer chains?

(b) do some periplasmic electron transfer proteins play a key role in acting to transfer electrons rapidly from one large membrane-bound complex to another?

The implications for Gram-positive organisms of these points are now considered. In connection with (a), an important point is whether such organization could occur in Gram-positive organisms. Indeed, DiSpirito and Hooper (1985) have argued that the restriction of chemolithotrophy to Gram-negative organisms may be a reflection of the need of the volume of the region provided by the periplasm for accommodation of the essential proteins; a region of equivalent volume may not exist in Gram-positive organisms. This argument ought to be generally applicable; therefore, other metabolic functions that occur in the periplasm ought to be absent from Gram-positive organisms. This appears to be the case in that periplasmic sugar-binding proteins have not been reported for Gram-positive organisms. It has also been noted that Gram-positive organisms may utilize methanol via a cytoplasmic NAD-dependent methanol dehydrogenase (Duine *et al.*, 1987; page 301, Chapter 6, this volume), whereas, as discussed earlier, Gram-negative organisms use the periplasmic, PQQ-dependent dehydrogenase (page 298, Chapter 6). As discussed in this chapter several of the electron transfer reactions of denitrification occur in the periplasm and, therefore, it might be predicted that Gram-positive organisms would be incapable of denitrification. In this context, it is interesting that *Bacillus licheniformis*, long regarded as a prototype denitrifier, is now argued to be incapable of denitrification (Pichinoty *et al.*, 1978).

As discussed earlier, the importance in electron transfer of relatively rapid motion of low molecular weight water-soluble electron transport proteins, either within the periplasm or over the periplasmic surface of the cytoplasmic membrane, is not clear. Accordingly, relatively little can be said about whether such motion of these proteins is a prerequisite for electron transfer and therefore operational in Gram-positive bacteria (Ferguson, 1987). The recognition of c-type cytochromes in *Bacillus licheniformis* (Woolley, 1987) means that the possibility of direct electron transfer between b-type cytochrome and cytochrome aa_3 discussed previously for this organism (Ferguson, 1987) should be reassessed. In this context it may be significant that cytochrome aa_3 isolated from the thermophilic organism PS3 (probably *Bacillus stearothermophilus*) and *Bacillus firmus RAB* has a tightly attached c-type haem on one of the constituent polypeptide chains (Ludwig, 1987). Whilst it is tempting to correlate this feature with the requirement that c-type cytochromes in Gram-positive organisms make good contacts with the cytoplasmic membrane, it must also be recognized that a similar organization is found for the oxidase from the Gram-negative organism *Thermus thermophilus* which, in common with the other two organisms mentioned, is an extremophile (see Chapter 5, this volume).

It is difficult at present to make a definite conclusion about the possible role of 'periplasmic' electron transport proteins in Gram-positive bacteria. The demonstration of c-type cytochromes, which are in some instances known to be located at the outer surface of the cytoplasmic membrane, amongst the Gram-positive bacteria means that consideration must at least be given to one of these proteins having similar functions to their counterparts in the periplasms of Gram-negative organisms. Indeed, if as mentioned in the Introduction and illustrated in Figure 3.1, the periplasm is filled with hydrated peptidoglycan, then the environments of peripheral membrane proteins on the outer surface of the cytoplasmic membrane may be similar in the Gram-positive and negative organisms. Whether chemolithotropy and other features of bacterial physiology that are associated with the periplasm in Gram-negative organisms are either absent from Gram-positive organisms or involve different pathways (e.g. oxidation of methanol, see above) remains to be finally determined.

3.7 A doubt over the accepted role for cytochrome c_2 as a periplasmic electron shuttle in *Rhodospirillacae*

As explained by Jackson in Chapter 7, energy transduction in the illuminated *Rhodospirillacae* depends on cyclic electron flow from and to a reaction centre via the ubiquinol-cytochrome c oxidoreductase. It has been thought that electrons are delivered from the latter to the former by cytochrome c_2, which would be relatively free to diffuse on the periplasmic surface of the membrane. However, as discussed by Ferguson (1987) recent experiments with a mutant of *Rb. capsulatus* that lacks specifically cytochrome c_2 show that electron transfer from the reductase to the reaction centre can continue relatively unimpaired. A similar situation may apply in *Pseudomonas* AM1 (now *Methylobacterium*). Mutants of this organism lacking cytochrome c have been obtained. Their cytoplasmic membranes are able to oxidize NADH or succinate using the same oxidase (cytochrome aa_3) as the wild type. The mutants are, however, unable to oxidize methanol, thus suggesting that c-type cytochromes are only necessary for connecting the periplasmic methanol dehydrogenase to the membrane-bound electron carriers (Anthony, 1982). This leads to the conclusion that bacterial electron transport between large membrane-bound complexes can occur at required rates without any requirement for an intermediate role of a periplasmic electron transfer protein. This returns us to the question of Gram-positive organisms: the results with the mutant of *Rb. capsulatus* and *Pseudomonas* AM1 suggest that lack of a periplasm may not be an impediment to the electron transfer process.

3.8 Conclusion

The periplasm is the location of a large number of electron transport reactions that are found amongst the bacteria. This has been realized only relatively recently. While many further examples will doubtless be identified, an important goal of future research will be identification of the electron transport pathways to and from the periplasm as well as elucidation of electron transport sequences within the periplasm. Such studies will correlate with continued investigation of the physical state of the periplasm. The possible role of a 'periplasm-like' region of Gram-positive organisms also requires to be clarified.

References

Alefounder, P. R. and Ferguson, S. J. (1980). *Biochem. J.* **192**, 231–240.
Alefounder, P. R. and Ferguson, S. J. (1981). *Biochem. Biophys. Res. Commun.* **98**, 778–784.
Alefounder, P. R., Greenfield, A. J., McCarthy, J. E. G. and Ferguson, S. J. (1983). *Biochim. Biophys. Acta* **724**, 20–39.
Anthony, C. (1982). 'The Biochemistry of Methylotrophs'. Academic Press, London.
Anthony, C. (1986). *Adv. Microbiol. Phys.* **27**, 113–210.
Badziong, W. and Thauer, R. K. (1980). *Arch. Microbiol.* **125**, 167–174.
Ballard, A. L. and Ferguson, S. J. (1987). *Biochem. Soc. Trans.* **15**, 937–938.
Bater, A. J. and Venables, W. A. (1977). *Biochim. Biophys. Acta* **468**, 209–226.
Beacham, I. R. (1979). *Int. J. Biochem.* **10**, 877–883.
Beardmore-Gray, M. and Anthony, C. (1984). *In* 'Microbial Growth on C₁ Compounds' (Eds. R. L. Crawford and R. S. Hanson), pp. 97–105, American Society for Microbiology, Washington DC.
Beardmore-Gay, M. and Anthony, C. (1986). *J. Gen. Microbiol.* **132**, 1257–1268.
Bell, G. R., LeGall, J. and Peck, H. D. (1974). *J. Bacteriol.* **120**, 994–997.
Blaurock, A. E., Stoeckenius, W., Oesterhelt, D. and Scherphof, G. L. (1976). *J. Cell. Biol.* **71**, 1–22.
Boogerd, F. C., Van Verseveld, H. W. and Stouthamer, A. H. (1981). *Biochim. Biophys. Acta* **638**, 181–191.
Brass, J. M., Higgins, C. F., Foley, M., Rugman, P. A., Birmingham, J. and Garland, P. B. (1986). *J. Bacteriol.* **165**, 787–794.
Burton, S. M., Byrom, D., Carver, M., Jones, G. D. D. and Jones, C. W. (1983). *FEMS Microbiol. Lett.* **17**, 185–190.
Brune, D. C. and Truper, H. G. (1986). *Arch. Microbiol.* **145**, 295–301.
Canters, G. W. (1986). *FEBS Lett.* **212**, 168–172.
Cole, J. A. (1968). *Biochim. Biophys. Acta* **162**, 356–368.
Cook, W. R., Macallister, T. J. and Rothfield, L. I. (1986). *J. Bacteriol.* **168**, 1430–1438.
De Vries, W., van Berchum, H. and Stouthamer, A. (1984). *Antonie van Leeuwenhoek. J. Microbiol.* **50**, 63–73.
De Vries, W., Niekus, H. G. D., van Berchum, H. and Stouthamer, A. H. (1984). *Arch. Microbiol.* **131**, 132–139.
De Vrij, W., van den Burg, B. and Konings, W. N. (1987). *Eur. J. Biochem.* **166**, 589–595.

DiSpirito, A. A., Taaffe, L. R. and Hooper, A. B. (1985). *Biochim. Biophys. Acta* **806**, 320–330.
Dokter, P., van Kleef, M. A. G., Frank, Jzn. J. and Duine, J. A. (1985). *Enzyme Microb. Technol.* **7**, 613–616.
Dorn, M., Anderseen, J. R. and Gottschalk, G. (1978). *Arch. Microbiol.* **119**, 7–11.
Duine, J. A., Frank, Jzn, J. and Jongejan, J. A. (1987). *Advances in Enzymology* **59**, 169–212. Interscience, J. Wiley, New York.
Ferguson, S. J. (1982). *Biochem. Soc. Trans.* **10**, 198–200.
Ferguson, S. J. (1986). *Trends in Biochem. Sci.* **11**, 351–353.
Ferguson, S. J. (1987). *Trends in Biochem. Sci.* **12**, 124–125.
Ferguson, S. J., Jackson, J. B. and McEwan, A. G. (1987). *FEMS Microbiol. Rev.* **46**, 117–143.
Fujita, T. and Sato, R. (1966). *J. Biochem.* **60**, 568–577.
Gray, K. A., Knaff, D. B., Husain, M. and Davidson, V. L. (1986). *FEBS Lett.* **207**, 239–242.
Guerlesquin, F., Bovier-Lapierre, G. and Bruschi, M. (1982). *Biochem. Biophys. Res. Commun.* **105**, 530–538.
Hackenbrock, C. R., Gupte, S. S. and Chazotte, B. (1985). *In* 'Achievements and Perspectives of Mitochondrial Research' (Eds. E. Quagliariello, E. C. Slater, F. Palmieri, C. Saccone and A. M. Kroon), pp. 83–101. Elsevier, Amsterdam.
Heppel, L. A. (1971). *In* 'Structure and Function of Biological Membranes' (Ed. L. I. Rothfield), pp. 222–247. Academic Press, New York.
Hobot, J. A., Carleman, E., Villiger, W. and Kellenberg, E. (1984). *J. Bact.* **160**, 143–152.
Hooper, A. B. and DiSpirito, A. A. (1985). *Microbiol. Rev.* **49**, 140–157.
Hopper, D. J., Jones, M. R. and Causer, M. J. (1985). *FEBS Lett.* **182**, 485–488.
Husain, M. and Davidson, V. L. (1985). *J. Biol. Chem.* **260**, 14626–14629.
Husain, M. and Davidson, V. L. (1986). *J. Biol. Chem.* **261**, 8577–8580.
Ingledew, W. J., Cox, J. C. and Halling, P. J. (1977). *FEMS Microbiol. Lett.* **2**, 193–197.
Jacobs, A. J., Kalra, V. K., Cavari, B. and Brodie, A. F. (1979). *Arch. Biochem. Biophys.* **194**, 531–541.
Kajie, S. and Anakaru, Y. (1986). *Eur. J. Biochem.* **154**, 457–463.
Kasprzak, A. A. and Steenkamp, D. J. (1983). *J. Bacteriol.* **156**, 348–353.
Knowles, C. J., Calcott, P. H. and Macleod, R. A. (1974). *FEBS Lett.* **49**, 78–83.
Kröger, A., Dorrer, E. and Winkler, E. (1980). *Biochim. Biophys. Acta* **589**, 118–136.
Lawton, S. A. and Anthony, C. (1985a). *Biochem. J.* **228**, 719–726.
Lawton, S. A. and Anthony, C. (1985b). *J. Gen. Microbiol.* **131**, 2165–2171.
LeGall, J. and Peck, H. D. Jr. (1987). *FEMS Microbiol. Rev.* **46**, 35–40.
Liu, M.-Y., Liu, M.-C., Payne, W. J. and LeGall, J. (1986). *J. Bacteriol.* **166**, 604–608.
Liu, M.-C. and Peck, H. D. (1981). *J. Biol. Chem.* **256**, 13159–13164.
Lorence, R. M., Yoshida, T., Findling, K. L. and Fee, J. A. (1981). *Biochem. Biophys. Res. Comm.* **99**, 591–599.
Ludwig, B. (1987). *FEMS Microbiol. Rev.* **46**, 41–56.
Martinkus, K., Kernelly, P. J., Rea, T. and Timkovich, R. (1980). *Arch. Biochem. Biophys.* **199**, 465–472.
Matsushita, K., Nonobe, M., Shinagawa, E., Adachi, O. and Ameyama, M. (1987). *J. Bacteriol.* **169**, 205–209.

Mauck, J. and Glaser, L. (1970). *Biochemistry* **9**, 1140–1147.
McEwan, A. G., Jackson, J. B. and Ferguson, S. J. (1984). *Arch. Microbiol.* **137**, 344–349.
McEwan, A. G., Greenfield, A. J., Wetzstein, H. G., Jackson, J. B. and Ferguson, S. J. (1985a). *J. Bacteriol.* **164**, 823–830.
McEwan, A. G., Wetzstein, H. G., Ferguson, S. J. and Jackson, J. B. (1985b). *Biochim. Biophys. Acta* **806**, 410–417.
Meijer, E. M., Van der Zwaan, J. W. and Stouthamer, A. H. (1979). *FEMS Microbiol. Lett.* **5**, 369–372.
Midgley, M. and Dawes, E. A. (1973). *Biochem. J.* **132**, 141–154.
Miki, K. and Okunuki, K. (1969). *Biochem.* **66**, 831–843.
Motteram, P. A. S., McCarthy, J. E. G., Ferguson, S. J., Jackson, J. B. and Cole, J. A. (1981). *FEMS Microbiol. Lett.* **12**, 317–320.
Odom, J. S. and Peck, H. D. Jr. (1984). *Ann. Rev. Microbiol.* **38**, 551–592.
Olson, T. C. and Hooper, A. B. (1983). *FEMS Microbiol. Lett.* **19**, 47–50.
Pankhania, I. P., Gow, L. A. and Hamilton, W. A. (1986). *FEMS Microbiol. Lett.* **35**, 1–4.
Pettigrew, G. W. and Moore, G. R. (1987). 'Cytochrome *c*: Biological Aspects', p. 160. Springer Verlag, Berlin.
Pichinoty, F., Garcia, J.-L., Job, C. and Durand, M. (1978). *Can. J. Microbiol.* **24**, 45–49.
Pickrill, B. C., Czechowski, M. H., Przybyla, A. E., Peck, H. D. Jr. and LeGall, J. (1986). *J. Bacteriol.* **167**, 722–725.
Prince, R. C., Baccarini-Melandri, A., Hauska, G., Melandri, B. A. and Crofts, A. R. (1975). *Biochimica Biophys. Acta.* **387**, 212–227.
Probst, I. and Schlegel, H. G. (1976). *Biochem. Biophys. Acta* **440**, 412–428.
Quilter, J. A. and Jones, C. W. (1984). *FEBS Lett.* **174**, 167–172.
Saraste, M. and Kuronen, T. (1978). *Biochim. Biophys. Acta* **513**, 117–131.
Sawada, E. and Satoh, T. (1980). *Plant and Cell Physiol.* **21**, 205–210.
Scholes, P. B., McLain, G. and Smith, L. (1971). *Biochemistry* **10**, 2072–2075.
Steenkamp, D. J. and Peck, H. D. (1981). *J. Biol. Chem.* **256**, 5450–5458.
Stock, J. B., Rauch, B. and Roseman, S. (1977). *J. Biol. Chem.* **252**, 7850–7861.
Tobari, J. and Harada, Y. (1981). *Biochem. Biophys. Res. Commun.* **101**, 502–508.
Urata, K., Shimada, K. and Satoh, T. (1982). *Plant and Cell Physiol.* **23**, 1121–1124.
Van der Western, H. M., Mayhew, S. G. and Veeger, C. (1978). *FEBS Lett.* **86**, 122–126.
Van Gemerden, H. (1984). *Arch. Microbiol.* **139**, 289–294.
Van Schie, B. J., Hellingwerf, K. J., van Dijken, J. P., Elferink, M. G. L., van Dijl, J. M., Kuenen, J. G. and Konings, W. M. (1985). *J. Bacteriol.* **163**, 493–499.
Vernon, L. P. and Mangum, J. H. (1960). *Arch. Biochem. Biophys.* **90**, 103–104.
Voordouw, G. and Brenner, S. (1986). *Eur. J. Biochem.* **159**, 347–351.
Weber, P. C., Salemme, F. R., Matthews, F. S. and Bethge, P. H. (1981). *J. Biol. Chem.* **256**, 7702–7704.
Weyer, K. A., Schäfer, W., Lottspeich, F. and Michel, H. (1987). *Biochemistry* **26**, 2909–2914.
Wood, P. M. (1978). *FEBS Lett.* **92**, 214–218.
Wood, P. M. (1983). *FEBS Lett.* **164**, 223–226.
Woolley, K. J. (1987). *Arch. Biochem. Biophys.* **254**, 376–379.
Wu, W.-P. (1986). *FEMS Microbiol. Lett.* **34**, 313–317.

4 Chemolithotrophy

P. M. Wood

4.1	**Introduction**	**184**
4.1.1	What is chemolithotrophy?	184
4.1.2	Reviews and sources of chemical data	185
4.1.3	The inorganic elements	186
4.1.4	Acid production and metal corrosion	188
4.2	**Biochemical features common to chemolithotrophs**	**189**
4.2.1	The underlying conventional electron transport chain	189
4.2.2	NADH synthesis by reversed proton flow	191
4.2.3	Generalizations concerning enzyme location	192
4.2.4	Why is CO_2 fixation important for chemolithotrophs?	193
4.2.5	Energy utilization and efficiency	194
4.3	**Oxidation of ammonia to nitrite**	**195**
4.3.1	Introduction	195
4.3.2	Oxidation of ammonia to hydroxylamine	196
	4.3.2.1 Evidence for hydroxylamine as an intermediate	196
	4.3.2.2 Ammonia oxidation as a monooxygenase reaction	197
	4.3.2.3 Molecular nature of ammonia monooxygenase	198
4.3.3	Oxidation of hydroxylamine to nitrite	199
4.3.4	Coupling of specialized enzymes to the conventional respiratory chain	200
	4.3.4.1 Electron donors to ammonia monooxygenase	200
	4.3.4.2 Electron flow by way of cytochrome c_{554}	200
	4.3.4.3 Integration of electron transport	201
4.3.5	Coupling to protonmotive force generation	202
	4.3.5.1 H^+/O stoichiometries	202
	4.3.5.2 Sensitivity to uncouplers	203
4.4	**Nitrite oxidation**	**203**
4.4.1	Introduction	203
4.4.2	Enzymology	203
	4.4.2.1 Redox centres implicated in nitrite oxidation	203
	4.4.2.2 Purified complexes	204
	4.4.2.3 Electron entry into the conventional chain	205
4.4.3	Coupling to generation of protonmotive force	205
	4.4.3.1 Location of nitrite oxidation	205
	4.4.3.2 Cytoplasmic release of protons as nitrite is oxidized	205
	4.4.3.3 Evidence for a proton-pumping cytochrome oxidase	206

         4.4.3.4   Integration of respiration 206
4.5   Oxidation of inorganic sulphur 208
4.5.1   Introduction 208
         4.5.1.1   Microbiology 208
         4.5.1.2   Chemistry 208
         4.5.1.3   Pathways of sulphur oxidation 212
         4.5.1.4   Transport of inorganic sulphur 213
         4.5.1.5   Molar growth yields 213
4.5.2   Sulphite oxidation 214
         4.5.2.1   Reversal of APS reductase 214
         4.5.2.2   Direct oxidation to sulphate 214
4.5.3   Sulphur oxidation 215
4.5.4   Sulphide oxidation 216
         4.5.4.1   Sulphur as product 216
         4.5.4.2   Reversal of sulphite reductase 216
4.5.5   Thiosulphate oxidation 217
         4.5.5.1   Cleavage by rhodanese 217
         4.5.5.2   Thiosulphate-combining enzyme 219
         4.5.5.3   Single-enzyme oxidation to sulphate 219
4.6   Oxidation of metal ions at acidic pH 220
4.6.1   Introduction 220
4.6.2   Electron flow from Fe(II) to molecular oxygen 221
4.6.3   Coupling to energy conversion 222
4.7   Manganese and iron oxidation at neutral pH 223
4.7.1   Microbiology 223
4.7.2   Energetics 224
References 225

4.1 Introduction

Chemolithotrophs are capable of growth with an inorganic oxidation as their source of energy. For example, ammonium ions may be oxidized to nitrite, elemental sulphur to sulphate, or ferrous iron to the ferric state. The oxidation may be an aerobic process coupled to reduction of molecular oxygen, or it may occur anaerobically with an electron acceptor such as nitrate or fumarate. Such organisms may also have other modes of growth, with light or organic compounds as the source of energy. This chapter describes the inorganic oxidations that serve as energy sources for specialized bacteria, and discusses how these oxidations are coupled to synthesis of ATP and NADH. The oxidation of hydrogen gas has little of an unusual nature because the first step in this process is direct formation of NADH; the hydrogen bacteria will therefore not be discussed in this chapter.

4.1.1 What is chemolithotrophy?

A century ago, Winogradsky made pioneering studies of the bacterial oxidation of reduced states of nitrogen, sulphur and iron. He came to two important conclusions: that for certain bacteria an inorganic oxidation provides a source of energy; and that such bacteria are chemoautotrophs and necessarily restricted to CO_2 as carbon source (see Rittenberg, 1969). Subsequent work has confirmed the first of his conclusions, but it has become clear that his second conclusion was mistaken. Indeed, the sulphur and iron bacteria used in his investigations have since been shown to have little or no ability to fix CO_2 (Rittenberg, 1969). In 1946, chemolithotrophy (as it had come to be known) was officially redefined as growth with an inorganic oxidation as energy source, irrespective of the source of carbon (Lwoff *et al.*, 1946). A chemolithotroph that fixes its own CO_2 may be more narrowly described as a chemolithoautotroph. These useful distinctions are still ignored in some microbiology texts which follow Winogradsky in treating chemolithotrophy as a synonym for chemoautotrophy.

4.1.2 Reviews and sources of chemical data

Energy conservation in chemolithotrophs provides a chapter in Dawes (1986) and is covered in similar depth in Jones (1982). Chemolithotrophy forms an important part of the text on geomicrobiology by Ehrlich (1981). 'Microbial Chemoautotrophy', edited by Strohl and Tuovinen (1984), lacks any description of sulphur oxidation, but is otherwise quite comprehensive. A number of aspects of chemolithotrophy (including ecological aspects) are covered in 'The Nitrogen and Sulphur Cycles' edited by Cole and Ferguson (1988). Hooper and DiSpirito (1985) discuss extracytoplasmic oxidation of respiratory substrates, and in so doing cover the major types of chemolithotroph (see also Chapter 3, this volume). Older sources include Kelly (1978), Aleem (1977) and Suzuki (1974).

The discussion of individual classes of chemolithotroph will include a brief statement of pK values for reactants and products at 25°C. The values given are taken from Smith and Martell (1976). These pK values indicate the pH ranges at which the chemical species will be uncharged and hence more likely to cross biological membranes without the aid of a carrier. They also enable chemical reactions to be written with the correct charge states and correct stoichiometry for proton involvement. In the literature, for example, equations concerning sulphide are often written in terms of the dianion, S^{2-}, which is only predominant at pH values above 13. The proportion of a minor species at a given pH can be calculated from the Henderson–Hasselbalch equation. It should be noted that pK values vary with temperature and are affected by high salt concentrations.

The literature on chemolithotrophs frequently gives values for redox potentials that are inappropriate for the pH in question or are simply wrong. Unless otherwise stated, midpoint redox potential values in this chapter are for pH 7.0 and are taken from Bard *et al.* (1985). In certain cases (mentioned in the text), redox potentials have been derived from ΔG_f^0 values for individual chemical species. The values have generally been rounded to the nearest 5 mV.

4.1.3   The inorganic elements

There are over 80 stable elements but only a much smaller number of classes of chemolithotroph. The periodic table starts with hydrogen, which has sufficient unique features as an inorganic energy source to be excluded from this chapter. The prime difference between molecular hydrogen and nearly all the inorganic reductants discussed in this chapter is that its redox potential is lower than that of $NAD^+[E_{m,7}(H^+/H_2) = -414\,mV;$ $E_{m,7}(NAD^+/NADH = -320\,mV]$. A hydrogenase catalyses the first step in hydrogen metabolism (page 31), after which energy transduction is indistinguishable from that in normal chemoorganotrophs.

Turning to the rest of the periodic table, the noble gases in Group O are too inert to be of interest, while the alkaline metals and alkaline earths in Groups I and IIA only exist in a single oxidation state in aqueous solution. In Group IIB (Zn, Cd, Hg), mercury undergoes a redox cycle in nature between Hg^0 and Hg^{II}, but is not known to serve as an energy source (Ehrlich, 1981). Of Group III (B, Al, Ga, In, Tl) there is likewise nothing to be said.

Group IV starts with carbon, which provides *organic* energy sources. Nevertheless, CO oxidation (page 36) is often regarded as an inorganic process ($E_{m,7}$ for $HCO_3^-/CO_{(g)} = -490\,mV$), because the carbon is only available for biosynthesis after it has been oxidized to CO_2. Below carbon in Group IV, the inorganic chemistry of silicon in the natural environment is restricted to silica and silicates (Ehrlich, 1981). Germanium and lead compounds have not been reported as energy sources, but stannous salts have been shown to be oxidized by *Thiobacillus ferrooxidans* (Ingledew, 1982).

The oxidative reactions of the Nitrogen Cycle (page 8) are primarily carried out by two important classes of chemolithotroph: the ammonia oxidizers ($NH_4^+ \rightarrow NO_2^-$) and the nitrite oxidizers ($NO_2^- \rightarrow NO_3^-$). Below nitrogen in Group V, phosphorus has a series of lower oxidation states which combine low redox potentials with stability in aqueous solution. The following potentials have been calculated from ΔG_f^0 and pK values in Bard *et al.* (1985): $E_{m,7}$ (hypophosphite/phosphine) $= -670\,mV$; $E_{m,7}$ (phosphite/hypophosphite) $= -900\,mV$; $E_{m,7}$ (phosphate/phosphite) $= -620\,mV$. Because the potentials are so low, phosphate remains the normal state in the terrestrial environment, although there are reports of lower oxidation states being formed biologically (Ehrlich, 1981). The status of the reduced states of phosphorus as energy sources is discussed by Ehrlich (1981). Surprisingly, there is no clear-cut example of bacterial growth coupled to oxidation of inorganic phosphorus, although an NAD^+-linked phosphite-oxidizing enzyme has been purified. Arsenic undergoes redox cycling in nature, but has not been reported as an energy source (Ehrlich, 1981). One explanation is the general toxicity of arsenate, which mimics phosphate in enzyme reactions but yields esters that hydrolyse spontaneously. The oxidation of antimony[III] to antimony[V] has been reported to serve as energy source for a bacterium from an ore deposit, as discussed by Ehrlich (1978, 1981).

In Group VI, reduction of O_2 to water provides the cornerstone for respiration: $E_{m,7}$ $(O_{2(g)}/H_2O) = +815\,mV$. The reverse reaction supplies reductant for oxygenic photosynthesis. However, there is no report of water oxidation being used as an energy source. Oxidation of hydrogen peroxide to molecular O_2 $(E_{m,7}$ $(O_{2(g)}/H_2O_2) = +280\,mV)$ is not known to occur in living organisms, except as part of the catalytic cycle of catalase. Sulphur is the second element in Group VI. The oxidative reactions of the Sulphur Cycle (page 8) are carried out by chemolithotrophic sulphur-oxidizing bacteria (e.g. *Thiobacillus, Beggiatoa*) and by photosynthetic sulphur bacteria (e.g. *Chlorobium, Chromatium*). Certain species of *Chromatium* have been shown to grow chemolithotrophically in the dark (see Section 4.5). For this reason the sulphur-oxidizing enzymes of chemolithotrophs will be discussed in conjunction with those of photosynthetic sulphur bacteria. Selenium is believed to pass through a similar redox cycle in nature, but has been little studied as an energy source (Ehrlich, 1981; Doran, 1982).

The halogens in Group VII have an extensive redox chemistry. However, at neutral pH, all redox potentials are either above that for O_2/H_2O or only marginally below. Two of the lowest are $E_{m,7}(ClO_3^-/ClO_2^-) = +710\,mV$ and E_m $(I_{2(aq)}/I^-) = +620\,mV$ (independent of pH).

This leaves the transition metals. Many transition metals have stable redox states that differ by a single electronic charge (e.g. Cu^I/Cu^{II}; Fe^{II}/Fe^{III}). In aqueous solution, ions in such states will exchange electrons with each other without the aid of enzymes, so establishing a common redox poise. There is no clear evidence that separate enzymes exist to accept electrons from different metal ions in solution, although this may well be the case for less reactive couples such as Mn^{II}/Mn^{IV}. The role of transition metals as energy sources will be discussed under two general headings: metal ion oxidation in acidic solution; and oxidation of divalent iron and manganese at neutral pH (pages 220 and 223).

The free metals are also possible energy sources. For example, oxidation of metallic iron has a low redox potential, $E_m(Fe^{2+}/Fe) = -440\,mV$. There seem to be no reports of electrons from a metal passing directly into an electron transport chain. Nevertheless, metal corrosion can still provide chemolithotrophs with an energy source, as will be described.

4.1.4   Acid production and metal corrosion

With the exception of nitrite oxidizers, the major classes of chemolithotroph gain energy from proton-releasing reactions, and therefore tend to lower the surrounding pH. The following equations are appropriate for a pH range from 5 to 7:

$$NH_4^+ + 1.5\,O_2 \rightarrow NO_2^- + H_2O + 2\,H^+$$

$$H_2S + 2\,O_2 \rightarrow SO_4^{2-} + 2\,H^+$$

$$Fe^{2+} + 0.25\,O_2 + 2.5\,H_2O \rightarrow Fe(OH)_3 + 2\,H^+$$

The acidification tends to be out of all proportion to the number of bacteria present, because chemolithotrophs consume large amounts of reductant in relation to net growth. Thus *Nitrosomonas* oxidizes at least 11.5 molecules of substrate for each carbon assimilated, while for *Nitrobacter* the ratio is at least 32, and is often much more (Ward, 1986).

Ammonia-oxidizing bacteria seem to have a requirement for free ammonia, for reasons discussed in Section 4.3. Over the pH range at which such organisms normally grow, the ratio of ammonia to ammonium ions falls by a factor of 10 per unit fall in pH (pH = $pK + \log_{10}([base]/[acid])$). In practice, growth stops when the pH falls below about 6. Ammonia oxidizers have been isolated from acidic soils, where they may exist in neutral microsites, but have not been successfully grown in acidic media (Robertson, 1982). However, for sulphur- and iron-oxidizing bacteria, many acidophilic strains are known. *Thiobacillus ferrooxidans* is a much-studied bacterium that uses both sulphur and iron compounds as energy sources at low pH (Ingledew, 1982). Its ability to convert iron pyrites (FeS_2) and other sulphide ores into solutions of metal ions in sulphuric acid has proved useful in ore leaching, but creates corrosion problems in mine drainage (Ehrlich, 1981). (The biochemistry of acidophily is discussed in Chapter 8.)

An important corrosion mechanism involves part of a metal surface acting as an anode at which the metal passes into solution (e.g. $Fe \rightarrow Fe^{2+} + 2e^-$), while an equivalent cathodic reaction occurs on *other* regions at a higher ambient redox poise (Hamilton, 1985). The cathodic reaction may be anaerobic or aerobic. At an anaerobic cathode, molecular H_2 is formed transiently $(2H^+ + 2e^- \rightarrow H_2)$, and H_2-oxidizing sulphate reducers are important in its reoxidation (page 128). The reduction of molecular O_2 at an aerobic cathode is non-biological $(0.5O_2 + H_2O + 2e^- \rightarrow 2OH^-)$. However, chemolithotrophs assist the corrosion indirectly by forming a biofilm over the anodic parts of the surface, beneath which O_2 becomes depleted. The anodic dissolution of steel, for example, yields ferrous ions, plus lesser quantities of non-ferrous metal ions such as Mn^{2+}, and low oxidation states of carbon, phosphorus and sulphur. The associated biofilm may include all the chemolithotrophs described below with the exception of nitrifiers (Hamilton, 1985). Their precipitated products (e.g. $Fe(OH)_3$) help to maintain a sharp oxic/anoxic boundary.

4.2 Biochemical features common to chemolithotrophs

4.2.1 The underlying conventional electron transport chain

It has become clear that chemolithotrophs possess a conventional respiratory chain with specialized components grafted onto it (see e.g. Ingledew, 1982; Kelly, 1982; Wood, 1987). Not all chemolithotrophs are aerobes (see Chapter 2 for anaerobic H_2-oxidizers), but virtually all those discussed in this chapter are capable of aerobic respiration. As described in Chapter 1, bacterial electron flow from NADH to O_2 involves membrane-bound complexes linked by an isoprenoid quinone and, optionally, by a soluble c-type cytochrome on the periplasmic face of the membrane. There is widespread indirect evidence for NADH dehydrogenase in chemolithotrophs, and it is clearly vital for reduction of NAD^+ to NADH by reverse electron flow (see below); but in no case has the complex been purified. The isoprenoid quinone has been found to be ubiquinone in all species where this has been investigated. The widespread reports of membrane-bound b- and c-type cytochromes suggest that a ubiquinone–cytochrome bc_1 complex is invariably present, but this has not been studied in isolation.

Soluble cytochromes c showing sequence homology with mitochondrial cytochrome c are denoted Class I cytochromes c (Ambler, 1980). Cytochrome c_{550} from *Nitrobacter agilis* and cytochrome c_{552} from *Nitrosomonas europaea* have been shown to belong to this Class (Wood, 1987). The *Nitrobacter* cytochrome has 66 residues in common with cytochrome c_2 from the photosynthetic bacterium *Rhodopseudomonas viridis*, while the

Table 4.1 Terminal oxidases with a-type haem in chemolithotrophs

Name	α-band maximum in reduced state nm	Redox potentials of haems mV	Apparent M_r of subunits kDa	Cu atoms per mole	References
Nitrobacter agilis cytochrome aa_3	605	240, 400 (pH 7.0)	31, 51	2–3	Yamanaka et al., 1981a Sewell et al., 1972
Nitrosomonas europaea cytochrome a_1	597		33, 50	1–2	Yamazaki et al., 1985
Thiobacillus ferrooxidans cytochrome a_1	597	420, 500 (pH 7.0) 610, 720 (pH 3.2)			Ingledew, 1982
Thiobacillus versutus (Th. A_2) cytochrome aa_3	607	210, 390 (pH 7.0)			Kula et al., 1982

Nitrosomonas protein is closest to cytochromes c_{551} from certain pseudomonads. It is evident from these sequences that chemolithotrophs are not particularly closely related. Soluble cytochromes c with molecular weight and redox potential similar to mitochondrial cytochrome c have been obtained from a variety of other chemolithotrophs (Ingledew, 1982; Lu *et al.*, 1984; Trudinger *et al.*, 1985).

Terminal oxidases with *a*-type haem (see Chapter 5) have been reported from many chemolithotrophs and the properties of a selection are summarized in Table 4.1. The designations a_1 and aa_3 have been based solely on the position of the reduced α-band (Poole, 1983). This distinction is not known to have any biochemical significance. In nitrite-oxidizers, however, cytochrome a_1 has a specialized role; as discussed in Section 4.4. Few reports of cytochromes o or d in the organisms described in this chapter. This may be because their mode of life requires the maximum possible stoichiometry of proton pumping, which is perhaps provided by the cytochrome c-cytochrome aa_3 pathway.

Terminal reductases for nitrate, nitrite and other electron acceptors enable certain chemolithotrophs to grow anaerobically (Kelly, 1982). These reductases have not been studied in detail, but appear to be similar to the analogous proteins in heterotrophs (Chapter 2).

Many studies have shown that electron transport and ATP synthesis can proceed separately from each other in chemolithotrophs and are consistent with their coupling via the protonmotive force (see Chapter 1, page 20):

$$ADP + P_i + 2H^+_{out} \rightarrow ATP + H_2O + 2H^+_{in}$$

The only clear evidence for substrate level phosphorylation is the APS reductase pathway for sulphite oxidation to sulphate (Section 4.5.2).

4.2.2 NADH synthesis by reversed proton flow

In Figure 4.1 the two ends of a conventional aerobic respiratory chain are placed on a scale of redox potential, together with two intermediate carriers, ubiquinone (UQ) and cytochrome c. The redox potential corresponding to glucose oxidation to bicarbonate (as a typical heterotrophic energy source) is more negative than that of the $NAD^+/NADH$ couple. Among the inorganic couples of prime importance, this is only true for molecular H_2. Sulphide oxidation to sulphate provides electrons at a higher average redox potential, while for the nitrite/ammonia and nitrate/nitrite couples the redox potential is higher even than for cytochrome c. These systems must be near the limit for an aerobic energy source. This is also true for ferrous iron oxidation at acid pH (Section 4.6).

The standard respiratory electron transport chain is fully reversible, with the exception of the terminal oxidase reaction which cannot be driven in the direction of O_2 evolution. Given a sufficient protonmotive force, electrons on intermediate carriers will be pushed back up the chain, and NAD^+ will become reduced:

$$2 \operatorname{cyt} c(Fe^{2+}) + NAD^+ + n H^+_{out} \rightarrow 2 \operatorname{cyt} c(Fe^{3+}) + NADH + (n-1) H^+_{in}$$

This process is termed 'reversed electron flow', and enables bacteria with low-grade energy sources to synthesize NADH (see page 24). The driving force is the reversed proton flow. The coupling of the inorganic oxidation to NADH biosynthesis has been shown to require the presence of a protonmotive force in all the major classes of chemolithotroph described in this chapter (Ingledew, 1982; Kelly, 1982; Wood, 1987). In nitrite oxidizers, the membrane potential is also a significant factor, as discussed in Section 4.4.

The proton-linked NAD/NADP transhydrogenase (page 27) has not been studied in chemolithotrophs.

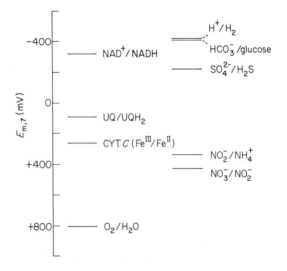

Fig. 4.1 Redox potentials for respiratory components and inorganic couples. All couples are shown conventionally as (oxidized/reduced), although, of course, the substrate is often the reduced form.

4.2.3 Generalizations concerning enzyme location

Hooper and DiSpirito (1985) have put forward the generalization that inorganic reductants are oxidized outside the cytoplasm. The rationale is that such oxidations nearly all liberate protons (see Section 4.1.4 and

Chapter 3); if these protons were to be released into the cytoplasm, they would oppose the production of a protonmotive force. It is true that the majority of the oxidations described below are extracytoplasmic, but there are exceptions. Hydrogen oxidation can be coupled to NAD^+ reduction by cytoplasmic enzymes (see page 31), while one pathway of sulphite oxidation includes a substrate-level phosphorylation (Section 4.5.2). Nitrite oxidation uses cytoplasmic release of protons to drive what would otherwise be an uphill reaction (Section 4.4.3). Other exceptions involve reactions catalysed by rhodanese and sulphite reductase, which are probably confined to the cytoplasm by their other biological roles (Sections 4.5.4 and 4.5.5).

Cytochromes c contain a haem prosthetic group exactly as in b-type cytochromes and many other haem proteins, with the distinction that the haem is attached by covalent bonds, formed by adding cysteine –SH groups across vinyl side chains of proto-haem (page 47). In bacteria, c-type cytochromes seem invariably to be located in the periplasm, or on the periplasmic side of the cytoplasmic membrane (see Chapter 3). Wood (1983) suggested that the reason for the existence of c-type cytochromes is that the covalent bonds prevent the haem from being lost into the surrounding medium. The corollary is that a soluble protein with c-type haem can be assumed to be periplasmic. A few apparent exceptions are discussed by Wood (1983); none has been confirmed by repeated study.

4.2.4 Why is CO_2 fixation important for chemolithotrophs?

The reason why chemolithotrophy is popularly assumed to be associated with CO_2 fixation is because this is substantially correct for certain inorganic reductants. All ammonia-oxidizing bacteria are dependent on CO_2 as major carbon source (Wood, 1987). Many strains can assimilate organic carbon to a limited extent, but any stimulation in the rate of growth is quite modest. The nitrite oxidizers are slightly more flexible, and some strains can use organic compounds as an *energy* source. However, heterotrophic growth is even slower than growth by nitrite oxidation, a doubling time of 65 h being reported in one case (Bock, 1978). Sulphur oxidizers are more versatile: some are autotrophs; some are heterotrophs (e.g. *Beggiatoa*); and some make use of both forms of carbon. The inorganic reductant with least association with autotrophy is hydrogen.

As a generalization, the dependence of chemolithotrophs on CO_2 correlates with how far removed their electron transport pathways are from conventional respiration. The least adaptation is required for H_2 oxidation, electrons being fed in at or near NADH. As the opposite extreme, the use of ammonia as an energy source implies three destinations for electrons from hydroxylamine: the terminal oxidase, the monooxygenase, and reversed electron flow to NAD^+. It is probable that the control mechanisms necessary to prevent the monooxygenase running out of reductant or the cells being poisoned by hydroxylamine, leave the electron transport chain incapable of any other mode of operation. Similar considerations must explain why ammonia-oxidizers do not oxidize their substrate all the way to nitrate.

For other chemolithotrophs the problem is not so obvious. If the whole electron transport chain is reversible, apart from the terminal oxidase, why should heterotrophic growth be slow? One factor of importance is the redox poise of the ubiquinone–cytochrome bc_1 region, which is crucial for efficient electron flow (McEwan *et al.*, 1985). This can be readily understood in terms of the Q-cycle presented in Chapter 1 (Fig. 1.29); if a side reaction leads to all components becoming reduced (or oxidized), the cycle is blocked, and it can only be restarted by a bypass outside the scheme.

4.2.5   Energy utilization and efficiency

The energy balance of CO_2-dependent chemolithotrophs has been discussed by Kelly (1978, 1982). Each CO_2 fixed by the Calvin cycle implies a consumption of 4 reducing equivalents as 2 NAD(P)H. These electrons are derived from the inorganic reductant; they are unavailable for passage *down* the respiratory chain and must instead be promoted back by reversed electron flow. The Calvin cycle requires 3 ATP per CO_2 converted to hexose, while biosynthesis with hexose as starting point is estimated to consume an additional 0.88 ATP per carbon (Kelly, 1978). Thus, if an inorganic oxidation releases x electrons and results in fixation of y CO_2, it follows that only $(x - 4y)$ electrons pass down the respiratory chain, while $3.88y$ ATP are used in biosynthetic reactions with CO_2 and NAD(P)H as starting-points. In the absence of substrate level phosphorylation, production of $3.88y$ ATP will require pumping of $7.76y$ protons. The parameter y can be measured by determining the maximal yield in g dry weight per mole of substrate (known as Y_{max}) and the percentage of dry weight attributable to carbon. The H^+/e^- ratio for the substrate can be measured, for instance with pulsed preparations of vesicles (page 64 and Chapter 9). One can then compare the total H^+ pumped by $(x - 4y)$ electrons with the $7.76y$ protons already accounted for. By using such calculations it can be shown that in many chemolithotrophs the major proton-requiring reaction is reversed electron flow.

A different type of calculation concerns the overall thermodynamic efficiency (Kelly, 1978). Fixation of CO_2 to the level of hexose has $\Delta G^{0\prime} = +495$ kJ.mol^{-1}. This is believed to account for about 80% of the energy budget of an autotroph. A measurement of CO_2 fixed per mole of substrate oxidized enables the thermodynamic efficiency to be estimated, provided

$\Delta G^{0\prime}$ for the substrate oxidation is known. With thiosulphate, recent Y_{max} values for different *Thiobacilli* range from 7.8 to 14.7 g dry weight per mole substrate oxidized (Kelly *et al.*, 1986). These values imply thermodynamic efficiencies ranging from 14 to 30%. For *T. ferrooxidans* growing by Fe^{2+} oxidation at low pH, $Y_{max} = 1.33$ g dry weight per mol substrate oxidized, implying an overall efficiency of 21% (Kelly, 1978). Ammonia and nitrite oxidation appear to be less efficient; Kelly (1978) cites values of 5.9 and 7.9% respectively.

4.3 Oxidation of ammonia to nitrite

4.3.1 Introduction

The bacteria that derive energy from oxidation of inorganic nitrogen fall into two classes: ammonia oxidizers ($NH_4^+ \rightarrow NO_2^-$) and nitrite-oxidizers ($NO_2^- \rightarrow NO_3^-$); the overall process is called nitrification (see page 8). The ammonia oxidizers are placed in genera prefixed *Nitroso-*. Five genera are distinguished by morphological factors (Watson *et al.*, 1981); almost all biochemical studies have used a single species, *Nitrosomonas europaea*. For all these bacteria, oxidation of ammonia is the only known energy source.

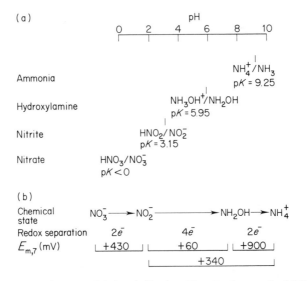

Fig. 4.2 pK values and redox potentials for nitrification. The pK values are for 25°C and zero ionic strength (Smith and Martell, 1976). Note that ΣnE is the same for different pathways, e.g. $(4 \times 60) + (2 \times 900) = 6 \times 340$.

All are primarily autotrophic, fixing CO_2 by the Calvin cycle. The bio-chemistry of ammonia oxidation as an energy source has recently been reviewed by Wood (1987), and Kuenen and Robertson (1988) provide an excellent review of the microbiology of nitrification. Earlier descriptions (other than those cited in Section 4.1.2) include Drozd (1980) and Hooper (1984).

Figure 4.2 shows the pK values of importance in nitrification. It can be seen that ammonia is predominantly in the cationic state at pH values of biological significance. However, the uncharged species is believed to be the active form, as will be explained. Its concentration in a 1 mM solution of NH_4^+ will be 10 μM at pH 7.2 and 1 μM at pH 6.2. Hydroxylamine (an intermediate in ammonia oxidation) is predominantly uncharged at neutral pH. Nevertheless, its polar nature makes it virtually insoluble in non-polar solvents, so it may only slowly diffuse across a biological membrane (Wood, 1987). Nitrite is anionic over the pH range of biological ammonia oxidation. The overall system for oxidation of ammonia to nitrite is summarized in Figure 4.3.

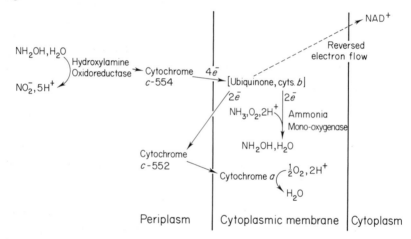

Fig. 4.3 Electron transport in *Nitrosomonas*.

4.3.2   Oxidation of ammonia to hydroxylamine

4.3.2.1   Evidence for hydroxylamine as an intermediate
The experimental findings are as follows:

(a) Lees (1952) demonstrated that exogenous hydroxylamine was oxi-dized by whole cells of *Nitrosomonas europaea*;

(b) a hydroxylamine-oxidizing enzyme has been isolated and shown to have hydrazine (H_2NNH_2) as an alternative substrate (Nicholas and Jones, 1960);

(c) a few years earlier, Hofman and Lees (1953) found that addition of hydrazine to ammonia-oxidizing cells led to an accumulation of hydroxylamine, a concentration of 80 µM being found after a 30 min incubation.

It can be concluded from these observations that hydroxylamine is intermediate in ammonia oxidation, and that it accumulates when hydrazine is added as a competitive inhibitor of its oxidation. However, there have been no reports of hydroxylamine being detected or trapped during normal ammonia oxidation, implying that the free concentration is very low.

4.3.2.2 Ammonia oxidation as a monooxygenase reaction

The oxidation of ammonia to hydroxylamine has a redox potential of + 900 mV at pH 7.0. This is distinctly more positive than for reduction of O_2 to water, thus ruling out a simple coupling to cytochrome oxidase.

There is good evidence that ammonia oxidation requires a supply of *reductant*. Hooper (1969) showed that addition of ammonia to starved cells resulted in a lag before O_2 uptake gathered speed, the lag being eliminated by a priming addition of hydroxylamine. With cell-free extracts, both NADH and hydroxylamine have been shown to act as priming agents (Suzuki *et al.*, 1976). Further evidence that reductant is necessary comes from studies with alternative substrates in place of ammonia, as discussed by Wood (1987).

Hollocher *et al.* (1981) incubated cells with ammonia and hydrazine, as in the experiment of Hofman and Lees (1953), but with $^{18}O_2$ in place of normal oxygen. They found that [^{18}O]-hydroxylamine accumulated, implying that molecular O_2 is a substrate for the ammonia-oxidizing reaction.

Suzuki *et al.* (1974) found the apparent K_m for NH_4^+ fell sharply with increasing pH, being 4 mM at pH 7.0 and 0.48 mM at pH 8.0. Expressed in terms of free NH_3, the K_m values were almost independent of pH, suggesting that NH_3 is the true substrate for oxidation. Similar results were obtained with a cell-free extract, ruling out the possibility that the K_m reflected a transport process into the cell. As further corroboration, it has since been discovered that the ammonia-oxidizing enzyme has a range of alternative substrates, all of them uncharged. Examples are methane, methanol, CO, ethene, bromoethane, benzene and phenol (Wood, 1987).

In summary, ammonia oxidation involves uncharged ammonia, molecular O_2 and reductant, and it yields hydroxylamine as product. A balanced equation requires the production of water, not directly demonstrated:

$$NH_3 + O_2 + 2[H] \rightarrow NH_2OH + H_2O$$

The enzyme responsible is known as ammonia monooxygenase.

The closest parallel to ammonia monooxygenase is found in methane-oxidizing bacteria. These organisms contain a methane monooxygenase which is likewise capable of oxidizing a wide range of non-polar organic substrates (Dalton, 1981; Anthony, 1986). Methane monooxygenase also oxidizes ammonia, although ammonium ions cannot provide an energy source for methanotrophs (Dalton, 1977; Wood, 1987). A particulate methane monooxygenase has the same sensitivity to chelating agents as ammonia monooxygenase (Dalton and Leak, 1985). A soluble form of methane monooxygenase has a very different active site, and NADH as electron donor (Dalton, 1981; Anthony, 1986).

4.3.2.3   Molecular nature of ammonia monooxygenase

Many general texts state that ammonia oxidation is brought about by an enzyme similar to cytochrome P-450. This is incorrect, as discussed by Wood (1987). Another common mistake in published pathways is to show ammonia oxidation as an essential part of the catalytic cycle of hydroxylamine oxidoreductase. This is disproved by the rapid oxidation of added hydroxylamine in the absence of ammonia (see above).

Ammonia monooxygenase is located in the membrane fraction from broken cells. Suzuki and co-workers have made studies with membranes separated from soluble components by gel filtration (Suzuki and Kwok 1981; Tsang and Suzuki, 1982) but this is as far as purification has been taken, without activity being lost.

Acetylene acts as an irreversible inhibitor (Hynes and Knowles, 1978) and incubation of whole cells with $[^{14}C]$-acetylene was found to lead to labelling of a single membrane polypeptide, which ran on SDS/polyacrylamide gels with an apparent M_r of 28 000 (Hyman and Wood, 1985). No soluble protein was labelled. This is the only information on the protein composition of the monooxygenase. It would seem that the enzyme attempts to oxidize acetylene, but forms such a reactive species that it becomes covalently modified and destroyed.

Ammonia oxidation is sensitive to near UV light (Hooper and Terry, 1974), resulting in an absence of nitrification in environments exposed to bright sunlight (Olson, 1981). A difference spectrum of irradiated cells relative to an untreated control showed a broad trough centred near 380 nm (Shears and Wood, 1985) and an absorption change with a similar profile resulted from adding an organic substrate to resting cells. The simplest explanation is that resting cells contain an oxygenated state of the enzyme with a broad near UV absorption band. The excited state resulting from UV light absorption must give rise to an active form of oxygen, which destroys the enzyme.

The sensitivity towards different metal chelators is consistent with an involvement of cuprous copper (Hooper and Terry, 1973). In the previous section, analogies were drawn with the membrane form of methane mono-oxygenase; this is synthesized only if copper is added to the growth medium (Prior and Dalton, 1985). Moreover, the copper enzyme tyrosinase has an oxygenated state with a near UV absorption band, and is sensitive in this form to near UV light (Solomon, 1981). These considerations led Shears and Wood (1985) to propose a speculative catalytic cycle, based on that for tyrosinase.

4.3.3   Oxidation of hydroxylamine to nitrite

The conversion of hydroxylamine to nitrite is a four-electron oxidation. The corresponding redox potential, $E_{m,7}(NO_2^-/NH_2OH_{(aq)})$, is $+60\,mV$. A second O-atom must be introduced, but the source remained ambiguous in early isotope studies, because the cells catalyse a rapid exchange of oxygen between nitrite and water. More recently, Andersson and Hooper (1983) have established that water contributes one O-atom to newly-synthesized nitrite, implying a mechanism such as:

$$E + H_2NOH \rightarrow E\text{–}NO^+ + 3H^+ + 4e^-$$

$$E\text{–}NO^+ + H_2O \rightarrow E + NO_2^- + 2H^+$$

Hydroxylamine oxidation is brought about by hydroxylamine dehydrogenase (also called hydroxylamine oxidoreductase EC 1.7.3.4), a soluble enzyme released on breakage of cells. This has an exceptionally complex structure, revealed through detailed studies by Hooper and co-workers (Hooper, 1984). Each α-subunit contains six c-type haems and a single P-460, and each β-subunit contains a single c-type haem (Terry and Hooper, 1981). The molecular mass is about 220 kDa, corresponding to the composition $\alpha_3\beta_3$. Thus each molecule contains no less than 21 c-type haems, plus three P-460. The redox potentials fall into four groups, the values of which are $+100$; 0; between -100 and -250; and $-320\,mV$ (Hooper, 1984).

The P-460 prosthetic group is covalently bound and physical studies point to it being an unusual form of haem (Lipscomb *et al.*, 1982; Andersson *et al.*, 1984). The spectroscopic absorption of P-460 is selectively removed if the oxidized enzyme is treated with hydrogen peroxide and this correlates with the enzyme losing its catalytic activity (Hooper *et al.*, 1983; Hooper, 1984). Hydroxylamine is normally capable of reducing about 40% of the total haem; this percentage is much decreased by hydrogen peroxide treatment (Hooper *et al.*, 1983). These findings indicate that P-460 is intimately connected with electrons from hydroxylamine; nevertheless, P-460 is not reduced by hydroxylamine, even under anaerobic conditions. This is explained by its redox potential of $-320\,\text{mV}$ (Hooper, 1984). There is EPR evidence for spin interactions between P-460 and c-type haem (Prince and Hooper, 1987). Either hydroxylamine binds between P-460 and a c-type haem, or hydroxylamine binding is stabilized by P-460.

4.3.4 Coupling of specialized enzymes to the conventional respiratory chain

4.3.4.1 Electron donors to ammonia monooxygenase

Where do electrons for the monooxygenase branch off from the underlying 'normal' electron transport chain? For uncharged ammonia as substrate, the monooxygenase reaction is:

$$NH_{3(aq)} + O_{2(g)} + 2H^+ + 2e^- \rightarrow NH_2OH_{(aq)} + H_2O$$

This reaction has $E_{m,7}$ of $+800\,\text{mV}$, about the same as that for oxygen reduction to water and so the enzyme could, in principle, accept electrons from anywhere in the respiratory chain.

The known electron donors fall into three classes. First is hydroxylamine and its analogues, acting by a pathway described below. Secondly, NADH acts as an electron donor and priming agent in isolated membranes (Tsang and Suzuki, 1982; Suzuki, 1974). With whole cells, organic substrates can be hydroxylated at a low rate without any added electron donor (Hyman and Wood, 1983, 1984); this can be explained by a significant level of endogenous respiration, which may be expected to proceed via NADH. On the other hand, reconstitutions show that the monooxygenase is kinetically competent in the absence of NADH (Suzuki and Kwok, 1981; Suzuki *et al.*, 1981). Tetra- and tri-methylhydroquinone constitute a third class of electron donor (Shears and Wood, 1986). These reduced quinones have redox potentials close to that of ubiquinone, and they have similar structures apart from the absence of the isoprenoid side chain. They presumably feed in electrons at the level of ubiquinone.

The provisional conclusion is that the monooxygenase accepts electrons from the ubiquinone–cytochrome b region of the chain, and that when NADH acts as donor, it does so via NADH dehydrogenase.

4.3.4.2 Electron flow by way of cytochrome c_{554}

Olson and Hooper (1983) have demonstrated that hydroxylamine dehydrogenase is released in the course of preparation of spheroplasts, implying a periplasmic location.

The addition of hydroxylamine to resting cells greatly stimulates the rate of hydroxylation of organic substrates by the ammonia monooxygenase

(Hyman and Wood, 1984). This system has been reconstituted by mixing membranes containing the monooxygenase with purified hydroxylamine dehydrogenase plus a cytochrome c_{554} (Suzuki and Kwok, 1981; Tsang and Suzuki, 1982). The reconstituted system carried out the complete oxidation of ammonia to nitrite, and also brought about organic conversions in the presence of hydroxylamine.

The properties of cytochrome c_{554} are unusual and complex (Andersson *et al.*, 1986). Four haems are present, attached to a 25 kDa polypeptide. At neutral pH three haems are in a low-spin state, as is usual in cytochromes, while the fourth is high spin. Physical studies indicate unique magnetic interactions between the haems which were sensitive to changes in pH. Only one redox potential has been reported, $E_{m,7} = +20\,\text{mV}$ (Miller and Wood, 1982).

4.3.4.3 Integration of electron transport

Ammonia oxidizers contain an underlying 'normal' electron transport chain (see Section 4.2.1). Hydroxylamine dehydrogenase (oxidoreductase) feeds in electrons, presumably via cytochrome c_{554}. The electrons could reach the conventional respiratory chain either at the level of ubiquinone and the b-type cytochromes, or at cytochrome c_{552} (a soluble Class I c-type cytochrome). The latter pathway would be less efficient, and would make it hard to understand the presence of cytochrome c_{554}. Mechanistic aspects of electron flow from cytochrome c_{554} to the ubiquinone–cytochrome b region are discussed by Wood (1987).

The point at which electrons branch off from the conventional chain for use by ammonia monooxygenase is probably also close to ubiquinone, as explained above.

An integrated scheme for electron transport is shown in Figure 4.3. When ammonia is added to resting cells, the initial source of reductant for the monooxygenase must be endogenous substrates, acting via NADH and (presumably) NADH:ubiquinone oxidoreductase. The oxidation is observed to gain speed, and then settles down to a steady rate, implying that hydroxylamine has reached a steady-state concentration. In this phase each oxidation of hydroxylamine to nitrite must result in exactly one new hydroxylamine being generated by the monooxygenase. A bleed-off of reductant is promoted back to NAD^+, for use in biosynthesis. This leaves slightly fewer than two electrons per hydroxylamine for passage to the terminal oxidase, The steady-state level of hydroxylamine is likely to be controlled by a feedback inhibition of the monooxygenase (Wood, 1987).

Under optimal growth conditions, 11.5 molecules of ammonia are oxidized per carbon assimilated (Ward, 1986). The reduction of CO_2 to the level of reduction of cell material ($=$ HCHO) requires four electrons. In the steady state, for each hydroxylamine oxidized, 0.35 electrons would have to be promoted back by reverse electron flow for NADH synthesis, while only 1.65 would pass down the chain to the terminal oxidase.

Ammonia-oxidizing bacteria produce small amounts of nitrous and nitric oxides in addition to nitrite, with a rising yield of N_2O as the O_2 tension is decreased (Goreau *et al.*, 1980; Lipschultz *et al.*, 1981). The labelling pattern obtained from incubation with $^{14}NH_4^+$ and $^{15}NO_2^-$ implied that N_2O is formed by reduction of nitrite, and not as an intermediate of nitrite formation (Poth and Focht, 1985). A soluble nitrite reductase with 'Type I' and 'Type II' copper has been characterized from *Nitrosomonas europaea* (DiSpirito *et al.*, 1985; Miller and Nicholas, 1985); its relationship to the pathways shown in Figure 4.3 is discussed in Wood (1987).

4.3.5   Coupling to protonmotive force generation

4.3.5.1 H^+/O stoichiometries
Two studies have shown that hydroxylamine and ammonia oxidation are coupled to proton translocation (Drozd, 1976; Hollocher *et al.*, 1982). The more detailed later study found that the H^+/O ratio fell as the substrate concentration was increased. For hydroxylamine, an H^+/O ratio of 4.0 was obtained at 1 mM, falling to 3.0 at 6 mM; ammonium ions gave a value of 3.1 at 1 mM, falling to 2.2 at 5 mM and 1.3 at 10 mM. The decline was attributed to the substrates acting as permeant amines. Extrapolation to zero substrate led to limiting H^+/O ratios of 4.4 for hydroxylamine and 3.4 for ammonium ions.

These ratios include the protons released for chemical reasons in the course of oxidation of hydroxylamine and ammonium ion to nitrite (1 per O_2 and 2 per 1.5 O_2, respectively). These 'stoichiometric' protons are released outside the cell and must therefore be subtracted (Wood, 1987). The limiting H^+/O ratios for protons from the cytoplasm then become 3.9 for hydroxylamine and 2.7 for ammonium ions.

Let us count protons for one ammonia being oxidized to nitrite, assuming hydroxylamine is in a steady state. Two electrons pass from hydroxylamine to the monooxygenase; nearly two electrons pass from hydroxylamine to the terminal oxidase; and 1.5 O_2 are consumed. The H^+/O ratio deduced for hydroxylamine implies that the two electrons passing to the terminal oxidase are coupled to removal of 3.9 protons from the cytoplasm. Therefore, about 4.2 protons must be removed from the cytoplasm per two electrons passing to the monooxygenase, in order to obtain the correct overall H^+/O ratio of 2.7. It should be emphasized that these experiments used O_2 pulses in the presence of a permeant anion, and may not give a true picture of ammonia oxidation proceeding at a steady rate.

4.3.5.2 Sensitivity to uncouplers

Ammonia oxidation is characteristically inhibited by uncouplers (Hooper and Terry, 1973). One explanation would be that the steps between cytochrome c_{554} and the monooxygenase are coupled to *inward* movement of protons, as discussed for nitrite oxidation in Section 4.4.3. However, this would be in marked disagreement with the H^+/O ratio just deduced. A key to a simpler explanation lies in the fact that hydroxylamine oxidation is stimulated by uncouplers (Aleem, 1977). This stimulation implies that, under normal conditions, the protonmotive force is slowing down electron flow from hydroxylamine to the terminal oxidase. The monooxygenase and terminal oxidase are competing acceptors (sinks) for electrons, so the balance between them must be carefully controlled. Uncouplers allow electrons to flow too freely to the terminal oxidase; this starves the monooxygenase of reductant, and hence ammonia oxidation becomes inhibited.

4.4 Nitrite oxidation

4.4.1 Introduction

Nitrite oxidizing bacteria are placed in genera prefixed *Nitro-*, three such genera being distinguished by morphological considerations (Watson *et al.*, 1981). Most biochemical studies have made use of *Nitrobacter agilis* or *N. winogradskyi*, regarded as synonymous by Watson *et al.* Some strains are obligate autotrophs, while others make use of organic carbon and can grow heterotrophically in the absence of nitrite.

Nitrite oxidation as an energy source has recently been reviewed by Wood (1987). Earlier descriptions include Aleem and Sewell (1984) and Bock (1978). Experiments of Hollocher *et al.* (1982) which failed to find evidence for proton translocation are discussed in Wood (1987).

As is shown in Figure 4.2, nitrite is a weak acid, with a pK value of 3.15. Nitrate, by contrast, is anionic at all pH values of biological interest. No acidophilic nitrite oxidizers are known, one reason being that nitrous acid decomposes spontaneously at low pH. The redox potential for the nitrate/nitrite couple ($+430 \, mV$) is even higher than that for nitrite formation from ammonia ($+340 \, mV$). A summary diagram (Fig. 4.4) is given on page 207.

4.4.2 Enzymology

4.4.2.1 Redox centres implicated in nitrite oxidation

Oxidation of nitrite to nitrate occurs without detectable intermediates. The extra O-atom in nitrate is derived from water (Aleem *et al.*, 1965):

$$NO_2^- + H_2O \rightarrow NO_3^- + 2H^+ + 2e^-$$

The enzyme responsible is known as nitrite oxidoreductase (strictly, nitrite: acceptor oxidoreductase). It is also capable of oxidizing formate, which is isoelectronic with nitrite but much more easily oxidized: $E_{m,7}(HCO_3^-/HCO_2^-) = -410\,mV$ (Cobley, 1984).

Cells and cell-free extracts of *Nitrobacter* are capable of anaerobic reduction of nitrate to nitrite. Aleem and Sewell (1984) and Sundermeyer-Klinger *et al.* (1984) conclude that nitrate reduction is brought about by nitrite oxidoreductase. In other words, nitrite oxidation is a reversible process. This is important for understanding the activating influence of the protonmotive force, discussed below. The reversibility also implies that the redox centres involved must have potentials close to that of the nitrate/nitrite couple ($E_{m,7} = +430\,mV$). Cytochrome aa_3 is universally agreed to be the terminal oxidase. The properties of cytochrome a_1 and a high potential molybdenum centre will now be described.

Absorption spectra of cells of nitrite oxidizers generally show a peak near 590 nm on reduction, attributed to cytochrome a_1. The protein responsible for this absorption has recently been purified, and the presence of a-type haem confirmed (Tanaka *et al.*, 1983). Redox titrations of isolated membranes revealed two components of cytochrome a_1, present in equal amounts, with potentials of $+350$ and $+100\,mV$ (Sewell *et al.*, 1972). Addition of nitrite led to an immediate partial reduction of cytochrome a_1 (Cobley, 1976b; Aleem, 1977). This was not prevented by uncouplers, unlike the cytochrome c reduction described below (Aleem, 1977).

Ingledew and Halling (1976) reported EPR studies of *Nitrobacter winogradskyi* in which fully oxidized membranes yielded a signal attributed to Mo(V), having a potential of $+340\,mV$. A role in nitrite oxidation was surmised from the fact that nitrate reductase is a molybdoenzyme, but no evidence was presented that the molybdenum was reduced by nitrite, and no further studies have been reported.

4.4.2.2   Purified complexes

Two very different purifications of nitrite oxidoreductase have been published. Tanaka *et al.* (1983) treated *Nitrobacter agilis* with the detergent Triton X-100 and extracted a particle containing cytochrome a_1 and c-type haem in about equal amounts. They denoted it cytochrome a_1c_1. It was composed of subunits of 55, 29 and 19 kDa, the c-type haem being bound to the 29 kDa subunit. Horse cytochrome c was reduced by the cytochrome a_1c_1 in the presence of nitrite, the reaction being inhibited by azide, cyanide and nitrate.

The second purification was reported by Sundermeyer-Klinger *et al.* (1984), who treated *Nitrobacter hamburgensis* with deoxycholate and obtained a complex with 116, 65 and 32 kDa subunits. Molybdenum was present, but only at a stoichiometry of about 0.12 per particle weight of 400 kDa. Haem a_1 was present, but c-type haem was absent. Nitrite was oxidized with ferricyanide as electron acceptor, but whether or not cytochrome c is the 'natural electron acceptor' has not been reported.

It is hard to judge whether these very different preparations reflect the use of different species or different techniques of purification. In both cases, the final activity on a protein basis was lower than for the starting material. Molybdenum and a-type haem may be lost during purification. An involvement of molybdenum is plausible, given that molybdenum is present in nitrate reductase. On the other hand, a recent report claims that mitochondrial cytochrome oxidase possesses some nitrite oxidase activity (Paitian *et al.*, 1985), suggesting that a nitrite oxidoreductase could evolve from an a-type terminal oxidase.

4.4.2.3 Electron entry into the conventional chain

The purified nitrite oxidoreductase readily donates electrons to Class I c-type cytochromes, while addition of nitrite to whole cells or inverted vesicles leads to reduction of c-type haem (see below). These observations suggest that nitrite oxidoreductase donates electrons *in vivo* to the Class I cytochrome, c_{550}. Hence its 'correct' name is nitrite:cytochrome c_{550} oxidoreductase.

4.4.3 Coupling to generation of protonmotive force

4.4.3.1 Location of nitrite oxidation

Nitrite is oxidized rapidly both by whole cells and by inverted vesicles (Cobley, 1976a,b). One of these preparations must involve nitrite and nitrate being transported across the cytoplasmic membrane. It might be thought that since nitrite oxidation leads to release of protons into the cytoplasm, as explained below, it must be located on the cytoplasmic side of the membrane. However, the aa_3-type terminal oxidase is able to abstract protons from the cytoplasm, despite having all its redox centres near the periplasmic face of the membrane (see Chapter 5). As things stand, it is not clear whether nitrite is oxidized on the periplasmic or cytoplasmic face of the membrane.

4.4.3.2 Cytoplasmic release of protons as nitrite is oxidized

The experimental findings are as follows. With whole cells plus nitrite, uncouplers decrease the rate of nitrite oxidation and decrease the degree of reduction of c-type cytochromes (Kiesow, 1967; Cobley, 1976b). With inverted membrane vesicles, the rate of nitrite oxidation is decreased by reagents that collapse the membrane potential, but not by reagents that prevent a pH gradient from forming across the membrane (Cobley, 1976a).

The simple coupling of nitrite oxidation to cytochrome c reduction is an endergonic process, $\Delta G^{0\prime}$ being $+30.5\,kJ.mol^{-1}$ (Wood, 1987). However, the reduction of cytochrome c by electrons from nitrite requires a release of two protons from water:

$$NO_2^- + H_2O + 2\ cyt\ c(Fe^{3+}) \rightarrow NO_3^- + 2H^+ + 2\ cyt\ c(Fe^{2+})$$

Wood (1987) has discussed how release of one or both of these protons into the *cytoplasm* makes nitrite oxidation a more favourable process. It was assumed that any nitrite and nitrate transport occurs by a simple nitrite/nitrate antiporter, and therefore involves no net transfer of protons or electric charge. The observed dependence of nitrite oxidation on the proton-motive force with whole cells, and on the membrane potential with inverted vesicles, is consistent with release of both protons into the cytoplasm. A release of both protons into the periplasm would lead to energization having no effect with whole cells, while release of one proton on each side is inconsistent with the simple dependence on the membrane potential with inverted vesicles.

4.4.3.3   Evidence for a proton-pumping cytochrome oxidase

The suggestion that nitrite oxidation is coupled to H^+ release into the cytoplasm was first presented by Cobley (1976b). He rejected the possibility that both protons were released into the cytoplasm, because this would provide no net proton translocation when coupled with a classical cytochrome oxidase reaction:

$$NO_2^- + H_2O \rightarrow NO_3^- + 2H^+{}_{in} + 2e^-$$

$$2H^+{}_{in} + 0.5O_2 + 2e^- \rightarrow H_2O$$

However, it has become clear since Cobley's work that in many bacteria O_2 reduction by cytochrome aa_3 is associated with translocation of $2H^+$ per $2e^-$ from cytoplasm, in *addition* to the $2H^+$ consumed in reduction of oxygen to water (see Chapter 5):

$$2\ cyt.\ c(Fe^{2+}) + 4H^+{}_{in} + 0.5O_2 \rightarrow 2\ cyt.\ c(Fe^{3+}) + 2H^+{}_{out} + H_2O$$

Wetzstein and Ferguson (1985) have used artificial electron donors to test proton-pumping by the terminal oxidase in whole cells. Despite technical difficulties, data were obtained implying that cytochrome aa_3 did translocate protons. Results obtained with the purified oxidase are discussed in Wood (1987).

4.4.3.4   Integration of respiration

The coupling of nitrite oxidation to ATP synthesis has been studied with inverted membrane vesicles; P/O rates of 0.8–0.9 were obtained by Aleem (1968), while Cobley (1976a) reported lower values of 0.2–0.35.

A scheme for electron transfer and proton translocation in *Nitrobacter* is presented in Figure 4.4. Nitrite oxidation is placed centrally in the membrane, to emphasize its uncertain location. Both protons are shown being released into the cytoplasm, as suggested above. A proton-pumping cytochrome oxidase will then permit a net translocation of $2H^+$ per $2e^-$. The predicted P/O ratio would approach 1.0, both with whole cells and with inverted vesicles.

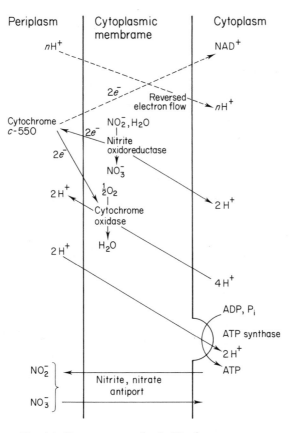

Fig. 4.4   Energy conservation in *Nitrobacter*.

4.5   Oxidation of inorganic sulphur

4.5.1   Introduction

4.5.1.1   Microbiology

Oxidation of reduced sulphur compounds forms an energy source for a range of obligate and facultative chemoautotrophs and mixotrophic bacteria. Most of the autotrophic sulphur-oxidizing bacteria are motile Gram-negative rods, assigned to the genus *Thiobacillus*. Table 4.2 provides a brief guide to the metabolic capabilities of the species mentioned below. In addition, *Sulfolobus* is a genus of sulphur-oxidizing Archaebacteria, isolated from acidic hot springs. Gliding sulphur bacteria with a predominantly heterotrophic metabolism are typified by the genus *Beggiatoa*. Further information on the microbiology of these and other sulphur-oxidizing bacteria can be found in 'The Prokaryotes' (Starr *et al.*, 1981). The interactions between different types of sulphur-oxidizing bacteria are discussed in Kuenen *et al.* (1985), and aspects of their ecology are reviewed by Jørgensen (1988).

The same oxidations are also brought about by photosynthetic sulphur bacteria, which use reduced states of sulphur as electron donors for NAD(P)H biosynthesis (see page 9). Many strains of purple sulphur bacteria can also grow in the dark with O_2 as electron acceptor (Kämpf and Pfennig, 1980; Kelly and Kuenen, 1984). This demonstrates that there is no clear-cut distinction with respect to sulphur oxidation between sulphur-oxidizing chemolithotrophs and sulphur-oxidizing phototrophs. For this reason the enzymology of sulphur oxidations in both classes of organism is discussed below. The Sulphur Cycle in nature is shown on page 8.

The enzymes involved in oxidation of inorganic sulphur by *Thiobacilli* have been reviewed and placed in a historical perspective by Kelly (1982, 1985, 1988), and Trüper and Fischer (1982) have reviewed the biochemistry of sulphur compounds as electron donors for bacterial photosynthesis.

4.5.1.2   Chemistry

The redox chemistry of sulphur has complexities not encountered with other substrates for chemolithotrophs (Figs 4.5 and 4.6). The element exists primarily in S_8 rings, which have a low solubility in water and readily form a separate phase of globules or colloidal sol. A two-electron reduction gives sulphide, a four-electron oxidation gives sulphite, and a further two-electron oxidation of sulphite yields sulphate. Sulphur and sulphite react together to form thiosulphate ($S_2O_3^{2-}$), which is the most stable reduced state at neutral pH. Other inorganic states include dithionate ($^-O_3S–SO_3^-$), trithionate ($^-O_3S–S–SO_3^-$) and higher thionates with chains of sulphur atoms

Table 4.2 Major species of *Thiobacilli*

Name	Preferred pH	Predominant carbon source	Donors in place of S compounds	Acceptors in place of O_2
T. denitrificans	Neutral	CO_2	—	NO_3^-, NO_2^-, N_2O
T. ferrooxidans	Acidic	CO_2; organic for some strains	Fe^{2+}; organic for some strains	—
T. intermedius	Acid tolerant	CO_2 or organic	—	—
T. neapolitanus	Neutral	CO_2	—	—
T. novellus	Neutral	CO_2 or organic	Organic	—
T. thiooxidans	Acidic	CO_2	—	—
T. versutus (*T. A₂*)	Neutral	CO_2 or organic	Organic	NO_3^-

Compiled from Kuenen and Tuovinen (1981) and Kelly (1985).

linking the two sulphonate groups. (Dithionate should not be confused with dithionite (^-O_2S–$SO_2$$^-$), indispensable in studies of redox proteins, but far too reactive to have any importance *in vivo*.) Figure 4.5 gives pK values and protonation states for the most important inorganic forms of sulphur.

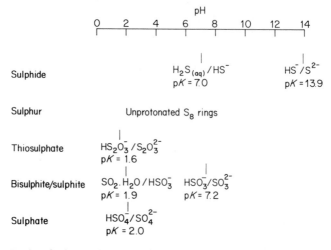

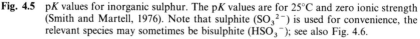

Fig. 4.5 pK values for inorganic sulphur. The pK values are for 25°C and zero ionic strength (Smith and Martell, 1976). Note that sulphite ($SO_3$$^{2-}$) is used for convenience, the relevant species may sometimes be bisulphite ($HSO_3$$^-$); see also Fig. 4.6.

The calculation of appropriate redox potentials is complicated by the fact that elemental sulphur is not formed in the thermodynamic standard state of rhombic sulphur. In many cases, sulphur accumulates as refractile globules. X-ray diffraction studies have shown these to consist of a supercooled liquid state, which can be formed non-biologically by condensation of sulphur vapour (Hageage *et al.*, 1970). Schedel and Trüper (1980) estimated the relative Gibbs energy of different forms of sulphur by comparing activation energies for reaction with cyanide (S + HCN → SCN$^-$ + H$^+$). They concluded that the intracellular sulphur sol formed by *T. denitrificans* had ΔG^0_f of about +30 kJ.g atom^{-1} relative to rhombic sulphur. A hydrophilic sulphur sol formed non-biologically had an excess Gibbs energy of about +10 kJ.g atom^{-1}. A reactive state of sulphur has a *higher* than normal redox potential for formation from sulphide, and a *lower* redox potential for oxidation to sulphite. Figure 4.6 gives redox potentials at pH 7.0, including values for rhombic sulphur and the intracellular sol of Schedel and Trüper (1980) as possible extremes.

Steudel *et al.* (1987) regard sulphur globules as micelles. The hydrophobic interior is predominantly S_8, with minor amounts of S_7 and S_9 preventing crystallization. An amphipathic coating is provided by polythionates, up to $^-O_3S-S_{11}-SO_3^-$ in size.

A number of non-enzymic reactions are important. Sulphur and sulphite react together to form thiosulphate $(S + SO_3^{2-} \rightarrow S_2O_3^{2-})$. At equilibrium, the ratio $[S_2O_3^{2-}]/[HSO_3^-]$ at pH 7.0 is 3×10^4 in the presence of rhombic sulphur, rising to 5×10^9 if the sulphur is in the reactive form reported by Schedel and Trüper (1980). Such high equilibrium constants imply that direct fission of thiosulphate requires an input of energy. The sulphide and sulphite redox states are both liable to autooxidation, which for sulphite often proceeds by chain reactions. A further complication is the tendency of sulphide to react with sulphur to form polysulphides. Enzyme-bound polysulphides may well be important in building up and breaking down S_8 rings.

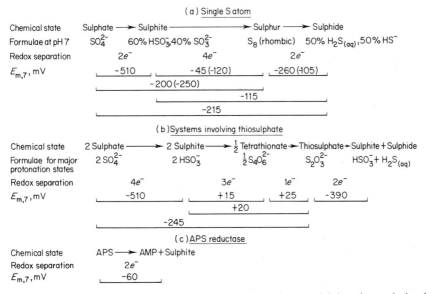

Fig. 4.6 Redox potentials at pH 7.0 for inorganic sulphur. The potentials have been calculated from ΔG_f^0 values in Thauer *et al.* (1977). (Bard *et al.* (1985) gives an older ΔG_f^0 value for thiosulphate, but is otherwise identical.) The redox potential for APS (adenosine phosphosulphate) is from Thauer *et al.* (1977). Values in parentheses are for a sulphur sol with $\Delta G_f^0 = +30\,\mathrm{kJ.mol^{-1}}$ (see text). For sulphite and sulphide, two protonation states are present at pH 7.0. Redox potentials near a pK value can be calculated by the following general method (Clark, 1960). If the redox equation on the acid side of the pK value is $\mathrm{ox} + ne^- + mH^+ \rightarrow \mathrm{red}$, with standard potential E_m, while on the alkaline side the equation becomes $\mathrm{ox} + ne^- + (m-1)H^+ \rightarrow \mathrm{red}$, then

$$E_m(\mathrm{pH}) = E_m - (m/n)(2.303\,RT/F).\mathrm{pH} + {}^*(1/n)(2.303\,RT/F).\log(1 + 10^{\mathrm{pH}-\mathrm{p}K})$$

If the equation at alkaline pH is $\mathrm{ox} + ne^- + (m+1)H^+ \rightarrow \mathrm{red}$, the plus sign marked * is changed to minus. (See also note in legend for Fig. 4.2.)

4.5.1.3   Pathways of sulphur oxidation

In the 1960s it was widely believed that bacterial sulphur oxidation took place by complex cyclic pathways involving polythionates. However, Kelly (1982) concludes a description of these proposals with the statement, 'I am firmly convinced that polythionates have no central role as free intermediates of inorganic sulphur oxidation in thiobacilli'. Such pathways will therefore not be discussed further. Only one enzyme of polythionate metabolism has been isolated; it synthesizes tetrathionate from thiosulphate.

The pathways proposed for different chemolithotrophs and phototrophs are summarized in Figure 4.7. The intermediate reductants will be discussed in the order: sulphite, sulphur, sulphide and thiosulphate. As will become apparent, each conversion can be catalysed in at least two different ways. Many *Thiobacilli* can also be grown with di-, tri- or higher thionates as reductant (Kelly, 1982).

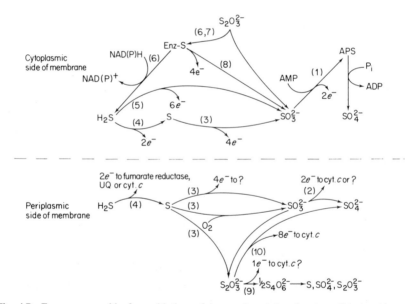

Fig. 4.7   Enzyme ensemble for oxidation of inorganic sulphur in chemolithotrophic and photosynthetic bacteria. Reaction of sulphur with sulphite to form thiosulphate is spontaneous (see text); all other reactions involve the enzymes described in text. Electrons produced on the cytoplasmic side of the membrane enter the electron transport chain at the level of ubiquinone/cytochrome *b*. (1) APS reductase; (2) 'sulphite oxidase'; (3) 'sulphur oxygenase' and other unknown enzymes; (4) two oxidation routes, donating electrons to the cytochrome chain at the level of cytochrome *c* or at the level of ubiquinone/cytochrome *b*; (5) 'reverse' sulphite reductase; (6) thiosulphate reductase; (7) rhodanese; (8) oxidation of sulphur bound to rhodanese by 'reverse' sulphite reductase; (9) thiosulphate-oxidizing enzyme (thiosulphate-combining enzyme); (10) thiosulphate-oxidizing enzyme.

4.5.1.4 Transport of inorganic sulphur

The transport of inorganic forms of sulphur across the cytoplasmic membranes of sulphur-oxidizing bacteria is poorly characterized. Sulphide is likely to diffuse across in the uncharged H_2S state. There is indirect evidence for transport of sulphur, despite its low solubility in water. Thus sulphide-oxidizing cyanobacteria have the photosynthetic electron transport chain located on internal thylakoid membranes, but they deposit sulphur outside the cell (Cohen *et al.*, 1975). Conversely, most purple sulphur bacteria store sulphur in intracytoplasmic globules, but are also able to make use of sulphur added to the growth medium (Trüper, 1978). The two pathways discussed below for sulphite oxidation are likely to be located on opposite sides of the cytoplasmic membrane. The choice between them is probably not dictated by the site of sulphite formation, implying a transport mechanism for sulphite or bisulphite ions. Finally, sulphate produced by the cytoplasmic pathway of sulphite oxidation must be transported out of the cell.

4.5.1.5 Molar growth yields

Studies of molar growth yields have particular significance for sulphur-oxidizing bacteria, because it is not generally possible to deduce which of the alternative pathways is in operation by assaying individual enzymes (Kelly, 1982, 1985). Comparisons of two sorts have proved instructive: a single species growing with different reductants, or different species growing with the same reductant. *T. versutus* provides a useful yardstick: the thiosulphate-combining enzyme is absent, and thiosulphate oxidation is believed to proceed solely by direct conversion to sulphate (Lu and Kelly, 1984b). APS reductase is also absent, implying that sulphite oxidation is a periplasmic process. The molar growth yields on thiosulphate and sulphide are similar (Kelly and Kuenen, 1984). Both these processes are 8-electron oxidations; that of thiosulphate is believed to be coupled to the conventional electron transport chain at the level of cytochrome *c* (Lu, 1986). The similar molar growth yield on sulphide suggests a similar point of coupling for the conversion of sulphide to sulphite. It would certainly not be explained by sulphur oxidation to sulphite being mediated by a non-energy conserving sulphur oxygenase.

    T. denitrificans was found to give higher growth yields on a range of substrates than other species including *T. versutus* (Justin and Kelly, 1978). This suggests that in *T. denitrificans* more steps are coupled to the conventional electron transport chain at the ubiquinone–cytochrome *b* level. This is probably essential for it to be able to grow as a facultative anaerobe, with reduced forms of nitrogen as electron acceptors.

4.5.2   Sulphite oxidation

4.5.2.1   Reversal of adenosine phosphosulphate reductase (Fig. 4.7, reaction 1)

For sulphite oxidation to sulphate, two mechanisms are well-established. In *T. thioparus* and *T. denitrificans*, sulphite undergoes an oxidative addition to AMP:

$$SO_3^{2-} + AMP \rightarrow APS + 2e^-$$

The product is adenylyl sulphate, also known as adenosine phosphosulphate (APS). The enzyme name, APS reductase, is derived from the reverse reaction, carried out by a very similar enzyme in sulphate respirers (Chapter 2). The purified enzyme from *T. thioparus* contained 1 FAD, 8–10 iron and 4–5 labile sulphides per polypeptide of 170 kDa (Lyric and Suzuki, 1970a). A similar enzyme has been obtained from *T. denitrificans* (Bowen *et al.*, 1966). Further reactions of APS liberate sulphate and result in synthesis of one high-energy phosphate bond by substrate-level phosphorylation (Peck, 1960; Kelly, 1982):

$$APS + P_i \rightarrow ADP + SO_4^{2-}$$

$$2\ ADP \rightarrow AMP + ATP$$

It has also been proposed that an ATP sulphurylase may yield ATP in place of ADP by use of pyrophosphate in place of phosphate (Aminuddin, 1980):

$$APS + PP_i \rightarrow ATP + SO_4^{2-}$$

APS reductase presumably has a cytoplasmic location, given its requirement for AMP. For the purified enzyme, ferricyanide and horse cytochrome c have been used as acceptors. The redox potential for the reaction ($E_{m,7} = -60\ mV$; Fig. 4.6) is consistent with a coupling to the conventional respiratory chain at the ubiquinone–cytochrome b level, as for succinate. However, the physiological acceptor is not known. APS reductase is also found in certain photosynthetic sulphur bacteria, where it is more intimately associated with the membranes and can be a haemoprotein (Trüper and Fischer, 1982).

4.5.2.2   Direct oxidation of sulphite to sulphate (Fig 4.7, reaction 2)

In several other species, sulphite is oxidized by a sulphite:cytochrome c oxidoreductase, loosely referred to as 'sulphite oxidase' (Kelly, 1982). This has been purified from *T. novellus* (Toghol and Southerland, 1983) and *T. versutus* (Lu and Kelly, 1984a). The purified enzyme has fluorescence spectra similar to those for sulphite oxidase from liver mitochondria, attributed to a molybdenum–pterin cofactor (Toghol and Southerland, 1983). Mitochondrial sulphite oxidase contains b-type haem; the bacterial enzyme differs in having c-type haem, with an α-band at 550–551 nm. The cytochrome c_{551} component has also been obtained in a form without enzymic activity (Yamanaka *et al.*, 1981b; Lu and Kelly, 1984a). A similar enzyme has been reported from a number of purple photosynthetic bacteria (Trüper and Fischer, 1982).

The presence of c-type haem suggests a periplasmic location (page 193), as has been confirmed for the related thiosulphate-oxidizing enzyme (see below). The purified enzyme donates electrons to soluble cytochrome c. This is the probable point of electron entry into the conventional chain *in vivo*.

A sulphite-oxidizing enzyme with more efficient energy conservation appears to be present in *T. denitrificans* (Sawhney and Nicholas, 1977). Electron flow from sulphite to nitrate in nitrate-grown cells was associated with the membrane fraction. An active preparation, solubilized with deoxycholate, contained flavin and haem and was capable of nitrate reduction with NADH or sulphite as electron donor. An involvement of APS reductase was ruled out by activity in the absence of AMP. Antimycin was inhibitory, suggesting that sulphite oxidation was coupled to a site prior to cytochrome b in the conventional chain.

4.5.3 Sulphur oxidation (Fig. 4.7, reaction 3)

In many species, sulphide oxidation results in the appearance of elemental sulphur. The elemental sulphur is extracellular for green photosynthetic bacteria and most growth regimes of *Thiobacilli* (Trüper, 1978; Buchanan and Gibbons, 1974). Sulphur may also be deposited on the outer membrane and cell wall (Gromova *et al.*, 1983). In *Beggiatoa*, apparently intracellular deposits of sulphur are separated from the cytoplasm by a conventional membrane, produced by deep intrusions of the bacterial periplasm (Larkin and Strohl, 1983). However, the sulphur globules of most purple sulphur bacteria are truly cytoplasmic, being separated from the cytoplasm by a unimolecular layer of protein similar to that around a gas vacuole (Trüper and Fischer, 1982; Nicholson and Schmidt, 1971).

A sulphur-oxidizing enzyme, 'sulphur oxygenase', formed the subject of several studies in the 1960s (Kelly, 1982). It brought about the following reaction, S_n being the oligomeric state of elemental sulphur:

$$S_n + O_2 + H_2O \rightarrow S_{n-1} + HSO_3^- + H^+$$

Reduced glutathione served as cofactor. The enzyme was purified and found to contain non-haem iron (Suzuki and Silver, 1966). The reaction takes place without energy conservation, making it a curiously wasteful (if not improbable) process. One possibility is that the enzyme acquires its ability to reduce molecular O_2 when separated from a physiological electron acceptor. The activity of any such enzyme *in vivo* must be controlled in some way, or sulphur would not be observed to accumulate. As a further objection, the sulphur oxygenase reaction requires the presence of molecular O_2. It therefore cannot mediate sulphur oxidation in photosynthetic sulphur bacteria (normally anaerobes), or in anaerobic cultures of *T. denitrificans*.

The location of sulphur oxidation is unknown but the simplest hypothesis is that oxidation takes place on the same side of the membrane as sulphur storage in globules.

4.5.4   Sulphide oxidation

4.5.4.1   Sulphur as product (Fig. 4.7, reaction 4)
The enzymes involved in sulphide oxidation to sulphur are poorly characterized. They appear to have different locations and sites of coupling in different organisms. The comments in the previous section on the location of sulphur oxidation apply equally to the location of sulphide oxidation to sulphur, with the proviso that newly formed sulphur *may* be transported across the membrane.

The anaerobe *Wolinella succinogenes* has recently been reported to use sulphide oxidation coupled to fumarate reduction as an energy source (Macy *et al.*, 1986; page 141). A plausible respiratory chain would consist of sulphide oxidation on the outer face of the membrane, inward movement of electrons (by unknown carriers) and uptake of protons from the cytoplasm as fumarate is reduced to succinate:

$$H_2S \rightarrow S + 2H^+_{out} + 2e^-$$

$$2e^- + 2H^+_{in} + \text{fumarate}^{2-} \rightarrow \text{succinate}^{2-}$$

In *Beggiatoa* and certain photosynthetic bacteria, aerobic sulphide oxidation is inhibited by HOQNO and antimycin A, suggesting a similar point of coupling with the conventional respiratory chain (prior to cytochrome *b*) (Knaff and Buchanan, 1975; Larkin and Strohl, 1983). A P/O ratio of 1.4 was measured for *T. novellus* (Cole and Aleem, 1973).

In other photosynthetic bacteria, sulphide oxidation is mediated by a soluble flavocytochrome *c*, which uses the Class I cytochrome *c* as electron acceptor (Trüper and Fischer, 1982). This implies a less efficient energy conservation. In *T. concretivorus*, sulphide oxidation to a membrane-bound form of sulphur is also believed to involve *c*-type cytochromes (Kelly, 1982).

4.5.4.2   Reversal of sulphite reductase (Fig. 4.7, reaction 5)
Schedel and Trüper (1979) have purified a protein from *T. denitrificans* that closely resembles sulphite reductase from *Desulfovibrio* and other sulphate

respirers (page 127). It contained siro-haem and iron–sulphur centres, and was capable of reducing sulphite to sulphide when supplied with artificial reductant. A similar enzyme was obtained from *Chromatium vinosum* grown photoautotrophically with sulphide as reductant (Trüper and Fischer, 1982). The authors proposed that the enzyme operates *in vivo* in the direction of sulphide oxidation. They were unable to test this hypothesis with the purified enzyme, because suitable electron acceptors were all reduced by sulphide non-enzymically. However, incubation of sulphide with an equivalent amount of oxidized enzyme led to its oxidation, with thiosulphate as the predominant product.

Oxidation of sulphide to sulphite is a proton-releasing reaction:

$$HS^- + 3H_2O \rightarrow SO_3^{2-} + 7H^+ + 6e^-$$

Since the development of a protonmotive force requires proton *removal* from the cytoplasm, Hooper and DiSpirito (1985) postulate a periplasmic location for 'reverse' sulphite reductase (see Section 4.2.3). The same line of reasoning would predict a cytoplasmic location for the sulphite reductase mode of operation in *Desulfovibrio* (Wood, 1978). This has been confirmed for *D. gigas* (Bell *et al.*, 1974). Whether the 'reverse' sulphite reductase in sulphide oxidizers has a different location remains to be seen.

4.5.5   Thiosulphate oxidation

4.5.5.1   Cleavage by rhodanese (Fig. 4.7, reactions 6,7,8)
The *reductive* cleavage of thiosulphate into sulphide and sulphite has been the subject of several studies. Peck (1960) showed that cell extracts of *T. thioparus* were able to metabolize thiosulphate when reduced glutathione was also present. He regarded the initial reaction as a cleavage to sulphide and sulphite, with reduced glutathione as reductant:

$$S_2O_3^{2-} + 2[H] \rightarrow HS^- + SO_3^{2-} + H^+$$

The enzyme responsible was named thiosulphate reductase. In a second study, Silver and Kelly (1976) demonstrated that the enzyme rhodanese (EC 2.8.1.1) from *Thiobacillus* A2 (later named *T. versutus*) catalysed the reaction of thiosulphate with dihydrolipoate to form sulphite and dihydrolipoate persulphide:

$$\text{lip}\begin{smallmatrix} S^- \\ \\ SH \end{smallmatrix} + S_2O_3^{2-} \rightarrow \text{lip}\begin{smallmatrix} S\text{--}S^- \\ \\ SH \end{smallmatrix} + SO_3^{2-}$$

This was detected chromatographically, but it rapidly decomposed into free sulphide and oxidized lipoate:

$$\text{lip}\!\!\begin{array}{c} \text{S–S}^- \\ \\ \text{SH} \end{array} \quad \rightarrow \quad \text{lip}\!\!\begin{array}{c} \text{S} \\ \\ \text{S} \end{array} \quad + \text{HS}^-$$

Schedel and Trüper (1980) found the activities of thiosulphate reductase and rhodanese to be equal in growing cells of *T. denitrificans*, strongly suggesting that they are different reactions of a common enzyme. Thus in the experiments of Peck (1960), glutathione would have acted in place of dihydrolipoate as S-atom acceptor. Nevertheless, it is not clear that reductive cleavage of thiosulphate is important during growth with thiosulphate oxidation as energy source. The redox potential for the reduction is $-400\,\text{mV}$ (Fig. 4.6), implying that one molecule of NAD(P)H per thiosulphate needs to be oxidized, regardless of the nature of the immediate H-atom donor. The reoxidation of sulphide to sulphur or sulphite releases electrons at *higher* redox potential (Fig. 4.6). The growth yield with thiosulphate would therefore be expected to be lower than with sulphide, contrary to experimental findings (Kelly, 1982).

One pathway that would avoid this problem would be for sulphur bound to rhodanese to be oxidized to sulphite by the reverse sulphite reductase or one of the sulphur-oxidizing enzymes:

$$\text{E–SH} + S_2O_3{}^{2-} \rightarrow \text{E–S–SH} + SO_3{}^{2-}$$

$$\text{E–S–SH} + 3H_2O \rightarrow \text{E–SH} + SO_3{}^{2-} + 6H^+ + 6e^-$$

The *overall* redox potential for thiosulphate oxidation to sulphite is only slightly less favourable than for free sulphur, being $+20\,\text{mV}$ as opposed to $-45\,\text{mV}$.

Schedel and Trüper (1980) reported that anaerobic oxidation of thiosulphate by *T. denitrificans* led to about half the sulphane sulphur appearing as a milky sol in the cytoplasm. They interpreted the sol as being caused by a direct cleavage of thiosulphate, without any reductive step. However, the reaction of rhodanese with thiosulphate forms a cysteine persulphide at the active site (Finazzi Agrò *et al.*, 1972) and it is hard to understand how this could be cleaved without expenditure of energy.

Rhodanese is an almost universal enzyme in microorganisms, plants and animals. It has recently been proposed to function in the biosynthesis of iron–sulphur centres (Cerletti, 1986). There is no evidence that the levels of rhodanese in *Thiobacilli* and photosynthetic sulphur bacteria increase during growth on thiosulphate. Further work is required before the true importance of rhodanese in thiosulphate oxidation can be assessed.

4.5.5.2 Thiosulphate-combining enzyme (Fig 4.7, reaction 9)

A widely distributed enzyme catalyses the conversion of thiosulphate into tetrathionate, with a soluble cytochrome c or ferricyanide as one-electron acceptor (Trudinger, 1967; Kelly, 1982):

$$2S_2O_3^{2-} \rightarrow S_4O_6^{2-} + 2e^-$$

The enzyme is referred to in the literature as thiosulphate-oxidizing enzyme (an ambiguous term; see next section) or thiosulphate-combining enzyme. It has been purified from *T. ferrooxidans* (Silver and Lundgren, 1968) and *T. thioparus* (Lyric and Suzuki, 1970b); other sources include thiosulphate-oxidizing strains of photosynthetic bacteria (Trüper and Fischer, 1982) and *Ps. aeruginosa* (Schook and Berk, 1979). The enzyme is reported to have no significant absorption above 300 nm, implying an absence of flavin or haem prosthetic groups. The reported molecular mass for different species ranges from 35 to 115 kDa (Lyric and Suzuki, 1970b; Schmitt *et al.*, 1981), and inhibitors include cyanide, sulphydryl reagents and sulphite (Lyric and Suzuki, 1970b; Kusai and Yamanaka, 1973; Schook and Berk, 1979).

The affinity of the enzyme for soluble c-type cytochromes as electron acceptor suggests a periplasmic location, but this has not been directly demonstrated. The further metabolism of tetrathionate is thought to involve non-enzymic reactions of disproportionation and hydrolysis, and also perhaps uncharacterized enzymes (Trüper, 1978; Kelly, 1982). These conversions eventually return the sulphur to S_1-states and thiosulphate. Such reactions would explain the frequent appearance of elemental sulphur in the medium when thiosulphate is being oxidized.

4.5.5.3 Single-enzyme oxidation to sulphate (Fig. 4.7, reaction 10)

A thiosulphate-oxidizing enzyme has been purified from *T. versutus* and intensively studied by Kelly and co-workers (Lu *et al.* 1985). The enzyme complex contained four proteins, denoted A, B, cytochrome c_{551} and cytochrome $c_{552.5}$. These were present in the stoichiometry 1:2:6:2, and their respective molecular masses were 16, 32, 43, 29 kDa. Cytochrome c_{551} was purified from the complex and contained four haems (two with $E_{m,7}$ of $+240$ and two $E_{m,7}$ of -115 mV). Cytochrome $c_{552.5}$ was also purified; each molecule contained two haems, which titrated at $+220$ mV and about -215 mV (Lu *et al.*, 1984). The complexity of the complete enzyme can be compared to that of hydroxylamine oxidoreductase (Section 4.3.3).

Component A was found to bind thiosulphate with a stoichiometry of 1:1, the binding being inhibited by sulphite. Activity was inhibited by sulphydryl reagents, which did not prevent binding. Both enzyme B and cytochrome c_{551} gave rise to fluorescence spectra similar to those for 'sulphite oxidase' (see above), indicating the presence of the same molybdenum cofactor.

Indeed, there is evidence that the cytochrome c_{551} component *is* the 'sulphite oxidase' enzyme; although when this is integrated into the enzyme complex, the ability to oxidize sulphite is lost (Lu and Kelly, 1984a).

The enzyme complex carried out the complete eight-electron oxidation of thiosulphate to sulphate, which could be coupled to reduction of an equivalent amount of horse cytochrome *c*. The mechanism is discussed by Lu *et al.* (1985); the central question is whether the S–S bond is cleaved in the initial 2-electron oxidation, or after oxidation has progressed to the level of dithionate ($^-O_3S-SO_3^-$).

The location of the thiosulphate multienzyme complex in *T. versutus* has been shown to be periplasmic (Lu, 1986). Cole and Aleem (1973) obtained a P/O ratio of 0.9 for thiosulphate oxidation by particles from *T. novellus*, while an H^+/O ratio of 1.9 was obtained by Drozd (1974) for whole cells of *T. neapolitanus*. These values correspond to electrons feeding in at the level of cytochrome *c*, plus H^+ uptake by a non-proton pumping cytochrome oxidase. Lu *et al.* (1984) discuss the possibility that some of the electrons from the complex may reach the conventional chain at the ubiquinone–cytochrome *b* level. A similar suggestion for hydroxylamine oxidoreductase is considered in Section 4.3.4.

A molybdenum requirement during growth with thiosulphate oxidation as energy source has been demonstrated for *T. versutus*, *Thiosphaera pantotropha* and *Paracoccus denitrificans* (Friedrich *et al.*, 1986).

4.6   Oxidation of metal ions at acidic pH

4.6.1   Introduction

Thiobacillus ferrooxidans provides the classic example of ferrous iron oxidation as an energy source for growth. It is an obligate acidophile, which grows optimally on Fe^{II} at a pH of 2.0. The biochemistry of Fe^{II} oxidation by *T. ferrooxidans* has been reviewed by Ingledew (1982). Iron oxidation serves as an energy source for a few other acidophiles, notably strains of *Sulfolobus*, but *T. ferrooxidans* is the only organism to have been studied in detail. The oxidation of iron at neutral or only slightly acid pH has a different character owing to the precipitation of ferric hydroxide (see Section 4.7).

T. ferrooxidans has been implicated in oxidation of several non-ferrous metal ions, including stannous (Sn^{II}), cuprous (Cu^I) and uranous (U^{IV}) ions. However, iron is required by the cells for synthesis of haem proteins and, as has been stated before, other transition metal ions tend to exchange electrons with iron to establish a common redox poise. Ingledew (1982) discusses claims that oxidations of non-ferrous metals are catalysed directly by the cells, but considers that the presence of iron in culture media or bound to harvested cells makes such claims difficult to prove. At present there is no evidence that non-ferrous metals are oxidized by a different pathway.

The ability of *T. ferrooxidans* to grow by oxidation of ferrous iron is markedly dependent on the choice of anion, with sulphate being far superior to phosphate or chloride (Ingledew, 1982). This phenomenon is partly related to growth at an acidic pH (see Chapter 8). However, the sulphate anion binds to both ferrous and ferric iron, to form $FeSO_{4(aq)}$ and $(FeSO_4)^+$ respectively. The precise binding constants depend on the ionic strength, but binding to ferric iron is generally about 100 times stronger than binding to ferrous (Sillén and Martell, 1964). This binding will lower the effective Fe^{3+}/Fe^{2+} redox potential from its standard value of $+770\,mV$ to about $+650\,mV$, a value found by Ingledew (1982) with equal concentrations of Fe^{II} and Fe^{III} in the growth medium. One reason why sulphate is more effective than other common anions is that it retains its double negative charge at pH 2.0.

4.6.2 Electron flow from Fe(II) to molecular oxygen

This is summarized in Figure 4.8. There are several lines of evidence for an extracellular oxidation of iron (Ingledew, 1982). Experiments with cell-free membranes indicate that only right-side-out vesicles have significant 'iron oxidase' activity. Furthermore, at the cytoplasmic pH of 6.5, iron would be rapidly autooxidized, and the cytoplasm would fill with precipitated ferric hydroxide.

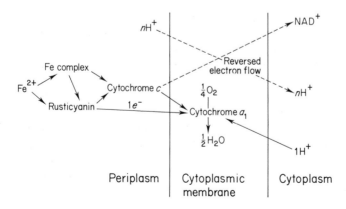

Fig. 4.8 Electron transport in *Thiobacillus ferrooxidans*.

The pathway of electron flow from ferrous ions in solution to the terminal oxidase remains uncertain, because of a plethora of one-electron carriers:

(a) A very broad EPR signal was abolished by multiple washing, implying an extracytoplasmic location (Ingledew and Cobley, 1980; Ingledew, 1982). It was similar to the EPR signal for phosvitin, an iron storage protein in egg yolks, and was attributed to a polynuclear complex of ferric iron atoms linked by hydroxyl groups. The iron atoms titrated over the range $+150$ to $+500$ mV in a redox titration.

(b) A 'blue' copper protein has been purified from *T. ferrooxidans* and named rusticyanin (Ingledew, 1982). The polypeptide, of 16.3 kDa, is associated with a single copper atom, and has EPR properties as for 'Type I' copper proteins such as azurin and plastocyanin. A redox potential of $+680$ mV was measured at pH 3.2.

(c) A soluble c-type cytochrome, c_{552}, has an E_m of $+640$ mV at pH 3.2.

Cytochrome c_{552} and rusticyanin interact with extracellular paramagnetic probes, confirming a periplasmic location. Cytochrome c_{552} is likely to be a Class I cytochrome, donating electrons to the terminal oxidase. The kinetics of electron transfer from Fe(II) in solution to rusticyanin are relatively slow (Lappin *et al.*, 1985), suggesting that bound iron atoms are important as an intermediate. Rusticyanin is likely to equilibrate with cytochrome c_{552}, but it may also donate electrons directly to the terminal oxidase. Impaired iron oxidase activity in mutants has been found to correlate with an absence of rusticyanin (Cox and Boxer, 1986).

The terminal oxidase of *T. ferrooxidans* has an α-band at 597 nm on reduction and has consequently been denoted cytochrome a_1 (see Chapter 5). Its haem redox potentials are $+725$ and $+610$ mV at pH 3.2, falling to more conventional values of $+500$ and $+420$ mV at pH 7.0 (Ingledew and Cobley, 1980).

4.6.3 Coupling to energy conversion

Iron-oxidizing cells of *T. ferrooxidans* maintain a cytoplasmic pH of 6.5 during growth at pH 2.0. The membrane potential for such cells is close to zero, implying an overall pmf of about -260 mV (Ingledew, 1982). This pmf is unusual in bacteria in being made up of only a pH gradient, which is used for ATP synthesis and for driving 'reverse' electron transport to NAD^+, to provide NADH for biosynthesis.

As discussed below, electron transport may, in theory, be coupled to proton translocation; however, the sole function of electron transport is best regarded as a way of consuming protons in the oxidase reaction, thus maintaining the internal pH at pH 6.5 by 'mopping up' the protons that enter by way of the ATP synthase down the 'environmentally produced' pH gradient.

At pH 2.0, it is not possible to measure H^+ pumping directly with a pH electrode. A proton:electron stoichiometry can therefore only be deduced by theoretical arguments. At a cytoplasmic pH of 6.5, the redox potential for reduction of O_2 to water is about $+850\,mV$. The E_h of the Fe^{3+}/Fe^{2+} couple in a chemostat can be as high as $+770\,mV$. Thus at first sight it might appear that significant proton translocation would be impossible. However, the true driving force is that for the reaction:

$$0.5O_2 + 2Fe^{2+} + 2H^+_{out} \rightarrow H_2O + 2Fe^{3+}$$

At pH 2, the (O_2/H_2O) redox potential is $+1100\,mV$ (O_2 at 0.21 atmospheres). If the ambient Fe^{3+}/Fe^{2+} redox potential is $+770\,mV$ (see above), ΔE for the reaction will be $+330\,mV$. Thus a Δp of $-260\,mV$ permits coupling to pumping of $1H^+$ per $1e^-$ from the cytoplasm, but a proton-pumping terminal oxidase with $2H^+/1e^-$ would not be possible.

T. ferrooxidans is sensitive to uncouplers. This is believed to be due to a collapse of cytoplasmic pH regulation, rather than an inhibitory effect on electron transport (Ingledew, 1982).

A scheme for electron transport is presented in Figure 4.8.

4.7 Manganese and iron oxidation at neutral pH

4.7.1 Microbiology

At neutral pH, Fe(II) autooxidizes rapidly. The orange encrustation associated with many 'iron bacteria' is mainly a deposition of colloidal $Fe(OH)_3$ formed non-biologically (Jones, 1986). However, in stratified eutrophic lakes, a zone of low O_2 separates the aerated epilimnion from the anaerobic hypolimnion. At low O_2 concentrations, autooxidation of Fe(II) is relatively slow, especially if slightly acid conditions prevail (Ghiorse, 1984). Certain iron bacteria are characteristically found in such zones, and in similar locations in soils and sediments. A more recent habitat is provided by steel structures (see Section 4.1.4).

These 'iron bacteria' can be cultured in redox gradients established by adding solid ferrous sulphide to tubes of mineral medium. In particular, strains of *Gallionella* grow in mineral media in the absence of significant amounts of organic carbon, provided ferrous iron is present, and uptake of label from [14C]-bicarbonate has been demonstrated (Ehrlich, 1981). The status of enzymic iron oxidation by *Sphaerotilus*, *Leptothrix* and other sheathed bacteria is discussed in Ehrlich (1981).

Manganous ions in solution at neutral pH are more stable than ferrous iron and do not autooxidize rapidly below pH 9.0 (Ghiorse, 1984). Ehrlich (1984) divides organisms that oxidize Mn(II) enzymically into three groups. Group I bacteria oxidize Mn^{2+} salts in solution, and have been isolated from soil and water. Group II organisms require the presence of solid MnO_2 which absorbs Mn^{2+} to form Mn_2O_3; they have mainly been isolated from the surface of deep-sea ferromanganese nodules. The organisms in Groups I and II include strains of *Arthrobacter*, *Vibrio* and *Leptothrix*. Group III organisms merely use Mn^{2+} as a peroxidase substrate, the aim being to remove H_2O_2. Reports of autotrophic growth coupled to Mn(II) oxidation are discussed and questioned in Ehrlich (1984).

A marine *Pseudomonas* strain has recently provided the first clear-cut evidence for bacterial growth coupled to manganese oxidation (Kepkay and Nealson, 1987). The bacteria were grown in continuous culture, and the extent of CO_2 fixation was shown to be proportional to manganese oxidation.

4.7.2   Energetics

The redox potential for oxidation of Fe^{2+} at neutral pH is dominated by the insolubility of ferric hydroxide (Smith and Martell, 1976):

$$Fe^{3+} + 3OH^- \rightleftharpoons Fe(OH)_3 \qquad K = 1.6 \times 10^{-39}$$

This equilibrium constant implies a maximal concentration of free Fe^{3+} of 1.6 mM at pH 2.0, falling to 1.6 µM at pH 3.0, and 1.6×10^{-18} M at pH 7.0. The total solubility of Fe(III) is somewhat increased by states such as $Fe(OH)^{2+}$, and may be dramatically increased by chelating anions. For manganese, the oxidized state (MnO_2) is essentially insoluble.

Provided the reduced state is present as the divalent cation, in the presence of the precipitated oxidized state, the redox potential for the reaction, $ox + ne^- + mH^+ \rightarrow red$, is given by:

$$E = E_m \, (ox_{(solid)}/red) + (RT/nF).\ln([H^+]^m/[red])$$

The redox potential thus depends on the concentration of the reduced form, and on the ambient pH. Figure 4.9 shows redox potentials for 10 mM and 10 µM concentrations of Fe^{2+} and Mn^{2+} over the pH range 5.0–8.0, as compared with the redox potential for reduction of O_2 to water. A plot for Mn_2O_3 oxidation by a Group II organism is also shown.

From Figure 4.9 it can be seen that Fe(II) oxidation is a much more favourable energy source at neutral pH than for *T. ferrooxidans* at pH 2.0. However, the combined problems of autooxidation and an insoluble oxidized state have prevented the biochemistry from being studied.

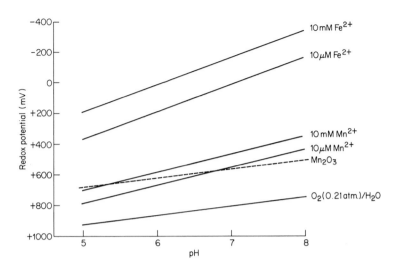

Fig. 4.9 Redox potentials for iron and manganese at pH 5.0–8.0. The figure shows redox potentials for unchelated Fe^{2+} and Mn^{2+} in equilibrium with the precipitated oxidized form, $Fe(OH)_3$ or MnO_2. The lines for Fe^{2+} are derived from $E_m(Fe^{3+}/Fe^{2+}) = +770\,mV$ and $K = 10^{-38.8}$ for $Fe(OH)_3 \leftrightarrow Fe^{3+} + 3\,OH^-$ (Smith and Martell, 1976). The lines for Mn^{2+} are derived from $E_m(MnO_2/Mn^{2+}) = +1230\,mV$. An additional line (dashed) shows the redox potential of the solid-state couple, Mn_2O_3 and MnO_2, derived from $E_m(MnO_2/Mn_2O_3) = +1010\,mV$. The lowest line is for reduction of O_2 at atmospheric pressure to liquid water; this is derived from $E_m(O_2/H_2O) = +1229\,mV$. The standard potentials are all from Bard *et al.* (1985).

Figure 4.9 also shows that Mn oxidation is a very marginal energy source. As the pH falls it becomes still more unfavourable; Mn^{2+} is therefore not among the ions oxidized by *T. ferrooxidans*. The addition of Mn^{2+} to whole cells of manganese-oxidizing bacteria in Groups I and II has been shown to result in synthesis of ATP, the reaction being blocked by cyanide and azide (Ehrlich, 1984). A reduction of *c*-type cytochromes has been observed, perhaps caused indirectly by an increased protonmotive force. For Group I organisms it is possible that either $Mn^{II} \rightarrow Enz–Mn^{III} + e^-$ or $Enz–Mn^{III} \rightarrow Mn^{IV} + e^-$ releases electrons at a more favourable redox potential, but there is little leeway if the other one-electron oxidation is to be coupled to reduction of O_2 to water.

References

Aleem, M. I. H. (1968). *Biochim. Biophys. Acta* **162**, 338–347.
Aleem, M. I. H. (1977). *Symp. Soc. Gen. Microbiol.* **27**, 351–381.

Aleem, M. I. H. and Sewell, D. L. (1984). *In* 'Microbial Chemoautotrophy', Ohio State University 8th Bioscience Colloquium (Eds. W. R. Strohl and O. H. Tuovinen), pp. 185–210. Ohio State University Press, Ohio.

Aleem, M. I. H., Hoch, G. E. and Varner, J. E. (1965). *Proc. Natl. Acad. Sci. USA* **54**, 869–873.

Ambler, R. P. (1980) *In* 'From Cyclotrons to Cytochromes' (Eds. N. O. Kaplan and A. Robinson), pp. 263–280. Academic Press, New York.

Aminuddin, M. (1980). *Arch. Microbiol.* **128**, 19–25.

Andersson, K. K. and Hooper, A. B. (1983). *FEBS Lett.* **164**, 236–240.

Andersson, K. K., Kent, T. A., Lipscomb, J. D., Hooper, A. B. and Münck, E. (1984), *J. Biol. Chem.* **259**, 6833–6840.

Andersson, K. K., Lipscomb, J. D., Valentine, M., Münck, E. and Hooper, A. B. (1986), *J. Biol. Chem.* **261**, 1126–1138.

Anthony, C. (1986). *Adv. Microb. Physiol.* **27**, 113–210.

Bard, A. J., Parsons, R. and Jordan, J. (Eds.) (1985). 'Standard potentials in aqueous solution'. IUPAC, Marcel Dekker, New York.

Bell, G. R., LeGall, J. and Peck, H. D. (1974). *J. Bacteriol.* **120**, 994–997.

Bock, E. (1978). *In* 'Microbiology 1978' (Ed. D. Schlessinger), pp. 310–314. American Society of Microbiologists, Washington DC.

Bowen, T. J., Happold, F. C. and Taylor, B. F. (1966). *Biochim. Biophys. Acta* **118**, 566–576.

Buchanan, R. E. and Gibbons, N. E. (Eds.) (1974). 'Bergey's Manual of Determinative Bacteriology, 8th Edn. Williams and Wilkins, Baltimore, USA.

Cerletti, P. (1986). *Trends Biochem. Sci.* **11**, 369–372.

Clark, W. M. (1960). 'Oxidation-reduction Potentials of Organic Systems', pp. 118–130. Waverly Press, Baltimore, Maryland.

Cobley, J. G. (1976a). *Biochem. J.* **156**, 481–491.

Cobley, J. G. (1976b). *Biochem. J.* **156**, 493–498.

Cobley, J. G. (1984). *In* 'Microbial Chemoautotrophy', Ohio State University 8th Bioscience Colloquium (Eds. W. R. Strohl and O. H. Tuovinen), pp. 169–183. Ohio State University Press, Ohio.

Cohen, Y., Jørgensen, B. B., Padan, E. and Shilo, M. (1975). *Nature* **257**, 489–492.

Cole, J. S. and Aleem, M. I. H. (1973). *Proc. Natl. Acad. Sci. USA* **70**, 3571–3575.

Cox, J. C. and Boxer, D. H. (1986). *Biotechnol. Appl. Biochem.* **8**, 269–275.

Dalton, H. (1977). *Arch. Microbiol.* **114**, 273–279.

Dalton, H. (1981). *In* 'Microbial Growth on C_1 Compounds', Proceedings of the 3rd International Symposium (Ed. H. Dalton), pp. 1–10. Heyden, London.

Dalton, H. and Leak, D. J. (1985) *In* 'Microbial Gas Metabolism' (Eds. R. K. Poole and C. S. Dow), pp. 173–200. Academic Press, London.

Dawes, E. A. (1986). 'Microbial Energetics'. Blackie, Glasgow.

DiSpirito, A. A., Taaffe, L. R., Lipscomb, J. D. and Hooper, A. B. (1985). *Biochim. Biophys. Acta* **827**, 320–326.

Doran, J. W. (1982). *Adv. Microb. Ecol.* **6**, 1–32.

Drozd, J. W. (1974). *FEBS Lett.* **49**, 103–105.

Drozd, J. W. (1976). *Arch. Microbiol.* **110**, 257–262.

Drozd, J. W. (1980). *In* 'Diversity of Bacterial Respiratory Systems' (Ed. C. J. Knowles), Vol. 2, pp. 87–111. CRC Press, Boca Raton, Florida.

Ehrlich, H. L. (1978). *Geomicrobiol. J.* **1**, 65–83.

Ehrlich, H. L. (1981). 'Geomicrobiology'. Marcel Dekker, New York.

Ehrlich, H. L. (1984). *In* 'Microbial Chemoautotrophy', Ohio State University 8th Bioscience Colloquium (Eds. W. R. Strohl and O. H. Tuovinen), pp. 47–56. Ohio State University Press, Ohio.

Finazzi Agrò, A., Federici, C., Giovagnoli, C., Cannella, C. and Cavallini, D. (1972). *Eur. J. Biochem.* **28**, 89–93.

Friedrich, C. G., Meyer, O. and Chandra, T. S. (1986). *FEMS Microbiol. Lett.* **37**, 105–108.

Ghiorse, W. C. (1984). *Ann. Rev. Microbiol.* **38**, 515–550.

Goreau, T. J., Kaplan, W. A., Wofsy, S. C., McElroy, M. B., Valois, F. W. and Watson, S. W. (1980). *Appl. Env. Microbiol.* **40**, 526–532.

Gromova, L. A., Karavaiko, G. I., Sevtsov, A. V. and Pereverzev, N. A. (1983). *Mikrobiologiya* **52**, 455–459.

Hageage, G. J., Eanes, E. D. and Gherna, R. L. (1970). *J. Bacteriol.* **101**, 464–469.

Hamilton, W. A. (1985). *Ann. Rev. Microbiol.* **39**, 195–217.

Hofman, T. and Lees, H. (1953). *Biochem. J.* **54**, 579–583.

Hollocher, T. C., Tate, M. E. and Nicholas, D. J. D. (1981). *J. Biol. Chem.* **256**, 10834–10836.

Hollocher, T. C., Kumar, S. and Nicholas, D. J. D. (1982). *J. Bacteriol.* **149**, 1013–1020.

Hooper, A. B. (1969). *J. Bacteriol.* **97**, 968–969.

Hooper, A. B. (1984). *In* 'Microbial Chemoautotrophy', Ohio State University 8th Bioscience Colloquium (Eds. W. R. Strohl and O. H. Tuovinen), pp. 133–167. Ohio State University Press, Ohio.

Hooper, A. B. and DiSpirito, A. A. (1985). *Microbiol. Rev.* **49**, 140–157.

Hooper, A. B. and Terry, K. R. (1973). *J. Bacteriol.* **115**, 480–485.

Hooper, A. B. and Terry, K. R. (1974). *J. Bacteriol.* **119**, 899–906.

Hooper, A. B., Debey, P., Andersson, K. K. and Balny, C. (1983). *Eur. J. Biochem.* **134**, 83–87.

Hyman, M. R. and Wood, P. M. (1983). *Biochem. J.* **212**, 31–37.

Hyman, M. R. and Wood, P. M. (1984). *Arch. Microbiol.* **137**, 155–158.

Hyman, M. R. and Wood, P. M. (1985). *Biochem. J.* **227**, 719–725.

Hynes, R. K. and Knowles, R. (1978). *FEMS Microbiol. Lett.* **4**, 319–321.

Ingledew, W. J. (1982). *Biochim. Biophys. Acta* **683**, 89–117.

Ingledew, W. J. and Cobley, J. G. (1980). *Biochim. Biophys. Acta* **590**, 141–158.

Ingledew, W. J. and Halling, P. J. (1976). *FEBS Letters* **67**, 90–93.

Jones, C. W. (1982). 'Bacterial Respiration and Photosynthesis'. Nelson, Walton-on-Thames, Surrey.

Jones, J. G. (1986). *Adv. Microb. Ecol.* **9**, 149–185.

Jørgensen, B. B. (1988). *In* 'The Nitrogen and Sulphur Cycles'. *S. G. M. Symp.* **42**, 31–63.

Justin, P. and Kelly, D. P. (1978). *J. Gen. Microbiol.* **107**, 123–130.

Kämpf, C. and Pfennig, N. (1980). *Arch. Microbiol.* **127**, 125–135.

Kelly, D. P. (1978). *In* 'Companion to Microbiology' (Eds. A. T. Bull and P. M. Meadow), pp. 363–386. Longman, London.

Kelly, D. P. (1982). *Phil. Trans. R. Soc. Lond.* **B298**, 499–528.

Kelly, D. P. (1985). *Microbiol. Sci.* **2**, 105–109.

Kelly, D. P. (1988). *In* 'The Nitrogen and Sulphur Cycles'. *S. G. M. Symp.* **42**, 65–98.

Kelly, D. P. and Kuenen, J. G. (1984). *In* 'Aspects of Microbial Metabolism and Ecology' (Ed. G. A. Codd), pp. 211–240. Academic Press, London.

Kelly, D. P., Mason, J. and Wood, A. P. (1987). *In* 'Microbial Growth on C_1 Compounds', Proceedings of the 5th International Symposium (Haren, Netherlands, August 1986).

Kepkay, P. E. and Nealson, K. H. (1987). *Arch. Microbiol.,* **148**, 63–67.

Kiesow, L. (1967). *Curr. Top. Bioenerg.* **2**, 195–233.

Knaff, D. B. and Buchanan, B. B. (1975). *Biochim. Biophys. Acta* **376**, 549–560.

Kuenen, J. G. and Tuovinen, O. H. (1981). *In* 'The Prokaryotes' (Eds. M. P. Starr, H. Stolp, H. G. Trüper, A. Balows and H. G. Schlegel), Vol. 1, pp. 1023–1036. Springer Verlag, Berlin.

Kuenen, J. G. and Robertson, L. A. (1988). *In* 'The Nitrogen and Sulphur Cycles'. *S. G. M. Symp.* **42**, 161–218.

Kuenen, J. G., Robertson, L. A. and van Gemerden, H. (1985). *Adv. Microb. Ecol.* **8**, 1–59.

Kula, T. J., Aleem, M. I. H. and Wilson, D. F. (1982). *Biochim. Biophys. Acta* **680**, 142–151.

Kusai, K. and Yamanaka, T. (1973). *Biochim. Biophys. Acta* **325**, 304–314.

Lappin, A. G., Lewis, C. A. and Ingledew, W. J. (1985). *Inorg. Chem.* **24**, 1446–1450.

Larkin, J. M. and Strohl, W. R. (1983). *Ann. Rev. Microbiol.* **37**, 341–367.

Lees, H. (1952). *Nature* **169**, 156–157.

Lipschultz, F., Zafiriou, O. C., Wofsy, S. C., McElroy, M. B., Valois, F. W. and Watson, S. W. (1981). *Nature* **294**, 641–643.

Lipscomb, J. D., Andersson, K. K., Münck, E., Kent, T. A. and Hooper, A. B. (1982). *Biochem.* **21**, 3973–3976.

Lu, W.-P. (1986). *FEMS Microbiol. Lett.* **34**, 313–317.

Lu, W.-P. and Kelly, D. P. (1984a). *J. Gen. Microbiol.* **130**, 1683–1692.

Lu, W.-P. and Kelly, D. P. (1984b). *In* 'Microbial Growth on C_1 Compounds', Proceedings of the 4th International Symposium (Eds. R. L. Crawford and R. S. Hanson), pp. 34–41. American Society for Microbiology, Washington DC.

Lu, W.-P., Poole, R. K. and Kelly, D. P. (1984). *Biochim. Biophys. Acta* **767**, 326–334.

Lu, W.-P., Swoboda, B. E. P. and Kelly, D. P. (1985). *Biochim. Biophys. Acta* **828**, 116–122.

Lwoff, A., van Niel, C. B., Ryan, F. J. and Tatum, E. L. (1946). *Cold Spring Harbor Symp. Quant. Biol.* **11**, 302–303.

Lyric, R. M. and Suzuki, I. (1970a). *Can. J. Biochem.* **48**, 344–354.

Lyric, R. M. and Suzuki, I. (1970b). *Can. J. Biochem.* **48**, 355–363.

Macy, J. M., Schröder, I., Thauer, R. K. and Kröger, A. (1986). *Arch. Microbiol.* **144**, 147–150.

McEwan, A. G., Cotton, N. P. J., Ferguson, S. J. and Jackson, J. B. (1985). *Biochim. Biophys. Acta* **810**, 140–147.

Miller, D. J. and Nicholas, D. J. D. (1985). *J. Gen. Microbiol.* **131**, 2851–2854.

Miller, D. J. and Wood, P. M. (1982). *Biochem. J.* **207**, 311–317.

Nicholas, D. J. D. and Jones, O. T. G. (1960). *Nature* **185**, 512–524.

Nicolson, G. L. and Schmidt, G. L. (1971). *J. Bacteriol.* **105**, 1142–1148.

Olson, R. J. (1981). *J. Marine Res.* **39**, 227–238.

Olson, T. C. and Hooper, A. B. (1983). *FEMS Microbiol. Lett.* **19**, 47–50.

Paitian, N. A., Markossian, K. A. and Nalbandyan, R. M. (1985). *Biochem. Biophys. Res. Commun.* **133**, 1104–1111.

Peck, H. D. (1960). *Proc. Natl. Acad. Sci. USA* **46**, 1053–1057.

Poole, R. K. (1983). *Biochim. Biophys. Acta* **726**, 205–243.

Poth, M. and Focht, D. D. (1985). *Appl. Env. Microbiol.* **49**, 1134–1141.
Prince, R. C. and Hooper, A. B. (1987). *Biochemistry* **26**, 970–974.
Prior, S. D. and Dalton, H. (1985). *J. Gen. Microbiol.* **131**, 155–163.
Rittenberg, S. C. (1969). *Adv. Microb. Physiol.* **3**, 159–196.
Robertson, G. P. (1982). *Phil. Trans. R. Soc. Lond.* **B296**, 445–457.
Sawhney, V. and Nicholas, D. J. D. (1977). *J. Gen. Microbiol.* **100**, 49–58.
Schedel, M. and Trüper, H. G. (1979). *Biochim. Biophys. Acta* **568**, 454–467.
Schedel, M. and Trüper, H. G. (1980). *Arch. Microbiol.* **124**, 205–210.
Schmitt, W., Schleifer, G. and Knobloch, K. (1981). *Arch. Microbiol.* **130**, 334–338.
Schook, L. B. and Berk, R. S. (1979). *J. Bacteriol.* **140**, 306–308.
Sewell, D. L., Aleem, M. I. H. and Wilson, D. F. (1972). *Arch. Biochem. Biophys.* **153**, 312–319.
Shears, J. H. and Wood, P. M. (1985). *Biochem. J.* **226**, 499–507.
Shears, J. H. and Wood, P. M. (1986). *FEMS Microbiol. Lett.* **33**, 281–284.
Sillén, L. G. and Martell, A. E. (Eds.) (1964). 'Stability constants of metal ion complexes'. *Chemical Society Special Publications* **17**, 159. The Chemical Society, London.
Silver, M. and Kelly, D. P. (1976). *J. Gen. Microbiol.* **97**, 277–284.
Silver, M. and Lundgren, D. G. (1968). *Can. J. Biochem.* **46**, 1215–1220.
Smith, R. M. and Martell, A. E. (1976). 'Critical Stability Constants. Vol. 4: Inorganic Complexes'. Plenum Press, New York.
Solomon, E. I. (1981). *In* 'Metal Ions in Biology' (Ed. T. G. Spiro), Vol. 3, pp. 41–108. John Wiley, New York.
Starr, M. P., Stolp, H., Trüper, H. G., Balows, A. and Schlegel, H. G. (Eds.) (1981) 'The Prokaryotes', Vol. 1. Springer-Verlag, Berlin.
Steudel, R., Holdt, G., Göbel, T. and Hazeu, W. (1987). *Angew. Chem. Int. Ed. Engl.* **26**, 151–153.
Strohl, W. R. and Tuovinen, O. H. (Eds.) (1984). 'Microbial Chemoautotrophy', Ohio State University 8th Bioscience Colloquium. Ohio State University Press, Ohio.
Sundermeyer-Klinger, H., Meyer, W., Warninghoff, B. and Bock, E. (1984). *Arch. Microbiol.* **140**, 153–158.
Suzuki, I. (1974). *Ann. Rev. Microbiol.* **28**, 85–101.
Suzuki, I. and Kwok, S-C. (1981). *Can. J. Biochem.* **59**, 484–488.
Suzuki, I. and Silver, M. (1966). *Biochim. Biophys. Acta* **122**, 22–33.
Suzuki, I., Dular, U. and Kwok, S. C. (1974). *J. Bacteriol.* **120**, 556–558.
Suzuki, I., Kwok, S.-C. and Dular, U. (1976). *FEBS Lett.* **72**, 117–120.
Suzuki, I., Kwok, S.-C., Dular, U. and Tsang, D. C. Y. (1981). *Can. J. Biochem.* **59**, 477–483.
Tanaka, Y., Fukumori, Y. and Yamanaka, T. (1983). *Arch. Microbiol.* **135**, 265–271.
Terry, K. R. and Hooper, A. B. (1981). *Biochem.* **20**, 7026–7032.
Thauer, R. K., Jungermann, K. and Decker, K. (1977). *Bacteriol. Rev.* **41**, 100–180.
Toghol, F. and Southerland, W. M. (1983). *J. Biol. Chem.* **258**, 6762–6766.
Trudinger, P. A. (1967). *Rev. Pure Appl. Chem.* **17**, 1–24
Trudinger, P. A., Meyer, T. E., Bartsch, R. G. and Kamen, M. D. (1985). *Arch. Microbiol.* **141**, 273–278.
Trüper, H. G. (1978). *In* 'The Photosynthetic Bacteria' (Eds. R. K. Clayton and W. R. Sistrom), pp. 677–690. Plenum Press, New York.
Trüper, H. G. and Fischer, U. (1982). *Phil. Trans. R. Soc. Lond.* **B298**, 529–542.
Tsang, D. C. Y. and Suzuki, I. (1982). *Can. J. Biochem.* **60**, 1018–1024.

Ward, B. B. (1986). *In* 'Nitrification' (Ed. J. Prosser). *S. G. M. Special Publications* **20**, 157–184. IRL Press, Oxford.

Watson, S. W., Valois, F. W. and Waterbury, J. B. (1981). *In* 'The Prokaryotes' (Eds. M. P. Starr, H. Stolp, H. G. Trüper, A. Balows and H. G. Schlegel), Vol. 1, pp. 1005–1022. Springer-Verlag, Berlin.

Wetzstein, H-G. and Ferguson, S. J. (1985). *FEMS Microbiol. Lett.* **30**, 87–92.

Wood, P. M. (1978). *FEBS Lett.* **95**, 12–18.

Wood, P. M. (1983). *FEBS Lett.* **164**, 223–226.

Wood, P. M. (1987). *In* 'Nitrification' (Ed. J. Prosser). *S. G. M. Special Publications* **20**, 39–62. IRL Press, Oxford.

Yamanaka, T., Kamita, Y. and Fukumori, Y. (1981a). *J. Biochem. (Tokyo)* **89**, 265–273.

Yamanaka, T., Yoshioka, T. and Kimura, K. (1981b). *Plant Cell Physiol.* **22**, 613–622.

Yamazaki, T., Fukumori, Y. and Yamanaka, T. (1985). *Biochim. Biophys. Acta* **810**, 174–183.

5 Bacterial cytochrome oxidases

R. K. Poole

5.1 Historical perspective and the contribution of studies
    on bacterial systems 232
5.2 Oxygen and its reduction products 235
5.3 Mitochondrial cytochrome oxidase: a brief survey 236
5.4 Cytochromes aa_3, caa_3 and c_1aa_3 238
5.4.1 Distribution 238
5.4.2 Purification and properties of the subunits of aa_3-type
      oxidases 240
5.4.3 The redox centres of aa_3-type oxidases 242
      5.4.3.1 Haems 242
      5.4.3.2 Copper 244
5.4.4 Redox properties of aa_3-type oxidases 245
5.4.5 Reaction with oxygen of aa_3-type oxidases 246
5.4.6 Chemiosmotic function of aa_3-type oxidases 248
5.4.7 Genetics and molecular biology of aa_3-type oxidases 249
5.5 Cytochrome o 249
5.5.1 Definition and distribution 249
5.5.2 Purification and properties of o-type oxidases 250
5.5.3 Spectral, potentiometric and ligand-binding
      properties of the haems of o-type oxidases 252
      5.5.3.1 General features 252
      5.5.3.2 *Escherichia coli* 252
      5.5.3.3 Other bacteria 255
5.5.4 Reactions with oxygen and catalytic activity o-type
      oxidases 256
5.5.5 Chemiosmotic aspects of o-type oxidases 258
5.5.6 Genetics of o-type oxidases 260
5.6 Cytochrome d 261
5.6.1 Definition, distribution and nomenclature 261
5.6.2 Purification and properties of the subunits 262
5.6.3 Optical properties of d-type oxidases 264
      5.6.3.1 Cytochrome d 264
      5.6.3.2 Cytochrome b_{558} 265
      5.6.3.3 The component resembling "cytochrome a_1" 265
5.6.4 EPR properties of d-type oxidases 266
5.6.5 Potentiometric studies of d-type oxidases 267

5.6.6   The reaction with oxygen and other ligands of
        d-type oxidases 268
5.6.7   Chemiosmotic function of *d*-type oxidases 272
5.6.8   Genetics and molecular biology of *d*-type oxidases 272
5.7   Cytochromes of the a_1-type **274**
5.7.1   Classification 274
5.7.2   Type I cytochromes a_1 278
**5.8   Other haemoproteins involved in the metabolism of
        oxygen and its partial reduction products** **278**
5.8.1   The cytochrome *o*-like haemoproteins of *Vitreoscilla* 278
5.8.2   *Pseudomonas* cytochrome cd_1 280
5.8.3   Hydroperoxidases 280
5.8.4   Superoxide dismutase 281
5.9   Conclusions and prospects **283**
References 284

5.1   Historical perspective and the contribution of studies on bacterial systems

The history of research on respiration and cytochromes (Keilin, 1966; Hempfling. 1979) and cytochrome oxidases in particular (Wikström *et al.*, 1981) is almost as long as the history of biochemistry itself (Lehninger, 1975). Even before Buchner discovered that alcoholic fermentation occurs in cell-free yeast extracts, MacMunn had discovered and reported the pigments 'myohaematin' (in muscle) and 'histohaematin' (in other tissues) by virtue of their four-banded absorption spectra (for references, see Keilin, 1966). These bands are now attributable to the partly superimposed α-, β- and γ-bands of cytochromes; one of the bands (613–593 nm) is cytochrome oxidase. Mac-Munn demonstrated the dependence of the visible bands on the reduced state, but made an important error in stating that myohaematin was the only pigment of pigeon breast muscle when, in fact, this tissue is very rich in both haemoglobin and cytochrome. This threw him into conflict with the influential biochemist, Hoppe-Seyler, who dismissed the existence of a 'special colouring matter' in pigeon muscle, and effectively inhibited further work. With little reference to MacMunn's work, Warburg proposed that cellular oxygen consumption is catalysed by an iron-containing enzyme, *der Atmungsferment*, which, in its divalent state, reacts with oxygen and is oxidized before being re-reduced by organic substances.

In 1925, Keilin reported that MacMunn's observations had been fundamentally correct but that the pigments he described were of much wider distribution than hitherto supposed and warranted a new name, cytochrome ('cellular pigment') (see Keilin, 1966). Of particular relevance was Keilin's discovery in 1925 that *Bacillus subtilis* (which he had used for feeding experiments with the fly *Gasterophilus* and in which he first observed cytochromes) displayed the same characteristic spectrum with absorption bands occupying the approximate positions: (a) 604 nm; (b) 564 nm; (c) 550 nm; and (d) 521 nm. Band (d) was shown to be a fused (β) band and the others were attributed to cytochromes 'a', 'b' and 'c'. In 1938, Keilin showed that the band of cytochrome a is not due to one component (a) but to two (a and a_3), one of which (a_3) is autooxidizable and combines with cyanide and carbon monoxide. Significantly, the position of the band of the carbon monoxide compound (593 nm) was similar to that of the band in Warburg's earlier (1926–1933) photochemical action spectrum. These results might have suggested that the autooxidizable component (a_3) was cytochrome oxidase itself (this term having replaced *Atmungsferment*) but Keilin and Hartree instead considered cytochrome oxidase to be separate from the cytochromes and possibly be a copper protein (Keilin, 1966).

(It is worth noting from its use in this context that the term 'cytochrome oxidase' does not imply that the enzyme itself is a cytochrome but only that it oxidizes a cytochrome. Thus in current usage, 'cytochrome oxidase o', for example, is preferable to 'cytochrome o oxidase', the latter implying that cytochrome o is oxidized by some other enzyme. In summary, cytochrome oxidases need not be cytochromes.)

By 1925, Keilin had completed extensive experiments on the cytochrome complement of yeasts and bacteria but the work had to be removed from a paper on cytochrome at the request of editors of the journal in order to shorten the manuscript! Subsequently, similar work was published from other laboratories. Keilin emphasized the need for a uniform terminology to describe the rather diverse components being reported in bacteria and his terminology is the basis of that used today (see pages 46–52). The recommended naming of the α-bands (reduced form) was:

Cytochrome	a_2		628–632 nm
Cytochrome	a	(and a_3)	603–605 nm
Cytochrome	a_1		587–592 nm
Cytochrome	b		562–564 nm
Cytochrome	b_1		557–560 nm
Cytochrome	c_1		553 nm
Cytochrome	c		550 nm

In 1932 and 1933, the absorption spectrum of *Acetobacter pasteurianum* was described by Kubowitz and Haas, and by Warburg's group (see Keilin, 1966). The composite β-band, the bands at 550, 553 and 563 nm, plus a very weak 589 nm band, were recorded. The bands disappeared on shaking the cell suspension and reappeared on standing; CO caused the 'yellow' band to be displaced from 589 to 592 nm whereas the other bands were unchanged. Cyanide addition to a suspension that was almost free of O_2 intensified the 589 nm band and caused the appearance of a new band at 639 nm. Subsequent aeration caused the 589 nm band to disappear and then slowly reappear, while the other bands, including the one at 639 nm, were unchanged. It was concluded that the 589 nm band was the ferrous form of the 'oxygen-transporting ferment' which combined with CO, underwent oxidation in the presence of cyanide and, in the ferric state, combined with cyanide to form a compound that was not easily reducible. At about this time, Negelein and Gerischer demonstrated c, b and a_2 ($= d$)-type cytochromes in *Azotobacter chroococcum* and showed that, on aeration, the bands of the first two disappeared whilst that of cytochrome a_2 moved from 632 to 647 nm. CO shifted the band of reduced cytochrome a_2 from 632 to 637 nm and it was concluded that this component was the 'oxygen-transporting ferment' (see Keilin, 1966). These studies of *Acetobacter* and *Azotobacter* were instrumental in Warburg waiving his objections to the concept of 'cytochrome' and accepting the view that it forms an integral part of respiratory catalysis. In particular, he identified bands a_1 and a_2 as the 'oxygen-transporting ferment', but still objected to the view that they should be called 'cytochromes', a term used previously to refer only to non-autooxidizable components.

Final proof of the co-identity of *der Atmungsferment* with ligand-reactive cytochromes such as a_3 (in yeast and mitochondria) and a_1, o and d (in bacteria), came from the elegant photochemical action spectra of Chance and his collaborators (Chance 1953a,b; Chance *et al.*, 1975), discussed in relevant sections of this chapter. He demonstrated the light sensitivity of the a_3-CO compound and showed that the photodissociation difference spectrum was identical to the 'static' (i.e. CO difference spectrum) and to the photochemical action spectrum. This set the stage for further structural and functional studies, notably the development of methods for cytochrome oxidase purification; the identification of cytochrome c oxidation as a "site" of oxidative phosphorylation; and the proof of a functional role for copper in the enzyme (for references see Wikström *et al.*, 1981).

Detailed structural studies of bacterial cytochrome oxidases were soon to develop but were given great impetus by the realization that the subunit composition of bacterial aa_3 types was much simpler than that of their mitochondrial counterpart (see Sections 5.3 and 5.4). Subsequently, oxidase complexes containing cytochrome o, cytochrome d and 'cytochrome a_1' have been purified from various bacteria in which their physiological roles were established. These oxidases are treated in turn in later sections of this Chapter.

Although the various oxidase classes have a propensity for occurrence in certain physiological groups, it is also clear that the nature of the oxidase(s) present is influenced by environmental conditions. The overall reaction catalysed by these oxidases appears similar in all cases. Each oxidase is a complex comprising the ligand-binding haem (which may be of the *b* ('*o*')-, *d*- or *a*-type), one or more further haems (of the *a*-, *b* or *c* types) and sometimes copper atoms (see Figs 1.20 and 1.21, pages 47 and 49). The complex catalyses electron transfer from a cytochrome or quinol to oxygen, generating H_2O, and the reaction usually contributes to energy-conservation by removal of internal protons (in reducing oxygen to water) and by being coupled (in some cases) to proton pumping across the bacterial cytoplasmic membrane. A fully rational explanation of why such structurally diverse, but functionally similar, oxidases occur in bacteria and may even coexist in the same organism, is lacking.

5.2 Oxygen and its reduction products

Oxygen has been 'chosen' by nature to act as the terminal oxidant of respiratory chains because it is 'the only element in the most appropriate physical state, with a satisfactory solubility in water and with desirable combinations of kinetic and thermodynamic properties' (P. George, see Naqui *et al.*, 1986). The electronic structure of the dioxygen molecule explains its relative kinetic inertness and the restriction on likely reduction products. The one-electron reduction of molecular oxygen is thermodynami-cally unfavourable, the redox potential of superoxide generation being $-0.33\,V$. It is for this reason that concerted two-electron transfers are the likely (but unproved) reductive pathways in the mechanisms of O_2-reducing enzymes:

$$\tfrac{1}{2}O_2 + H^+ + e^- \rightleftharpoons \tfrac{1}{2}H_2O_2,\ E_0 = +0.3\,V\,(pH\,7)$$

$$\tfrac{1}{2}H_2O_2 + H^+ + e^- \rightleftharpoons H_2O,\ E_0 = +1.35\,V\,(pH\,7)$$

Cytochrome oxidases and other oxygen-activating enzymes employ the unpaired electrons of transition metal ions to catalyse oxygen reduction. As shall be seen in Section 5.3, the best-studied of these enzymes, cytochrome *c* oxidase, employs a binuclear [Fe(II),Cu(I)] centre to catalyse oxygen reduc-tion by transfer of two electrons, but in the case of bacterial oxidases of the a_1, *o* and *d* types such binuclear centres have not been firmly identified. Copper may (e.g. cytochrome *o*) or may not (e.g. cytochrome *d*) be involved, but all these enzymes are haemoproteins. The mechanism of action of bacterial aa_3-type oxidases is presumed, and has in part been demonstrated, to resemble that of the mitochondrial enzyme.

5.3   Mitochondrial cytochrome oxidase: a brief survey

The cytochrome oxidase of mitochondria in eukaryotes catalyses the reaction:

$$4H^+ + O_2 + 4\,cyt\ c^{2+} \rightarrow 2H_2O + 4\,cyt\ c^{3+}$$

This is the last step in the series of membrane-bound redox reactions in which reducing equivalents are passed from $NADH_2$, succinate and other reductants to oxygen, with conservation of the redox energy of the overall reaction. The reaction involves coupling of the one-electron oxidoreduction of cytochrome c to the four-electron reduction of oxygen to water, in such a way that the generation of potentially damaging, free partial-reduction products (e.g. superoxide and peroxide) is minimized.

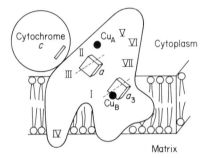

Fig. 5.1   Structure and topography of mitochondrial cytochrome c oxidase. The shape of the protein mass is derived from electron microscopic analysis of a crystal form obtained by deoxycholate extraction. The topography of the subunits (I–VII) is based on chemical and immunological labelling. The square planar structures are the a haems, and the circles, the two copper atoms. Cu_B and haem a_3 are in very close proximity, probably coupled by a bridging ligand. Electron paramagnetic resonance (EPR) studies with dysprosium-EDTA as probe suggest that haem a is close to the cytoplasmic side of the membrane, whilst haem a_3 is less accessible. Optical and EPR studies of vesicles and multilayers show that both haems are oriented with their planes perpendicular to the membrane plane. Estimates of haem c–haem a (25A) and haem a–haem a_3 (10A) have been made. For further details, see Wikström *et al.* (1981). The structure of haem a is given in Fig. 1.20 (page 47).

The eukaryotic enzyme is an integral membrane protein comprising 7 to 13 subunits of M_r 4000–36 000. The isolated enzyme is believed to be a dimer of M_r 300 000. *In situ*, the enzyme is asymmetrically disposed in the membrane (Fig. 5.1). Several of the polypeptide chains have amino acid sequences in which stretches of largely hydrophobic residues are interspersed with hydrophylic regions, consistent with the view that such polypeptides traverse the membrane, probably more than once. It is especially notable, in the context of bacterial oxidases, that the largest subunits (I–III) are synthesized on mitochondrial ribosomes, whereas the other, smaller subunits are made extramitochondrially. It is these three largest subunits that appear to constitute the functional 'core' of the enzyme and which are analogous to the two or three subunits (only) that comprise bacterial oxidases of this type. The functions of the extra, cytoplasmically encoded polypeptides are unclear, but they may be involved in the coupling of electron transport and proton pumping, and in regulation adapted to the specific requirements of different tissues, stages of development and adaptation (Kadenbach *et al.*, 1983).

The oxidase has long been considered to contain four metal centres, i.e. two haems *a* and two copper atoms. Recently, zinc, magnesium and a third copper have also been found (Einarsdottir and Caughey, 1985; Steffens *et al.*, 1987). Cytochrome *a* is low spin, having two imidazole nitrogens as axial ligands in the ferric and ferrous states; it is responsible for up to 75% of the 605 nm absorbance of the reduced enzyme and contributes about 50% (with a_3) to the Soret absorbance. Cu_A, in the Cu(II) state, is EPR-detectable and also makes significant contributions to the optical spectrum at 830 nm and possibly 615 nm. The ligand-binding site comprises haem a_3 and Cu_B, neither of which is EPR-detectable in the oxidized resting state, probably because the two metals (Cu_B and Fe) are magnetically and physically coupled, possibly by a sulphur atom. Subunit I is thought to contain this critical metal pair plus haem *a*. Subunit II is the cytochrome *c*-binding site and contains at least one copper atom, possibly Cu_A (Holm *et al.*, 1987). The finding of a third copper leads, however, to the attractive idea (Steffens *et al.*, 1987) that it is this copper that is bound to subunit II and that all four redox-active metal centres are bound to subunit I. This is consistent with the view that the *Thermus thermophilus* oxidase lacks subunit II and consists of a large subunit containing Cu_A, Cu_B, and haems *a* and a_3 (Yoshida *et al.*, 1984; Hon-Nami and Oshida, 1984; Section 5.4.2).

Electrons enter the oxidase at haem *a*; Cu_A is reduced next, being in rapid redox equilibrium with haem *a*. Much more slowly, are reduced Cu_B and haem a_3 and it is to the latter that oxygen binds (for a review, see Naqui *et al.*, 1986). Studies at very low temperatures have suggested that the initial product is an 'oxy' intermediate. Its spectral similarity to the Fe^{2+}.CO compound suggests the structure $Fe^{2+}-O_2 Cu_B^+$, but the possibility of a partial charge transfer from iron to the bound dioxygen, as postulated for oxyhaemoglobin (Thomson, 1977) must be considered (Clore *et al.*, 1980). Electron transfers from Cu_B and haem a_3 give a bound peroxide intermediate (Compound B). Significantly, at physiological temperatures, the reaction is complete within 1 ms and, when oxygen is bound to the binuclear (a_3–Cu_B) site, electron transfer from Cu_A is 10^3 to 10^4 times faster than in the absence of the ligand.

Quite recently, a 'pulsed' or peroxidatic intermediate of cytochrome oxidase has been described when the fully reduced or resting enzyme is exposed to O_2. The pulsed form is a more efficient catalyst; in particular the internal electron transfer from the electron-accepting couple (Cu_A and cytochrome a) to the oxygen-reactive couple (Cu_B and cytochrome a_3) is greatly enhanced. The pulsed form has characteristic optical (420 and 655 nm absorbances) and EPR ($g = 5$) characteristics but structural information on the presumptive more 'open' configuration is at present inconclusive. The absence of a sulphur atom bridging the Fe of a_3 and Cu_B, or the existence of ferryl (Fe IV) iron in a_3 have been proposed as alternative bases for the pulsed (= peroxidatic) species (Naqui *et al.*, 1986).

The most striking feature of the catalytic role of cytochrome oxidase is the homology with other haemoproteins, notably haemoglobin (with respect to oxycompound formation) and peroxidases (which structurally and functionally resemble intermediate states in the oxidase reaction).

The role of cytochrome oxidase in mitochondrial energy transduction has been controversial. The key issue has been whether in addition to the involvement of 'scalar' protons during reduction by cytochrome c of oxygen, the oxidase transports protons against the gradient formed during catalysis. The overwhelming opinion, now supported by Mitchell (see Mitchell *et al.*, 1985) is that cytochrome oxidase is indeed such a proton pump. The molecular mechanism remains obscure, but subunit III has been implicated as a proton ionophore or channel. Further information on this much studied enzyme can be found from Table 5.1 and in the references cited by Brunori and Wilson (1982) and Naqui *et al.* (1986).

5.4 Cytochromes aa_3, caa_3 and c_1aa_3

5.4.1 Distribution

These are the oxidases that resemble most closely the mitochondrial cytochrome c oxidase (EC 1.9.3.1). Nevertheless, there are two outstanding, distinguishing features of the bacterial aa_3-types: first, their relatively simple subunit composition; secondly, the association of a c- or c_1-type cytochrome with the oxidases of thermophiles and certain other bacteria. A claim that cytochromes a, c and o comprise a novel oxidase complex in the thermophile PS3 has been refuted (Poole, 1981; Poole *et al.*, 1985a). Cytochrome aa_3 was noted (see Keilin, 1966) in the earliest spectroscopic recordings of a bacterium (*Bacillus subtilis*), and is now known to occur in most physiological groups of prokaryotes, including cyanobacteria (Peschek, 1981), often in association with cytochrome c.

Table 5.1 Major sources in the field of cytochrome oxidases

Authors and year	Coverage
Degn et al. (Eds.) (1978)	Terminal oxidases, particularly cyanide-insensitive systems, in bacteria and eukaryotes.
Jones and Poole (1985)	Survey of cytochrome structure and function with emphasis on methods of analysis.
Keilin (1966)	Posthumous, fascinating history of studies on respiration and cytochromes
King et al. (Eds.) (1979)	Mostly mitochondrial aa_3-type oxidases.
Knowles (Ed.) (1980)	Two volumes describing selected topics in bacterial respiratory systems.
Lemberg and Barrett (1973)	Whole field of cytochrome biochemistry.
Wikström et al. (1981)	Cytochrome aa_3-type oxidases only.
Various authors (1985)	Up to date, authoritative reviews constituting entire edition of *Journal of Inorganic Biochemistry* **23**, numbers 3 and 4.

Oxygen deprivation sometimes causes the replacement of cytochrome aa_3 by o; e.g. in *Paracoccus denitrificans*, *Rhodopseudomonas sphaeroides* (for references, see Poole, 1983), *Bacillus stearothermophilus* PS3 (Sone et al., 1983a; B. S. Baines and R. K. Poole, unpublished) and other bacteria (Jones, 1977). In *Paracoccus denitrificans*, the two oxygen-terminated respiratory chains branch at ubiquinone (Kučera et al., 1984; Parsonage et al., 1986). These replacements cannot be rationalized with current knowledge of the properties of the oxidases. *Methylophilus methylotrophus* is similar at first sight to these bacteria, cytochrome aa_3 being produced under oxygen-excess conditions (carbon limitation), and being replaced by cytochrome o under conditions of oxygen limitation (carbon excess). Further investigation showed, however, that cytochrome aa_3 is also replaced by cytochrome o under conditions of nitrogen limitation (i.e. carbon- or oxygen-excess conditions), suggesting that in this organism carbon status may be at least as important as oxygen status (Cross and Anthony, 1980). As many investigations of regulation of oxidases have considered only two growth conditions (carbon or oxygen limitation), it is possible that the conclusions drawn for *M. methylotrophus* may also apply to a wider range of bacteria and that oxygen may not be the most important regulator of oxidase synthesis.

5.4.2   Purification and properties of the subunits of aa_3-type oxidases

The major impetus to studies of the bacterial aa_3-type oxidases was the discovery of their simple subunit composition (Ludwig and Schatz, 1980). In all species from which the oxidases have been purified, only two or three subunits have been identified (Table 5.2), in contrast to the 7 or 13 found in the mitochondrial enzyme (Section 5.3). Full details are given in the excellent review by Ludwig (1987). Purification of the hydrophobic complexes from cytoplasmic membranes requires solubilization (e.g. with Triton X-100, dodecyl maltoside or Brij-35; Baines and Poole, 1985). In addition to more conventional procedures, affinity chromatography with yeast cytochrome c as ligand has been exploited in the cases of *P. denitrificans* (Ludwig, 1986), *Rhodobacter sphaeroides* (Azzi and Gennis, 1986) and *Bacillus subtilis* (de Vrij *et al.*, 1986).

The two subunits of the *Paracoccus* oxidase resemble the two largest subunits in the mitochondrial oxidase in their hydrophobicity, immunological cross-reactivity (Ludwig, 1982) and sequence homology (Steffens *et al.*, 1983). Subunit II antibody from *Paracoccus* also cross-reacts with subunits II and III of the *Rhodobacter* enzyme (Ludwig, 1987).

Further comparison with the mitochondrial enzyme, in which subunits I to III are encoded by the organelle, suggests the possibility of a third subunit in the *Paracoccus* enzyme (as occurs in some other bacteria; Table 5.2). Although subunit III of the mitochondrial oxidase is easily detached and a similarly loose association might cause loss of a subunit during purification of the *Paracoccus* oxidase, extensive studies on the effects of various solubilization and purification protocols on the latter have not revealed a third subunit. Most importantly, a reconstituted two-subunit complex is functional in electron transport and proton translocation, albeit with a stoichiometry of only $0.6 \, H^+/e^-$. Resolution of this problem may come from future studies on expression of cloned oxidase genes (page 249). The arrangement of the subunits in the membrane of *Paracoccus* has been little explored. The intact (?) enzyme free from phospholipid has a molecular mass of 79–85 000, showing that the isolated enzyme is a monomer and accounting satisfactorily for the presence of the two subunits in a 1:1 stoichiometry (see Ludwig, 1987).

Three subunits have been reported in the caa_3-type oxidase of the thermophile PS3 (Sone and Yanagita, 1982; Baines *et al.*, 1984a) despite earlier claims of a single subunit oxidase (Sone *et al.*, 1979) probably caused by aggregation of subunits during heating to 95°C for denaturation. Subunits I (55–56 K), II (28–38 K) and III (22 K) may be homologous to the three mitochondrially encoded subunits in eukaryotes and, indeed, an immunological cross-reactivity with yeast subunit II has been reported (Ludwig, 1980). Subunit II bears the haem c, making its homology with a

Table 5.2 Properties of Purified Oxidases of the aa_3 type

Organism	Haems			Copper			Subunits		Proton translocation		References
	type	α-band (reduced; nm)	nmol haem a/mg	Cu:haem a ratio	nmol /mg	No.	M_r(k)	Ratio	→H⁺/e	DCCD inhibition	
Bacillus subtilis	aa_3	601	15–17			3	57, 37, 21		no		De Vrij et al. (1983)
Bacillus firmus RAB	caa_3	605, 551	17 (plus 0.43 mol haem c/mol haem a)	1	22	3	56, 40, 14[a]				Kitada and Krulwich (1984)
Thermus thermophilus HB8	c_1aa_3	604, 550	21.3	1	23–26	2	55, 33[a]	1:1	0.8–0.9	no	Yoshida and Fee (1984); Yoshida et al. (1984); Hon-Nami and Oshima (1984)
PS3 (thermophilic Bacillus sp.)	caa_3	605, 550	15–16 (plus 8–8.9 nmol haem c/mg)	1	14–18	3	55–56, 28–38, 22[a]	1:1:1	0.45–1.4	yes	Baines et al. (1984a); Baines et al. (1985); Sone and Yanagita (1982); Sone (1986); Sone and Yanagita (1984); Sone and Hinkle (1982)
Paracoccus denitrificans	aa_3	605	27	1	31–35	2	45, 28	1:1	0.6	no	Ludwig and Schatz (1980); Ludwig (1982, 1986); Solioz et al. (1982); Püttner et al. (1983)
Pseudomonas AM1	aa_3	604	14.9	1 or 0.5[b]		2	50, 30				Fukumori et al. (1985a,b)
Rhodobacter sphaeroides	aa_3	606	14			3	45, 37, 35		none		Gennis et al. (1982)
Thiobacillus novellus	aa_3	602		1		2	32, 23				Yamanaka et al. (1979); Yamanaka and Fujii (1980)
Nitrobacter agilis	aa_3	606	20	1.6		2	51[c], 31	1:1	none		Yamanaka et al. (1981); Sone et al. (1983); Sato et al. (1983)
Nitrosomonas europaea	aa_3?	597		0.7		2	50, 33				Yamanaka et al. (1985)
Micrococcus luteus	aa_3	601	17.4	1.33	23.2	3	47, 31, 19	1:1:1	?		Artzatbanov et al. (1987)

[a] Carries haem c.
[b] Copper-limited growth.
[c] Amino acid composition similar to bovine (Yamanaka and Fukumori, 1981).

mitochondrial subunit II rather surprising. In another thermophile, *Thermus thermophilus* HB8, the oxidase also contains *c*-type haem which is associated with the smaller (33 K) of the two polypeptides found. Yoshida *et al.* (1984) have suggested that the four redox centres of the oxidase *per se* (two haems *a*, two coppers) are all bound in the larger (55 K) subunit. The argument is based in part on the inability to resolve the *caa*$_3$ complex into more than two subunits, coupled with the finding that the purified smaller subunit contains no haem *a* or Cu. The extent of similarity between these two thermophile oxidases remains to be established, but it is likely that the smallest of the PS3 subunits will be suspected of being a contaminating peptide. Similar arguments have been rampant in the mitochondrial field for some years (Saraste, 1983).

Data on the oxidase subunits from *Bacillus subtilis* (numbering 3), *Nitrobacter agilis* (2), *Thiobacillus novellus* (2) and *Pseudomonas* AM1 (2) are summarized in Table 5.2. Although the oxidase from the cyanobacterium *Anacystis nidulans* has not been purified, Trnka and Peschek (1986) have probed with antisera to the *Paracoccus* oxidase and demonstrated two subunits of $M_r = 55$ K and 32 K.

In summary, structural and immunological evidence support the idea that the bacterial and mitochondrial oxidases are evolutionarily continuous. The basic redox and chemiosmotic functions of the oxidases are performed by a core structure of two or three subunits. These are the only subunits in the bacterial enzymes, but in eukaryotes these mitochondrially synthesized subunits are supplemented by many others (cytoplasmically synthesized) of unknown function. There is direct evidence for the presence of the haems and coppers on the mitochondrially synthesized subunits I and II (Winter *et al.*, 1980) and there are also similarities in the amino acid sequence of the latter with the copper metalloprotein, plastocyanin (Steffens and Buse, 1979).

5.4.3   The redox centres of *aa*$_3$-type oxidases

5.4.3.1   Haems

In the case of *P. denitrificans*, the *a*-type haems give rise to peaks at 424 and 602 nm for the oxidized enzyme, and at 445 and 605 nm for the reduced enzyme. The absorption coefficient $\varepsilon_{(605-630 \text{ red-ox})}$ is $11.7 \text{ m}M^{-1} \text{cm}^{-1}$. Only haem *a*$_3$ binds CO, causing a shift of the γ-band to 434 nm. In the thermophile PS3 and *Thermus* oxidases, the cytochrome signals again closely resemble their mitochondrial counterparts but the purified enzymes also contain a cytochrome *c* (Table 5.2). Cold-lability of the purified PS3 enzyme leads to spectral changes at low temperatures that can be mistaken for the presence of a *b*-type cytochrome (Baines *et al.*, 1984a).

As in the case of the mitochondrial oxidase, the two haems are readily distinguishable by their ligand-binding properties. In the reduced form, cytochrome a_3 binds CO, cyanide and other ligands, and in the oxidized form, cyanide. Cytochrome a does not bind these ligands (e.g. Fig. 5.2; Poole, 1981).

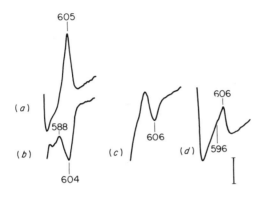

Fig. 5.2 Difference spectra of cytochrome aa_3 in membranes of the thermophile PS3, and the effects of ligands. (*a*) is the dithionite-reduced *minus* oxidized difference spectrum and (*b*) is the CO + reduced *minus* reduced difference spectrum; (*c*) is the analogous CN^- + reduced *minus* reduced difference spectrum; in (*d*), both cuvettes were oxidized by aeration and 5 mM-KCN added to each. The sample (front) cuvette was then treated with dithionite. This effectively eliminates cytochrome a_3 from the difference spectrum, since this component is not readily reduced when cyanide-liganded. All spectra were recorded in 2 mm pathlength cuvettes at 77 K and 1 nm.s^{-1}. The bar represents ΔA 0.03 (a and d) or 0.016 (b and c). Protein concentration was 7 mg.ml^{-1} for (a) and (b) and 3.5 mg.ml^{-1} for (c) and (d).

EPR spectra confirm the essential similarity between the *Thermus* and mitochondrial oxidases (Fee *et al.*, 1987). Both spectra show low-spin haem signals ($g \sim 3.3$, 2.2, 1.4) assigned to cytochrome a. However, two weak resonances ($g \sim 1.92$, 1.85) have not been firmly assigned, although they resemble those from an oxygenated state of the mitochondrial oxidase (Shaw *et al.*, 1978). The most striking difference between mitochondrial and *Thermus* oxidases is the presence in the latter of a distinct $g \sim 3.3$ signal, similar to that assigned to cytochrome c_1 in the mitochondrial bc_1 complex suggesting that the *c*-type cytochrome in the *Thermus* oxidase should be similarly named.

The orientation of the EPR-detectable haem(s) can be determined by centrifuging membrane vesicles onto a Mylar support and examining the angular dependence of the signals as the flattened vesicles are rotated in the EPR cavity. Such experiments (Erecinska *et al.*, 1979) show that the haem planes in *Paracoccus* membranes are normal to the membrane plane as they are in mitochondria.

Mössbauer spectroscopy can provide direct information on the electronic structure of the a_3–Cu_B site (Kent *et al.*, 1982). All three haems in the *Thermus* enzyme (c_1, a, a_3) are distinguishable. Reduced a_3 is high spin and five-coordinate as might be expected of its ligand-binding role, but work on the electronic structure of oxidized a_3 is incomplete (see Fee *et al.*, 1987). Interpretation of the structure of the oxidized cyano complex is equivocal, as indeed it is for the mitochondrial oxidase. The EPR properties of the cyano complex of cytochrome aa_3 resemble the native enzyme in revealing only $Cu_A(II)$ and cytochrome a. However, the cyano complexes of ferric haems generally exhibit EPR signals characteristic of low-spin haems. The absence of an EPR signal from a_3 has been attributed to either *anti*ferromagnetic coupling between $Cu_B(II)$ and $a_3(III)$ (Tweedle *et al.*, 1978) or ferromagnetic coupling (Thomson *et al.*, 1981). However, the availability of high-quality Mössbauer spectra from the *Thermus* enzyme, made possible by [57]Fe enrichment after growing the organism in the presence of the isotope, has clearly shown *ferro*magnetic coupling between a_3 and Cu_B (Kent *et al.*, 1982, 1983).

Resonance Raman spectroscopy relies on laser excitation of the individual haems to probe their vibrational properties. Such a study of the PS3 oxidase has shown close similarities of the haem environment with that in the mitochondrial oxidase (Ogura *et al.*, 1984). For example, excitation of cytochrome c at 514.5 nm gave spectra almost identical to those of horse heart cytochrome c, suggesting minimal interaction between the tightly bound cytochrome c and aa_3. Specific excitation of the two similar haems a is more problematic, but preferential excitation of a and a_3 has shown that the PS3 (Ogura *et al.*, 1984; Sone *et al.*, 1986), *Thermus* (Ogura *et al.*, 1983; Babcock and Callahan, 1983) and mitochondrial enzymes show close similarities. In particular,

(a) the core size of the metalloporphyrin (centre-to-N distance);
(b) the position of a band attributed to the formyl (C=O) stretch in cytochrome a_3; and
(c) a band attributed to a Fe–His stretching frequency in the reduced protein are almost identical.

5.4.3.2   Copper

EPR studies of *Paracoccus* membranes (Albracht *et al.*, 1980) show that the copper signal (Cu_A) is indistinguishable from that in submitochondrial particles. The $g = 2$ copper signals are partly obscured by tightly bound manganese of unknown function (Seelig *et al.*, 1981). EPR studies of the

Thermus purified oxidase show a characteristic signal ($g \sim 2.14$, 2.0) assigned to $Cu_A(II)$ (Fee *et al.*, 1987).

For most bacterial aa_3-type oxidases, analysis of Cu and haem are taken to indicate a 1:1 ratio. EPR studies of *Paracoccus* membranes, however, indicated a Cu:haem *a* ratio of well over 1 under many growth conditions (Albracht *et al.*, 1980). Recent measurements on the purified *Paracoccus* oxidase, using inductively coupled plasma atomic emission spectroscopy (Steffens *et al.*, 1986, 1987), also suggest > 2 Cu:aa_3, as do atomic absorption studies of the *N. agilis* enzyme and *Micrococcus* enzymes (Table 5.2). The functions of the third coppers are not known. A broad absorption band at 830 nm in the spectrum of the oxidized enzyme from *Paracoccus* has been attributed to copper (Ludwig and Gibson, 1981) by analogy with the mitochondrial enzyme. An 830 nm band is lost from *Paracoccus* when cells are grown in a chemostat with growth-limiting concentrations of copper (Hubbard *et al.*, 1986). Similar signals emanate from the oxidases of *Nitrobacter agilis* and of *Pseudomonas* AM1 (Ludwig, 1987) and, at 800 nm, in *Thermus* (Fee *et al.*, 1987).

Growth of *Pseudomonas* AM1 in copper-depleted medium, a strategy adopted by earlier authors (see Poole, 1983), results in synthesis of an oxidase lacking the 820 nm absorption band and having a haem *a*:copper ratio of only 0.59 (Fukumori *et al.*, 1985a). Even so, the enzyme retains catalytic function, suggesting that one copper, possibly Cu_B, is sufficient. Further studies on this phenotypically modified oxidase are warranted.

The environment of the haem irons and coppers in the *Paracoccus* protein has been probed by X-ray absorption fine structure (EXAFS) spectroscopy (Powers *et al.*, 1981). Despite difficulties in unequivocally interpreting such spectra, it seems that the three-dimensional organization of the redox centres and their amino acid-derived ligands is very similar in both the *Paracoccus* and mitochondrial oxidases. A more 'rigid' coordination environment was noted in the case of *Paracoccus*.

5.4.4 Redox properties of aa_3-type oxidases

The aa_3-type oxidases are cytochrome *c* oxidases, their physiological electron donor being a cytochrome *c*. In some cases, however, it is not clear which soluble or membrane-bound cytochrome *c* is the donor. The well-characterized *Paracoccus* cytochrome c_{550}, for example (see Ludwig, 1987), is a poorer donor than horse cytochrome *c*, for which the *Paracoccus* oxidase has a particularly high affinity. In the case of the thermophiles PS3 and *Thermus thermophilus*, horse cytochrome *c* exhibits a lower affinity than either the 'natural' or other bacterial cytochromes *c*, even though the *Thermus* c_{552} closely resembles mitochondrial cytochrome *c* in molecular weight, pI, water solubility and topological location (Hon-Nami and Oshima, 1977).

Interpretation of the rates of oxidation of exogenous cytochrome c by thermophilic oxidases is complicated by the presence of a tightly bound c or c_1. Thus the periplasmic c_{552} of *Thermus* is effectively oxidized by the c_1aa_3 complex, suggesting an analogy with the $c_1 \rightarrow c \rightarrow aa_3$ pathway of mitochondria. Various cytochromes c share a common binding site on the oxidase (Yoshida and Fee, 1984). Electrons from TMPD or soluble cytochromes c are transferred to cytochrome c_1 and thence rapidly to aa_3, suggesting that the bound c_1 is the physiological donor to the a haems.

Rates of reaction between cytochromes c from various sources and the *Paracoccus*, *Thiobacillus* and *Nitrobacter* oxidases, have been used to propose evolutionary relationships and suggest that the *Paracoccus* enzyme is closest in evolution terms to the eukaryotic oxidases (Yamanaka and Fukumori, 1981).

A number of studies have demonstrated the similarity of the potentiometric properties of the bacterial cytochromes a and a_3 to their mitochondrial counterparts (Table 5.3).

Table 5.3 Potentiometric properties of the components of cytochrome aa_3-type oxidases

Organism	Component	E_m (mV) [pH]	Reference
Bacillus alcalophilus			
(i) wild type	a	+230 [7], +240 [9][a]	Lewis *et al.* (1981)
	a_3	+390 [7], +240 [9]	
(ii) non-alkalophilic strain	a	+100 [7]	
Thiobacillus A2	a	+210 [7], +190 [8]	Kula *et al.* (1982)
	a_3	+390 [7], +345 [8]	
Thermophile PS3	a	+190 [7]	Poole *et al.* (1983a)
	a_3	+340 [7]	
	c	+229 [7]	
Thermus thermophilus	a	+270 [7]	Yoshida and Fee (1985)
	a_3	+360 [7]	

[a] Normal cytoplasmic pH.

5.4.5 Reaction with oxygen of aa_3-type oxidases

Both mitochondrial and bacterial aa_3-type oxidases oxidize cytochrome c (Section 5.4.4) and mediate electron transfer to oxygen with rapid reoxidation of the enzyme ($t_\frac{1}{2} \leqslant 3$ to 6 ms). Water is the product. The oxygen affinities of the bacterial enzymes (K_m, 4 to 8 µM; see Poole, 1983) are lower than some values reported for mitochondria (cf. Poole *et al.*, 1979a) but are probably underestimated, being measured with a membrane-covered oxygen electrode.

Although there is little reason to doubt that the aa_3-type enzymes are indeed oxidases and can support measured respiration rates of intact cells, there have been few attempts to confirm the oxidase function by the classical technique of photochemical action spectroscopy, which has been used to demonstrate a_3 as an oxidase in the thermophile PS3 (Poole *et al.*, 1982a) (Fig 5.3) and in *B. subtilis* (Edwards *et al.*, 1981).

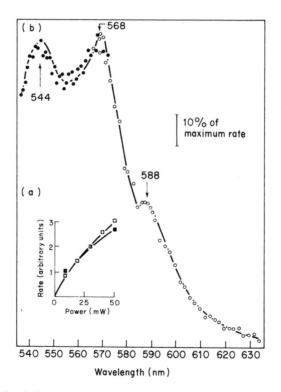

Fig. 5.3 Photochemical action spectrum of the thermophile PS3. (*a*) Irradiation of a cell suspension at 570 nm (–■–; rhodamine 110) and at 590 nm (–□–; rhodamine 6G). Output of the dyes was varied by adjustment of power output of the argon-ion pump laser. (*b*) Irradiation from 536–572 nm used rhodamine 110 (–●–; mean of two spectra), that from 567–634 nm used rhodamine 6G, both at 30 mW. Spectra were normalized by comparison of data obtained at 571 nm. Respiration rate is plotted on the ordinate in arbitrary units; the O_2 concentration in the liquid phase was 3.3 µM.

There have been few studies of the reaction mechanisms of bacterial aa_3-type oxidases to date, but the data available confirm that the similarity to the mitochondrial enzyme extends to the nature of intermediates observed in the oxygen reaction. Thus, photolysis of the CO compound of the PS3 oxidase at subzero temperatures and in the presence of O_2 is followed by O_2 binding to give a photoinsensitive intermediate, which is spectrally similar to the $a_3(II)$ CO compound (Poole *et al.*, 1982a) but probably analagous to the 'Compound A' ($a_3(II)O_2$) observed in mitochondria. At slightly higher temperatures, another intermediate, similar to Compound B, is formed as a result of electron transfer from the enzyme to O_2 (Sone *et al.*, 1984). After flash photolysis of the CO compound at room temperatures, two kinetically and spectrally distinct components are observed, almost identical to those of the mitochondrial enzyme (Ludwig and Gibson, 1981). A species equivalent to the 'pulsed form' has also been described (Reichardt and Gibson, 1983).

5.4.6   Chemiosmotic function of aa_3-type oxidases

Several bacterial cytochrome oxidases of the aa_3 type act as proton pumps. Reconstitution of purified *Paracoccus* oxidase into phospholipid vesicles allows demonstration of an uncoupler-sensitive acidification of a weakly buffered external medium on pulsing with reduced cytochrome *c* (Solioz *et al.*, 1982). In marked contrast to the mitochondrial situation, DCCD neither binds to the subunits nor inhibits proton translocation (Püttner *et al.*, 1983). The measured $\rightarrow H^+/e^-$ ratio of ~ 0.6 contrasts with a value of 0.93 measured with whole cells (Van Verseveld *et al.*, 1981). Possible explanations for the discrepancy include the purification and reconstruction of an incomplete oxidase (e.g. one lacking a third subunit with a role in proton translocation; the most probable explanation, see page 249) or the choice of the inappropriate (i.e. horse heart) cytochrome *c* as donor.

In the case of PS3, the purified and reconstituted oxidase has given H^+/e^- ratios between 0.45 and 1.4, which are partly sensitive to DCCD (e.g. Sone and Yanagita, 1984). Ratios near unity have been observed in analagous experiments with *Thermus* oxidase, providing the donor used is *Thermus* c_{552} (Yoshida and Fee, 1984). DCCD does not inhibit H^+ pumping, consistent with the lack of a subunit resembling the subunit III of the mitochondrial enzyme (Hon-Nami and Oshima, 1984). In contrast, no proton pumping could be observed in the case of the *Nitrobacter* cytochrome aa_3 (Sone *et al.*, 1983b). The failure to observe proton pumping by the *Nitrobacter* oxidase, even under conditions where the *Thermus* enzyme exhibited such pumping, is not consistent with a scheme proposed by Ferguson (1982) which has, however, the attraction of allowing $3H^+$ (and 3 positive charges) to be translocated per two electrons flowing.

5.4.7 Genetics and molecular biology

Taber (1974) isolated mutants of *Bacillus subtilis* defective in cytochrome oxidase but the genetic loci concerned were not mapped. Subsequently, W. S. James and F. Gibson (pers. comm.) identified a mutation (*cox*) resulting in loss of spectroscopically detectable cytochrome aa_3 and mapping at about 130′ on the chromosomal map between *met*C and *pyr*D. The finding that cytochrome aa_3 is necessary for streptomycin accumulation (see Arrow and Taber, 1986) will facilitate screens for further cytochrome aa_3-deficient mutants.

Rapid progress in this area can be expected. Rather surprisingly, most progress has been reported, not with *Bacillus subtilis* or a bacterium with a well-established genetic base in which aa_3-deficient mutants were previously isolated (Willison *et al.*, 1981), but with *Paracoccus denitrificans* and *Bradyrhizobium japonicum* (O'Brian and Maier, 1987). A *Paracoccus* gene bank has been constructed in an inducible expression vector. Gene expression in *E. coli*, detected immunologically (Paetow *et al.*, 1985; Panskus *et al.*, 1986), and DNA sequencing has revealed an amino acid sequence with extensive homologies to subunit II of the mitochondrial oxidase (Ludwig, 1987). A further gene 1.5 kb downstream of the subunit II gene locus has been partially sequenced and shown to be extensively homologous to subunit III of the mitochondrial oxidase (Saraste *et al.*, 1986). Clearly, this points to the possibility of a third, hitherto unrecognized subunit which is probably involved in proton pumping.

Synthetic oligonucleotide probes have recently been used to clone two loci from the chromosomal DNA of *P. denitrificans* (Raitio *et al.*, 1987). One locus contains four or five genes, two of which encode oxidase subunits II and III respectively. A distinct locus codes for subunit I. All three subunits show extensive homology with the corresponding subunits of the mitochondrial oxidase. The most conserved protein is COI (subunit I), which probably has 12 transmembrane segments and may bind the *a*-type haems and Cu_B (Holm *et al.*, 1987). The protein products of four other, adjacent reading frames are unknown. *In vitro*, site-directed mutagenesis is indicated to relate conserved sequences to function but this will require a suitable expression system for the oxidase; *E. coli* is probably unable to synthesize haem *a*.

5.5 Cytochrome *o*

5.5.1 Definition and distribution

A CO-binding pigment, first described in *Staphylococcus aureus* as having the spectral properties of a protohaem-containing enzyme, was later shown by Chance and colleagues to be a terminal oxidase in this and other bacteria (for a review, see Poole, 1983). It was designated (Castor and Chance, 1959) cytochrome *o* ('o' for oxidase), although current recommendations favour the term *b'* to indicate its haem type and ligand-binding property. The term has been used and misused to describe a multitude of pigments with similar CO spectra, even though the crucial criterion of oxidase function has been rarely demonstrated.

It should be noted that the name 'cytochrome *o*' reflects *sensu stricto* the enzyme's function (oxidase) and not its complement of redox-active centres. This is in line with designating, for example, the mitochondrial oxidase as 'cytochrome *c* oxidase' and not '$aa_3Cu_ACu_BCu_x$'. Although the term 'cytochrome *co*' has proved useful in distinguishing those oxidases that contain both haems *b* and *c* from those containing only haem *b* as the sole haem, it may be premature to call the members of the latter group '*bo*' oxidases (Wood, 1984). Clear distinctions between the ligand-binding haem *b* and the second haem *b* in the complex have not yet been described. In those cases where the presence of two dissimilar *b*-type haems is suspected, the term 'cytochrome *o* complex' may be preferable and does not suggest that copper atoms are not an essential component. Here, unless specified to the contrary, cytochrome *o* is used to mean the oxidase complex, without stipulating its composition.

Cytochrome *o* is probably the most widely distributed of bacterial oxidases (Poole, 1983; Poole *et al.*, 1985a), occurring in most physiological groups, most commonly (as in *E. coli*) terminating one of two or more respiratory chains (Poole and Ingledew, 1987; Anraku and Gennis, 1987). Immunological similarity between the cytochromes *o* in a number of bacteria has been demonstrated (Kranz and Gennis, 1985). An *o*-like cytochrome has recently been described in the archaebacterium *Sulfolobus acidocaldarius* (Anemüller *et al.*, 1985), extending the distribution further to another 'primary kingdom' (Woese, 1981). CO-binding *b*-type cytochromes, superficially resembling cytochrome *o*, have been observed in some eukaryotic microbes and in higher eukaryotes, but in only a few cases have such pigments been proven to be oxidases (see Edwards, 1984; Mendis and Evans, 1984 and references therein).

5.5.2   Purification and properties of *o*-type oxidases

With the apparently notable exception of *Vitreoscilla* 'cytochrome *o*' (see page 278), cytochromes *o* are integral membrane proteins that, until recently, proved refractory to purification. In *E. coli* and most other bacteria, cytochrome *o* is accompanied by other oxidases, further frustrating

Table 5.4 Properties of purified cytochrome o-containing complexes

Organism	Subunits (kd)	M_r of native enzyme (kd)	Haem b (nmol/mg)	Haem c (nmol/mg)	copper (nmol/mg)	λ max (γ, CO + reduced minus reduced)	References
Azotobacter vinelandii	28–29		1.6	3.6(c_4)	n.d.	414–416 nm	Yang (1986), Wong and Jurtshuk (1984); Jurtshuk et al. (1981)
Escherichia coli	55, 33	n.d.	19.5[b]	n.d.	16.8	416 nm	Kita et al. (1984a)
	[a]55, 34, 22, 17	140	17	n.d.	n.d.	419 nm	Matsushita et al. (1984)
	51, 28.5, 18, 12.7						Kranz and Gennis (1983)
Methylophilus methylotrophus	23.8 (contains haem c)	124 (2 mols of each subunit)		1 mol/subunit	n.d.	416 nm	Froud and Anthony, (1984)
	31.5 (contains haem b)			1 mol/subunit			
PS3	47–49	Unknown	20.2		(1.7)	412 nm	Baines et al. (1984a); Baines and Poole (1985)
Pseudomonas aeruginosa	29, 21, 11.5, 9.5		5.2[b]	+		415 nm	Matsushita et al. (1982)
			+	+		418.5 nm	Yang (1982)
Rhodopseudomonas capsulata	54–65		7.7 (1 mol/dimer)	–		CO-insensitive	Hüdig and Drews (1982a,b)
Rhodopseudomonas palustris	30.5, 25.5, 12.2, 9.5	–	+	+(c_2)		414 nm	King and Drews (1976)
Thermus	36, 16		18.4		(3.9)	419 nm	Ranger et al. (unpublished; cited in Fee et al. 1987)

[a] Uncorrected for retardation coefficients in acrylamide.
[b] CO-binding haem only.
n.d. not detectable.

purification. Thus, Kita *et al.* (1984a) have used an *E. coli* strain with higher cytochrome *o* levels conferred by an F′, whilst Kranz and Gennis (1983) and Matsushita *et al.* (1984) used a cytochrome *d*-deficient mutant (see Section 5.6.8). Table 5.4 summarizes the properties of the surprisingly diverse preparations from several species; only a few further comments are needed.

Most preparations contain, in addition to the CO-binding *b*-type haem, a second haem *b* or *c*, possibly constituting a binuclear site for O_2 reduction. The distribution of haems among the non-identical subunits is unknown in the case of the *E. coli* enzyme, but in cytochrome '*co*' haem *c* is on the smaller subunit. The presence of more than one *b*-type haem in the complexes from thermophilic bacteria is uncertain. Copper has been detected in the cytochrome *o* complex from *E. coli* by atomic absorption (Kita *et al.*, 1984a) and EPR spectroscopy (Hata *et al.*, 1985). However, optical studies have not revealed a band near 830 nm (Poole *et al.*, 1979b, 1983b,c) as seen in the mitochondrial enzyme, and further studies on the environment of the copper in the protein are required. *Chromatium vinosum* lacks the potential cytochrome oxidases *d* (Section 5.6) and *aa₃* (Section 5.4) but, despite the presence of two CO-binding, *o*-like cytochromes, no Cu(II) signal was detectable by EPR (Wynn *et al.*, 1985).

5.5.3   Spectral, potentiometric and ligand binding properties of the haems of *o*-type oxidases

5.5.3.1   General features

Since the functional, ligand-binding haem of cytochrome *o* is a *b*-type, its reduced minus oxidized spectrum in unpurified preparations is generally masked by other cytochromes *b*, and the CO difference spectrum remains the most useful guide to its presence, although there is considerable variation in the positions of the diagnostic α-(565–575 nm), β-(531–543 nm) and γ-(410–421 nm) bands of the CO adduct (Poole, 1983). Nevertheless, due to loss of absorbance of the reduced form, these bands, as well as the troughs (in the difference spectra), can still be partly masked by signals from other CO-binding haemoproteins, notably cytochromes *c*, *a* and high-spin type *B* haemoproteins. The photodissociation spectrum (Fig. 5.4; see also Sone *et al.*, 1983) which exploits the light-sensitivity of the carbonmonoxy, ferrous form can aid identification, since the CO complexes of potentially interfering CO pigments are generally either not readily photodissociable or rapidly reform after photolysis at low temperatures.

5.5.3.2   *Escherichia coli*

The purified cytochrome *o* preparations from *E. coli* each reveal a split α-band in the reduced state with maxima at 555–558 nm and 562–563 nm. It is tacitly assumed that the two maxima arise from two different haems but it is possible that the two peaks constitute a split α-band. Both bands titrate with the same midpoint potential in the purified preparation (+125 mV, pH 7.4; Kita *et al.*, 1984a) and in membranes (+260 mV, pH 7.0; Reid and Ingledew, 1979). The latter authors concluded, however, that a lower potential cytochrome *b* (+80 mV) was the oxygen-binding haem on the basis of its CO-binding behaviour.

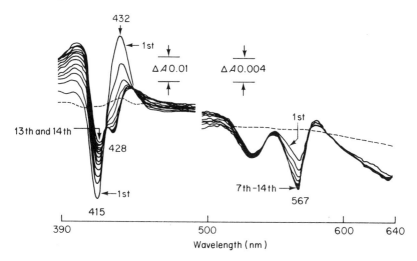

Fig. 5.4   The reaction of cytochrome *o* with oxygen to form 'Compound A'. The spectrum of a suspension of CO-liganded, reduced *E. coli* cells was scanned and subtracted digitally from all subsequent scans. The reaction was initiated at −105°C by photolysis of the CO complex; the numbering of successive scans is with reference to the first after the flash. Scanning proceeds from left to right at 3.5 nm s^{-1}. The dashed line is the pre-photolysis baseline. The first scan is the photodissociation spectrum and is the inverse of the conventional CO + reduced *minus* reduced difference spectrum.

Second-order finite difference analysis of the CO-liganded complex purified from *E. coli* suggests that the lower wavelength component (∼ 555 nm) is more affected by CO (Kita *et al.*, 1984a). Such spectra need cautious interpretation, however, since the band intensities in higher order spectra are strongly dependent on the natural band widths of the components.

The CO difference and photodissociation spectra of *E. coli* cytochrome *o*, in their shape and the sharpening at 77 K, are suggestive of low-spin haem (Wood, 1984). Conversely, the high γ/α ratio of the CO difference spectrum and, in the absolute spectrum of the oxidized form, the presence of a faint shoulder at 630 nm and continuous absorption from 460 to 530 nm, point to the presence of some high-spin form, as occurs in most haemoproteins that react with oxygen. EPR spectra of the purified air-oxidized cytochrome *o* (b_{562}–*o* complex) show a high-spin signal at $g = 6.0$, plus signals at $g = 3.0$ and 2.26 (Hata *et al.*, 1985). Signals near $g = 2.0$ were attributed to Cu(II). The $g = 6$ signal decreased on addition of KCN, whilst the rhombicity of the low-spin signals increased. On this basis, the high-spin ($g = 6$) signal was assigned to the ligand-binding species. Subsequent resonance Raman studies have confirmed the existence of both low-(b_{562}) and high-spin ligand-binding haems in the oxidized and reduced b_{562}–*o* complex (Uno *et al.*, 1985); existence of both spin-states is reminiscent of cytochrome *c* oxidase, in which a_3 is high-spin and *a* is low-spin (page 237).

Table 5.5 Possible relationships of the measured redox potentials of *b*-type cytochromes in *E. coli* to spectrally distinguishable forms

Component	Observed absorption maxima in reduced form (nm)	E_m (mV)[a] (a)	(b)	(c)	(d)	(e)
b_{556}	555–556	+129	—	+80	+46	+35
b_{558}[b]	556–558.6	−43	—	−50	−75	—
b_{562}–*o* complex	(555–556.8) (562–563.5)	+196	+125	+260	+174	+165

[a] The midpoint potentials were measured at room temperature and pH 7.0 (in the absence of CO) by: (a) Hackett and Bragg (1983); (b) Kita *et al.* (1984a); (c) Reid and Ingledew (1979); (d) van Wielink *et al.* (1982); (e) Lorence *et al.* (1984a); neither the component names given, nor their assignments, are necessarily those used in the papers cited.
[b] Readily solubilized and probably not identical to the spectrally similar component of the cytochrome *bd* complex (page 265).

There have been many, often conflicting, attempts to unravel the spectral and potentiometric properties of the *b*-type cytochromes of *E. coli*. The recent purification of the cytochrome *o* complex and its potentiometric titration (Kita *et al.*, 1984a) has clarified the contribution(s) of the complex to the pool of *b*-type cytochromes and thus aided enumeration of the remaining cytochromes *b* (Table 5.5).

The reaction of CO with cytochrome *o* has been studied at subzero temperatures, where the recombination that follows photolysis of the CO adduct is sufficiently slow to monitor with conventional dual-wavelength scanning apparatus. Observation at 432 *minus* 444 nm reveals pseudo-first-order kinetics (activation energy 34.6 kJ mol^{-1}), whilst at 415–444 nm clearly biphasic kinetics, of unknown origin, are seen (see Poole, 1983). The dependence of CO recombination rates on temperature is broadly similar to that of cytochrome aa_3, both enzymes showing sluggish CO recombination below −90°C, in contrast to the kinetics of myoglobin, haemoglobin and *E. coli* cytochrome *d* (page 268). The slow CO binding to cytochrome aa_3 has been attributed to the combination of CO with oxidase copper (Alben *et al.*, 1981), but no such explanation can yet be definitely offered for cytochrome *o*. Cyanide also binds to *E. coli* cytochrome *o*. Indeed, the greater sensitivity of this oxidase, relative to cytochrome *d*, is a distinguishing feature (Pudek and Bragg, 1974). Azide inhibits ubiquinol oxidase activity of the purified cytochrome b_{562}–*o* complex, as do Zn^{2+} and Cd^{2+}, but their reactions with the oxidase have not been further studied. The sensitivity of cytochrome *o* to these agents has been exploited in the design of selective media for mutants lacking cytochrome *d* (page 273).

5.5.3.3 *Other bacteria*

The extraordinary range of redox potentials claimed for bacterial cytochromes *o* (see Poole, 1983) may reflect an intrinsic diversity of the structure and function of these oxidases, but it is just as likely that the potentials are in some cases wrongly attributed. Evidence that a *b*-type cytochrome is a ligand-binding oxidase such as might be obtained (e.g. Reid and Ingledew, 1979) from testing the dependence of the potential on CO (an inappropriate criterion for carboxydobacteria; Cypionka *et al.*, 1985) or from making measurements on a purified preparation, is often lacking (Poole, 1983). Similar comments apply to assignments of the α-absorption maxima of the reduced forms. Coupled spectrum deconvolution and potentiometric analysis of *Proteus mirabilis* (van Wielink *et al.*, 1983) reveal two bands at 556 and 563 nm of high potential (142–143 mV). This is consistent with the assignment to a cytochrome *o* complex of a split α-peak as for *E. coli*. In *Methylophilus methylotrophus*, a component with a much higher potential ($+260$ mV), seen as a peak at 558 nm at 77 K, has been attributed to cytochrome *o* (as has a similar component in *E. coli* in Table 5.5). The assignment is based on (a) the absence of other cytochrome oxidases, and (b) the co-induction (in carbon-excess conditions) of the new spectrally-distinguishable, high potential component together with a great increase in turnover number for azide-sensitive oxidation of ascorbate + TMPD (Cross and Anthony, 1980). The properties of the pure cytochrome *co* confirmed this assignment (Froud and Anthony, 1984; see also page 308).

   The CO-insensitive *o*-like haemoprotein of *Rps. capsulata* shown in Table 5.4 requires further comment. In this organism, cytochrome *c* oxidase activity has been attributed to an unusual *b*-type cytochrome that is distinguished from classical cytochrome *o* by its inability to bind CO and by its very high midpoint potential; this is $+410$ mV in membranes (Zannoni *et al.*, 1974) or $+385$ mV for the purified oxidase (Hüdig and Drews, 1982a,b). A similar oxidase has been reported in the phytopathogen *Ps. cichorii* (Zannoni, 1982), which also contains a second *b*-type oxidase (E_m $+250$ mV) responsible for cyanide-resistant but CO-sensitive respiration. Which, if either, of these cytochromes should be called cytochrome *o* is a matter that can be resolved only by unequivocal demonstration of the oxidase role(s) of the components.

There are few reports of the EPR properties of *o*-type cytochromes, but the oxidized, purified protein from *Az. vinelandii* (Yang, 1986) exhibits a high-spin ($g = 6.0$) signal.

5.5.4   Reactions with oxygen and catalytic activity of *o*-type oxidases

The appearance of peaks at 412–420, 534–544, and/or 560–572 nm in photochemical action spectra is generally accepted as firm evidence for cytochrome *o* as an oxidase (for a recent example, see Poole *et al.*, 1982a and Fig. 5.3). Such experiments do not rule out the possibility, however, that a CO-binding *b*-type cytochrome is upstream of a CO-insensitive oxidase. Similar comments apply to the claim that a CO-binding cytochrome *c* is an oxidase in *Beneckea* (see Wood, 1984). Identification of cytochrome *o* has also been made on kinetic grounds in *E. coli* and *Haemophilus influenzae*; a component of the cytochrome *b* pool was shown in stopped-flow experiments to be kinetically competent to support the observed respiration rates. K_m values between 1.8 and 6.5 µM have generally been attributed to cytochrome *o*, but much lower values were reported for *E. coli* (0.1–0.3 µM; Rice and Hempfling, 1978), *Salmonella typhimurium* (0.74 µM; Laszlo *et al.*, 1984), and *Beneckea natriegens* (see Poole, 1983).

More recently, direct confirmation of the reaction of *o*-type cytochromes with O_2 has come from photolysis of the CO compound of the oxidase in the presence of O_2 at subzero temperatures. In *E. coli* (Poole *et al.*, 1979b,c; Poole and Chance, 1981), and subsequently *Vitreoscilla* (De Maio *et al.*, 1983) and *Acetobacter pasteurianum* (Williams and Poole, 1988), it was shown that the post-photolysis (photodissociation) difference spectrum resembles the conventional (inverted) CO difference spectrum, being the difference between the unliganded oxidase and the pre-photolysis carbon-monoxy ferrous form used as reference (Figs 5.4, 5.5). Ligand binding was revealed by diminution of the peak (attributed to the free ferrous form) and of the trough (due to prior loss of the CO-liganded form). The ligand that binds to the ferrous form after photolysis may be O_2 or CO, but the two are distinguishable by their kinetics and photosensitivity. In the case of *E. coli* cytochrome *o*, for example, the half-times of the apparent first-order kinetics of binding of O_2 and CO at 432–444 nm are 6.8 and 47 min at $-105°C$, respectively (Poole *et al.*, 1979b). Secondly, the binding of CO is fully reversible by photolysis, whereas O_2 binding is light-insensitive (Poole *et al.*, 1979c).

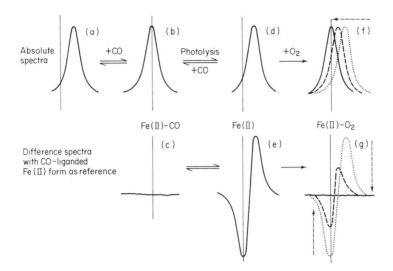

Fig. 5.5 Spectral changes associated with CO binding, photolysis, and O_2 binding of a cytochrome oxidase. Spectra (*a*), (*b*), (*d*) and (*f*) are absolute spectra (recorded, for example, with purified protein using buffer as reference, or with a turbid suspension using a suitably light-scattering material as reference). Spectra (*c*), (*e*) and (*g*) are difference spectra with spectrum (*b*) as reference. The solid vertical lines indicate the absorption maximum of the CO-liganded form (e.g. 415 nm for cytochrome *o*). Reaction of the reduced form (*a*) with CO causes a shift of the band (e.g. 430 nm for cytochrome *o*) to shorter wavelengths. Photolysis of the CO-liganded form (*b*) reforms the ferrous species (*d* or *e*), which, on reaction with O_2, results in re-formation (*f* or *g*) of the oxygenated species with a spectrum similar to (*b*) or (*c*). The band widths and shapes are arbitrary and equal absorbance coefficients for the reduced and CO-liganded forms are assumed. Note that wavelength increases from left to right in all spectra shown.

Such experiments have provided the following information on the O_2 reactions of cytochrome *o* in *E. coli.*

(a) Oxygen reacts with a ferrous haem *b* to form a species that, in the Soret region, is spectrally indistinguishable from the CO compound. There is evidence that in the α/β region, the compound resembles the ferric rather than the carbonmonoxy form. By analogy with the O_2 compound of cytochrome aa_3, this (first?) intermediate may be termed compound A.

(b) With reference to the following equation, in which a possible dihaem structure of the oxidase complex is shown, the forward velocity constant K_{+1} has been shown to be approximately 3 orders of magnitude lower than the corresponding value for cytochrome aa_3 (Chance *et al.*, 1975):

$$(Fe^{2+}).Fe^{2+} + O_2 \underset{K_{-1}}{\overset{K_{+1}}{\rightleftharpoons}} \underset{Compound\ A}{(Fe^{2+}).Fe^{2+}.O_2} \xrightarrow{electron\ transfer} ?$$

Furthermore, the dissociation constant k_d is 7–15 μM between -90 and $-101°C$, but 480 μM for cytochrome aa_3. There is good agreement between the values reported for the cytochrome o of *E. coli* (Poole *et al.*, 1979b) and o of *Vitreoscilla* (De Maio *et al.*, 1983).

(c) Cells from oxygen-limited cultures of *E. coli* give a photodissociation spectrum with a peak at 436, not 432 nm. Based on the insensitivity of the CO complex to irradiation with a HeNe laser (line at 632.8 nm), which selectively photolyses the CO compound of cytochrome d, the 436 nm band was attributed to an o-like cytochrome (Poole and Chance, 1981). The 'cytochrome o_{436}' reacts with oxygen at subzero temperatures approximately tenfold faster than does cytochrome o_{432}, consistent with its synthesis under O_2-limited conditions.

In marked contrast to those oxygenated intermediates of *E. coli* cytochrome o and certain aa_3 types, stabilized only at subzero temperatures, compounds attributed to oxygenated species and readily observed at room temperatures have been reported for three bacterial cytochrome o-like haemoproteins. The *Vitreoscilla* o-like haemoglobin is covered below (page 278). Earlier, Iwasaki (1966) had described a CO-binding pigment in *Acetobacter suboxydans*, which exhibits peaks (in the CO-liganded form) at 423, 544 and 574 nm. It is possible that the 423 nm band, which is at an unusually long wavelength for cytochrome o, may be influenced by the CO complex of a cytochrome a, but the finding that the presumptive oxygen compound also absorbs at this wavelength, and the realization that haemoglobins do occur in bacteria (see below) demand a reassessment of the nature and properties of this pigment.

More recently, Matsushita *et al.* (1982) have noted the formation of an 'oxygenated' species of a purified o-type cytochrome from *Ps. aeruginosa*. This form had a diminished γ-peak at 413 nm, intermediate between the positions of the oxidized and reduced forms, with α and β forms similar to the ferric form; no other information is available.

5.5.5 Chemiosmotic aspects of o-type oxidases

Measurements of respiration-driven proton translocation made on intact cells from aerobically grown *E. coli* cultures (Lawford and Haddock, 1973) suggested a H^+/O ratio of 2.26. This has been taken to represent an integer value of 2.0 for the oxidation of succinate, together with a contribution from respiration of endogenous NADH-linked substrates. A H^+/O ratio of 2 for succinate oxidation allows postulation of models of respiration-driven proton translocation that need invoke neither protonmotive Q-cycles nor a proton-pumping oxidase. In the redox loop model shown in Figure 5.6, succinate dehydrogenase reduces quinone on the cytoplasmic face of the membrane; quinol is reoxidized by an electron carrier on the periplasmic face, with the concomitant release of $2H^+$. The electrons are then transferred to O_2 and the scalar oxidase protons are consumed from the cytoplasm.

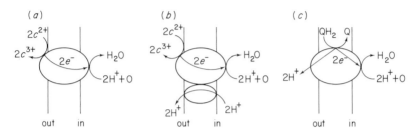

Fig. 5.6 Models for $\Delta\psi$ and ΔpH generation by cytochrome oxidases, after Matsushita *et al.* (1984). (*a*) Mitchell's early proposal for mitochondrial cytochrome aa_3; $\Delta\psi$ is generated by vectorial electron flow from reduced cytochrome c (c^{2+}) on the outside of the membrane to the inner face; ΔpH is generated solely by proton consumption during oxygen reduction on the inner face of the membrane. (*b*) Mitochondrial and bacterial (e.g. *P. denitrificans*, PS3, *Th. thermophilus*) aa_3-types acting as proton pumps in addition to the features shown in *a*. (*c*) Cytochrome oxidase *o* from *E. coli*; vectorial electron flow from ubiquinol (QH_2) results in $\Delta\psi$ generation. ΔpH is generated by scalar reactions leading to H^+ appearance on the outside of the membrane and H^+ consumption during oxygen reduction (see also Fig. 1.29, page 68).

More recent studies have exploited the purified oxidase (Kita *et al.*, 1982; Matsushita *et al.*, 1983, 1984). Proteoliposomes made by detergent dilution or dialysis in the presence of oxidase and phospholipids, followed by freeze-thawing and sonication, generate a proton electrochemical gradient ($\Delta\bar{\mu}_{H^+}$) of -115 to $-140\,mV$ (interior negative and alkaline) during oxidase turnover, but with low respiratory control ratios (about 1.5). In the additional presence of the purified *lac* permease, such proteoliposomes can transport lactose against a concentration gradient in a direction consistent with inward translocation in the intact cell (Matsushita *et al.*, 1983). Similarly, proteoliposomes containing the cytochrome *o* complex, ubiquinone-8 and pyruvate oxidase, generate a transmembrane potential of at least $-120\,mV$ (negative inside) (Carter and Gennis, 1985). By using a glass electrode to measure changes in external pH and the fluorescence of vesicle-entrapped 8-hydroxy-1,3,6-pyrenetrisulphonate to measure changes in internal pH, Matsushita *et al.* (1984) also showed that oxidation of quinol (a hydrogen donor) is accompanied by proton release on the external surface and consumption on the internal surface with an H^+/O ratio of 0.9–1.4.

However, oxidation of the electron donor, N,N,N',N'-TMPD resulted in little change in external pH. These results and the confirmatory data of Hamamoto *et al.* (1985), are consistent with the view that cytochrome *o* in *E. coli* does not catalyse vectorial proton translocation but generates a pH gradient from scalar (non-vectorial) proton-consuming and -releasing reactions on opposite aspects of the membrane. Cytochrome *o* constitutes the only site of $\Delta\mu_{H^+}$ generation in the span of the respiratory chain between D-lactate dehydrogenase or ubiquinol and oxygen, whereas between NADH and oxygen, there is an additional site upstream of quinone (Matsushita and Kaback, 1986). Such a scheme was earlier suggested by Poole and Haddock (1975).

These findings contrast with the consensus of opinion on the mitochondrial aa_3-type oxidase and with the report (Hüdig and Drews, 1984) that an *o*-type oxidase from *Rps. capsulata* functions as a true proton-pumping system on incorporation into proteoliposomes.

5.5.6   Genetics of *o*-type oxidases

Although 'TMPD-negative' mutants, defective in the ability of cytochromes *o* and a_1 to oxidize cytochromes $c_4 + c_5$, were reported in *Az. vinelandii* by Hoffman *et al.* (1979), there have been few reported attempts to isolate such mutants. The task of isolating a cytochrome *o*-deficient mutant has been facilitated by the availability of *E. coli* mutants already defective in cytochrome *d* (page 273). Thus, a mutant lacking both oxidases was found by Au *et al.* (1985) amongst a number of mutants. These mutants had been enriched by ampicillin killing, were unable to grow aerobically on the non-fermentable substrates succinate and lactate, but retained the ability to grow anaerobically with nitrate. Spectroscopic and immunological studies show the mutant to be deficient in cytochrome *o*. The reduced *minus* oxidized spectrum was devoid of cytochrome b_{562} and the CO-difference spectrum lacked the γ-bands characteristic of cytochrome *o*. A redox titration in the presence of CO showed the highest potential component to be missing. Co-transduction experiments map the lesion, designated *cyo*, between *pho*R and *acr*A at about minute 10 on the *E. coli* chromosome. This mapping is consistent with the finding that merodiploids containing the F13 F'-factor (*lac-pur*E) have twofold amplified levels of the b_{562}–*o* complex (Kita *et al.*, 1985). The *cyo* gene, which encodes the cytochrome *o* complex, has been cloned (Au *et al.*, 1987) and sequencing is in progress (R. B. Gennis *et al.*, unpublished; see Anraku and Gennis, 1987).

Mutants have been isolated by a similar procedure but in a different cytochrome *d*-deficient strain and without enrichment by ampicillin (R. K. Poole, F. Gibson, G. B. Cox and H. D. Williams, unpublished). Such mutants have been shown to retain some spectroscopically detectable cytochrome *o*, yet are unable to grow on succinate. Genes complementing two mutants have been cloned from a cosmid library and the gene products analysed in an *in vitro* coupled transcription/translocation system. Molecular weight similarities with the smaller subunits of the cytochrome *o* complex are evident, but further work is required to characterize the gene products and predict their amino acid sequences.

5.6 Cytochrome *d*

5.6.1 Definition, distribution and nomenclature

Cytochrome *d* was discovered in 1928 in *E. coli* and *Shigella dysenteriae* (early papers are cited by Poole, 1983). Its current nomenclature replaced 'cytochrome a_2' on identification of its prosthetic group as a chlorin, i.e. a porphyrin with one pyrrole ring saturated. Recently, a detailed structure of the haem has been proposed based on spectroscopic studies of the haem from the purified cytochrome *bd* oxidase complex of *Escherichia coli* (Timkovich *et al.*, 1985).

Cytochrome *d* is widely distributed, especially in Gram-negative heterotrophs. It is rare in Gram-positive heterotrophs, phototrophs and chemolithotrophs. It is often found in the same organism as cytochrome *o* (as in *E. coli*; Poole and Ingledew, 1987; Anraku and Gennis, 1987) but rarely with cytochrome aa_3. Interestingly, it has been reported in anaerobes, where its high O_2 affinity may endow the cell with a 'respiratory protection mechanism' (see Poole, 1983, for references). Antigenically identical *d*-type oxidases are found in a number of Gram-negative bacteria (Kranz and Gennis, 1985). Although a cytochrome *d* complex has recently been purified from *Photobacterium phosphoreum* (Section 5.6.2) the vast majority of published work relates to *E. coli* and it is this system that is described here unless explicitly stated otherwise.

All attempts to purify cytochrome *d* from bacteria have resulted in the isolation of an oxidase complex that contains haem *d*, identified by its characteristic spectral properties, and a *b*-type haem with a spectrum indicative of low-spin haem *b* (λ_{max} at 558 nm for the α-band of the reduced form at low temperatures). In addition, the purified complexes show an absorbance in the reduced state near 590–595 nm, originally but erroneously attributed to 'cytochrome a_1', but now believed to be a high-spin *b*-type haem with an unusually distinct α-band at this wavelength. It seems appropriate, therefore, to describe the oxidase as a cytochrome *bd* complex. The *d*-haem(s) bind oxygen, CO and other ligands, whilst the function of the *b*-type haems is electron transfer to haem *d*, probably in the order $b_{558} \to$ 'a_1' $\to d$.

It has frequently been reported that cytochrome *d* appears during oxygen-limited growth. This is but one, however, of a number of 'inducing' conditions (Poole, 1983; Ingledew and Poole, 1984) which include attainment of the late exponential or stationary phases of growth, growth on glucose, anaerobic growth, growth with cyanide or under copper-limited or sulphate-limited conditions but not iron-limited conditions (Hubbard *et al.*, 1986). In *E. coli*, the biosynthesis of cytochrome *o* is repressed when cytochrome *d* is induced by lowering the dissolved O_2 tension (Kranz and Gennis, 1983). A mutant isolated by Johnson and Bragg (1985) exhibits control by growth substrate of cytochrome *d* formation; the components of the cytochrome *d* complex are present during growth on glycerol or glucose but not on succinate or other substrates.

5.6.2 Purification and properties of the subunits

An oxidase complex containing cytochrome *d* has been purified from *E. coli* and *Photobacterium phosphoreum* (Table 5.6). With the exception of an earlier purification of the *Ph. phosphoreum* complex (Watanabe *et al.*, 1979) in which the same molecular weight was observed in SDS–PAGE or on gel filtration, suggesting the existence of only one polypeptide, the other reports describe two dissimilar subunits. Reduced *minus* oxidized difference spectra show the presence of *b*- and *d*-type cytochromes and a peak resembling 'cytochrome a_1'. CO binds to haem *d* (as does O_2) and to at least part of the cytochrome *b*, with an α peak near 560 nm in both the *E. coli* (Miller and Gennis, 1983; Kita *et al.*, 1984b) and *Ph. phosphoreum* (Konishi *et al.*, 1986) preparations. Cytochrome *o* (Section 5.5.3) has not been reported as a contaminant of these preparations and so it is possible that the CO-binding to cytochrome *b* results from some denaturation during purification. The purified complexes lack copper, quinones and non-haem iron, and oxidize ubiquinol-1 and TMPD. Inhibition of quinol oxidation is achieved only with concentrations of cyanide, azide, Zn^{2+} or HQNO far in excess of those required for inhibition of quinol oxidation by purified cytochrome *o* (see papers cited in Table 5.6).

The following genetic and immunochemical studies have identified subunit I as the site of the *b* haem with a λ_{max} at 558 nm. Antibodies to subunit I inhibit ubiquinol oxidation but not TMPD oxidation, which is inhibited by antisubunit II antibodies (Kranz and Gennis, 1984). Furthermore, the *cyd*B phenotype (page 273) is the lack of subunit II and cytochromes *d* and 'a_1', whilst subunit I and b_{558} are retained.

Cloning the gene for subunit I (see page 273) results in overproduction of

Table 5.6 Properties of cytochrome *d*-containing oxidase complexes

Property	*E. coli*			*Photobacterium phosphoreum*
	Reid and Ingledew (1980) Finlayson and Ingledew (1985)	Miller and Gennis (1983)	Kita *et al.* (1984b)	Konishi *et al.* (1986)
Subunits				
M_r of native enzyme	43 K, 70 K 150 K	43 K[a], 57 K	26 K, 51 K 77 K (assumed)	41 K, 54 K 160 K
pI	4.8	5.3	—	—
Redox centres (nmol mg protein^{-1})				
Fe	—	25.5 to 34.1	26.6	22.5
Haem *b*	7.1	18.9	12.3	10.2
Haem *d*	—	None found	9.54	—
α-absorbance peaks (nm) (77 K) (reduced *minus* oxidized)	558, 595, 629	558.8, 591, 624.5	558 (594[b]), 624	560, 590, 625
CO-difference spectra (nm) (room temperature)	420(p), 433(t), 444(t), 625(p), 647(p)	420(p), 430.5(t), 444(t) 622(t), 642(p)	420(p), 430(t), 442(t), 560(t), 622(t), 642(p)	418(p), 434(t), 560(t), 620(t), 639(p)

[a] 28 K in 12.5% acrylamide.
[b] 77 K value not given, 594 nm at room temperature.
(p) peak: (t) trough.

cytochrome b_{558} and has facilitated purification of subunit I (Green *et al.*, 1986). Its M_r (57 000) is the same as that reported previously (Table 5.6).

5.6.3   Optical properties of *d*-type oxidases

5.6.3.1   Cytochrome *d*

The spectral properties of the chlorin haem (Timkovich *et al.*, 1985) of cytochrome *d* are distinctive and were noted in the pioneering studies of Warburg's group (see Poole, 1983). Reduced preparations show an absorbance maximum at 628 to 632 nm (Fig. 1.21, page 49); this is shifted by about 5 nm further to the red in the presence of CO. Aeration of cell or membrane suspensions containing cytochrome *d* results in a further shift of the peak from 646 to 652 nm. It has been widely assumed that this latter, narrow band arises from the ferric form (Keilin, 1966) but more recent evidence suggests that it should be attributed to a stable 'oxygenated' compound of cytochrome *d* (Pudek and Bragg, 1976a; Poole *et al.*, 1982b, 1983b; Hata *et al.*, 1985). Early evidence that the Soret band is weak and diffuse has been supported by:

(a) photodissociation spectra at 4 K (Poole *et al.*, 1982c);
(b) selective photolysis by a He-Ne laser of the CO complex of cytochrome *d* (Poole and Chance, 1981);
(c) lack of correlation between the intensities of the α and putative Soret bands in photodissociation spectra in the absence or presence of oxygen (Poole and Chance, 1981);
(d) the assignment of a band at 448 nm to 'cytochrome a_1', not *d* (Poole *et al.*, 1981).

A further absorption band at 675–680 nm is also attributable to a form of cytochrome *d*. The signal appears during the course of oxidation of 'cytochrome d_{650}' at subzero temperatures (Poole *et al.*, 1983c) and it can also be observed in reduced *minus* oxidized difference spectra. It is now believed to represent a peroxy form of cytochrome *d* (Poole and Williams, 1988).

In the presence of oxygen, cyanide binds to the reduced form of cytochrome *d*, eliminating its α-band (Pudek and Bragg, 1974). Cyanide also reacted with the form that absorbs near 650 nm or with some 'invisible' form, intermediate between the reduced and 650 nm-absorbing forms. A current hypothesis (Poole *et al.*, 1983a; Section 5.7) is that this cyanide-binding form is the ferric species which, like other oxidized cytochromes, is devoid of distinct α features and is, in that sense, 'invisible'.

CO causes a red shift in the position of the α-band of the reduced form near 630 nm (Table 5.6). The reaction with CO is unusual in that the recombination of photodissociated CO with the ferrous form is too rapid to be observed by conventional scanning apparatus even at 77 K (Poole *et al.*, 1982c; Section 5.6.6).

A number of small nitrogen-containing ligands react with or bind to cytochrome *d*. Thus, NO_3^-, NO_2^-, $N_2O_3^{2-}$ (trioxodinitrate) and NO all react with cytochrome *d* in *E. coli* membranes, giving diminution and shifting of the 630 nm peak to 641–645 nm (Meyer, 1973; Hubbard *et al.*, 1983, 1985). The rapid reduction of trioxodinitrate is of special interest since this compound has long been postulated as an intermediate in microbial denitrification; the reaction probably occurs between cytochrome *d* and nitroxyl ion, produced in the self-decomposition:

$$HN_2O_3^- \rightarrow HNO + NO_2^-$$

Thus, it seems likely that all four compounds result in formation of the nitrosyl species.

The finding that NO_3^- and its reduction products react with cytochrome *d* may explain the red-shifted peak of cytochrome *d* in nitrate-grown cells (e.g. Haddock *et al.*, 1976). The *Az. vinelandii* (see Poole, 1983) and *Ph. phosphoreum* (Watanabe *et al.*, 1979) cytochromes *d* demonstrate similar spectral shifts on binding NO_2^- and various nitrogen ligands. In none of these instances has the identity of the presumptive nitrosyl compound been confirmed.

A 'cytochrome *d*' in *Thiobacillus denitrificans* and some other bacteria has been reported to absorb, in the reduced state, at 610–620 nm (see Poole, 1983). They have not been further characterized.

5.6.3.2   Cytochrome b_{558}

The *b*-type cytochrome of the oxidase complex accounts for the α-absorbance at 560–562 nm in absolute (reduced) and difference (reduced *minus* oxidized) spectra at room temperature. The β-band is at about 531–532 nm (room temperature) and the γ-band at 429–430 nm. The α and β band positions are confirmed by potentiometric resolution of the reduced *minus* oxidized spectrum (Koland *et al.*, 1984a). Low temperature (77 K) shifts all bands 1–4 nm to the blue, most diagnostically that of the α-band of the low spin cytochrome *b* to 558 nm. An extinction coefficient for the cytochrome *b* has been obtained by study of purified subunit I (Green *et al.*, 1986).

5.6.3.3   The component resembling 'cytochrome a_1'

Intact cells, membranes and 'soluble' subcellular fractions from *E. coli* show, in reduced *minus* oxidized difference spectra, a broad band between 585 and 590 nm. The component has been called cytochrome a_1 because of its superficial similarity to the oxidase in *Acetobacter* (Section 5.7.1) but its role in *E. coli* and many other bacteria where it frequently coexists with cytochrome *d* is unclear.

Both soluble and membrane-bound forms of the 'a_1-like' haemoprotein exist in *E. coli* (Poole *et al.*, 1985a,b, 1986); the latter co-purifies with the cytochrome *bd* complex from *E. coli* (Table 5.6) but its function is unknown. Reduced (absolute) or reduced *minus* oxidized (difference) spectra of cells, membranes or the purified oxidized complex show a distinct shoulder on the red side of the γ-peak of cytochrome *b* (e.g. Kita *et al.*, 1984b), whilst the CO difference spectra in this region are characteristically 'W-shaped', with troughs at about 430 and 442–444 nm (room temperature). The longer wavelength features have been assigned to cytochrome *d*, ignoring the substantial evidence that cytochrome *d* has only a very weak γ-band. The a_1- like component, therefore, is perhaps responsible for this band near 440 nm. Interestingly, the preparation of Kita *et al.* (1984b), which is claimed to have only 3% 'contamination' by the a_1-like species, has the γ-peak (reduced minus oxidized) at 429 nm and only a shoulder at higher wavelengths, whilst the preparation of Miller and Gennis (1983) shows the γ-peak at 432 nm, possibly by fusion of the 429 nm band with an intense band at a higher wavelength. The soluble a_1-like component (haemoprotein b_{590}) also has intense Soret features (Poole *et al.*, 1986).

The reduced *minus* oxidized spectrum of the a_1-like component has been resolved from a set of spectra of the purified complex recorded at different solution potentials (Koland *et al.*, 1984a; Lorence *et al.*, 1986). The α-band is at about 595 nm, the β-band is at 560 nm and there is a trough at 645 nm indicative of a high spin haem *b*, as in peroxidase. It has been called 'cytochrome b_{595}' (Lorence *et al.*, 1986), following the example of the similar soluble component, haemoprotein b_{590} (page 274). It makes a significant contribution to the haem *b* content of the complex, since the extinction coefficient of the pure cytochrome b_{558} (22 000 M^{-1} cm^{-1}) is twice as high as the apparent value (10 800 M^{-1} cm^{-1}) for total *b*-type cytochromes in the complex (Green *et al.*, 1986). A 1:1 stoichiometry of b_{558}:'b_{595}' with two mols of cytochrome *d* has been proposed (Lorence *et al.*, 1986).

5.6.4   EPR properties of *d*-type oxidases

Membranes of *Azotobacter vinelandii* exhibit EPR signals due to high-spin ferric haem at $g = 6.2$, 5.86 and 5.48 (Dervartanian *et al.*, 1973); these resolve into two components, one rhombically distorted ($g = 6.24$, 5.5) and a dominant axial resonance ($g = 5.94$) (Kauffman *et al.*, 1975). These signals were attributed to cytochrome *d*, being sensitive to the ligands cyanide and CO. Miller and Gennis (1983) drew similar conclusions regarding the *E. coli* oxidase.

Recently, however, Hata *et al.* (1985) have described the EPR properties of the *bd* complex in membranes from *E. coli* and in a purified form. High spin signals near $g = 6$, with rhombic signals ($g = 6.2, 5.6$) resembling those in *Az. vinelandii*, as well as weak low-spin signals ($g = 2.5, 2.3$) were detected. The low-spin signals were altered by ligands such as azide and cyanide, and assigned to cytochrome *d*. Consistent with this assignment are the E_m values of the high spin ($g = 6$) and the low spin ($g = 2.5$) signals which are in good agreement respectively with the potentials for cytochromes b_{558} and *d* (Section 5.6.5 and Table 5.7). This assignment leaves unanswered the origin of the high-spin rhombic signals, which, after photolytic activation of the reaction with O_2 at low temperature (page 268), appear (i.e. become oxidized) *before* the $g = 6$ (axial) signal (Kumar *et al.*, 1985). They were at first attributed to conformational change of oxidized cytochrome b_{558} (Hata *et al.*, 1985), but Poole and Williams (1987) and, subsequently, Hata-Tanaka *et al.* (1987) suggested that they might arise from the 'cytochrome a_1-like' b_{595} component.

The low-spin signals assigned to cytochrome *d* were more intense on oxidation in the absence of oxygen than in aerated samples (Hata *et al.*, 1985), in agreement with the proposal (Poole *et al.*, 1983b) that cytochrome *d* readily forms an oxygenated species in which the haem *d* remains in the ferrous state and spin-inactive.

5.6.5   Potentiometric studies of *d*-type oxidases

The midpoint potential of cytochrome *d* in membranes, measured at the absorbance maximum of the reduced form, and with purified complexes, are in good agreement (Table 5.7) but the latter are influenced by the detergent used to solubilize the complex and by pH (Lorence *et al.*, 1984b). Even at potentials as high as $+440$ mV, the 650 nm form does not appear unless oxygen is present (Pudek and Bragg, 1976b) or unless the sample is cycled through a preliminary phase of air oxidation, followed by substrate reduction (Hendler and Schrager, 1979). Under the latter condition, the absorption bands at about 630 and 650 nm exhibit very unusual and complex behaviour which could be interpreted only by invoking concerted four-electron transport by cytochrome(s) *d*. An alternative explanation (Poole *et al.*, 1983b) is that the 650 nm species does not arise directly from oxidation of the ferrous state to ferric but is attributable to an oxy-ferrous form (see next section).

The associated cytochromes b_{558} and 'a_1' have lower potentials. The midpoint potential of cytochrome b_{558} purified from an over-producing strain and incorporated into phospholipid vesicles ($+160$ mV) is similar to the values (Table 5.7) obtained in membranes. In other purified preparations, not reconstituted, much lower potentials have been recorded. The midpoints reported for 'cytochrome a_1' (cytochrome b_{595}) are in the range $+113$ to $+160$ mV (Table 5.7).

Table 5.7   Midpoint potentials (mV, pH 7.0, $n = 1$) of the components of the cytochrome *d*-containing oxidase complex of *E. coli*, measured optically

d	Cytochrome b_{558}	'a_1'	Preparation	Reference
$+260$	$+165$–196	$+147$	Membranes	Pudek and Bragg (1976b)
$+280$	$+250$	$+260^{(b)}$	Membranes	Reid and Ingledew (1979)
		$+160$		
$+260$	$+180$	$+150$	Membranes	Lorence *et al.* (1984b)
$+240$	$+10^{(a)}$	n.g.	Purified	Kita *et al.* (1984b)
$+232$	$+61$	$+113$	Purified	Koland *et al.* (1984a)
n.g.	$+160$	n.g.	Purified	Green *et al.* (1986)
	$+90^{(c)}$			

[a] pH 7.4.
[b] Higher value probably due to interference from cytochrome *d*.
[c] In cholate.
n.g. not given.

5.6.6   The reaction with oxygen and other ligands of *d*-type oxidases

Photochemical action spectroscopy has clearly demonstrated an oxidase function for cytochrome *d* in *E. coli*, *Az. vinelandii* and other bacteria, although the CO complex is unusually photostable. Stopped-flow studies have also shown that cytochrome *d* in *E. coli* and *Haemophilus* reacts with O_2 sufficiently rapidly to be considered a terminal oxidase (see Poole, 1983).

High affinity for oxygen (K_m values in the range 0.024–0.38 μM) has been reported for the *E. coli* cytochrome *d* (Rice and Hempfling, 1978; Kita *et al.*, 1984b), although a surprisingly high K_m (32 μM) has been claimed by Sánchez Crispin *et al.* (1979) in cells grown anaerobically with nitrate. This may be a special case, since unusual multiphasic kinetics were observed for such cells in stopped-flow experiments (Haddock *et al.*, 1976) and the reduced form of cytochrome *d* in nitrate-grown cells has an unusually long λ_{max} (Section 5.6.3.1). There is general agreement, however, that cytochrome *d* is a high-affinity oxidase which may serve to maintain low free-oxygen concentrations necessary for nitrogen fixation or other oxygen-labile processes.

The primary events in O_2-binding to ferrous cytochrome *d* have been studied using the low-temperature trapping and ligand-exchange procedures (triple-trapping) (see page 256). In these experiments, a reduced preparation of cells or membranes is reacted with CO and then, at about $-25°C$ in the presence of ethylene glycol (as antifreeze), O_2 is added before rapid trapping of the sample at $-78°C$. A small proportion of the carbonmonoxy ferrous cytochrome d is oxidized during this procedure. Initiation of the reaction at $-130°C$ by photolysis of the remaining CO adduct has been studied spectrophotometrically and by EPR spectroscopy. At this temperature, the first optically detectable species has a sharp, symmetrical band at 650–652 nm (where the reference spectrum is the CO-liganded form) (Fig. 5.7). This species ('cytochrome d_{650}') has spectral properties generally attributed to the *oxidized* cytochrome, but a consideration of the available optical (Poole *et al.*, 1983b,c), EPR (Kumar *et al.*, 1985) and resonance Raman spectroscopic (Poole *et al.*, 1982b) data has led to the hypothesis that cytochrome d_{650} in *E. coli* is a ferrous oxygenated form and an early (first?) intermediate in the oxidation of cytochrome d. The oxidized ferric form should be equated with the 'invisible species' (d_x or $d*$) postulated by others (see Poole, 1983), having little characteristic absorbance in the red region of the spectrum and being analogous to, say, methaemoglobin. This is the converse of previous proposals, which invoke the existence of a hypothetical, invisible intermediate between the ferrous and ferric forms (see Section 5.6.3.1). It provides an explanation of the curious potentiometric and kinetic behaviour of the 650 nm form (Section 5.6.5). It also rationalizes the ability of CO to react with d_{650} in which, in this proposal, the haem is formally ferrous. The hypothesis is supported by resonance Raman spectroscopy (Poole *et al.*, 1982b) of aerated cell suspensions or solubilized enzyme in which the d_{650} form is quite stable. During laser excitation at 647.1 nm, the spectrum shows resonances at 1078 to 1105 cm^{-1}, attributable to the O–O stretching frequency of the oxidase–oxygen adduct and reminiscent of oxyhaemoglobin and oxymyoglobin. Subsequently, further EPR evidence has suggested that the form of cytochrome d predominating in air-oxidized preparations is in the ferrous state (Hata *et al.*, 1985; page 266).

The d_{650} form is readily prepared by aeration of cells, membranes or the purified oxidase complex. Demonstration of its nature as an oxygenated species ($Fe^{3+}.O_2^-$ or $Fe^{2+}O_2$) (though the literature still abounds with descriptions of its 'oxidized' state) prompts speculation of the role of this species in oxygen metabolism. In particular, the appearance of cytochrome d under conditions of lowered O_2 availability (page 262) suggests that cytochrome d may play a role in oxygen transport or storage.

The presence of cytochrome b_{558} in the purified oxidase complex makes it a likely immediate reductant of cytochrome d itself. Although cytochrome b_{558} is reduced by ubiquinol, it does not reduce oxygen (Green *et al.*, 1986). The progress of oxidation of the oxygenated form in

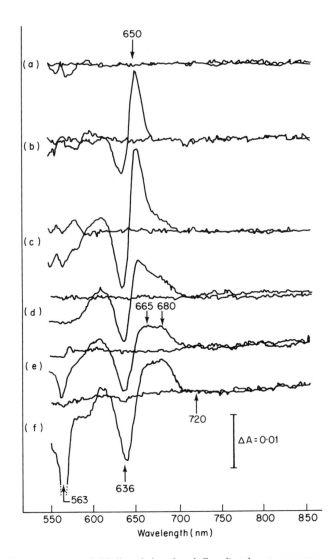

Fig. 5.7 Flash photolysis of CO-liganded, reduced *E. coli* at low temperatures and observation in the 550–860 nm range. In (*a*), the sample was freeze-trapped in the absence of O_2 and its spectrum scanned and stored in the digital memory of the dual-wavelength spectrophotometer at $-132°C$. This spectrum was subtracted from a subsequent pre-photolysis spectrum to yield the reduced +CO *minus* reduced + CO baseline, or from the spectrum recorded immediately after photolysis (30 flashes of a 200 J Xenon lamp). In (*b*) to (*f*), the procedure was identical except that O_2 (approx. $400 \,\mu M$) was present and spectra were scanned at $-132°C$ (*b*), $-97°C$ (*c*), $-88°C$ (*d*), $-80°C$ (*e*) or $-70°C$ (*f*). In each case, the reference wavelength was 720 nm and the scan speed approx. $1.4 \, nm \, s^{-1}$. Protein concentration was $11.8 \, mg \, ml^{-1}$.

membranes, observed in 'snapshots' during the reaction at low temperature (Poole *et al.*, 1983c), shows sequential oxidation of two optically distinguishable *b*-type cytochromes. Since the reaction was initiated with a He–Ne laser at 633 nm, intended to photolyse selectively the CO compound of cytochrome *d*, neither of these is thought to be cytochrome *o*. However, as noted above, parallel EPR experiments show (Kumar *et al.*, 1985) that the axial *g*6 signal, assigned to b_{558} by Hata *et al.* (1985), appears *after* the development of the flanking rhombic signals. Since the latter are not believed by these authors to be oxidized cytochrome *d* (Fe III), a further component must be invoked. Is the associated cytochrome a_1-like haemoprotein a candidate?

Recent low-temperature studies on *Acetobacter pasteurianus* strain NCIB 6428 (Williams and Poole, 1986, 1987) have revealed generally similar behaviour to that described above in *E. coli*, except for the apparent oxidation of an a_1-like component during cytochrome *d* oxidation and preceding cytochrome *b* oxidation. Subsequently optical studies of *E. coli* membranes, in which the reaction of oxygen with cytochrome *d* was initiated by laser photolysis of the CO-liganded form, clearly showed that oxidation of cytochrome b_{595} precedes oxidation of the low-spin haem *b* (Poole and Williams, 1987).

In both organisms loss of absorbance of the 650 nm form is also accompanied by increased absorbance at 675 to 680 nm (Poole, 1983; Poole *et al.*, 1983c). Reaction of the ferric haem *d* with either hydrogen peroxide or ethyl hydrogen peroxide also results in the formation of such a species (Poole and Williams, 1988), strongly suggesting its identity as a peroxy compound.

No significant absorbance changes occur between 720 and 860 nm in low-temperature kinetic experiments over a wide temperature range, where changes in the oxidation states of the cytochrome *d* complex are thought to occur (Poole *et al.*, 1983c). This is consistent with, but does not prove, the absence of copper from the purified complex (page 262).

The presence of cytochrome *d* is widely held to correlate with cyanide resistance (Ashcroft and Haddock, 1975; Green and Gennis, 1983; see, however, Akimenko and Trutko, 1984). It has been suggested that cyanide binds to a form of "invisible" cytochrome *d* (d_x or d*) intermediate between the reduced form and that absorbing near 650 nm (see Poole, 1983, for references). The cyanide-binding form is likely, however, to be the ferric species (Poole *et al.*, 1983b).

Carbon monoxide gives characteristic spectral changes in cytochrome *d* (page 264). Reports that CO could react with the 650 nm form (attributed to the oxidized form) may be rationalized by the hypothesis that this is an oxygenated species from which CO can displace oxygen. The reaction of CO with ferrous cytochrome d is unusual in that photolysis of the CO-liganded form (in membranes) at temperatures down to $-130°C$ does not elicit a photodissociation spectrum. The explanation (Poole *et al.*, 1982c) is that, as in myoglobin and haemoglobin, CO recombination is too rapid at these temperatures to be detected and photolysis at 4–20 K is required to observe the photodissociated form and subsequent CO recombination without recourse to rapid kinetic spectroscopy. This fast recombination is unique among cytochrome oxidases and may be related to the absence of copper, which, in the mitochondrial oxidase and possibly cytochrome o (page 255), may offer a binding site for the photodissociated CO (Alben *et al.*, 1981).

5.6.7   Chemiosmotic function of d-type oxidases

Unequivocal measurements of the H^+/O ratio for a cytochrome d-terminated respiratory chain functioning in intact cells lacking another oxidase have not been made, but can be expected in view of the availability of cytochrome o-deficient mutants of *E. coli* (page 260).

A simple model that does not invoke a pump or Q-cycle is one that places the oxygen-reacting site on the cytoplasmic aspect of the membrane, with the oxidase functioning as the electron-carrying arm of a conventional proton-translocating loop and the quinone acting as the hydrogen-carrying arm (see Figs 1.29a and 5.6). Unfortunately, no direct data on the location of the O_2 reduction site are available, although a cytochrome d–NO compound can be shown to be closer to the cytoplasmic aspect of the membrane than to the periplasmic aspect by EPR studies utilizing the interaction of the paramagnetic centre with an extrinsic paramagnetic probe (W. J. Ingledew and N. Bradbury, unpublished).

The purified cytochrome d-containing complexes from *E. coli* (Kita *et al.*, 1984b; Koland *et al.*, 1984b; Miller and Gennis, 1985) and *Ph. phosphoreum* (Konishi *et al.*, 1986) have been reconstituted into proteoliposomes and shown to generate an uncoupler-sensitive transmembrane potential of upto 160–180 mV (negative inside). In the *E. coli* system, net proton translocation (H^+/O ratios of about 1.7) has recently been demonstrated and interpreted in terms of the model outlined above, in which spatially separated proton-consuming and -releasing reactions account for the stoichiometry.

5.6.8   Genetics and molecular biology

Cytochrome d is generally found in the same organism together with other oxidases, especially of the o- and possibly a_1-types; therefore, isolation of cytochrome *d* mutants requires a screen that distinguishes lesions in the cytochrome *d* complex from those in other oxidases and their respiratory chains. Three general approaches have been adopted, namely:

(a) reduction of dyes believed to interact specifically with the *d* complex;
(b) growth in the presence of respiratory inhibitors to which cytochrome *d* may be relatively resistant; and
(c) direct spectroscopic screening for the distinctive optical properties of cytochrome *d*.

However, mutants of *Az. vinelandii* isolated by McInery *et al.* (1984) as being unable to reduce tetrazolium red proved to have *elevated* cytochrome *d* levels and decreased activity of the cytochrome *o*- and a_1-terminated branches.

Green and Gennis (1983) devised a screening procedure that permitted the isolation of mutants unable to oxidize N,N,N'N'-TMPD. One such mutant studied in detail lacked both subunits of the oxidase and the spectral signals attributed to haems *b*, *d* and the a_1-like haem. The gene, designated *cyd*, mapped near minute 17 on the *E. coli* chromosome. Taking advantage of the increased sensitivity of cytochrome *d*-deficient mutants to cyanide and azide, localized mutagenesis was used to select new mutants, sensitive to azide (Green *et al.*, 1984a). Two classes of mutations, *cyd*A and *cyd*B, were obtained. Both map in the same 7-kilobase locus as does the original *cyd* mutation. The *cyd*A phenotype is the lack of all three chromophores of the complex, as judged by spectrophotometry, and the absence of subunits I and II of the oxidase complex (Kranz *et al.*, 1983). *Cyd*B strains lack the cytochrome a_1-like component and subunit II (43 K), but have a normal amount of cytochrome b_{558} and retain subunit I. On this basis, subunit I has been proposed as the cytochrome b_{558} component of the complex.

Green *et al.* (1984b) have isolated from the Clarke and Carbon *E. coli* DNA bank two plasmids carrying the structural genes for the cytochrome *bd* complex, and have subcloned a 5.4 kb fragment into pBR322. Colonies containing cyd^+ plasmids had a yellow-green colour arising from the cytochrome *d* chromophore. Cyd^+ subclones code for two polypeptides with mobilities identical to subunits I and II of the purified oxidase complex. The DNA sequence of the cloned *cyd* locus has been determined (R. B. Gennis *et al.*; see Anraku and Gennis, 1987) and used to deduce the amino acid sequence of subunits I and II. Both appear to be transmembranous proteins probably crossing the membrane several times.

A further locus involved in the expression of components of the cytochrome *d* complex has been identified. It maps at 18.9 min and has been named *cyd*C (Georgiou *et al.*, 1987).

5.7 Cytochromes of the a_1-type

5.7.1 Classification

Despite its germinal role in the development of understanding of cyto-chromes in general, little is known about cytochrome a_1. Indeed, in recent years more has been learned about a_1-like haemoproteins that have proved to be *not* a_1 than about the increasingly small number of pigments that still warrant this historic name. The classic experiments of Warburg's group on *Acetobacter pasteurianum* (now *Ac. pasteurianus*) revealed a weak band at 589 nm, which disappeared on shaking in air and reappeared on standing. Warburg concluded that the band belonged to the ferrous form of the 'oxygen-transporting ferment', which bound carbon monoxide, underwent autooxidation in the presence of cyanide and, in the ferric form, combined with cyanide to form a complex that was not readily reducible. The band of the reduced form at 589 nm was sufficiently removed from that of cyto-chrome aa_3 in *Bacillus* to warrant a distinguishing subscript. Subsequently, any haemoprotein that shows a band near this wavelength (and the range has grown to encompass wavelengths between about 585 and 596 nm) has been termed a_1 without consideration of the other properties, notably auto-oxidizability, described by Warburg, Keilin, Chance and others. It has been proposed elsewhere (Poole *et al.*, 1985b) that the cytochrome a_1-like haemo-proteins of bacteria be classified along the lines in Table 5.8. Excluded are even more components that could be wrongly identified as cytochrome a_1, like oxyhaemoglobin and sirohaem (Poole, 1983). The spectrum of *E. coli* membranes in Figure 1.21 (page 49) illustrates a typical 'false' cytochrome 'a_1'.

Class I comprises the 'true' cytochromes a_1 which have been proved to be terminal oxidases and whose properties are covered in Section 5.7.2. Even here, the demonstration of haem a is rare. Class II is represented by a unique enzyme comprising a cytochrome a_1-like component and cytochrome c_1. Because it is not autooxidizable but is believed to mediate electron transfer from nitrite to cytochromes c and aa_3, it is not discussed further here. Class III is represented by the soluble haemoprotein b_{590} of *E. coli* and probably by many other bacterial a_1-like haemoproteins (Poole, 1983). The *E. coli* component (Fig. 5.8) is a hydroperoxidase and, although evidence has been obtained that it may react with oxygen (C. A. Appleby and R. K. Poole, unpublished) this does not distinguish it from other peroxidases. The binding of CO to a cytochrome a_1-like pigment in intact cells (Poole *et al.*, 1981) is probably attributable to the soluble form (Baines *et al.*, 1984b; Poole *et al.*, 1986) rather than the d-associated form (Lorence *et al.*, 1986). The a_1-like 'cytochrome b_{595}' that is membrane-bound and copurifies with the

Table 5.8 The diversity of cytochrome a_1-like haemoproteins in bacteria (modified from Poole et al., 1985b)

Proposed class	Example	Haem type	Location	Function	Reference
I	Acetobacter (some strains)	a	m	Oxidase	See Table 5.9
	Thiobacillus ferrooxidans	n.d.	m	Oxidase	See Table 5.9
II	Nitrobacter agilis	a (with c)	m	Nitrite dehydrogenase	Tanaka et al. (1983)
III	E. coli	b	s	Hydroperoxidase	Baines et al. (1984b), Poole et al. (1984, 1985b)
	Rhizobium sp.	b	s	Hydroperoxidase	C. A. Appleby, H. D. Williams and R. K. Poole (unpublished)
	Corynebacterium	n.d.	s	Oxygenase?	Jurtshuk et al. (1975)
	Halobacterium halobium	b	s	Hydroperoxidase	Fukumori et al. (1985b)
IV	E. coli	b	m	Electron donor to cytochrome d?	see Sections 5.6.6, 5.7.2

m = membrane-bound; s = soluble; n.d. = not determined

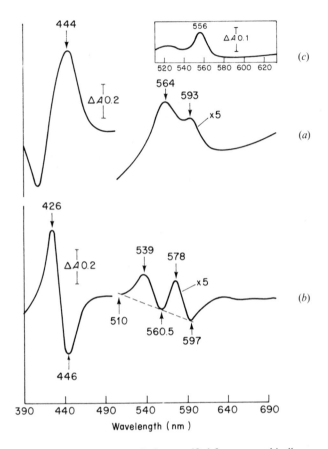

Fig. 5.8 Difference spectra of haemoprotein b_{590} purified from anaerobically grown *E. coli*. Spectrum (*a*) (top) is the computed difference between the spectra of native (ferric) and dithionite-reduced samples recorded at room temperature (1 cm light path). Spectrum (*b*) (bottom) is the computed difference between the spectra of a dithionite-reduced sample and the same sample bubbled with CO, recorded under the same conditions as for (*a*). Spectrum (*c*) (inset) is the reduced *minus* oxidized difference spectrum of the pyridine haemochrome of the sample. The spectral band width was 1.0 nm throughout and the scan speed was 1 nm s^{-1}.

cytochrome *bd* complex may or may not be identical with class III components. The spectral differences are small, but extensively washed membranes do not exhibit the catalase activity expected of haemoprotein b_{590} and the redox potential of cytochrome b_{595} is higher than that of the soluble form (R. K. Poole, unpublished). Still unconfirmed is the single report that an *a*-type cytochrome functions as an oxidase in *E. coli* (Edwards *et al.*, 1981). In *Paracoccus denitrificans*, a CO-binding haemoprotein absorbing in the reduced form at 590 nm has been assigned to 'cytochrome a_1', but there is

Table 5.9 Cytochromes a_1 believed to be terminal oxidases, i.e. Type I (for further information see Poole, 1983)

Bacterium	λ_{max} of reduced form (nm)	Features in CO difference spectra (nm)	Peaks in photochemical action spectra (nm)	Reference
Acetobacter pasteurianus (Smith strain) Ac. peroxydans (Keilin strain)	441	427(p), 442(t)	427–428, 548, 591–592	Castor and Chance (1955, 1959) Meyer and Jones (1973) Chance (1953a,b)
Acinetobacter	594	610(p)?	595	Ensley and Finnerty (1980)
Az. vinelandii	440, 588–590		(415–435), 585[b], 589	Castor and Chance (1959) Edwards et al. (1981)
Beneckea natriegens	588–595 432–436	Not well resolved	434, 588	Weston and Knowles (1974)
Proteus vulgaris	430	418(p), 437(t)	432, 591[a]	Castor and Chance (1955, 1959)
Thiobacillus[c] ferrooxidans	439–446, 597	426(p), 440(t)		Ingledew and Cobley (1980); Ingledew (1982); de Fonseka et al. (1980)
Ps. carboxydovorans	596		422, 595	Cypionka and Meyer (1983a,b)

[a] Small shoulder on o peak.
[b] Displacement from 590 nm possibly due to overlapping bands.
[c] There is also kinetic evidence for an oxidase role in this organism.

little evidence that this is a functional oxidase (van Verseveld *et al.*, 1983). The *a*-type cytochromes of certain carboxydobacteria show an α-band in the reduced state at 596 nm (Cypionka and Meyer, 1983a). In *Ps. carboxydovorans*, a peak at 595 nm in the photochemical action spectrum has been attributed to cytochrome a_1 (Cypionka and Meyer, 1983b).

5.7.2   Type I cytochromes a_1

Table 5.9 lists a number of bacteria believed to have cytochrome a_1 as a terminal oxidase, offers the main spectral features and gives an indication of the evidence for an oxidase role. Only a few further comments are warranted. The early results on *Acetobacter* remain the most convincing; the action spectra of Chance showed cytochrome a_1 to be the only oxidase. Attempts to extend this work have been frustrated by loss of the strain used. Subsequent studies with strains containing cytochromes *o*, *d* and an a_1-like component (Williams and Poole, 1986, 1987) have revealed similarities in kinetic behaviour to the *E. coli* cytochromes, except that there is evidence for electron donation from a cytochrome a_1-like pigment to cytochrome *d* (see also Section 5.6.6).

In *T. ferrooxidans* (Ingledew, 1982), cytochrome a_1 terminates an Fe^{2+}-oxidizing respiratory chain comprising multiple cytochromes *c*, cytochrome *b* and two coppers. Two *a*-type haems are distinguishable by their redox potential and kinetics. An interesting feature is that copper does not appear to be involved in the oxidase reaction since kinetic experiments show that, at $-20°C$, copper oxidation is 20-fold slower than cytochrome a_1 oxidation, and no optical changes occur in the 650–900 nm range at $-80°C$ (Ingledew and Cobley, 1980).

The nomenclature of an *a*-type cytochrome oxidase recently purified from *Nitrosomonas europaea* (Yamazaki *et al.*, 1985) poses problems. The reduced form has a peak at 595 nm, indicative of an a_1-type. However, the composition of the enzyme resembles aa_3-type oxidases from other bacteria. There are two subunits (M_r 50 000 and 33 000), binding two mol of haem *a* and 1 or 2 copper atoms. The lack of absorbance at 655 or 830 nm suggests that Cu_A may be absent. The other unusual feature is the ready dissociation of the CO compound in air to give the oxidized enzyme and CO.

5.8   Other haemoproteins involved in metabolism of oxygen and its partial reduction products

5.8.1   The cytochrome *o*-like haemoproteins of *Vitreoscilla*

Vitreoscilla is a filamentous bacterium in the Beggiatoa family. It is a strict aerobe and has long been assumed to possess cytochrome *o* as terminal oxidase. A water-soluble cytochrome *o*-like haemoprotein was purified and extensively characterized (for early references, see Poole, 1983). The protein is a dimer consisting of two identical 13 000-dalton subunits and two protohaem IX molecules. Non-haem iron, copper and flavin are absent. The reduced enzyme binds CO to give an absorption spectrum similar to membrane-bound cytochromes *o* but, in marked contrast to these, oxygen (in the presence of NADH as reductant) gives a stable 'oxygenated' form, with peaks at 578, 545 and 425 nm (with a persulphate-oxidized sample as reference). The oxygenated species is a predominant steady-state species in respiring cells and in partially purified preparations oxidizing NADH. Infrared spectroscopy (Choc *et al.*, 1982) has revealed that dioxygen is bound to the haem in a bent end-on geometry, as in oxyhaemoglobins and oxymyoglobins. Oxygen titrations have revealed two interconvertible oxygenated species: one is a dimeric compound (compound D), $(b^{2+}b^{2+})_2$–O_2; the second is the oxygenated compound, $(b^{2+}b^{2+})$–O_2.

Recent work (Orii and Webster, 1986) has utilized photodissociation of the oxygenated form to show that the oxygenated species reforms from the ferrous haemoprotein and O_2, without the generation of compound D, to give an oxygen compound that is stable and in which the haem iron does not become oxidized. The suggestion from these results, namely that this haemoprotein is not a true oxidase but an oxygen carrier or storage protein, receives support from:

(a) the finding that a true membrane-bound cytochrome *o* with the characteristics of an oxidase was identified by photolysis of CO-liganded intact cells at $-100°C$ (De Maio *et al.*, 1983);

(b) the identification of a second potential oxidase, cytochrome *d* (cited in Wakabayashi *et al.*, 1986); and

(c) the homology of the amino acid sequence of the soluble form with the helical regions of several animal and plant globins (Wakabayashi *et al.*, 1986).

The alignment is consistent with this 'cytochrome *o*' being a structural homologue of eukaryotic haemoglobins. It shows maximum sequence homology (24%) with lupin leghaemoglobin and is to be called *Vitreoscilla* Hb hereafter. Its present function is unclear, but it is notable that the cellular content of the protein (and other haems) increases dramatically under hypoxic conditions, suggesting that it serves as an oxygen storage-trap to enable the organism to survive in oxygen-poor environments. Alternatively, if localized near the bacterial plasma membrane it could serve to facilitate oxygen transfer to the membrane-bound terminal oxidases. Indeed, expression of *Vitreoscilla* Hb in *E. coli* increased cell yield and respiration rates, especially at low dissolved oxygen concentrations (Khoshla and Bailey, 1988). A soluble haemoglobin-like molecule has also been reported in cultured *Rhizobium* (Appleby, 1969) and appears to be of similar molecular size to *Vitreoscilla* Hb.

5.8.2   Pseudomonas cytochrome cd_1

One of the most extensively studied, but perhaps atypical, oxidases in bacteria is the ferrocytochrome c_{551}:oxygen oxidoreductase (EC 1.9.3.2) of *Pseudomonas aeruginosa*. This haemoprotein catalyses the reduction of O_2 to water using electrons from cytochrome c_{551} or the blue copper protein, azurin. It is atypical of cytochrome oxidases in that its dominant role *in vivo* is probably the single-electron reduction of NO_2^- to NO, and is also water-soluble. Its synthesis requires nitrate and is enhanced during anaerobic growth. This enzyme has been reviewed in detail (Meyer and Kamen, 1982; Poole, 1983) and will, consequently, be treated only in summary here.

The enzyme is associated with the inner surface of the cytoplasmic membrane but is largely solubilized on spheroplast disruption and is relatively easily purified. Recent estimates of molecular weight suggest values of 120 000–130 000 for the intact dimer, which bears 2 haems *c* and 2 haems d_1. The haems are clearly distinguishable by their optical, magnetic and functional characteristics. EPR signals from copper have been attributed to adventitious metal.

The reaction of the oxidase with various ligands has been intensively studied. Carbon monoxide, cyanide, NO and oxygen all bind to haem d_1. Interestingly, photolysis of the CO-liganded ferrous oxidase at very low temperatures (5–30 K) is followed by rapid recombination as occurs in myoglobin and cytochrome *d* (page 272). This fits with the absence of copper from cytochrome oxidases of the *d*- and cd_1-type and the belief that CO-binding to cytochrome aa_3 (and *o*?) is sterically hindered by copper (pages 252 and 272).

Haem d_1 is rapidly oxidized by oxygen, whereas haem *c* oxidation is sluggish, indicating rate-limiting intramolecular electron transfer. Oxygenated intermediates have been observed by stopped-flow spectrophotometry.

5.8.3   Hydroperoxidases

Catalases and peroxidases, collectively referred to as 'hydroperoxidases', are responsible for removing the deleterious hydrogen peroxide and alkyl peroxides. Their activities are fundamentally similar: peroxidases reduce peroxide in single electron transfer from a reduced donor, such as cytochrome *c*:

$$2e^- + 2H^+ + H_2O_2 \rightarrow 2H_2O$$

Catalase disproportionates peroxide, the oxidative power of one peroxide molecule being used to oxidize another in an electron pair transfer or dismutation of hydrogen peroxide:

$$2H_2O_2 \rightarrow 2H_2O + O_2$$

Although not directly involved in electron transport to dioxygen, it is clear that at least one of the cytochrome a_1-like haemoproteins in *E. coli* is a hydroperoxidase (page 274). Furthermore, it has already been shown that peroxide is a product of O_2 reduction by cytochrome *o* of the myxobacterium *Vitreoscilla* (for a survey see Poole, 1983) and so hydroperoxidases may have a role in completing O_2 reduction in other bacteria. Indeed, the synthesis of hydroperoxidases in *E. coli* is linked with that of certain components of the electron transport chain and is independent of O_2 or H_2O_2 *per se* (Hassan and Fridovich, 1978).

Certain microbial catalases are similar to the mammalian enzyme in comprising four identical subunits, each carrying one *b*-type haem, but there is also considerable diversity in the microbial and, in particular, bacterial enzymes (Table 5.10). The catalases of *E. coli* are the best-characterized enzymes of this type. HP-I (Table 5.10) is constitutive and present even in anaerobically grown cells, whereas HP-II is synthesized during growth in which electron transport occurs to a terminal acceptor such as O_2 or NO_3^- (Hassan and Fridovich, 1978). The soluble a_1-like haemoprotein b_{590} (Fig. 5.8; Baines *et al.*, 1984b; Poole *et al.*, 1984, 1986) and probably the 'a_1b' complex of Barrett and Sinclair (1967) share several features in common with HP-I (Poole *et al.*, 1986). A membrane-associated protein with a *b*-type haem, exhibiting catalase and peroxidase activities and having a subunit molecular weight of 70 000 has a pI of 4.7 (Kranz and Gennis, 1982; Kranz *et al.*, 1984) close to the value (4.4) of the pure haemoprotein b_{590}. In addition, glucose considerably suppresses the activity of catalase and peroxidase (Hassan and Fridovich, 1978) and the amount of the membrane-bound catalase (Kranz *et al.*, 1984). There is thus strong circumstantial evidence that all these proteins are closely related and may, indeed, be identical. No intracellular substrate for the peroxidase activity of HP-I has yet been identified (Claiborne and Fridovich, 1979) and the primary role of this and the related proteins is perhaps catalatic. There have been recent advances in genetic studies of *E. coli* and *S. typhimurium* catalases; the most recent literature has been reviewed by Poole and Ingledew (1987).

The cytochrome *c* peroxidase of *Ps. aeruginosa* (Table 5.10) is well studied. It catalyses the peroxidation of *c*-type cytochromes and azurin, a copper protein, and is unusual amongst hydroperoxidases in having 2 *c*-type haems in a single polypeptide chain. The high-spin low-potential haem is the target of H_2O_2 in the peroxidatic reaction cycle. Further details are given by Ellfolk *et al.* (1983).

5.8.4 Superoxide dismutase

This enzyme (SOD) plays an important role in defence against the biological production of the superoxide radical (O_2^-) by catalysing its disproportiona-

Table 5.10 Properties of selected bacterial hydroperoxidases

Enzyme	Source	M_r (no. of subunits)	Haem(s)	References
Catalase (with peroxidase activity)	*Comamonas compransoris*	150 K (2)	$2 \times b$	Nies and Schlegel (1982)
Hydroperoxidase I (*o*-dianisidine peroxidase and catalase activities)	*E. coli* B	337 K (4)	$2 \times b$	Claiborne and Fridovich (1979)
Hydroperoxidase II (predominantly catalase)	*E. coli* B	312 K (4)	$2 \times b$ (plus unidentified haems?)	Claiborne *et al.* (1979)
Cytochrome *c* peroxidase	*Ps. aeruginosa*	53.5 K (1)	$2 \times c$	Ellfolk *et al.* (1983) and refs therein

tion according to the equation:

$$2H^+ + 2O_2^- \rightarrow O_2 + H_2O_2$$

The peroxide may in turn be scavenged by hydroperoxidases (Section 5.8.3).

All SODs studied are metalloproteins in which the active site is reduced by one O_2^-, then reoxidized by another O_2^-. Whereas a copper- and zinc-containing enzyme is confined to eukaryotes, bacteria possess iron or manganese SODs (for a review, see Fridovich, 1981). The iron-containing enzymes generally contain one atom of the metal per subunit of M_r 18–24 K. Structures of the enzymes from *Ps. ovalis* and *E. coli* have recently been described at high resolution (2.9Å) (e.g. Stallings *et al.*, 1983). The bacterial manganese SODs tend to be dimeric although, again, exceptions occur. The amino acid sequences of the yeast and bacterial manganese SODs show considerable homology and the subunit sizes, amino acid sequences and oligomeric structures also show similarities to the iron SODs.

5.9 Conclusions and prospects

Keilin's realization that 'cytochrome' comprised a number of chemically related, but distinct, compounds was soon expanded to the concept that the functions of 'cytochrome oxidase' may be served in bacteria by a number of different haemoproteins. Thus, although some prokaryotes possess a cytochrome *c* oxidase having a catalytic 'core' of two or three subunits, with structures and functions similar to their mitochondrial counterparts, many bacteria can synthesize additional cytochrome oxidase complexes in which haems of the *b*, *d* or d_1 types replace haem a_3 as ligand-binding redox centre (for a brief survey see Anraku and Gennis, 1987).

Correlations are emerging between the nature of the oxidase(s) synthesized and the taxonomic position of the organism and its growth conditions. In some, but not all, cases the appearance of a particular oxidase can be rationalized, knowing its molecular and functional properties. For example, the fact that *E. coli* cytochrome *d* is synthesized under oxygen-limited conditions or in the presence of cyanide is explicable in terms of the oxidase's high affinity for O_2 and cyanide-resistance, respectively. Similarly, aa_3-type oxidases lacking one or more of the copper atoms appear to be synthesized under copper-limited conditions. Other correlations can be less readily interpreted but this is likely to reflect the still-preliminary information on the structure and function of the enzyme. There is especial ignorance of the mechanisms that regulate electron flow between branches of respiratory chains which are terminated by different oxidases. Are the criteria involved in the capacity for disposal of excess reducing equivalents or the efficiency of associated energy conservation?

The techniques of physical biochemistry and molecular genetics may be anticipated to make important contributions to studies of bacterial cytochrome oxidases in the near future. Spectroscopic techniques such as resonance Raman, magnetic circular dichroism, EPR and increasingly sophisticated optical approaches have their part to play, but their potential usefulness is currently limited by ignorance of some of the most fundamental aspects of organization, notably unequivocal enumeration and structure determination of the redox active centres (haems and copper), as well as the immediate environment of these centres, particularly the axial ligands to the haems and the copper ligands. Physical techniques hold great promise for a better understanding of the catalytic mechanism but, again, such studies are in their infancy. Only in the case of aa_3-type oxidases can a scheme be proposed for oxygen reduction in which the oxygen-reactive centres have been identified and enumerated (haem a_3 and Cu_B) and pathways of intramolecular electron transfer (from haem a and Cu_A) mapped. Cloning of genes involved in cytochrome oxidases is under way in a number of laboratories and the nucleotide sequence of structural genes can soon be expected to yield a primary protein structure with its inherent clues to organization in the membrane. Site-directed mutagenesis is now making important contributions to structure–function correlations for other membrane-bound proteins of energy transduction, notably the *lac* permease and the ATPase of *E. coli* and will surely do so in the case of the oxidases. These complex enzymes present a considerable challenge to the whole arsenal of techniques of biochemists and microbiologists and the eventual understanding of oxygen reduction and energy conservation by such enzymes will be a major achievement in bioenergetics.

References

Akimenko, V. K. and Trutko, S. M. (1984). *Arch. Microbiol.* **138**, 58–63.
Alben, J. O., Moh, P. P., Fiamingo, F. G. and Altschuld, R. A. (1981). *Proc. Natl. Acad. Sci. USA* **78**, 234–237.
Albracht, S. P. J., van Verseveld, H. W., Hagen, W. R. and Kalkman, M. L. (1980). *Biochim. Biophys. Acta* **593**, 173–186.
Anemüller, S., Lübben, M. and Schäfer, G. (1985). *FEBS Lett.* **193**, 83–87.
Anraku, Y. and Gennis, R. B. (1987). *Trends Biochem. Sci.* **12**, 262–266.
Appleby, C. A. (1969). *Biochim. Biophys. Acta* **172**, 88–105.
Arrow, A. S. and Taber, H. W. (1986). *Antimicrob. Agents Chemother.* **29**, 141–146.
Artzatbanov, V., Müller, M. and Azzi, A. (1987). *Archiv. Biochem. Biophys.* **257**, 476–480.

Ashcroft, J. R. and Haddock, B. A. (1975). *Biochem. J.* **148**, 349–352.

Au, D. C.-T., Lorence, R. M. and Gennis, R. B. (1985). *J. Bacteriol.* **161**, 123–127.

Au, D. C.-T. and Gennis, R. B. (1987). *J. Bacteriol.* **169**, 3237–3242.

Azzi, A and Gennis, R. B. (1986). *Methods Enzymol.* **126**, 138–145.

Babcock, G. T. and Callahan, P. M. (1983). *Biochemistry* **22**, 2314–2319.

Baines, B. S. and Poole, R. K. (1985). *In* 'Microbial Gas Metabolism: Mechanistic, Metabolic and Biotechnological Aspects' (Eds. R. K. Poole and C. S. Dow). Academic Press, London.

Baines, B. S., Hubbard, J. A. M. and Poole, R. K. (1984a). *Biochim. Biophys. Acta* **766**, 438–445.

Baines, B. S., Williams, H. D., Hubbard, J. A. M. and Poole, R. K. (1984b). *FEBS Lett.* **171**, 309–314.

Barrett, J. and Sinclair, P. (1967). Abstr. 7th Intern. Cong. Biochem., Tokyo, H-107, p. 907.

Brunori, M. and Wilson, M. T. (1982). *Trends Biochem. Sci.* **7**, 295–299.

Carter, K. and Gennis, R. B. (1985). *J. Biol. Chem.* **260**, 10986–10990.

Castor, L. N. and Chance, B. (1955). *J. Biol. Chem.* **217**, 453–465.

Castor, L. N. and Chance, B. (1959). *J. Biol. Chem.* **234**, 1587–1592.

Chance, B. (1953a). *J. Biol. Chem.* **202**, 383–396.

Chance, B. (1953b). *J. Biol. Chem.* **202**, 397–406.

Chance, B., Saronio, C. and Leigh, J. S. (1975). *J. Biol. Chem.* **250**, 9226–9237.

Choc, M. G., Webster, D. A. and Caughey, W. S. (1982). *J. Biol. Chem.* **257**, 865–869.

Claiborne, A. and Fridovich, I. (1979). *J. Biol. Chem.* **254**, 4245–4252.

Claiborne, A., Malinowski, P. and Fridovich, I. (1979). *J. Biol. Chem.* **254**, 11664–11668.

Clore, G. M., Andreasson, L.-E., Karlsson, B., Aasa, R. and Malmström, B. G. (1980). *Biochem. J.* **185**, 139–154.

Cross, A. R. and Anthony, C. (1980). *Biochem. J.* **192**, 429–439.

Cypionka, H. and Meyer, O. (1983a). *Archiv. Microbiol.* **135**, 293–298.

Cypionka, H. and Meyer, O. (1983b). *J. Bacteriol.* **156**, 1178–1187.

Cypionka, H., Reijnders, W. N. M., van Wielink, J. E., Oltmann, L. F. and Stouthamer, A. H. (1985). *FEBS Microbiol. Lett.* **27**, 189–193.

De Fonseka, K., Reid, G. A. and Ingledew, W. J. (1980). *Fed. Proc.* **39**, 6.

Degn, H., Lloyd, D. and Hill, G. C. (Eds.) (1978). *Proc. 11th FEBS Meeting* **vol. 49.** Pergamon Press, Oxford.

De Maio, R. A., Webster, D. A. and Chance, B. (1983). *J. Biol. Chem.* **258**, 13768–13771.

DerVartanian, D. V., Iburg, L. K. and Morgan, T. V. (1973). *Biochim. Biophys. Acta* **305**, 173–178.

De Vrij, W., Azzi, A. and Konings, W. N. (1983). *Eur. J. Biochem.* **131**, 97–103.

De Vrij, W., Poolman, B., Konings, W. N. and Azzi, A. (1986). *Methods Enzymol.* **126**, 159–173.

Edwards, C. (1984). *FEMS Microbiol. Lett.* **21**, 319–322.

Edwards, C., Beer, S., Siviram, A. and Chance, B. (1981). *FEBS Lett.* **128**, 205–207.

Einarsdottir, O. and Caughey, W. S. (1985). *Biochem. Biophys. Res. Comm.* **129**, 840–847.

Ellfolk, N., Rönnberg, M., Aasa, R., Andréasson, L-E. and Vänngård, T. (1983). *Biochim. Biophys. Acta* **743**, 23–30.

Ensley, B. D. and Finnerty, W. R. (1980). *J. Bacteriol.* **142**, 859–868.

Erecínska, M., Wilson, D. F. and Blasie, J. K. (1979). *Biochim. Biophys. Acta* **545**, 352–364.

Fee, J. A., Kuila, D., Mather, M. W. and Yoshida, T. (1987). *Biochim. Biophys. Acta* (in press).

Ferguson, S. J. (1982). *FEBS Lett.* **146**, 239–243.

Finlayson, S. D. and Ingledew, W. J. (1985). *Biochem. Soc. Trans.* **13**, 632–633.

Fridovich, I. (1981). In 'Oxygen and Living Processes. An Interdisciplinary Approach' (Ed. D. L. Gilbert), pp. 250–272. Springer-Verlag, New York.

Froud, S. J. and Anthony, C. (1984). *J. Gen. Microbiol.* **130**, 2201–2212.

Fukumori, Y., Nakayama, K. and Yamanaka, T. (1985a). *J. Biochem.* **98**, 1719–1722.

Fukumori, Y., Fujiwara, T., Okada-Takahashi, Y., Mukohata, Y. and Yamanaka, T. (1985b). *J. Biochem.* **98**, 1055–1061.

Gennis, R. B., Casey, R. P., Azzi, A. and Ludwig, B. (1982). *Eur. J. Biochem.* **125**, 189–195.

Georgiou, C. D., Fang, H. and Gennis, R. B. (1987). *J. Bacteriol.* **169**, 2107–2112.

Green, G. N. and Gennis, R. B. (1983). *J. Bacteriol.* **154**, 1269–1275.

Green, G. N., Kranz, R. G., Lorence, R. M. and Gennis, R. B. (1984a). *J. Biol. Chem.* **259**, 7994–7997.

Green, G. N., Kranz, J. E., and Gennis, R. B. (1984b). *Gene* **32**, 99–106.

Green, G. N., Lorence, R. M. and Gennis, R. B. (1986). *Biochemistry* **25**, 2309–2314.

Hackett, N. R. and Bragg, P. D. (1983). *J. Bacteriol.* **154**, 708–718.

Haddock, B. A., Downie, J. A. and Garland, P. B. (1976). *Biochem. J.* **154**, 285–294.

Hamamoto, T., Carrasco, N., Matsushita, K., Kaback, H. R. and Montal, M. (1985). *Proc. Natl. Acad. Sci. USA* **82**, 2570–2573.

Hassan, H. M. and Fridovich, I. (1978). *J. Biol. Chem.* **253**, 6445–6450.

Hata, A., Kirino, Y., Matsuura, K., Itoh, S., Hiyama, T., Konishi, K., Kita, K. and Anraku, Y. (1985). *Biochim. Biophys. Acta* **810**, 62–72.

Hata-Tanaka, A., Matsuura, K., Hoh, S. and Anraku, Y. (1987). *Biochim. Biophys. Acta* **893**, 289–295.

Hempfling, W. P. (Ed.) (1979). 'Benchmark Papers in Microbiology', Vol. 13. Dowden, Hutchinson & Ross, Stroudsberg, Pennsylvania.

Hendler, R. W. and Shrager, R. I. (1979). *J. Biol. Chem.* **254**, 11288–11299.

Hoffman, P. S., Morgan, T. V. and DerVartanian, D. V. (1979). *Eur. J. Biochem.* **100**, 19–27.

Holm, L., Saraste, M. and Wikström, M. (1987). *EMBO J.* **6**, 2819–2823.

Hon-Nami, K. and Oshima, T. (1977). *J. Biochem.* **8**, 769–776.

Hon-Nami, K. and Oshima, T. (1984). *Biochemistry* **23**, 454–460.

Hubbard, J. A. M., Hughes, M. N. and Poole, R. K. (1983). *FEBS Lett.* **164**, 241–243.

Hubbard, J. A. M., Hughes, M. N. and Poole, R. K. (1985). In 'Microbial Gas Metabolism: Mechanistic, Metabolic and Biotechnological Aspects' (Eds. R. K. Poole and C. S. Dow), pp. 231–236. Academic Press, London.

Hubbard, J. A. M., Lewandowska, K. B., Hughes, M. N. and Poole, R. K. (1986). *Arch. Microbiol.* **146**, 80–86.

Hubbard, J. A., Poole, R. K. and Hughes, M. N. (1986). *Biochem. Soc. Trans.* **14**, 1214–1215.

Hüdig, H. and Drews, G. (1982a). *FEBS Lett.* **146**, 389–392.

Hüdig, H. and Drews, G. (1982b). *Z. Naturforsch.* **37c**, 193–198.

Hüdig, H. and Drews, G. (1984). *Biochim. Biophys. Acta* **765**, 171–177.

Ingledew, W. J. (1982). *Biochim. Biophys. Acta* **683**, 89–117.
Ingledew, W. J. and Cobley, J. G. (1980). *Biochim. Biophys. Acta* **590**, 141–158.
Ingledew, W. J. and Poole, R. K. (1984). *Microbiol. Rev.* **48**, 222–271.
Iwasaki, H. (1966). *Pl. Cell Physiol.* **7**, 199–216.
Johnson, P. L. and Bragg, P. D. (1985). *FEMS Microbiol. Lett.* **26**, 185–189.
Jones, C. W. (1977). *In* 'Microbial Energetics' (Eds. B. A. Haddock and W. A. Hamilton), pp. 23–59. Cambridge University Press, Cambridge.
Jones, C. W. and Poole, R. K. (1985). *In* 'Methods in Microbiology' (Ed. G. Gottschalk), Vol. 18, pp. 285–328. Academic Press, London.
Jurtshuk, P., Mueller, T. J. and Acord, W. C. (1975). *CRC Crit. Rev. Microbiol.* **3**, 399–468.
Jurtshuk, P., Mueller, T. J. and Wong, T. Y. (1981). *Biochim. Biophys. Acta* **637**, 374–382.
Kadenbach, B., Ungibauer, M., Jarausch, J., Büge, U. and Kuhn-Nentwig, L. (1983). *Trends Biochem. Sci.* **8**, 398–400.
Kauffman, H. F., DerVartanian, D. V., van Gelder, B. F. and Wampler, J. (1975). *J. Bioenerg.* **7**, 215–221.
Keilin, D. (1966). 'The History of Cell Respiration and Cytochrome'. Cambridge University Press, Cambridge.
Kent, T. A., Munck, E., Dunham, W. R., Filter, W. F., Findling, K. L., Yoshida, T. and Fee, J. A. (1982). *J. Biol. Chem.* **257**, 12489–12492.
Kent, T. A., Young, L. J., Palmer, G., Fee, J. A. and Munck, E. (1983). *J. Biol. Chem.* **258**, 8543–8546.
Khoshla, C. and Bailey, J. E. (1988). *Nature* **331**, 633–635.
King, M.-T. and Drews, G. (1976). *Eur. J. Biochem.* **68**, 5–12.
King, T. E., Orii, Y., Chance, B. and Okunuki, K. (Eds.) (1979). 'Cytochrome Oxidase'. Elsevier/North Holland Biomedical Press, Amsterdam.
Kita, K., Kasahara, M. and Anraku, Y. (1982). *J. Biol. Chem.* **257**, 7933–7935.
Kita, K., Konishi, K. and Anraku, Y. (1984a). *J. Biol. Chem.* **259**, 3368–3374.
Kita, K., Konishi, K. and Anraku, Y. (1984b). *J. Biol. Chem.* **259**, 3375–3381.
Kita, K., Murakami, H., Oya, H. and Anraku, Y. (1985). *Biochem. Int.* **10**, 319–326.
Kitada, M. and Krulwich, T. A. (1984). *J. Bacteriol.* **158**, 963–966.
Knowles, C. J. (Ed.) (1980). 'Diversity of Bacterial Respiratory Systems', Vols. I and II. CRC Press, Boca Raton, Florida.
Koland, J. G., Miller, M. J. and Gennis, R. B. (1984a). *Biochemistry* **23**, 1051–1056.
Koland, J. G., Miller, M. J. and Gennis, R. B. (1984b). *Biochemistry* **23**, 445–453.
Konishi, K., Ouchi, M., Kita, K. and Horikoshi, I. (1986). *J. Biochem.* **99**, 1227–1236.
Kranz, R. G. and Gennis, R. B. (1982). *J. Bacteriol.* **150**, 36–45.
Kranz, R. G. and Gennis, R. B. (1983). *J. Biol. Chem.* **258**, 10614–10621.
Kranz, R. G. and Gennis, R. B. (1984). *J. Biol. Chem.* **259**, 7998–8003.
Kranz, R. G. and Gennis, R. B. (1985). *J. Bacteriol.* **161**, 709–713.
Kranz, R. G., Baraassi, C. A., Miller, M. J., Green, G. N. and Gennis, R. B. (1983). *J. Bacteriol.* **156**, 115–121.
Kranz, R. G., Baraassi, C. A. and Gennis, R. B. (1984). *J. Bacteriol.* **158**, 1191–1194.
Kučera, I., Křivánková, L. and Dadák, V. (1984). *Biochim. Biophys. Acta* **765**, 43–47.
Kula, T. J., Aleem, M. I. H. and Wilson, D. F. (1982). *Biochim. Biophys. Acta* **680**, 142–151.
Kumar, C., Poole, R. K., Salmon, I. and Chance, B. (1985). *FEBS Lett.* **190**, 227–231.

Laszlo, D. J., Fandrich, B. L., Sivaram, A., Chance, B. and Taylor, B. L. (1984). *J. Bacteriol.* **159**, 663–667.

Lawford, H. G. and Haddock, B. A. (1973). *Biochem. J.* **136**, 217–220.

Lehninger, A. L. (1975). 'Biochemistry', 2nd Edn., Worth Publishers, New York.

Lemberg, R. and Barrett, J. (1973). 'Cytochromes'. Academic Press, London and New York.

Lewis, R. J., Prince, R. C., Dutton, P. L., Knaff, D. B. and Krulwich, T. A. (1981). *J. Biol. Chem.* **256**, 10543–10549.

Lorence, R. M., Green, G. N. and Gennis, R. B. (1984a). *J. Bacteriol.* **157**, 115–121.

Lorence, R. M., Miller, M. J., Borochov, A., Faiman-Weinberg, R. and Gennis, R. B. (1984b). *Biochim. Biophys. Acta* **790**, 148–153.

Lorence, R. M., Koland, J. G. and Gennis, R. B. (1986). *Biochemistry* **25**, 2314–2321.

Ludwig, B. (1980). *Biochim. Biophys. Acta* **594**, 177–189.

Ludwig, B. (1982). *In* 'Electron Transport and Oxygen Utilization' (Ed. C. Ho), pp. 293–296. Elsevier North Holland, Amsterdam.

Ludwig, B. (1986). *Methods Enzymol.* **126**, 153–159.

Ludwig, B. (1987). *FEMS Microbiol. Rev.* (in press).

Ludwig, B. and Schatz, G. (1980). *Proc. Natl. Acad. Sci. USA* **77**, 196–200.

Ludwig, B. and Gibson, Q. H. (1981). *J. Biol. Chem.* **256**, 10092–10098.

Matsushita, K. and Kaback, H. R. (1986). *Biochemistry* **25**, 2321–2327.

Matsushita, K., Shinagawa, E., Adachi, O. and Ameyama, M. (1982). *FEBS Lett.* **139**, 255–258.

Matsushita, K., Patel, L., Gennis, R. B. and Kaback, H. R. (1983). *Proc. Natl. Acad. Sci. USA* **80**, 4889–4893.

Matsushita, K., Patel, L. and Kaback, H. R. (1984). *Biochemistry* **23**, 4703–4714.

McInerny, M. J., Holmes, K. S., Hoffman, P. and DerVartanian, D. V. (1984). *Eur. J. Biochem.* **141**, 447–452.

Mendis, A. H. W. and Evans, A. A. F. (1984). *Comp. Biochem. Physiol., Part B*, **78**, 729.

Meyer, D. J. (1973). *Nature (Lond.) New Biol.* **245**, 276–277.

Meyer, D. J. and Jones, C. W. (1973). *FEBS Lett.* **33**, 101–105.

Meyer, T. E. and Kamen, M. D. (1982). *Adv. Protein Chem.* **35**, 105–212.

Miller, M. J. and Gennis, R. B. (1983). *J. Biol. Chem.* **258**, 9159–9165.

Miller, M. J. and Gennis, R. B. (1985). *J. Biol. Chem.* **260**, 14003–14008.

Mitchell, P., Mitchell, R., Moody, A. J., West, I. C., Baum, H. and Wrigglesworth, J. M. (1985). *FEBS Lett.* **188**, 1–7.

Naqui, A., Chance, B. and Cadenas, E. (1986). *Ann. Rev. Biochem.* **55**, 137–166.

Nies, D. and Schlegel, H. G. (1982). *J. Gen. Appl. Microbiol.* **28**, 311–319.

O'Brian, M. R. and Maier, R. J. (1987). *Proc. Natl. Acad. Sci. USA* **84**, 3219–3223.

Ogura, T., Hon-Nami, K., Oshima, T., Yoshikawa, S. and Kitagawa, K. (1983). *J. Am. Chem. Soc.* **105**, 7781–7783.

Ogura, T., Sone, N., Kagawa, K. and Kitagawa, T. (1984). *Biochemistry* **26**, 2826–2831.

Orii, Y. and Webster, D. A. (1986). *J. Biol. Chem.* **261**, 3544–3547.

Paetow, B., Panskus, G. and Ludwig, B. (1985). *J. Inorg. Biochem.* **23**, 183–186.

Panskus, G., Steinrücke, P., Paetow, B. and Ludwig, B. (1986). Abstr. 4th EBEC, Prague, p. 103.

Parsonage, D., Greenfield, A. J. and Ferguson, S. J. (1986). *Arch. Microbiol.* **145**, 191–196.

Peschek, G. A. (1981). *Biochim. Biophys. Acta* **635**, 470–475.

Peschek, G. A., Schmetterer, G., Lauritsch, G., Nitschmann, W. H., Kienzl, P. F. and Muchl, R. (1982). *Arch. Microbiol.* **131**, 261–265.

Poole, R. K. (1981). *FEBS Lett.* **133**, 250–259.

Poole, R. K. (1983). *Biochim. Biophys. Acta* **726**, 205–243.

Poole, R. K. and Chance, B. (1981). *J. Gen. Microbiol.* **126**, 277–287.

Poole, R. K. and Haddock, B. A. (1975). *Biochem. J.* **152**, 537–546.

Poole, R. K. and Ingledew, W. J. (1987). In '*Escherichia coli* and *Salmonella typhimurium*: Cellular and Molecular Biology' (Eds. F. C. Heidhardt, J. L. Ingraham, K. B. Low, B. Magasanik, M. Schaechter and H. E. Umbarger), Vol. 1; pp. 170–200. Am. Soc. Microbiol., Washington DC.

Poole, R. K. and Williams, H. D. (1987). *FEBS Lett.* **217**, 49–52.

Poole, R. K. and Williams H. D. (1988). *FEBS Lett.* (in press).

Poole, R. K., Lloyd, D. and Chance, B. (1979a). *Biochem. J.* **184**, 555–563.

Poole, R. K., Waring, A. J. and Chance, B. (1979b). *Biochem. J.* **184**, 379–389.

Poole, R. K., Waring, A. J. and Chance, B. (1979c). *FEBS Lett.* **101**, 56–58.

Poole, R. K., Scott, R. I. and Chance, B. (1981). *J. Gen. Microbiol.* **125**, 431–438.

Poole, R. K., Scott, R. I., Baines, B. S., Salmon, I. and Lloyd, D. (1982a). *FEBS Lett.* **150**, 281–285.

Poole, R. K., Baines, B. S., Hubbard, J. A. M., Hughes, M. N. and Campbell, N. J. (1982b). *FEBS Lett.* **150**, 147–150.

Poole, R. K., Sivaram, A., Salmon, I. and Chance, B. (1982c). *FEBS Lett.* **141**, 237–241.

Poole, R. K., Kumar, C., Salmon, I. and Chance, B. (1983b). *J. Gen. Microbiol.* **129**, 1335–1344.

Poole, R. K., Salmon, I. and Chance, B. (1983c). *J. Gen. Microbiol.* **129**, 1345–1355.

Poole, R. K., van Wielink, J. E., Baines, B. S., Reijnders, W. N., Salmon, I. and Oltmann, L. F. (1983a). *J. Gen. Microbiol.* **129**, 2163–2173.

Poole, R. K., Baines, B. S., Curtis, S. J., Williams, H. D. and Wood, P. M. (1984). *J. Gen. Microbiol.* **130**, 3055–3058.

Poole, R. K., Baines, B. S., Hubbard, J. A. M. and Williams, H. D. (1985a). In 'Microbial Gas Metabolism: Mechanistic, Metabolic and Biotechnological Aspects (Eds. R. K. Poole and C. S. Dow). Academic Press, London, Orlando.

Poole, R. K., Baines, B. S. and Williams, H. D. (1985b). *Microbiol. Sci.* **2**, 21–24.

Poole, R. K., Baines, B. S. and Appleby, C. A. (1986). *J. Gen. Microbiol.* **132**, 1525–1539.

Powers, L., Chance, B., Ching, Y. and Angiolillo, P. (1981). *Biophys. J.* **34**, 465–498.

Pudek, M. R. and Bragg, P. D. (1974). *Archiv. Biochem. Biophys.* **164**, 682–693.

Pudek, M. R. and Bragg, P. D. (1976a). *FEBS Lett.* **62**, 330–333.

Pudek, M. R. and Bragg, P. D. (1976b). *Archiv. Biochem. Biophys.* **174**, 546–552.

Püttner, I., Solioz, M., Carafoli, E. and Ludwig, B. (1983). *Eur. J. Biochem.* **134**, 33–37.

Raitio, M., Jalli, T. and Saraste, M. (1987). *EMBO. J.* **6**, 2825–2833.

Reichardt, J. K. V. and Gibson, Q. H. (1983). *J. Biol. Chem.* **258**, 1504–1507.

Reid, G. A. and Ingledew, W. J. (1979). *Biochem. J.* **182**, 465–472.

Reid, G. A. and Ingledew, W. J. (1980). *FEBS Lett.* **109**, 1–4.

Rice, C. W. and Hempfling, W. P. (1978). *J. Bacteriol.* **134**, 115–124.

Sánchez Crispin, J. A., Dudourdieu, D. and Chippaux, M. (1979). *Biochim. Biophys. Acta* **547**, 198–210.

Saraste, M. (1983). *Trends Biochem. Soc.* **8**, 139–142.

Saraste, M., Raitio, M., Jalli, T. and Peramaa, A. (1986). *FEBS Lett.* **206**, 154–156.

Sato, M., Tamala, N., Kakiuchi, K., Fukumori, Y., Yamanaka, T., Kasai, N. and Kakudo, M. (1983). *Biochem. International* **7**, 345–352.

Seelig, A., Ludwig, B., Seelig, J. and Schatz, G. (1981). *Biochim. Biophys. Acta* **636**, 162–167.

Shaw, R. W., Hansen, R. E. and Beinert, H. (1978). *J. Biol. Chem.* **253**, 6637–6640.

Solioz, M., Carafoli, E. and Ludwig, B. (1982). *J. Biol. Chem.* **257**, 1579–1582.

Sone, N. (1986). *Methods Enzymol.* **126**, 145–153.

Sone, N. and Hinkle, P. C. (1982). *J. Biol. Chem.* **257**, 12600–12604.

Sone, N. and Yanagita, Y. (1982). *Biochim. Biophys. Acta* **682**, 216–226.

Sone, N. and Yanagita, Y. (1984). *J. Biol. Chem.* **259**, 1405–1408.

Sone, N., Ohyama, T. and Kagawa, Y. (1979). *FEBS Lett.* **106**, 39–42.

Sone, N., Kagawa, Y. and Orii, Y. (1983a). *J. Biochem.* **93**, 1329–1336.

Sone, N., Yanagita, Y., Hon-Nami, K., Fukumori, Y. and Yamanaka, T. (1983b). *FEBS Lett.* **155**, 150–154.

Sone, N., Naqui, A., Kumar, C. and Chance, B. (1984). *Biochem. J.* **221**, 529–533.

Sone, N., Ogura, T. and Kitagawa, T. (1986). *Biochim. Biophys. Acta* **850**, 139–145.

Stallings, W. C., Powers, T. B., Pattridge, K. A., Fee, J. A. and Ludwig, N. L. (1983). *Proc. Natl. Acad. Sci. USA* **80**, 3884–3888.

Steffens, G. J. and Buse, G. (1979). *Hoppe-Seyler's Z. Physiol. Chem.* **360**, 613–619.

Steffens, G. C. M., Buse, G., Oppliger, W. and Ludwig, B. (1983). *Biochem. Biophys. Res. Commun.* **116**, 335–340.

Steffens, G. C. M., Biewald, R. and Buse, G. (1986). *4th Eur. Bioenerg. Conf. Abstr.* (Prague), p. 191.

Steffens, G. C. M., Biewald, R. and Buse, G. (1987). *Eur. J. Biochem.* **164**, 295–300.

Taber, H. (1974). *J. Gen. Microbiol.* **81**, 435–444.

Tanaka, Y., Fukumori, Y. and Yamanaka, T. (1983). *Archiv. Microbiol.* **135**, 265–271.

Thomson, A. J. (1977). *Nature (Lond.)* **265**, 15–16.

Thomson, A. J., Johnson, M. K., Greenwood, C. and Gooding, P. E. (1981). *Biochem. J.* **193**, 687–697.

Timkovich, R., Cork, M. S., Gennis, R. B. and Johnson, P. Y. (1985). *J. Am. Chem. Soc.* **107**, 6069–6075.

Trnka, M. and Peschek, G. A. (1986). *Biochem. Biophys. Res. Commun.* **136**, 235–241.

Tweedle, M. F., Wilson, L. J., Garcia-Iñiguez, L., Babcock, G. T. and Palmer, G. (1978). *J. Biol. Chem.* **253**, 8065–8071.

Uno, T., Nishimura, Y., Tsuboi, M., Kita, K. and Anraku, Y. (1985). *J. Biol. Chem.* **260**, 6755–6760.

Van Verseveld, H. W., Krab, K. and Stouthamer, A. H. (1981). *Biochim. Biophys. Acta* **635**, 525–534.

Van Verseveld, H. W., Braster, M., Boogerd, F. C., Chance, B. and Stouthamer, A. H. (1983). *Archiv. Microbiol.* **135**, 229–236.

Van Wielink, J. E., Oltman, L. F., Leeuwerik, F. J., de Hollander, J. A. and Stouthamer, A. H. (1982). *Biochim. Biophys. Acta* **681**, 177–190.

Van Wielink, J. E., Reijnders, W. N. M., Oltman, L. F., Leeuwerik, F. J. and Stouthamer, A. H. (1983). *Arch. Microbiol.* **134**, 118–122.

Wakabayashi, S., Matsubara, H. and Webster, D. A. (1986). *Nature* **322**, 481–483.

Watanabe, H., Kamita, Y., Nakamura, T., Takimoto, A. and Yamanaka, T. (1979). *Biochim. Biophys. Acta* **547**, 70–78.

Weston, J. A. and Knowles, C. J. (1974). *Biochim. Biophys. Acta* **333**, 228–236.

Wikström, M., Krab, K. and Saraste, M. (1981). 'Cytochrome Oxidase, a Synthesis'. Academic Press, London.

Williams, H. D. and Poole, R. K. (1986). *Biochem. Soc. Trans.* **14**, 1215–1216.

Williams, H. D. and Poole, R. K. (1987). *J. Gen. Microbiol.* **133**, 2461–2472.

Williams, H. D. and Poole, R. K. (1988). *Curr. Microbiol.* **16**, 227–280.

Willison, J. C., Haddock, B. A. and Boxer, D. H. (1981). *FEMS Lett.* **10**, 249–255.

Winter, D. B., Bruyninckx, W. J., Foulke, F. G., Grinich, N. P. and Mason, H. S. (1980). *J. Biol. Chem.* **255**, 11408–11414.

Woese, C. R. (1981). Archaebacteria. *Scientific American*, June, 94–106.

Wong, T.-Y. and Jurtshuk, P. (1984). *J. Bioenerg. Biomicrob.* **16**, 477–489.

Wood, P. M. (1984). *Biochim. Biophys. Acta* **768**, 293–317.

Wynn, R. M., Kämpf, C., Gaul, D. F., Choi, W. K., Shaw, R. W. and Knaff, D. B. (1985). *Biochim. Biophys. Acta* **808**, 85–93.

Yamanaka, T. and Fujii, K. (1980). *Biochim. Biophys. Acta* **591**, 53–62.

Yamanaka, T. and Fukumori, Y. (1981). *Plant & Cell Physiol.* **22**, 1223–1230.

Yamanaka, T., Fujii, K. and Kamita, Y. (1979). *J. Biochem.* **86**, 821–824.

Yamanaka, T., Kamita, Y. and Fukumori, Y. (1981). *J. Biochem.* **89**, 265–273.

Yamanaka, T., Fukumori, Y., Yamazaki, T., Kato, H. and Nakayama, K. (1985). *J. Inorg. Biochem.* **23**, 273–277.

Yamazaki, T., Fukumori, Y. and Yamanaka, T. (1985). *Biochim. Biophys. Acta* **810**, 174–183.

Yang, T. (1982). *Eur. J. Biochem.* **121**, 335–341.

Yang, T. (1986). *Biochim. Biophys. Acta* **848**, 342–351.

Yoshida, T. and Fee, J. A. (1984). *J. Biol. Chem.* **259**, 1031–1036.

Yoshida, T. and Fee, J. A. (1985). *J. Inorg. Biochem.* **23**, 279–288.

Yoshida, T., Lorence, R. M., Choc, M. G., Tarr, G. E., Findling, R. L. and Fee, J. A. (1984). *J. Biol. Chem.* **259**, 112–123.

Zannoni, D., Baccarini-Melandri, A., Melandri, B. A., Evans, E. H., Prince, R. C. and Crofts, A. R. (1974). *FEBS Lett.* **48**, 152–155.

Zannoni, D. (1982). *Archiv. Microbiol.* **133**, 267–273.

6 Quinoproteins and energy transduction

C. Anthony

6.1   Introduction 293
6.2   Pyrrolo-quinoline quinone, PQQ 294
6.2.1   General properties 294
6.2.2   The natural occurrence of PQQ, and PQQ as a growth factor   296
6.3   Quinoprotein dehydrogenases 298
6.3.1   Methanol dehydrogenase 298
        6.3.1.1   Introduction and general properties 298
        6.3.1.2   Interaction with cytochrome c 299
        6.3.1.3   NAD-dependent, PQQ-containing methanol
                  dehydrogenase 301
6.3.2   Alcohol dehydrogenases 301
6.3.3   Methylamine dehydrogenase 302
        6.3.3.1   Introduction and general properties 302
        6.3.3.2   Interaction with amicyanin and cytochromes 303
6.3.4   Glucose dehydrogenase 305
        6.3.4.1   Introduction and general properties 305
        6.3.4.2   The electron acceptor for glucose dehydrogenase . 306
6.3.5   Other quinoprotein dehydrogenases and oxidases 307
6.4   Interactions of quinoproteins with electron transport chains,
      and energy transduction 307
6.4.1   Electron transport and energy transduction from methanol
        dehydrogenase 307
6.4.2   Energy transduction from other alcohol dehydrogenases ... 310
6.4.3   Electron transport and energy transduction from methylamine
        dehydrogenase 310
6.4.4   Electron transport and energy transduction from glucose
        dehydrogenase 311
6.5   Why have quinoproteins? 312
References 313

6.1   Introduction

Quinoproteins are dehydrogenases having pyrrolo-quinoline quinone (PQQ)

293

as prosthetic group. This separate chapter is devoted to the quinoproteins because their recent appearance on the scene has precluded discussion of them in previous textbooks, and because their interactions with electron transport chains are both unusual and varied. Although the most thoroughly characterized quinoproteins have so far been bacterial in origin, PQQ has now been found in all the main types of living organism and, as is the case with other prosthetic groups and coenzymes, it may well be a vitamin in higher organisms.

The first quinoproteins to be described and shown to have unusual prosthetic groups were the dehydrogenases for methanol (Anthony and Zatman, 1964, 1967a,b) and for glucose (Hauge, 1964). Most dehydrogenases that do not use NAD(P) as coenzyme have riboflavin derivatives (FMN or FAD) as their prosthetic group and that of glucose dehydrogenase was first thought to be a novel flavin derivative. Methanol dehydrogenase was, by contrast, clearly not a flavoprotein. It appeared to have either a novel pteridine derivative or some completely novel compound as prosthetic group. Duine and Frank and their colleagues have subsequently shown that both these dehydrogenases and other dehydrogenases have the same PQQ prosthetic group, and they have coined the name 'quinoprotein' to include them all (see Table 6.1 for complete list and references). In all cases, the PQQ is bound sufficiently tightly to the dehydrogenase apoprotein for the entire reaction cycle to occur on a single enzyme molecule. In this respect it is analogous to the flavin prosthetic group of flavoproteins and unlike coenzymes such as NAD which require two separate enzymes for the complete reaction cycle.

By contrast with flavoprotein dehydrogenases, which usually react with membrane-soluble quinones (e.g. ubiquinone), quinoproteins react with a variety of different electron acceptors. They are usually periplasmic, or arranged in membranes so that their active sites catalyse reactions with substrates in the periplasm. After extraction from bacteria, quinoproteins are usually assayed *in vitro* using an artificial electron acceptor such as phenazine methosulphate.

This chapter describes the PQQ prosthetic group itself, the quinoprotein dehydrogenases containing the PQQ, and their interactions with electron transport chains to generate a protonmotive force.

6.2 Pyrrolo-quinoline quinone

6.2.1 General properties

The prosthetic group of methanol dehydrogenase (MDH) was the first PQQ to be isolated and characterized. The absorption spectrum of MDH was clearly different from that of flavoproteins, and denaturation of MDH released a reddish-brown, low-molecular-weight, polar-acidic molecule with a typical green fluorescence, having an excitation maximum at 365 nm and a fluorescence maximum at 470 nm (Anthony and Zatman, 1967a,b). Release of this fluorescent chromophore from the MDH occurred concomitantly with loss of activity over a wide range of denaturing conditions. The excitation and emission characteristics of the fluorescence of the prosthetic group are typical of pteridines and it was originally concluded that the novel prosthetic group of MDH might be an unusual pteridine. The first demonstration that this was not the case was published by Duine and Frank and their colleagues. Using a wide range of chemical and physical techniques they showed that it is a multicyclic ring compound with two uncoupled aromatic protons, an inner ring orthoquinone, two nitrogen atoms and one or more carboxyl groups (Duine *et al.*, 1978, 1979, 1980; Westerling *et al.*, 1979). These proposals were all consistent with the structure based on X-ray diffraction analysis of a crystalline acetonyl derivative of the presumed prosthetic group extracted from whole cells of *Pseudomonas* TP-1 by Salisbury et al. (1979). The trivial name 'methoxatin' has been proposed for the prosthetic group but the more usual name pyrrolo-quinoline quinone (PQQ) emphasizes the functional importance of the orthoquinone part of the structure (Fig. 6.1).

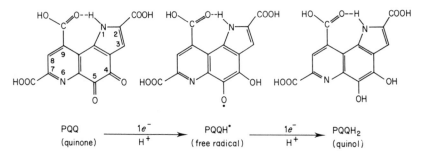

Fig. 6.1   The prosthetic group of methanol dehydrogenase. The full name of PQQ is 2,7,9-tricarboxy-lH-pyrrolo [2,3-f] quinoline-4,5-dione. The trivial name 'methoxatin' was initially proposed (Salisbury *et al.*, 1979) but the abbreviation PQQ (Pyrrolo-Quinoline Quinone) emphasizes the functional importance of the orthoquinone part of the structure (Duine and Frank, 1981). Adducts with water, methanol, acetaldehyde, acetone and ammonia are formed by addition at C_5. The midpoint redox potential of the $PQQ/PQQH_2$ couple is $+90\,mV$ at pH 7.0 and $+419\,mV$ at pH 2.0, indicating that PQQ is a $2e^-/2H^+$ redox carrier (Duine *et al.*, 1981); see also Faraggi *et al.* (1986). The spectra of PQQ are given on page 41 and in Anthony (1986).

Comprehensive reviews of the biochemistry, chemistry and physical properties of PQQ have been published recently, so these topics will not be considered in detail here (see Anthony, 1986; Duine *et al.*, 1987a).

Although it has not been possible to reconstitute active MDH from the isolated PQQ prosthetic group plus apoenzyme, such a reconstitution has been achieved using the apoenzyme of glucose dehydrogenase. This assay system has been used for studies of the relationship between structure and function of PQQ and its derivatives (Duine *et al.*, 1987a; Ameyama *et al.*, 1985b; Shinagawa *et al.*, 1986). Of special interest in considering the structure of PQQ is the question of how it binds to apoenzyme and how it is involved in electron (or hydrogen) transport. From a chemical point of view, binding could occur by interaction of the carboxylic acid groups or the quinone moiety with amino acid residues in the protein. Most evidence suggests that the quinone moiety does not take part in binding but that the carboxyl groups are probably involved, possibly in conjunction with magnesium or calcium ions. An important property of the orthoquinone structure is its ability to form adducts at the C-4 and C-5 positions, but whether or not such adduct formation is involved in the catalytic cycle of quinoproteins is not known (see Anthony, 1986; Duine *et al.*, 1987a). Similarly, it is not known if the ability of PQQ to foster the oxidation of aldehydes, ketones, thiols and amino acids has any relevance to its role as an enzyme prosthetic group (Oshiro *et al.*, 1983; Itoh *et al.*, 1984, 1985). The PQQ prosthetic group of methylamine dehydrogenase probably lacks carboxyl groups and is covalently bound (see page 302).

6.2.2   The natural occurrence of PQQ, and its role as a growth factor

Table 6.1 gives a list of bacterial quinoproteins which have PQQ as their prosthetic group. As mentioned above, PQQ has also been shown to be the prosthetic group of oxidoreductases from all sorts of microbe and from plants and animals. In addition to these sources, PQQ is found in extracts of lab media and in bacterial culture fluids and this can result in stimulation of PQQ-requiring bacteria by those that overproduce it (Ameyama *et al.*, 1984a,b,c, 1985c,d; Shimao *et al.*, 1984; Duine *et al.*, 1986). Bacteria whose growth is stimulated by PQQ are those that form the apoenzyme of a quinoprotein but are unable to synthesize PQQ, which must therefore be provided as a growth factor. The first example of this was *Pseudomonas* sp.VM15C which needs PQQ (provided as such, or provided by growth with a PQQ-producing symbiont) in order to produce a poly (vinyl alcohol)-degrading enzyme essential for growth on this substrate (Shimao *et al.*, 1984). A similar case is that of *Pseudomonas testosteroni* which needs PQQ in order to synthesize the alcohol dehydrogenase responsible for rapid growth on that substrate (Groen *et al.*, 1986). Some bacteria with the alternative routes for glucose metabolism are still able to oxidize glucose by GDH if PQQ is added to reconstitute active holoenzyme. Examples of such bacteria

Table 6.1 Bacterial quinoprotein dehydrogenases. This list of references is not exhaustive; it aims to refer to first descriptions, demonstrations of quinoprotein nature of the enzymes and reviews. The most recent reviews of quinoproteins are those by Duine et al. (1986, 1987a,b) and recent reviews of dehydrogenases for methanol and methylamine are by Anthony (1982, 1986). Further references to the dehydrogenases and their coupling to electron transport and energy transduction are given in the text.

Enzyme	Organism	References
Methanol dehydrogenase	Gram-negative methylotrophic bacteria	Anthony and Zatman (1964, 1967a,b); Duine and Frank (1979); Salisbury et al. (1979); Westerling et al. (1979); Duine et al. (1980); Anthony (1982, 1986)
Alcohol dehydrogenases	Pseudomonas aeruginosa	Groen et al. (1984)
	Pseudomonas testosteroni (a quinohaemoprotein)	Groen et al. (1986)
	Gluconobacter suboxydans	Adachi et al. (1978, 1982)
Methylamine dehydrogenase	Gram-negative methylotrophic bacteria	Eady and Large (1968); de Beer et al. (1980); Anthony (1982)
Glucose dehydrogenase	Acinetobacter	Hauge (1964); Duine et al. (1979, 1982); Geiger and Gorisch (1986); Dokter et al. (1987)
	Pseudomonas spp.	Matsushita et al. (1980a); Duine et al. (1983)
	Escherichia coli	Ameyama et al. (1986); Matsushita et al. (1987)
	Gluconobacter suboxydans	Ameyama et al. (1981)
Glycerol dehydrogenase	Gluconobacter industruis	Ameyama et al. (1985a)
Aldehyde dehydrogenase	Acetic acid bacteria	Ameyama and Adachi (1982)
Polyethylene glycol dehydrogenase	Flavobacterium sp. plus Pseudomonas sp.	Kawai et al. (1985)
Methylamine oxidase	Arthrobacter P1	Duine et al. (1987a)

include enteric bacteria (Neijssel *et al.*, 1983; Hommes *et al.*, 1984; Matsu-shita *et al.*, 1986) and some species of *Pseudomonas, Agrobacterium, Rhizo-bium* and *Acinetobacter* (van Schie *et al.* 1985, 1987). Particularly interesting is that PQQ becomes a cofactor for growth of *E. coli* on glucose when the normal route for glucose uptake and metabolism (the phosphotransferase system) is lost by mutation (Hommes *et al.*, 1984). Further examples of this and related phenomena are discussed in the review by Duine *et al.* 1986).

6.3 Quinoprotein dehydrogenases

6.31 Methanol dehydrogenase

6.3.1.1 Introduction and general properties

This dehydrogenase (MDH) is responsible for the oxidation of methanol to formaldehyde in methylotrophs growing on methane or methanol. Full references to the original work quoted below are given in the extensive reviews of Anthony (1982, 1986) and Duine *et al.* (1986, 1987a,b); and an excellent review of the molecular biology of methanol oxidation has been published by de Vries (1986).

MDH constitutes up to 15% of the soluble bacterial protein and is located exclusively in the periplasm together with its electron acceptor (cytochrome c_L), the concentrations of both of these proteins being about 0.5 mM. When bacterial extracts are prepared by sonication some MDH is found associated with membrane fractions; it is not known if MDH binds to the outside of the periplasmic membrane but there seems to be no obvious necessity for it to do so. A study of the biosynthesis and assembly of MDH suggests that a precursor form of the enzyme is made ($M_r = 1500$ larger than mature subunit) and that processing of this form occurs together with export of the apoprotein subunits, followed by assembly of the subunits and prosthetic group in the periplasm to form the mature holoenzyme (Davidson *et al.*, 1985). This necessity for having PQQ in the periplasm presumably accounts for the substantial amounts of PQQ found in the culture fluids of methylo-trophic bacteria.

Most methanol dehydrogenases are dimers of identical subunits of mol-ecular weight (M_r) of 60 000 and most are basic proteins. MDH is assayed at its usual pH optimum (pH 9.0) using an artificial electron acceptor such as phenazine methosulphate and ammonia as activator although it is unlikely that this activator is required *in vivo* because whole cells do not appear to require it and it is not always essential—particularly in experiments in which the natural electron acceptor (cytochrome c_L) is used *in vitro* (Anthony, 1986; Duine *et al.*, 1986).

MDH oxidizes a wide range of primary alcohols; the Km for methanol is usually low (about 20 µM) and affinity for the enzyme often decreases with increasing size of the alcohol. A common feature of methanol dehydrogenases is their ability to catalyse the oxidation of formaldehyde (in the hydrated gem-diol form) to formate. This must be prevented during growth because the formaldehyde must enter the bacteria in order to provide the carbon substrate for cell synthesis and because oxidation of formaldehyde by MDH is energetically wasteful. A protein is present in all methylotrophs tested that offers a solution to this problem. This 'modifier protein' (M-protein) alters the substrate specificity of MDH by altering its affinity for its substrates. The affinity for some alcohols is increased but the affinity for formaldehyde is markedly decreased: sufficient to stop formaldehyde oxidation (Ford *et al.*, 1985; Page and Anthony, 1986). The regulation of methanol and formaldehyde oxidation during the continuous culture of methylotrophs has been studied extensively by Cornish *et al.* (1984), by Greenwood and Jones (1986) and by Jones *et al.* (1987).

Methanol dehydrogenase has a characteristic absorption spectrum due to its prosthetic group, with a peak at 345 nm and a shoulder at about 400 nm. It is usually isolated as the yellow/green partially reduced enzyme, and the slight variations in spectra from one enzyme to another are probably due to varying proportions of partially and fully reduced enzyme present. Non-covalent bonds are involved in attachment of PQQ to the apoenzyme, and boiling or treatment with acid or alkali is sufficient to release it. It has not been possible to reconstitute active dehydrogenase from PQQ plus apoenzyme although this has been possible using glucose dehydrogenase.

During the catalytic cycle the orthoquinone (PQQ) becomes fully reduced to the orthoquinol ($PQQH_2$) by substrate; this fully reduced form is then oxidized to PQQ in two steps by way of the free radical ($PQQH^{\bullet}$) (Fig. 6.1). Although the C-4 and C-5 of the orthoquinone are able to form adducts with alcohols, amines, etc., there is no evidence that these are involved in catalytic activity of this dehydrogenase. Ideas relating to the catalytic mechanism of MDH have been reviewed recently by Duine *et al.* (1987a).

6.3.1.2   Interaction with cytochrome *c*

A full description of the *c*-type cytochromes of methylotrophs and their interactions is presented in Anthony (1986). In summary, there are always at least two soluble, periplasmic cytochromes. These are functionally distinct but spectrally almost indistinguishable; they were therefore named according to their isoelectric points. Cytochrome c_H usually has a high isoelectric point and is now known to correspond to the typical soluble *c*-type cytochromes found in mitochondria and many bacteria. Cytochrome c_L occurs only in methylotrophs; it has a low isoelectric point (pH 4.0–5.0) and its mass

(M_r = 17 000–21 000) is about twice that of typical soluble c-type cytochromes. Cytochrome c_L is a typical cytochrome c in most other respects; it has a single polypeptide chain, bearing a single haem, having histidine and methionine as its two axial ligands. It has a high midpoint redox potential (250–310 mV), characteristic absorption spectrum and is low spin in both oxidized and reduced states (O'Keeffe and Anthony, 1980a; Cross and Anthony, 1980a; Beardmore-Gray et al., 1982).

Cytochrome c_L is the physiological electron acceptor for MDH, as shown when using the pure proteins from a number of different methylotrophs (Beardmore-Gray et al., 1983). This has also been confirmed by isolating mutants lacking this cytochrome but containing all other cytochromes and dehydrogenases. These mutants are unable to oxidize or grow on methanol but are unimpaired with respect to oxidation of all other growth substrates including methylamine (Nunn and Lidstrom, 1986a,b). Unfortunately, the rate of reduction of cytochrome c_L by methanol catalysed by MDH is less than 1% of that occurring during cellular respiration. This suggests either that some other extra component is required in the system or that one or more of the proteins is damaged during extraction and purification. It is certain that some component involved in reduction of cytochrome by MDH in partially purified extracts is sensitive to oxygen, but anaerobic preparation of the proteins does not lead to markedly higher rates of reaction (see Duine et al., 1986; Dijkstra et al., 1988a).

In considering the interaction between MDH and cytochrome c_L it is necessary to consider the autoreduction of this cytochrome seen in all examples of cytochrome c_L that have been studied (O'Keeffe and Anthony, 1980b; Anthony, 1986). In the absence of MDH, as the pH is raised cytochrome c_L becomes autoreduced at high rates in the absence of any electron donor in an intramolecular reaction. In the presence of MDH this same intramolecular reduction of cytochrome c also occurs but at a much lower (more physiological) pH value which is usually pH 7.0, but is pH 4.0 in the acidophilic methylotroph *Acetobacter methanolicus* (Elliott and Anthony, 1988).

Whether or not this phenomenon of autoreduction is also involved in electron transfer between MDH and cytochrome c_L is not yet certain (see Anthony, 1982, 1986; Dijkstra et al., 1988b), but at the concentrations of MDH and cytochrome c_L present in the periplasm, the rate of MDH-induced autoreduction is similar to the overall rate of respiration. It is the rate of electron transfer from methanol dehydrogenase to the oxidized form of the cytochrome c_L that is the rate-limiting step in electron transport from methanol (Elliott and Anthony, 1988). It has been calculated that the concentrations of both MDH and cytochrome c_L are of the order of 0.5 mM in the periplasm (Beardmore-Gray et al., 1983). It is possible that the two proteins completely cover most of the outer surface of the bacterial membrane, and that the marked change in environment occurring on release of the two proteins on cell disruption may contribute to their low activity when measured *in vitro* (see Chapter 3, page 175 for further discussion of this point). Electron transport and energy transduction from MDH is discussed on pages 307 to 310.

6.3.1.3 NAD-dependent, PQQ-containing methanol dehydrogenase

When grown on methanol, the Gram-positive methylotroph, *Nocardia* sp.239, does not contain a classical methanol dehydrogenase. Instead, preliminary reports indicated that it contains a multienzyme complex consisting of three components: the methanol dehydrogenase; NAD-dependent aldehyde dehydrogenase; and NADH dehydrogenase. It was concluded that this novel methanol dehydrogenase transfers the reducing equivalents, derived from methanol, directly to its associated NADH dehydrogenase via a mechanism in which NAD and PQQ are involved (Duine *et al.*, 1984). This is so unusual that considerably more confirmatory results and more information are required before any assessment of its significance can be made.

6.3.2 Alcohol dehydrogenases

These enzymes are similar in many respects to methanol dehydrogenases, but they are unable to oxidize methanol at substantial rates. The first of these to be described was purified from membranes of an unidentified organism which was thought (erroneously) to be an *Acinetobacter* sp. and which has since been lost (Duine *et al.*, 1981; Beardmore-Gray and Anthony, 1983). A similar, but soluble, dehydrogenase has now been well characterized from *Pseudomonas aeruginosa* and shown to differ from MDH in using higher primary amines as activator and in being a large monomeric protein (M_r, 101 000) having two molecules of PQQ per enzyme molecule (Groen *et al.*, 1984).

A second type of alcohol dehydrogenase isolated from *Pseudomonas testosteroni* grown on butanol, differs more markedly from MDH in that it is a monomer and amine activators are not required (Groen *et al.*, 1986). It was isolated as the apoenzyme, activity being dependent on addition of PQQ in the presence of calcium ions. As well as a single molecule of PQQ, the enzyme contained one molecule of haem *c* which is reduced by substrate in the holoenzyme containing PQQ but not in the apoenzyme. This quinohaemoprotein, alcohol dehydrogenase, may be similar to the membrane-bound alcohol dehydrogenase from *Gluconobacter* which has been crystallized in a complex together with cytochrome *c* (Adachi *et al.*, 1982).

6.3.3   Methylamine dehydrogenase

6.3.3.1   Introduction and general properties

This enzyme, which is also called 'primary amine dehydrogenase', catalyses the oxidation of methylamine to formaldehyde and ammonia in the presence of its electron acceptor, amicyanin, or an artificial electron acceptor such as phenazine methosulphate. This is the first step in the oxidation of methylamine during growth on this substrate by some Gram-negative methylotrophic bacteria (see Anthony, 1982, for review of methylamine dehydrogenase).

Methylamine dehydrogenase is located exclusively in the periplasm (Burton *et al.*, 1983), together with its electron acceptor, amicyanin, and soluble *c*-type cytochromes (Lawton and Anthony, 1985b; Husain *et al.*, 1986). It has been suggested that it is the lack of a periplasmic space in Gram-positive bacteria that has led to the alternative copper-containing amine oxidase system in these methylotrophs (Duine *et al.*, 1987b).

Methylamine dehydrogenase differs from methanol dehydrogenase in its mode of binding its prosthetic group which is covalently bound PQQ or, rather, a closely related derivative of it (de Beer *et al.*, 1980; Kenney and McIntire, 1983). Mass spectral data indicate that the prosthetic group lacks the three carboxyl groups and is covalently bound to the small subunit of the apoprotein by a cysteine thio-ether (via a methylene bridge) and a serine ether linkage (McIntire and Stults, 1986). The dehydrogenase consists of two light subunits (about 13 000 daltons), each of which carries one molecule of prosthetic group, plus two heavy subunits (about 40 000 daltons). Amino acid sequence studies of the light subunit showed that the PQQ is covalently bound to two residues but these could not be identified (Ishii *et al.*, 1983). The dehydrogenase is induced during growth on methylamine and consists of up to 5% of the soluble protein of the bacteria. The dehydrogenase from obligate methylotrophs oxidizes a narrow range of primary amines, whereas facultative methylotrophs such as *Methylobacterium* AM1 oxidize a wide range of primary aliphatic amines and diamines.

The dehydrogenase is greenish-yellow, with a peak due to the prosthetic group at 430 nm and a shoulder at 460 nm. When reduced by addition of methylamine the peak at 430 nm diminishes and a new peak at 325 nm appears. The reduced form of the enzyme-bound prosthetic group is fluorescent with an excitation maximum at about 330 nm and an emission maximum at 380 nm. In these characteristics the methylamine dehydrogenase differs markedly from methanol dehydrogenase. This difference is presumably due to the slightly different form of PQQ and its different binding to the apoenzyme. The enzyme reaction occurs by way of a ping-pong mechanism, some of the proposed intermediates being identifiable by their spectra. Perhaps the most important question about the mechanism of the dehydrogenase is whether or not the *o*-quinone of the PQQ acts merely as an electron acceptor or whether the amine substrate binds to the PQQ as previously suggested (Anthony, 1982). The recent crystallization and X-ray investigation of this enzyme should lead to a resolution of this question (Vellieux *et al.*, 1986).

6.3.3.2   Interaction with amicyanin and cytochromes

The physiological electron acceptor for methylamine dehydrogenase is a type I blue copper protein (or cuproprotein) first discovered in *Methylobacterium* AM1 by Tobari and called amicyanin (Tobari and Harada, 1981; Tobari, 1984). This conclusion is supported by evidence from work with at least five very different methylotrophs including *Methylobacterium* AM1, a pink facultative methylotroph (previously called *Pseudomonas* AM1); organism 4025, an obligate methylotroph (closely related to *Methylophilus methylotrophus*) that produces so much amicyanin that it turns blue; and *Paracoccus denitrificans*, a facultative autotroph (see Table 6.2 for references). By contrast with methanol dehydrogenase and its proposed acceptor (cytochrome c_L), the rate of electron transfer from methylamine dehydrogenase to amicyanin is rapid and sufficient to account for the respiration rate in whole bacteria (Lawton and Anthony, 1985a; see Chapter 3, page 174). Most methylotrophs that contain amicyanin also contain, in some growth conditions, a second blue copper protein of uncertain function which is usually called azurin.

Type I blue copper proteins are small, water-soluble proteins that contain one copper atom per polypeptide chain, the ligands to the copper being two histidines, one cysteine and one methionine residue; they are blue in the oxidized form and colourless when reduced (page 45). It has been suggested that these proteins should be called cupredoxins by analogy with the iron–sulphur proteins, the ferredoxins. The cupredoxins act, like cytochrome *c*, as soluble carriers in electron transport having redox potentials that are similar to those of the high potential cytochome *c*, with which they are usually able to react. Amino acid sequence data suggest four classes, but these do not necessarily correlate with any known functions of the proteins (Ambler and Tobari, 1985). The proposed classes are plastocyanin, pseudoazurin, amicyanin, and azurin; the latter differing from the other three in having a disulphide bridge. At least one cupredoxin that acts as electron acceptor for methylamine dehydrogenase falls into the azurin sequence class rather than the expected amicyanin class; and it has been suggested by Ambler and Tobari (1985) that the name 'amicyanin' should be reserved exclusively for cupredoxins in the same sequence class as that from *Methylobacterium* AM1. Although logical, this poses problems because the amino acid sequence is one of the last features of a new protein to be characterized and it is likely that blue copper proteins able to accept electrons from methylamine dehydrogenase will always tend to be called amicyanins. This author has used the name 'amicyanin' for cupredoxins reacting with methylamine dehydrogenase and 'azurin' for other cupredoxins in methylotrophic bacteria in Table 6.2 which lists the cupredoxins. Like the soluble *c*-type cytochromes, bacterial cupredoxins are always located in the periplasm (see, e.g. Lawton and Anthony, 1985b; Husain and Davidson, 1985; Husain *et al.*, 1986; and see Chapter 3).

Table 6.2   The blue copper proteins (cupredoxins) of methylotrophs

Organism	Mol.wt	pI	$E_{m,7}$(mV)	λmax(nm)
Amicyanin				
Methylobacterium AM1 [a,b,c,k]	11 700	9.3	280	596
Methylomonas J [a,b]	—	7.7	—	613
Organism 4025 [d]	11 500	5.3	294	620
P. denitrificans [e,f,g,h]	15 000	4.8	—	595
Thiobacillus versutus [i]	13 800	4.7	261	596
'Azurin'				
Methylobacterium AM1 [a,b,c]	13 000	9.4	310	593
Methylomonas J [a,b]	—	9.6	—	620
Organism 4025 [d]	12 500	9.4	323	625
P. denitrificans [j]	13 800	4.6	230	595

Note: The proteins are grouped here according to their function: amicyanins are electron acceptor for methylamine dehydrogenase and 'azurins' are other cupredoxins also found in these methylotrophs. The spectrum of a cupredoxin (or cuproprotein) is given in Fig. 1.19, page 45; the structure of the prosthetic group is given in Fig. 1.18. page 45.
(*a*) Tobari and Harada (1981); (*b*) Tobari (1984); (*c*) Ambler and Tobari (1985); (*d*) Lawton and Anthony (1985a,b); (*e*) Husain and Davidson (1985, 1986a,b); (*f*) Husain *et al.* (1986); (*g*) Gray *et al.* (1986); (*h*) Lim *et al.* (1986); (*i*) van Houwelingen *et al.* (1985); (*j*) Martinkus *et al.* (1980); (*k*) Fukomori and Yamanaka (1987).

It should be emphasized that, although amicyanin is the best (usually only) electron acceptor for methylamine dehydrogenase and is sometimes induced to extraordinary high concentrations on methylamine (Lawton and Anthony, 1985b), it is not detectable in all methylotrophs growing on methylamine by way of methylamine dehydrogenase. The picture is confused by the large variation in amounts of amicyanin found in those bacteria that do produce it, by variations in amounts of total blue copper proteins brought about by varying copper concentrations during growth, and by variations in the ease of dissociation of copper from the amicyanin when studied *in vitro*. Methylamine dehydrogenase is sometimes able to react with the periplasmic cytochrome *c* found in methylotrophs (see Chandrasekar and Klapper, 1986; Fukumori and Yamanaka, 1987), but whether or not this has physiological significance is not known; it is clearly the most likely candidate for electron acceptor when amicyanin is absent. (See page 310 for discussion of electron transport and energy transduction from methylamine dehydrogenase.)

6.3.4   Glucose dehydrogenase

6.3.4.1   Introduction and general properties
Fermentative bacteria such as *E. coli*, having the glycolytic pathway for conversion of glucose to pyruvate usually take up glucose by way of a phosphotransferase system in which phosphorylation of glucose to glucose 6-phosphate occurs concomitantly with its uptake from the growth medium (Chapter 8, page 381). By contrast, aerobic bacteria, such as pseudomonads and acetic acid bacteria, which do not have the glycolytic pathway, oxidize glucose to the lactone of gluconic acid by way of glucose dehydrogenase (GDH). This is usually membrane-bound and is probably always arranged so that its catalytic site is available for periplasmic oxidation of glucose to gluconic acid which is further oxidized or released into the growth medium (Midgely and Dawes, 1973; Dawes, 1981; Lessie and Phibbs, 1984).

A clear summary of the role of glucose dehydrogenase (GDH) is hindered by the fact that some bacteria, such as *Acinetobacter* species, do not further metabolize their product (gluconic acid) and so are unable to grow upon glucose. Others, such as *E. coli* and other enteric bacteria, form the apoenzyme but are unable to synthesize PQQ, and so are unable to oxidize glucose by this route unless PQQ is provided (see page 296).

GDH was first isolated and described by Hauge (1960, 1961, 1964) and shown to be a quinoprotein by Duine *et al.* (1979); and the GDHs from a number of different bacteria have subsequently been solubilized with Triton from membranes, purified to homogeneity and shown to have similar properties (Table 6.1). The membrane GDHs from *E. coli*, *K. aerogenes*, *Gluconobacter suboxydans*, *Acetobacter aceti*, *Pseudomonas aeruginosa* and *Acinetobacter calcoaceticus* are all sufficiently closely related to cross-react with antibodies produced against the GDH of *Pseudomonas fluorescens*

(Matsushita *et al.*, 1986). The electron acceptor used *in vitro* for assays of GDH is usually phenazine methosulphate or ferricyanide, and the pH optimum for activity depends on the electron acceptor and the source of the enzyme. In most cases, the substrate range is broad and includes hexoses, pentoses and also disaccharides. The molecular weight is usually 80–90 000. Compared with other quinoproteins GDH is a relatively small proportion of the bacterial protein (e.g. 0.1%) but this is compensated for by a very high turnover number (e.g. $4000\,s^{-1}$). The absorption spectrum of GDH is similar to that of methanol dehydrogenase, having a peak due to PQQ at about 345 nm (Duine *et al.*, 1987a).

PQQ can be readily removed from GDH to form the apoenzyme, and active holoenzyme reconstituted by addition of PQQ or analogues. This has provided an enzymic assay for PQQ and a tool for investigating the structural requirements for an active prosthetic group (Duine *et al.*, 1986, 1987a; Shinagawa *et al.*, 1986). Reconstitution of the holoenzyme usually requires magnesium ions (Shinagawa *et al.*, 1986; van Schie *et al.*, 1987). In *Acinetobacter calcoaceticus* (formerly *Bacterium anitratum*) GDH exists in at least two forms, one of which is periplasmic and the other membrane-bound, the proportion depending on growth conditions (see Duine *et al.*, 1982). The significance of this is uncertain because it has been demonstrated that in *Acinetobacter calcoaceticus* 79–39 the sonication process removes some GDH from membrane (about half is soluble after sonication) (Beardmore-Gray and Anthony, 1986). When sphaeroplasts were formed very little GDH (less than 10%) and no cytochrome *b* was present in the periplasmic fraction and all the GDH and cytochrome *b* in the sphaeroplasts was in the membrane fraction produced after careful lysis.

6.3.4.2   The electron acceptor for GDH

The natural electron acceptor for GDH is still a matter of debate. The soluble form of the enzyme first studied by Hauge was closely associated with a soluble, autoxidizable cytochrome *b* and it was suggested that a second soluble factor may also be involved (Hauge, 1960, 1961). By contrast, in *Gluconobacter suboxydans* (now called *Acetobacter suboxydans*) some indirect evidence suggested that a CO-binding cytochrome *c* may be involved (as in methanol oxidation), whereas other evidence implicated cytochrome *b* (Daniel and Erickson, 1969; Daniel, 1970; Ameyama *et al.*, 1981; Matsushita *et al.*, 1981). In *Pseudomonas aeruginosa* the purified GDH has been shown to react with the artificial electron acceptor ubiquinones, but the rate of reaction with the 'natural' ubiquinone-9 was very slow (Matsushita *et al.*, 1980a,b, 1982).

This variety of suggestions is perhaps inevitable when studying a membrane-bound respiratory enzyme. What is certain is that in *Acinetobacter calcoaceticus* (and probably in other bacteria) electrons flow into the electron transport chain at the level of the quinone pool which, in turn, is oxidized by the cytochrome *b* and the *o*-type oxidase (Beardmore-Gray and Anthony, 1986) (see page 311).

6.3.5 Other quinoprotein dehydrogenases and oxidases

Other quinoprotein dehydrogenases, which have not been well characterized are listed in Table 6.1. Perhaps the most important of these is the glycerol dehydrogenase from acetic acid bacteria; this enzyme is likely to be responsible for oxidation of other polyhydric alcohols, a well-known activity of acetic acid bacteria. Table 6.1 also refers to the amine oxidases from *Arthrobacter* P1, which is similar to those from *Aspergillus niger* and mammalian sources. For many years it has been assumed that these copper-containing enzymes also contained a pyridoxal phosphate prosthetic group but it appears that this is not so: they contain two copper atoms and one covalently bound PQQ-like prosthetic group (Duine *et al.*, 1987a).

6.4 Interactions of quinoproteins with electron transport chains, and energy transduction

6.4.1 Electron transport and energy transduction from methanol dehydrogenase

A comprehensive description of electron transport components, electron transport systems, proton translocation and ATP synthesis, together with full references to the original work has been published previously (Anthony, 1986; Anthony and Jones, 1987) and the significance of the energetics of methanol oxidation to the general physiology of methylotrophs is more fully discussed in Anthony (1982).

During growth on conventional multicarbon substrates almost all electron transport is by way of the NADH or succinate dehydrogenases. During growth on methanol, by contrast, 50–90% of electron transport is by way of the 'methanol oxidase' system and the characteristics of this system clearly have an important influence on the bioenergetics of methylotrophs. This is also true for bacteria growing on methane, in which NADH is used almost exclusively for the initial hydroxylation of methane to methanol.

All bacteria oxidizing methanol have cytochromes *b* and *c*, the cytochrome *c* usually being at much higher concentrations than the *b*-type cytochromes; this is related to the fact that the primary electron acceptor from methanol dehydrogenase is the unusual cytochrome c_L (page 300).

Methylotrophs also have a second, small, more basic c-type cytochrome (cytochrome c_H) which is usually similar in all respects to the soluble cytochrome c that mediates between the cytochrome bc_1 complex and the terminal cytochrome oxidase of mitochondria and many other bacteria. It presumably has the same function in methylotrophs during oxidation of NADH and succinate. The methanol dehydrogenase and both c-type cytochromes are periplasmic proteins.

The terminal oxidase is either cytochrome aa_3 or an o-type oxidase and both may be present in a single organism, the relative amounts depending on growth conditions (Cross and Anthony, 1980b; see page 255). The cytochrome aa_3 from *Methylobacterium* AM1 appears to be a typical oxidase of this class (see page 245) but it does not appear to pump protons (Sone *et al.*, 1987) and it oxidizes cytochrome c_H much faster than cytochrome c_L (Fukumori *et al.*, 1985). The o-type oxidase from *Methylophilus methylotrophus* has four subunits; there are two subunits of a large c-type cytochrome and two of a b-type cytochrome which, because it reacts rapidly with carbon monoxide and because it is derived from the o-type oxidase has been called the cytochrome o component (Froud and Anthony, 1984a,b). The name now used for the oxidase is cytochrome co (see Chapter 5, page 250). It also oxidizes cytochrome c_H 50 times faster than cytochrome c_L and it is reasonable to assume that the cytochrome c_H is the usual electron donor to oxidases in methylotrophs.

Figure 6.2 shows the most likely path of electron transport from methanol to oxygen, drawn here as a redox arm mechanism with the two protolytic reactions on the opposite side of the membrane linked by vectorial electron transfer. This proposes that two protons are translocated for every methanol oxidized to formaldehyde, and this value has been measured in *M. methylotrophus*, in which either cytochrome co or cytochrome aa_3 is the predominant oxidase (Dawson and Jones, 1981a,b). It should be noted that an alternative arrangement of the oxidase in the membrane could result in both protolytic reactions occurring on its periplasmic side linked by a proton-pumping oxidase. It has not yet proved possible to distinguish experimentally between these two mechanisms. Interestingly, whole cells of *Paracoccus denitrificans* exhibit a $\rightarrow H^+/O$ quotient of 4 for the oxidation of methanol, a result that indicates a combined redox arm/proton pump mechanism (with the two protolytic reactions linked by vectorial electron transfer through a proton-pumping cytochrome oxidase aa_3) (van Verseveld *et al.*, 1981). It has recently been confirmed (by DNA sequencing) that this oxidase does have a proton pumping subunit (see page 249).

The relationship between the phosphorylation potential and the two components of the protonmotive force at various pH values has been measured by Dawson and Jones (1982) during the oxidation of methanol by

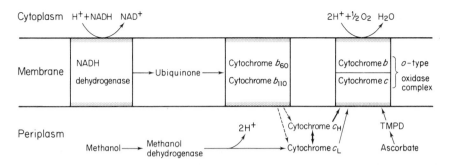

Fig. 6.2 The electron transport chain of *M. methylotrophus* grown under conditions of O₂-limitation. This figure is based on Froud and Anthony (1984a,b). This electron transport chain also operates in NH₃-limited conditions and, to some extent, in batch culture and continuous culture under conditions of methanol limitation. Solid lines indicate reactions that have been demonstrated in these bacteria, the relative thickness indicating the probable relative importance of the routes for oxidation of the soluble cytochromes *c*. Dotted lines are alternative routes for oxidation of cytochrome *b* which have not yet been investigated. About 40% of the cytochrome c_L is firmly attached to the membrane and this may form a cytochrome bc_L complex analogous to the cytochrome bc_1 complex in mitochondria and *Paracoccus denitrificans*. It should be noted that the methanol dehydrogenase and soluble cytochromes *c* are shown having a periplasmic location; this does not preclude the possibility that they are loosely bound to the periplasmic surface of the bacterial membrane. TMPD can probably donate electrons directly to all the *c*-type cytochromes and not only to the cytochrome *c* component of the oxidase as drawn here for convenience. The *o*-type oxidase is now usually called cytochrome *co*.

M. methylotrophus. They concluded that this organism is able to sustain a protonmotive force of up to −165mV during respiration with methanol as substrate. This was composed of a Δψ and ΔpH, the values for which depended on the external pH value, ΔpH being maximal at low external pH values and decreasing to zero at an external pH of 7.0 which is the same as the internal pH value of the bacteria, a value that is rather lower than is usual for neutrophiles. Either the Δψ or the ΔpH value alone was shown to be fully competent to drive ATP synthesis, and a Δ*Gp* of up to −46 kJmol⁻¹ was sustained during oxidation of methanol at pH 7.0.

Energy conservation *in vivo* has also been examined in *M. methylotrophus* by measuring steady-state concentrations of adenine nucleotides and inorganic phosphate in whole respiring cells, and the changes that occurred during the first few seconds following addition of oxidizable substrates. The conclusion drawn from this work was that the phosphorylation of ADP to ATP is coupled to translocation of about four protons and that the oxidation of methanol and NADH is accompanied by the synthesis of about 0.5 and 1.4–2.0 molecules of ATP respectively (Patchett *et al.*, 1985). This suggestion is consistent with the measured ATP production during the oxidation of methanol and NADH by membrane vesicle isolated from *Methylobacterium* AM1 (Netrusov and Anthony, 1979).

6.4.2   Energy transduction from other alcohol dehydrogenases

It is probable, by analogy with MDH, that alcohol dehydrogenases catalyse electron transfer from the alcohol to a cytochrome *c*. Whether or not the haem *c* or cytochrome *c* bound to some of these dehydrogenases replaces the cytochrome c_L is not known but it appears likely. In this case, coupling of alcohol oxidation by these enzymes to ATP synthesis is likely to be entirely analogous to that in MDH systems.

6.4.3   Electron transport and energy transduction from methylamine dehydrogenase

As described above (page 303), the evidence from work with at least five very different methylotrophs supports the view that amicyanin is the primary electron acceptor from methylamine dehydrogenase. It has been suggested that the cytochrome c_H might also act as electron acceptor in *Methylobacterium* AM1 (Fukumori and Yamanaka, 1987) and in organism W3A1 (Chandrasekar and Klapper, 1986). However, because amicyanin is usually induced to higher levels during growth on methylamine, as well as being the best acceptor *in vitro*, whenever amicyanin is present it is probable that it does act as the primary electron acceptor; in its absence, cytochrome c_H is clearly the most likely alternative candidate.

Neither of the two oxidases described in methylotrophs (cytochrome aa_3 and cytochrome *co*) is able to oxidize amicyanin, although azurin may be oxidized relatively slowly. Most of the periplasmic *c*-type cytochromes are able to react with the blue copper proteins, so precluding a definite statement about the sequence of electron flow in organisms containing at least two copper proteins plus at least two cytochromes in their periplasm. The oxidases of *Methylobacterium* AM1 (cytochrome aa_3) and *Methylophilus methylotrophus* (cytochrome *co*) oxidize cytochrome c_H in preference to cytochrome c_L (the electron acceptor for methanol dehydrogenase) (see pages 299–301). Furthermore, mutants lacking cytochrome c_L, although unable to grow on methanol, are able to grow well on methylamine (Nunn and Lidstrom, 1986a,b). It is thus unlikely that cytochrome c_L plays any role in the normal oxidation of methylamine, and the diagram below shows the most likely sequence in electron transport (solid arrows indicate likely reactions; dashed lines indicate reactions that occur *in vitro* but which may not be important *in vivo*):

Methanol → MDH → Cytochrome c_L → Cytochrome c_H → Oxidase → Oxygen

Methylamine → MNDH → Amicyanin → 'Azurin'

The proton translocation measured during respiration with methylamine in *Methylobacterium* AM1 is consistent with the proposal that, like 'methanol oxidase', the 'methylamine oxidase' system does not involve the midchain *b*-type cytochromes and that the protonmotive force is established by way of a protonmotive redox arm as shown in Figure 6.2 for methanol (O'Keeffe and Anthony, 1978). This would indicate that the P/O ratio for methylamine oxidation by way of methylamine dehydrogenase is the same as that operating during oxidation of methanol, and is always one or less.

6.4.4 Electron transport and energy transduction from glucose dehydrogenase

The interaction of glucose dehydrogenase (GDH) with the electron transport chain has been studied most extensively in bacteria lacking cytochrome *c* (*Acinetobacter* species and enteric bacteria), but the conclusions drawn are almost certainly valid for all bacteria having this dehydrogenase. In an investigation of this topic in *Acinetobacter calcoaceticus*, no evidence was obtained to support a previous suggestion that the soluble form of the dehydrogenase and the soluble cytochrome *b* associated with it are involved in the oxidation of glucose (Beardmore-Gray and Anthony, 1986; see page 306). It was concluded that glucose, succinate and NADH are all oxidized by way of the same membrane *b*-type cytochromes and cytochrome oxidases (cytochrome *o* and cytochrome *d*). The *b*-type cytochromes form a binary complex with the *o*-type oxidase and no communication occurs between the electron transport chains for the three substrates at the cytochrome level. It was concluded that the ubiquinone pool mediates electron transport from both the glucose and NADH dehydrogenases. In some conditions the quinone pool facilitated communication between the 'glucose oxidase' and 'NADH oxidase' electron transport chains, but in normal conditions these chains were kinetically distinct. This conclusion is illustrated in Figure 6.3. This proposal is supported by the reconstitution of a PQQ-dependent 'glucose oxidase' respiratory chain of *E. coli*. Proteoliposomes, reconstituted from GDH and the *o*-type oxidase and *E. coli* phospholipids containing ubiquinone-8, were able to oxidize glucose in the presence of added PQQ and to establish a protonmotive force (inside negative) (Matsushita *et al.*, 1987). That GDH is able to contribute to a pmf has also been demonstrated in whole cells and vesicles of *E. coli*, *Pseudomonas aeruginosa* and *Acinetobacter* species by measuring the membrane potential during respiration with glucose, and by measuring transport of glutamate, lactose and alanine (van Schie *et al.*, 1985; Pronk *et al.*, 1986).

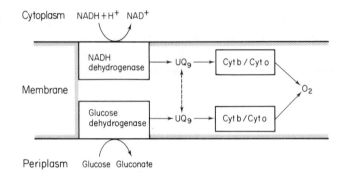

Fig. 6.3   The oxidation of glucose and NADH in *Acinetobacter calcoaceticus* (based on Beardmore-Gray and Anthony, 1986). The components in boxes are firmly membrane-bound. The cytochrome *b*/cytochrome *o* complexes are probably arranged in identical orientations with respect to the membrane, but they differ in their spacial relationship with respect to the two dehydrogenases. These have substrate-binding sites on opposite sides of the membrane but the NADH dehydrogenase may span the membrane. The dotted line indicates that in some conditions the ubiquinone pool (UQ$_9$) may mediate interactions between the two kinetically distinct chains for oxidation of NADH and glucose.

6.5   Why have quinoproteins?

There is no obvious answer to this question. The obvious alternative would be flavoproteins catalysing the same reactions and, as pointed out in a more comprehensive discussion of this question, there are examples of flavoprotein and quinoprotein dehydrogenases performing similar functions (Duine *et al.*, 1986). This is not the case for methanol and methylamine dehydrogenases and this may be significant; both these enzymes react with electron acceptors having a high redox potential. Electron transport chains for these substrates are thus unique in being the only chains in which membrane quinones and midchain *b*-type cytochromes are not involved in the oxidation of an organic compound. This may be because the redox potential of PQQ ($+90$ mV at pH 7.0) can be more readily modified than that of flavins to a suitably high potential for reaction with the cytochromes or blue copper proteins. This still leaves the question of why these enzymes are involved in such high potential 'oxidase systems'. The answer to this is likely to lie in a better understanding of the bioenergetics of growth of these bacteria.

An important feature of quinoprotein dehydrogenases that may be relevant to this question is their location. Most (or all) flavoprotein dehydrogenases are membrane-bound and react on the inner face of the membrane, whereas all quinoproteins are periplasmic or have their active sites on the periplasmic face of the bacterial membrane. This might dictate that processing of the enzymes, involving addition of PQQ, occurs to some extent at least on this face of the membrane and this might be more straightforward with PQQ than with flavins.

In summary, although it appears that the most suitable answer to the question 'Why have quinoproteins' remains, 'Why not?', it is certain that the search for a better answer will continue to provide even more satisfying biochemical contributions to the bioenergetics of bacteria than those that have been described in this chapter.

References

Adachi, O., Tayama, K., Shinagawa, E., Matsushita, K. and Ameyama, M. (1978). *Agric. Biol. Chem.* **42**, 2045–2056.

Adachi, O., Shinagawa, E., Matsushita, K. and Ameyama, M. (1982). *Agric. Biol. Chem.* **46**, 2859–2863.

Ambler, R. P. and Tobari, J. (1985). *Biochem. J.* **232**, 451–457.

Ameyama, M. and Adachi, O. (1982). *Methods Enzymol.* **89**, 491–497.

Ameyama, M., Shinagawa, E., Matsushita, K. and Adachi, O. (1981). *Agric. Biol. Chem.* **45**, 851–861.

Ameyama, M., Hayashi, M., Matsushita, K., Shinagawa, E. and Adachi, O. (1984a). *Agric. Biol. Chem.* **48**, 561–565.

Ameyama, M., Shinagawa, E., Matsushita, K. and Adachi, O. (1984b). *Agric. Biol. Chem.* **48**, 2909–2911.

Ameyama, M., Shinagawa, E., Matsushita, K. and Adachi, O. (1984c). *Agric. Biol. Chem.* **48**, 3099–3107.

Ameyama, M., Shinagawa, E., Matsushita, K. and Adachi, O. (1985a). *Agric. Biol. Chem.* **49**, 1001–1010.

Ameyama, M., Nonobe, M., Hayashi, M., Shinagawa, E., Matsushita, K. and Adachi, O. (1985b). *Agric. Biol. Chem.* **49**, 1227–1231.

Ameyama, M., Shinagawa, E., Matsushita, K. and Adachi, O. (1985c). *Agric. Biol. Chem.* **49**, 699–709.

Ameyama, M., Shinagawa, E., Matsushita, K. and Adachi, O. (1985d). *Agric. Biol. Chem.* **49**, 853–854.

Ameyama, M., Nonobe, M., Shinagawa, E., Matsushita, K., Takimoto, K. and Adachi, O. (1986). *Agric. Biol. Chem.* **50**, 49–57.

Anthony, C. (1982). 'The Biochemistry of Methylotrophs'. Academic Press, London.

Anthony, C. (1986). *Adv. Microbial. Physiol.* **27**, 113–210.

Anthony, C. and Zatman, L. J. (1964). *Biochem. J.* **92**, 614–627.

Anthony, C. and Zatman, L. J. (1967a). *Biochem. J.* **104**, 953–959.

Anthony, C. and Zatman, L. J. (1967b). *Biochem. J.* **104**, 960–969.

Anthony, C. and Jones, C. W. (1987). *In* 'Microbial Growth on C-1 Compounds' (Eds. H. W. van Verseveld and J. A. Duine), pp. 195–202. Martinus Nijhoff, Dordrecht.

Beardmore-Gray, M. and Anthony, C. (1983). *J. Gen. Microbiol.* **129**, 2979–2983.

Beardmore-Gray, M. and Anthony, C. (1986). *J. Gen. Microbiol.* **132**, 1257–1268.

Beardmore-Gray, M., O'Keeffe, D. T. and Anthony, C. (1982). *Biochem. J.* **207**, 161–165.

Beardmore-Gray, M., O'Keeffe, D. T. and Anthony, C. (1983). *J. Gen. Microbiol.* **129**, 923–933.

de Beer, R., Duine, J. A., Frank, J. and Large, P. J. (1980). *Biochim. Biophys. Acta* **622**, 370–374.

Burton, S. M., Byrom, D., Carver, M., Jones, G. D. D. and Jones, C. W. (1983). *FEMS Microbiol. Lett.* **17**, 185–190.

Chandrasekar, R. and Klapper, M. H. (1986). *J. Biol. Chem.* **261**, 3616–3619.

Cornish, A., Nicholls, K. M., Scott, D., Hunter, B. K., Aston, W. J., Higgins, I. J. and Saunders, J. K. M. (1984). *J. Gen. Microbiol.* **130**, 2565–2575.

Cross, A. R. and Anthony, C. (1980a). *Biochem. J.* **192**, 421–427.

Cross, A. R. and Anthony, C. (1980b). *Biochem. J.* **192**, 429–439.

Daniel, R. M. (1970). *Biochim. Biophys. Acta* **216**, 328–341.

Daniel, R. M. and Erickson, S. K. (1969). *Biochim. Biophys. Acta* **180**, 63–67.

Davidson, V. L., Neher, J. W. and Cecchini, G. (1985). *J. Biol. Chem.* **260**, 9642–9447.

Dawes, E. A. (1981). *In* 'Continuous Culture of Cells' (Ed. P. H. Calcott), Vol. 2. CRC Press, Boca Raton, Fla., USA. pp. 1–38.

Dawson, M. J. and Jones, C. W. (1981a). *Biochem. J.* **194**, 915–924.

Dawson, M. J. and Jones, C. W. (1981b). *Eur. J. Biochem.* **118**, 113–118.

Dawson, M. J. and Jones, C. W. (1982). *Arch. Microbiol.* **133**, 55–61.

Dijkstra, M., Frank, J. and Duine, J. A. (1988a). *FEBS Letts* **227**, 198–202.

Dijkstra, M., Frank, J., van Wielink, J. E. and Duine, J. A. (1988b). *Biochem. J.* **251**, 467–474.

Dokter, P., Frank, Jzn., J. and Duine, J. A. (1986). *Biochem. J.* **239**, 163–167.

Duine, J. A. and Frank Jzn., J. (1979). *Biochem. J.* **187**, 221–226.

Duine, J. A., Frank, J. and Westerling, J. (1978). *Biochim. Biophys. Acta* **524**, 277–287.

Duine, J. A., Frank Jzn., J. and van Zeeland, J. K. (1979). *FEBS Letts* **108**, 443–446.

Duine, J. A., Frank, J. and Verweil, P. E. J. (1980). *Eur. J. Biochem.* **108**, 187–192.

Duine, J. A., Frank, J. and Verweil, P. E. J. (1981). *Eur. J. Biochem.* **118**, 395–399.

Duine, J. A., Frank, Jzn., J. and van der Meer, R. (1982). *Arch. Microbiol.* **131**, 27–31.

Duine, J. A., Frank, Jzn., J. and Jongejan, J. A. (1983). *Anal. Biochem.* **133**, 239–243.

Duine, J. A., Frank, J. and Berhout, M. P. J. (1984). *FEBS Letts* **168**, 217–221.

Duine, J. A., Frank, Jzn., J. and Jongejan, J. A. (1986). *FEMS. Microbiol. Revs.* **32**, 165–178.

Duine, J. A., Frank, Jzn., J. and Jongejan, J. A. (1987a). *Adv. in Enzymology.* **59**, 169–212. John Wiley, New York.

Duine, J. A., Frank, Jzn., J. and Dijkstra, M. (1987b). *In* 'Microbial Growth on C-1 Compounds' (Eds. H. W. van Verseveld and J. A. Duine), pp. 105–112. Martinus Nijhoff, Dordrecht.

Eady, R. R. and Large, P. J. (1968). *Biochem. J.* **106**, 245–255.

Elliott, E. J. and Anthony, C. (1988). *J. Gen. Microbiol.* **134**, 369–377.

Faraggi, M., Chandrasekar, R., McWhirter, R. B. and Klapper, M. H. (1986). *Biochem. Biophys. Res. Comm.* **139**, 955–960.

Ford, S., Page, M. D. and Anthony, C. (1985). *J. Gen. Microbiol.* **131**, 2173–2182.

Froud, S. J. and Anthony, C. (1984a). *J. Gen. Microbiol.* **130**, 2201–2212.

Froud, S. J. and Anthony, C. (1984b). *J. Gen. Microbiol.* **130**, 3319–3325.

Fukumori, Y. and Yamanaka, T. (1987). *J. Biochem.* **101**, 441–445.

Fukumori, Y., Nakayama, K. and Yamanaka, T. (1985). *J. Biochem.* **98**, 1719–1722.

Geiger, O. and Gorisch, H. (1986). *Biochemistry* **25**, 6043–6048.

Gray, K. A., Knaff, D. B., Husain, M. and Davidson, V. L. (1986). *FEBS Letts* **207**, 239–242.

Greenwood, J. A. and Jones, C. W. (1986). *J. Gen. Microbiol.* **132**, 1247–1256.

Groen, B. W., Frank, J. Jzn. and Duine, J. A. (1984). *Biochem. J.* **223**, 921–924.

Groen, B. W., van Kleef, M. A. G. and Duine, J. A. (1986). *Biochem. J.* **234**, 611–615.

Hauge, J. G. (1960). *Biochim. Biophys. Acta* **45**, 250–262.

Hauge, J. G. (1961). *Arch. Biochem. Biophys.* **94**, 308–318.

Hauge, J. G. (1964). *J. Biol. Chem.* **239**, 3630–3639.

Hauge, J. G. and Haberg, P. A. (1964). *Biochim. Biophys. Acta* **81**, 251–256.

Hommes, R. W. J., Postma, P. W., Neijssel, O. M., Tempest, D. W., Dokter, P. and Duine, J. A. (1984). *FEMS Microbiol. Letts* **24**, 329–333.

van Houwelingen, T., Canters, G. W., Stobbelaar, G., Duine, J. A., Frank, Jzn., J. and Tsugita, A. (1985). *Eur. J. Biochem.* **153**, 75–80.

Husain, M. and Davidson, V. L. (1985). *J. Biol. Chem.* **260**, 14626–14629.

Husain, M. and Davidson, V. L. (1986a). *Biochemistry*, **25**, 2431–2436.

Husain, M. and Davidson, V. L. (1986b). *J. Biol. Chem.* **261**, 8577–8580.

Husain, M., Davidson, V. L. and Smith, A. J. (1986). *Biochemistry*, **25**, 2431–2436.

Ishii, Y., Hase, T., Fukumori, Y., Matsubara, H. and Tobari, J. (1983). *J. Biochem.* **93**, 107–119.

Itoh, S., Kato, N., Ohshiro, Y. and Agawa, T. (1984). *Tetrahedron Lett.* **25**, 4753–4756.

Itoh, S., Kato, N., Ohshiro, Y. and Agawa, T. (1985). *Chem. Lett.* 135–136.

Jones, C. W., Greenwood, J. A., Burton, S. M., Santos, H. and Turner, D. L. (1987). *J. Gen. Microbiol.* **133**, 1511–1519.

Kawai, F., Yamanaka, H., Ameyama, M., Shinagawa, E., Matsushita, K. and Adachi, O. (1985). *Agric. Biol. Chem.* **49**, 1071–1076.

Kenney, W. C. and McIntire, W. (1983). *Biochemistry* **22**, 3858–3868.

Lawton, S. A. and Anthony, C. (1985a). *Biochem. J.* **228**, 719–726.

Lawton, S. A. and Anthony, C. (1985b). *J. Gen. Microbiol.* **131**, 2165–2171.

Lessie, T. G. and Phibbs, P. V. (1984). *Ann. Rev. Microbiol.* **38**, 359–387.

Lim, L. W., Mathews, F. S., Husain, M. and Davidson, V. L. (1986). *J. Mol. Biol.* **189**, 257–258.

McIntire, W. S. and Stults, J. T. (1986). *Biochem. Biophys. Res. Comm.* **141**, 562–568.

Martinkus, K., Kennelly, P. J., Rea, T. and Timkovitch, R. (1980). *Arch. Biochem. Biophys.* **199**, 465–472.

Matsushita, K., Ohno, Y., Shinagawa, E., Adachi, O. and Ameyama, M. (1980a). *Agric. Biol. Chem.* **44**, 1505–1512.

Matsushita, K., Yamada, M., Shinagawa, E., Adachi, O. and Ameyama, M. (1980b). *J. Biochem.* **88**, 757–764.

Matsushita, K., Yayama, K., Shinagawa, E., Adachi, O. and Ameyama, M. (1981). *FEMS. Microbiol. Letts* **10**, 267–270.

Matsushita, K., Ohno, Y., Shinagawa, E., Adachi, O. and Ameyama, M. (1982). *Agric. Biol. Chem.* **46**, 1007–1011.

Matsushita, K., Shinagawa, E., Inoue, T., Adachi, O. and Ameyama, M. (1986). *FEMS. Microbiol. Letts* **37**, 141–144.

Matsushita, K., Nonobe, M., Shinagawa, E., Adachi, O. and Ameyama, M. (1987). *J. Bacteriol.* **169**, 205–209.

Midgely, M. and Dawes, E. A. (1973). *Biochem. J.* **132**, 141–154.

Neijssel, O. M., Tempest, D. W., Postma, P. W., Duine, J. A. and Frank, Jzn., J. (1983). *FEMS Microbiol. Letts* **20**, 35–39.

Netrusov, A. I. and Anthony, C. (1979). *Biochem. J.* **178**, 353–360.

Nunn, D. N. and Lidstrom, M. E. (1986a). *J. Bacteriol.* **166**, 581–590.

Nunn, D. N. and Lidstrom, M. E. (1986b). *J. Bacteriol.* **166**, 591–597.

O'Keeffe, D. T. and Anthony, C. (1978). *Biochem. J.* **170**, 561–567.

O'Keeffe, D. T. and Anthony, C. (1980a). *Biochem. J.* **192**, 411–419.

O'Keeffe, D. T. and Anthony, C. (1980b). *Biochem. J.* **190**, 481–484.

Ohshiro, Y., Itoh, S., Kurokawa, K., Kato, J., Hirao, T. and Agawa, T. (1983). *Tetrahedron Lett.* **24**, 3465–3468.

Page, M. D. and Anthony, C. (1986). *J. Gen. Microbiol.* **132**, 1553–1563.

Patchett, R. A., Quilter, J. A. and Jones, C. W. (1985). *Arch. Microbiol.* **141**, 95–102.

Pronk, J. T., van Schie, B. J., van Dijken, J. P. and Kuenen, J. G. (1986). *Ant. van Leeuv.* **51**, 560–561.

Salisbury, S. A., Forrest, H. S., Cruse, W. B. T. and Kennard, O. (1979). *Nature* **280**, 843–844.

van Schie, B. J., Hellingwerf, K. J., van Dijken, J. P., Elferink, M. G. L., van Dijl, J. M., Kuenen, J. G. and Konings, W. N. (1985). *J. Bacteriol.* **163**, 493–499.

van Schie, B. J., de Mooy, O. H., Linton, J. D., van Dijken, J. P. and Kuenen, J. G. (1987). *J. Gen. Microbiol.* **133**, 867–875.

Shimao, M., Yamamoto, H., Ninomiya, K., Kato, N., Adachi, O., Ameyama, M. and Sakazawa, C. (1984). *Agric. Biol. Chem.* **48**, 2873–2876.

Shinagawa, E., Matsushita, K., Nonobe, M., Adachi, O., Ameyama, M., Ohshiro, O., Itoh, S. and Kitamura, Y. (1986). *Biochem. Biophys. Res. Comm.* **139**, 1279–1284.

Sone, N., Sekimachi, M., Fukumori, Y. and Yamanaka, T. (1987). *J. Biochem.* **102**, 481–486.

Tobari, J. and Harada, Y. (1981). *Biochem. Biophys. Res. Comm.* **101**, 502–508.

Tobari, J. (1984). *In* 'Microbial Growth on C-1 Compounds' (Eds R. L. Crawford and R. S. Hanson), pp. 106–112. American Society for Microbiology, Washington DC.

van Verseveld, H. W., Krab, K. and Stouthamer, A. H. (1981). *Biochim. Biophys. Acta* **635**, 525–534.

Vellieux, F. M. D., Frank, Jzn., J., Swarte, M. B. A., Groendijk, H., Duine, J. A., Drenth, J. and Hol. W. G. J. (1986). *Eur. J. Biochem.* **154**, 383–386.

de Vries, G. E. (1986). *FEMS Microbiol. Revs* **39**, 235–258.

Westerling, J., Frank, Jzn., J. and Duine, J. A. (1979). *Biochem. Biophys. Res. Commun.* **87**, 719–724.

7 Bacterial photosynthesis

J. Baz Jackson

7.1 Introduction **318**
7.1.1 Why study bacterial photosynthesis? 319
7.1.2 An overview of bacterial photosynthesis in species of
Rhodopseudomonas and *Rhodobacter* 320
7.2 Electrogenic reactions in the photosynthetic reaction centre .. **324**
7.2.1 Isolation and structure of photosynthetic reaction centres .. 325
7.2.2 Primary photochemistry 328
7.2.3 The position and orientation of prosthetic groups within
the reaction centre 330
7.2.4 Electron transfer in and out of the reaction centre 332
7.2.4.1 The donor side of the reaction centre 332
7.2.4.2 The acceptor side of the reaction centre 333
7.2.5 Electrogenic processes in the photosynthetic reaction centre
revealed by electrochromism 334
7.2.5.1 Electrochromism as a probe of membrane potential 336
7.2.5.2 Electrochromic absorbance changes attributable to
transmembrane electric potential generation by the
photosynthetic reaction centre 338
7.2.5.3 Two-step generation of $\Delta\psi$ by photosynthetic
reaction centres 338
7.2.6 Membrane potential generation in reaction centres
reconstituted into artificial membranes..................... 342
7.2.7 Electrogenic components in the photosynthetic reaction
centre revealed by experiments with artificial membranes .. 345
7.2.8 Electrical activity of reaction centres embedded in
monolayers on solid supports 346
7.2.9 Direct recording of the electrical activity of photosynthetic
reaction centres *in vivo* 347
7.2.10 General comments on the electrogenic activity of the
photosynthetic reaction centre 349
7.3 Electrogenic reactions in the cytochrome b/c_1 complex **351**
7.3.1 Structure of the cytochrome b/c_1 complex 351
7.3.2 The mechanism of electron transport through the
cytochrome b/c_1 complex 352
7.3.2.1 Redox poise and electron transfer through the
cytochrome b/c_1 complex 353

7.3.2.2   A modified protonmotive Q-cycle for electron
          transfer through the cytochrome *b/c*₁ complex 354
7.3.3   Electrogenic activity in the cytochrome *b/c*₁ complex 357
7.3.4   Protolytic reactions associated with the cytochrome *b/c*₁
        complex 360
7.4   Charge recombination during ATP synthesis 361
7.4.1   Charge recombination through the ATP synthase as
        revealed by electrochromic absorbance changes 362
7.4.2   Are there interactions between the components of the
        photosynthetic electron transport chain and the ATP synthase
        other than those mediated by the protonmotive force? 364
7.4.3   The mechanism of ATP synthesis 366
**7.5   Concluding remarks: the protonmotive current and bacterial
        growth** 367
References 370

7.1   Introduction

Photosynthesis is the process in living organisms in which light energy is used to drive metabolic rearrangements such as the assimilation of carbon substrates. ATP is produced from ADP in the process known as photophosphorylation in which light-driven electron transport is used to establish a protonmotive force which subsequently provides the potential to drive ATP synthesis. When photosynthesis involves the fixation of carbon dioxide, for example in plants and some bacteria, then a strong reductant (NADH or NADPH) is required in addition to ATP. In higher plants, algae and the cyanobacteria water is the electron donor for reduction of NADP to NADPH, oxygen being released from the water as molecular oxygen (hence oxygenic photosynthesis). The energy conserved in the light reactions is used to overcome the unfavourable difference in redox potential between the O_2/H_2O couple and the NADP/NADPH couple. Phototrophic bacteria other than the cyanobacteria cannot use water as electron donor and so must carry out anoxygenic photosynthesis. During reduction of carbon dioxide by these bacteria, inorganic electron donors such as reduced sulphur compounds provide the necessary reducing power for reduction of NADP to NADPH. Most of these bacteria can also grow photoheterotrophically using reduced carbon compounds such as sugars and organic acids as a convenient source of cellular carbon. In these circumstances a supply of inorganic exogenous reductant is unnecessary. Often the phototrophic bacteria simultaneously assimilate reduced organic compounds and fix carbon dioxide, in which case the organic substrate provides the reducing power for fixation of the CO_2.

It should be noted that it was once thought that the purple and green sulphur bacteria are obligate photolithotrophs, using only inorganic reductants, and that, by contrast, the purple non-sulphur bacteria are usually photoheterotrophs, using only organic reductants (see page 3). However, as more becomes known of the physiology of these bacteria, the demarcations with respect to exogenous reductant are becoming blurred and the physiological distinctions between the various families of phototropic bacteria less well defined.

An outline of the photosynthetic systems in the photosynthetic bacteria is given in Chapter 1 (pages 9–11, 53–56, 59–62 and Figs. 1.4, 1.9, 1.22, 1.25 and 1.31), and a comprehensive account of all aspects of photosynthesis in plants and bacteria is given by Govindjee (1982).

7.1.1 Why study bacterial photosynthesis?

The bioenergetics of the purple non-sulphur photosynthetic (or phototrophic) bacteria are understood in detail. The reasons for this are to be found in a coincidence of properties which make the organisms amenable to study and in the supposition that the photosynthetic bacteria form a good model system for other organisms and organelles. It will be evident from other chapters in this book that the basic principles governing the bioenergetics of electron transport, ATP synthesis and solute translocation are similar in diverse species of bacteria. The similarity is believed to extend also to the mitochondrial and thylakoid membranes of higher animals and plants. This chapter deals with charge separation reactions in the photosynthetic bacteria: they can be uniquely studied by special techniques not generally applicable to other bacteria.

It will become apparent that one of the major experimental advantages of the phototrophic bacteria is that, using controlled illumination regimes, their energy conversion reactions can be monitored on a rapid time-scale. Electron transport processes, for example, can be observed in the range 10^{-12} seconds to one second. Charge separations within and across membranes (electrogenic reactions) are thought to be at the heart of energy conservation during electron transport-ATP synthesis and these too can be recorded with very high time resolution in the photosynthetic bacteria. This has led to a unique kinetic approach to the problem of energy conservation. A description of these fast charge separation and recombination reactions and attempts to correlate them with component processes in electron transport and ATP synthesis will be the objective of this chapter. Details of the electron transport reactions will only be summarized as far as necessary for an understanding of stable charge separation. It is hoped to persuade the reader that the findings are of relevance to the bioenergetics of other organisms.

7.1.2 An overview of bacterial photosynthesis in species of *Rhodopseudomonas* and *Rhodobacter*

The wealth of information available on bacterial photosynthesis is confined to only a handful of species. In kinetic analysis, *Rhodobacter sphaeroides* (until recently known as *Rhodopseudomonas sphaeroides*) and *Rb. capsulatus* (formerly *Rps. capsulata*) reign supreme. These two species are the prime subjects of the chapter. Some recent, outstanding structural work on the photosynthetic reaction centres of *Rps. viridis* has elevated the status of this organism (Deisenhofer *et al.*, 1984, 1985a,b) and honourable mention should be made of *Rhodospirillum rubrum*, the favourite of the 1960s which still makes a valuable contribution. There has been renewed interest in other related species, some of which have been only recently discovered, in the expectation that differences from the 'standard' organisms will reveal new insights into general principles of photosynthesis.

The taxonomic relationships of the photosynthetic bacteria are complex (see Imhoff *et al.*, 1984) and are not of immediate relevance. However, a few words about the physiology of the bacteria are in order (Vignais *et al.*, 1985). An important feature is their astounding versatility; indeed *Rb. capsulatus* has been described as the most versatile of all procaryotes (and therefore of all living organisms) (Madigan and Gest, 1979). It can grow:

(a) photoautotrophically (fixing CO_2 in the light);
(b) photoheterotrophically (utilizing reduced organic compounds in the light)
(c) chemoautotrophically (in the dark, fixing CO_2 using free enthalpy released in the reaction between H_2 and O_2);
(d) chemoheterotrophically (in the dark using organic compounds both as a source of carbon intermediates and as a source of reductant for respiratory electron flow to oxygen or other acceptors);
(e) fermentatively (anaerobically in the dark with appropriate substrates; this is a complex issue (see McEwan *et al.*, 1985; Schultz and Weaver, 1982)).

In view of this versatility, the complexity of the electron transport pathways in photosynthetic bacteria is not surprising. Figure 7.1 summarizes these pathways in *Rb. capsulatus*. It can be seen that the electron transport system shares properties in common with other non-photosynthetic bacteria, for example, the extensive branching at the level of the ubiquinone pool. Almost invariably for studies on the mechanism of photosynthetic energy transduction, the organisms are grown anaerobically under photoheterotrophic conditions. Even restricted to this mode of growth the organisms express several of the electron transport branches not directly concerned with photosynthesis, for example, the cytochrome c oxidase, alternative oxidase and nitrous oxide reductase (Kaufman *et al.*, 1982; McEwan *et al.*, 1985a,b). However, it is not difficult for the experimenter to eliminate the adventitious pathways and be left with the simplest of electron transport chains: the cyclic, photosynthetic electron transport shown in the shaded area in Figure 7.1.

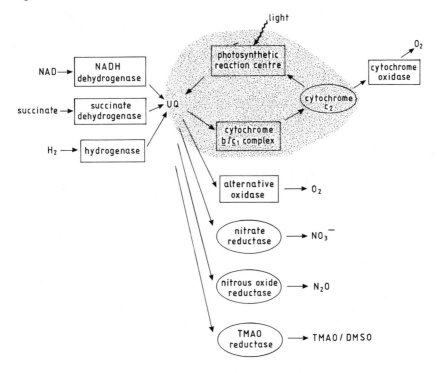

Fig. 7.1 The electron transport pathways of *Rb. capsulatus*. Square boxes depict membrane-bound complexes; ellipses depict soluble proteins located in the periplasm. Not all components are induced under all culture conditions. The shaded area represents the cyclic, photosynthetic electron transport pathway. Note that the cytochrome bc_1 complex in the membrane, and the periplasmic cytochrome c_2 are involved both in photosynthesis in the light (anaerobic) and in aerobic respiration in the dark. UQ = ubiquinone; TMAO = trimethylamine N-oxide; DMSO = dimethylsulphoxide. The anaerobic electron pathways are reviewed by McEwan *et al.* (1987), and hydrogen metabolism by Vignais *et al.* (1985).

The cyclic photosynthetic electron transport pathway is discussed on pages 9–11 and 322–324. The outline in Figure 7.1 serves to set the perspective. *Rb. capsulata*, *Rb. sphaeroides*, and probably *Rhs. rubrum*, have very similar photosynthetic electron transport pathways; *Rps. viridis* has many features in common with them, but at least some differences in the nature of its *c*-type cytochromes. Discussion is confined mainly to the two species of *Rhodobacter*.

The cyclic, photosynthetic electron transport chain is comprised of two large, membrane-bound protein complexes, the photosynthetic reaction centre complex and the cytochrome b/c_1 complex, together with ubiquinone and cytochrome c_2 (Fig. 7.2). The reaction centre catalyses a light-driven cytochrome c_2/ubiquinone oxidoreductase activity and the cytochrome b/c_1 complex catalyses ubiquinol/cytochrome c_2 oxidoreductase activity: hence the term 'cycle'. Photons are absorbed and channelled to the photosynthetic reaction centres by membrane proteins known as the 'light harvesting' or 'antenna' complexes (Fig. 7.2). Most of the organism's bacteriochlorophyll and carotenoid is, in fact, located in these complexes (see Chapter 1, pages 53–56). For reviews of the structure of the light-harvesting complexes and the mechanism of light absorption and energy-transmission to the reaction centre, see Drews (1985), Cogdell (1986), Thornber (1986) and Zuber (1986a,b).

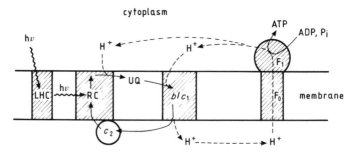

Fig. 7.2   Bacterial photophosphorylation. This diagram is a generalized picture based on *Rb. sphaeroides*; arrangements of components may differ slightly in other bacteria such as *Rps. viridis* (see, e.g. Table 7.1, Figs 7.4, 7.5). The heavy lines define the membrane and protein components; wavy lines depict the delivery of quanta of light to the reaction centre *via* the carotenoids and antenna chlorophylls of the light-harvesting complexes; thin lines show the reactions of electron transport and ATP synthesis and dashed lines show the path of protons as suggested by the chemiosmotic hypothesis [NB: this merely indicates 'pathways'; the stoichiometry of the process is discussed later in the text (and see Fig. 7.20)]. LHC, light-harvesting complex; RC, reaction centre; b/c_1, cytochrome bc_1, complex; F_1F_0, ATP synthase; c_2, cytochrome c_2.

Photons arriving at the photosynthetic reaction centre initiate the primary photochemical reaction which results in the generation of a strong reductant and a strong oxidant. It is their recombination through the cyclic electron transport chain that provides the Gibbs Free Energy to be used ultimately in the synthesis of ATP and other energy-requiring reactions. The scheme in Figure 7.2 summarizes the process of energy transduction in bacterial photophosphorylation within the framework of the chemiosmotic hypothesis (see also Chapters 1 and 9). The cyclic electron transport system is depicted as a proton translocator which when driven by light leads to the development of a protonmotive force (Δp) across the membrane. Protons are driven back across the membrane by Δp through an F_0F_1-ATP synthase with concomitant synthesis of ATP. The photosynthetic bacteria provide an ideal system with which to analyse component reactions in this model and to seek out its deficiencies.

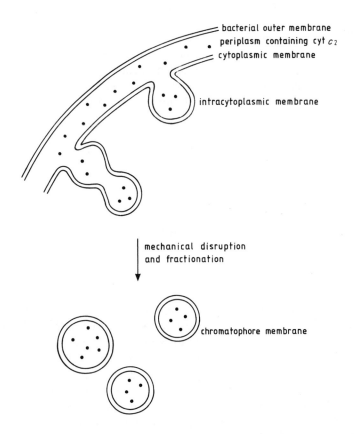

bacterial outer membrane
periplasm containing cyt c_2
cytoplasmic membrane

intracytoplasmic membrane

mechanical disruption
and fractionation

chromatophore membrane

Fig. 7.3 Membrane structure in some species of the photosynthetic bacteria and the formation of chromatophores. Membrane proteins that faced into the cytoplasm in the whole organism will face outwards from the chromatophores which will contain periplasmic proteins. In this context it should be noted that the periplasm includes the 'contents' of the invaginated membrane system. For more information on membrane structure see Remsen (1978) and Sprague and Varga (1986).

The membrane which anchors the components of photosynthetic electron transport and which provides the dielectric material across which charges are separated during energy conservation in the bacteria is the 'intracytoplasmic membrane'. This membrane takes different forms in different species and even within a single species can be influenced strongly by environmental conditions (for reviews see Remsen, 1978; Sprague and Varga, 1986). The intracytoplasmic membrane of *Rps. viridis* has a lamellar appearance. In *Rb. capsulatus*, *Rb. sphaeroides* and *Rhs. rubrum* grown under phototrophic conditions the intracytoplasmic membrane appears to arise as invaginations of the cytoplasmic membrane and takes the form of vesiculated tubules extending into the cytoplasm; the consensus view is that the whole intracytoplasmic membrane system is contiguous with the cytoplasmic membrane (see Fig. 7.3). Upon mechanical disruption of these organisms by sonication or French Press-treatment the intracytoplasmic membrane fragments and then re-seals to form vesicles, called chromatophores, which have a predominantly inverted orientation with respect to the intact cell membrane (see Fig. 7.3; Prince *et al.*, 1975: Lommen and Takemoto, 1978). The chromatophores, which can be easily separated by differential centrifugation, are fully capable of photosynthetic electron transport and ATP synthesis and are a useful source of material for experimental investigation.

7.2   Electrogenic reactions in the photosynthetic reaction centre

The progress made during the last ten years towards understanding the structure and mechanism of photosynthetic reaction centres in the purple bacteria must rank as one of the most impressive achievements in bioenergetics. It will provide the framework for the understanding of electron transport in other bacteria and of photochemistry in higher plants. Evidence that the photosynthetic reaction centre is electrogenic is to be summarized in this section. It will be concluded that the reaction centre is an electric potential generator which is active in the formation of a protonmotive force across the bacterial membrane, as summarized in Figure 7.2. Details of reaction centre structure and photochemistry are not given here, but several excellent reviews offer further information (Dutton, 1986; Parson and Ke, 1982; Okamura *et al.*, 1982).

Early work had established that the crucial reaction in photosynthesis, the primary photochemical event, was the oxidation of a bacteriochlorophyll species called *P* (Clayton, 1980). The reaction proceeds, after light activation, from the excited singlet state, $P*$, an unknown carrier, X, being the presumed primary electron acceptor:

$$PX \xrightarrow{h\nu} P*X \to P^+X^-$$

Optical absorbance changes mainly attributable to the formation of P^+ established that this reaction was complete in less than 10^{-6} s. The immediate objective is to look at this reaction in a little more detail; to establish the chemical nature of P and X; and to outline how the reaction takes place. The results of both structural and kinetic measurements will be drawn together to provide a rather convincing picture of how, in the presence of light, the reaction centre can oxidize cytochrome c and reduce ubiquinone.

7.2.1   Isolation and structure of photosynthetic reaction centres

The new era began in 1967 when R. K. Clayton developed a procedure for the isolation and purification of photosynthetic reaction centres from *Rb. sphaeroides* (Clayton and Wang, 1971). The detergent-solubilized reaction centres were capable of primary photochemistry (the generation of P^+X^-) but were free of other electron transport components and of the dominating light-harvesting pigments which had previously obscured detailed investigations of the reaction in membrane preparations. The reaction centres comprised three subunits which were called H, M. and L (for heavy, medium and light) on the basis of their electrophoretic mobility on denaturing gel electrophoresis (Okamura *et al.*, 1974). It was found that for each mole of reaction centre in *Rb. sphaeroides* there were 4 moles of bacteriochlorophyll *a*, 2 moles of bacteriophaeophytin *a*, 1 mole of non-haem iron and 2 moles of ubiquinone; although up to one mole of quinone can be lost during isolation (Table 7.1). There is a similar reaction centre composition in other organisms, such as *Rb. capsulatus* and in *Rhs. rubrum*.

In *Rps. viridis* the situation is slightly different. Here a *c*-type cytochrome with four covalently linked haem groups is tightly associated with the reaction centre; bacteriochlorophyll *b* and bacteriophaeophytin *b* replace the bacteriochlorophyll *a* and bacteriophaeophytin *a*, respectively; and a menaquinone replaces one of the ubiquinones. The subunit and pigment compositions of these reaction centres are summarized in Table 7.1.

Advances in understanding of the photochemical events through application of modern biophysical techniques (see below) has provided the impetus for a considerable research effort into the protein chemistry of photosynthetic reaction centres. This has culminated in the recent sequence data and X-ray crystallographic analysis of Diesenhofer and Michel and colleagues on reaction centres from *Rps. viridis*. The genes encoding the reaction centres H, L and M polypeptides have been cloned and their primary structure has been determined by DNA sequence analysis (Deisenhofer *et al.*, 1985a,b; Michel *et al.*, 1985).

The production of crystals suitable for X-ray analysis required the development of new techniques, since membrane proteins have been notoriously difficult to crystallize. The breakthrough with *Rps. viridis* reaction centres was to use amphiphilic molecules like heptane-1,2,3 triol during crystallization by salt precipitation. This gave photochemically active crystals that were suitable for X-ray structural analysis to below 3 Å resolution. The results provide the first model of an integral membrane protein at nearly atomic resolution and will prove an important precedent for the study of other membrane proteins, including those from non-photosynthetic sources (Michel, 1982, 1983; Zinth *et al.*, 1983; Diesenhofer *et al.*, 1984, 1985a,b).

Table 7.1 Composition of photosynthetic reaction centres. The structures of these components of the reaction centres are given in Chapter 1. Although more is known about the structure of the reaction centre of *Rps. viridis* (see Figs 7.4 and 7.5), more is known about function from studies with *Rb. sphaeroides* (see Figs 7.2, 7.13, 7.19, 7.20).

	Rb. sphaeroides[a,b,c,d] *Rb. capsulatus*[e] *Rhs. rubrum*[a,d,f]	*Rps. viridis*[g,h]
Protein subunits	L, M, H	L, M, H, cyt *c*
Pigments per RC	4 bacteriochlorophyll *a* 2 bacteriophaeophytin *a*	4 bacteriochlorophyll *b* 2 bacteriophaeophytin *b*
Iron per RC	1	1
Quinone per RC	2 ubiquinone	1 ubiquinone 1 menaquinone

References: [a] Okamura *et al.* (1982); [b] Strayley *et al.* (1973); [c] Feher (1971); [d] Feher and Okamura (1978); [e] Prince and Youvan (1987); [f] van der Rest and Gingras (1974); [g] Thornber *et al.* (1980); [h] Prince *et al.* (1976).

The reaction centre is a transmembrane protein (Fig. 7.4). There are similarities in the way in which the L and the M subunits are folded and significant rotational symmetry can be recognized in the LM subunit complex. In both, five long helical segments can be seen to span the membrane and a further membrane spanning helix is provided by the H-subunit. A highly hydrophobic cylindrical surface provided by the helical segments probably identifies the region of the protein which interacts with the membrane lipids. On either side of the membrane the L and M subunits form two flat surfaces, where short helices provide intimate contact with the globular structures of the H subunit on one side (cytoplasmic) and the cytochrome *c* subunit on the other (periplasmic). The largest dimension of the protein, between the outer edges of the peripheral globular structures is approximately 130 Å. Figure 7.4 shows a very low-resolution, summary representation of the reaction centre protein in the membrane.

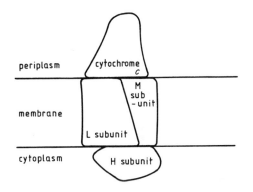

Fig. 7.4   The subunit structure of the reaction centre of *Rhodopseudomonas viridis*. This is adapted from the stereo-ribbon representations in Deisenhofer *et al.* (1984, 1985a). The bound cytochrome *c* on the periplasmic face is probably equivalent to the periplasmic cytochrome c_2 in Fig. 7.2. For details of the position of prosthetic groups, see Fig. 7.5.

Apart from the fact that *Rb. capsulatus* and *Rb. sphaeroides* lack the tightly bound cytochrome *c* subunit, their reaction centre structure is likely to be very similar to that of *Rps. viridis*. Thus the known sequences of the L and M subunits show strong (50–60%) homologies and the bacteriochlorophyll and non-haem iron binding ligands are conserved (Williams *et al.*, 1984; Youvan *et al.*, 1984; Deisenhofer *et al.*, 1985a; Michel 1985). Crystals of reaction centres from *Rb. sphaeroides*, compared with those from *Rps. viridis* by a method known as 'molecular displacement' confirm the homology at 3.6 Å resolution (Chang *et al.*, 1986).

The location of the prosthetic groups within the reaction centre of *Rps. viridis* has been ascertained by X-ray crystallographic analysis (Fig. 7.5). The bacteriochlorophyll *b*, bacteriophaeophytin *b*, non-haem iron and menaquinone are all bound non-covalently by the M and L subunits. A discussion of the relative position and orientation of these prosthetic groups within the protein will follow the description of the electron transport pathway through the reaction centre (page 330).

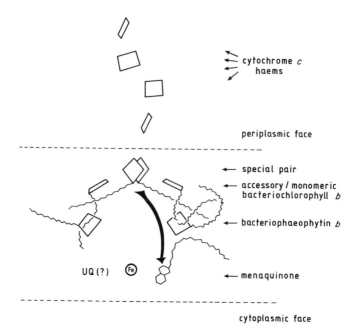

cytochrome *c*
haems

periplasmic face

←— special pair

←— accessory / monomeric
bacteriochlorophyll *b*

←— bacteriophaeophytin *b*

UQ (?)   (Fe)

←— menaquinone

cytoplasmic face

Fig. 7.5   Position and orientation of prosthetic groups in the photosynthetic reaction centre of *Rps. viridis*. Taken from the stereo representation in Deisenhofer *et al.* (1984). The plane of the cytochrome haems on the periplasmic face of the reaction centre, and the bacteriochlorophyll *b* and bacteriophaeophytin *b* are illustrated as flat squares. The bacteriochlorophyll 'special pair' is sometimes depicted as *P* and sometimes as [BChl]$_2$. The phytyl side chains of the bacteriochlorophyll and bacteriophaeophytin, and the isoprenoid side chain of the menaquinone, are also depicted. The menaquinone corresponds to the Q_A in Figs 7.7 and 7.20, and UQ is the ubiquinone designated Q_B in these figures; it comes from the quinone pool to which it is released after reduction. The dashed lines show the approximate position of the membrane (NB: its orientation is upside-down compared with Fig. 7.2). The shaded arrow indicates the probable route of electron transfer. Reaction centres of other photosynthetic bacteria, such as *Rb. sphaeroides* are probably very similar in principle, except that the *c*-type cytochrome is not as firmly bound. It should be noted that the crystal structure of the reaction centre of *Rps. viridis* provides us with the best structural information at present, whereas the system from *Rb. sphaeroides* has provided us with most of our information on electron and proton flow; ideas from the two approaches can probably be extrapolated from one organism to the other but this may not always be the case.

7.2.2   Primary photochemistry

Changes in the near infrared absorption spectrum of chromatophore membranes and isolated reaction centres had established that the formation of P^+ during illumination was due to the oxidation of a distinct species of bacteriochlorophyll. The unpaired electron on P^+ gives rise to a signal in the electron spin resonance spectrum. The linewidth of the signal is narrower than that expected of a bacteriochlorophyll monomer and this gave rise to the view that P^+ is a dimer in which the unpaired electron is shared between a "special pair" of two bacteriochlorophyll molecules (Norris *et al.*, 1971). This was later supported by other optical and magnetic resonance techniques (see Clayton, 1980). By the time the crystal structure was announced, it was generally recognized that the 'special pair' of bacteriophyll molecules loses an electron to form a radical dimeric cation $[BChl]_2^+$, during photochemistry.

Solvent extraction and reconstitution experiments showed rather convincingly that the electron acceptor, X, is a tightly bound molecule of quinone called Q_A. This is ubiquinone in *Rb. sphaeroides* and *Rb. capsulatus* and menaquinone in *Rps. viridis*. Upon receiving an electron from P during photochemistry, the quinone is reduced to an unusually stable anionic semiquinone, Q_A^- which has a characteristic ultraviolet absorption spectrum, and an e.p.r. spectrum which appears to be shifted from its expected position by the adjacent paramagnetic iron atom in the reaction centre (Okamura *et al.*, 1975; Clayton & Straley, 1970).

The identity of components in the reaction centre which we had described as:

$$PX \xrightarrow{h\nu} P^*X \rightarrow P^+X^-$$

can now be pencilled in:

$$[BChl]_2Q_A \xrightarrow{h\nu} [BChl]_2^*Q_A \rightarrow [BChl]_2^+Q_A^-$$

The advent of spectrophotometers capable of following photochemical conversions in the picosecond time-scale enabled several research groups working in the United States to discover intermediates in this reaction (Parson and Ke, 1982; Clayton, 1980; Rockley *et al.*, 1975; Kaufman *et al.*, 1975). A few picoseconds after the activation of reaction centres with a laser pulse, an absorption spectrum was recorded which was different from that of the well-known spectrum of P^+X^- measured in the nanosecond or microsecond time region. The new spectrum, called P^F (F for fast) had features which were characteristic of both bacteriochlorophyll and bacteriophaeophytin (Fajer *et al.*, 1975). In fact, a comparison with pigment spectra in organic solvents suggested that P^F was generated as the result of electron transfer from the bacteriochlorophyll special pair to bacteriophaeophytin, i.e. $[BChl]_2^+BPh^-$. Because P^F is only transient, it was difficult to test this idea with complementary techniques. However, using photosynthetic reaction centres from *Rps. viridis* and *Chromatium vinosum*, which retain bound *c*-type cytochrome during preparation, the intermediate was trapped in its reduced form during long periods of continuous illumination in the presence of sodium dithionite, and its bacteriophaeophytin character was confirmed by e.p.r. (Tiede *et al.*, 1976).

By making measurements at wavelengths specific to P^F and to P^+ as a function of time, it was possible to follow kinetically the electron transfer from BPh$^-$ to (presumably) Q_A (Rockley *et al.*, 1975; Kaufman *et al.*, 1975). It turned out that the spectrum of P^F disappeared at the same rate that P^+ was generated and so, very probably, we have:

$$[BChl]_2 \ BPh \ Q_A \xrightarrow{hv} [BChl]_2* \ BPh \ Q_A$$

$$[BChl]_2* \ BPh \ Q_A \xrightarrow{4 \ ps} [BChl]_2{}^+ \ BPh^-Q_A$$

$$[BChl]_2{}^+ \ BPh^- \ Q_A \xrightarrow{200 \ ps} [BChl]_2{}^+ \ BPh \ Q_A{}^-$$

This scheme can be accepted as a minimal model for the light-driven reaction $PX \rightarrow P^+X^-$, although it accounts for only two of the four bacteriochlorophyll molecules and only one of the bacteriophaeophytins in the reaction centre. It is now appropriate to see how the recent crystal structure has provided a spectacular visual corroboration of the kinetic model.

7.2.3   The position and orientation of prosthetic groups within the reaction centre

A diagrammatic representation of the prosthetic groups in the *Rps. viridis* reaction centre is shown in Figure 7.5. The first point of interest is that the 'special pair' can be recognized: two of the four bacteriochlorophyll *b* molecules appear to interact very strongly with one another; their pyrrole rings I are stacked, one on top of the other, and the acetyl groups of rings I are in direct contact with the Mg atoms of the other bacteriochlorophyll *b* in the pair (Fig. 7.6).

With each bacteriochlorophyll *b* of the special pair, another monomeric bacteriochlorophyll *b* is associated, and each of these is, in turn, in contact with a bacteriophaeophytin *b* (Fig. 7.5). The organization is such that there is an approximate twofold rotational symmetry relating one bacteriochlorophyll of the special pair, one of the monomeric bacteriochlorophylls and one bacteriophaeophytin. The non-haem iron atom is very close to this axis of symmetry. The menaquinone of the reaction centre (Q_A) is within van der Waals' contact of one of the bacteriophytophytins and breaks the twofold symmetry as shown in Figure 7.5. The probable pathway for electron transport from the special pair to Q_A via one of the bacteriophaeophytin molecules is shown in Figure 7.5. In principle, there is sufficient interaction between neighbouring pigments to facilitate electron transport. The symmetrical pathway through the other bacteriophaeophytin appears to be defunct. The reason for this organization is not yet clear. Another feature of this arrangement of pigments is that each of the monomeric bacteriochlorophyll molecules is situated intriguingly close to the juncture between the special pair and the bacteriophaeophytin. There is one school of thought for the involvement, on the picosecond time-scale, of one of the bacteriochlorophyll molecules as an intermediate electron acceptor between the special pair and the bacteriophaeophytin (Shuvalov and Duysens, 1986). Others have disputed this role as a true intermediate and suggest that the role of the monomeric bacteriochlorophyll may be to lower the overall energy requirements for electron transport by interacting with the special pair (Wasielewski and Tiede, 1986).

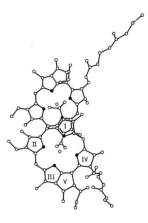

Fig. 7.6   The bacteriochlorophyll *b* special pair. Before its nature was known, this special pair was known as *P* (see Fig. 7.8); it is usually referred to as [BChl]$_2$. Ring numbers are indicated on one bacteriochlorophyll *b*. Note the close overlap of ring I of the two tetrapyrroles of the bacteriochlorophylls. For convenience, the phytyl chains are truncated in this diagram. Taken from Deisenhofer *et al.* (1984). See Fig. 1.22, page 54 for the structure of bacteriochlorophylls.

There is no doubt that there will be important new developments in the study of reaction centre photochemistry in the late 1980s. The crystal structure has opened up a new dimension for analysis and the role played by the amino acid residues in the protein has now become an accessible problem. Techniques for site-directed mutagenesis of reaction centres of *Rb. capsulatus* are already well advanced and could provide new insights into the mechanism of photochemistry (Youvan *et al.*, 1985).

7.2.4   Electron transport in and out of the reaction centre

7.2.4.1   The donor side of the reaction centre

The photosynthetic reaction centre of *Rb. sphaeroides* and *Rb. capsulatus* is a cytochrome c_2/ubiquinone oxidoreductase (see Figs 7.1 and 7.2). Cytochrome c_2 is a water-soluble peripheral membrane protein which shares strong homologies with mammalian cytochrome c (Dickerson *et al.*, 1976). Experiments with isolated reaction centres show convincingly that reduced cytochrome c_2 is the electron donor to $[BChl]_2^+$ generated during illumination (Overfield *et al.*, 1979). Transiently bound cytochrome c_2 appears to transfer electrons to the reaction centre with a $t_{\frac{1}{2}} \simeq 2\,\mu s$ but otherwise in reconstituted reaction centre membrane vesicles the process is slower, being limited by cytochrome diffusion or by the binding kinetics (Overfield and Wraight, 1980a,b). In intact cells and chromatophores of *Rb. sphaeroides* there have been several attempts to characterize kinetically the electron transfer between cytochrome c_2 and the reaction centre (Dutton *et al.*, 1975; Dutton and Prince, 1978; Bower *et al.*, 1979a,b; Overfield *et al.*, 1979; Crofts and Wraight, 1983). It is hoped that this will lead to useful information about the electron transport reaction itself from transiently bound cytochrome c_2 to the reaction centre and about the supposed diffusion of cytochrome c_2 from the cytochrome b/c_1 complex to the reaction centre at the membrane interface. It will become apparent (page 339) that the former process is particularly relevant to the electrogenic capabilities of the reaction centre. It should be recognized, however, that absorbance changes attributed to cytochrome c_2 also include a contribution from membrane-bound cytochrome c_1 in the b/c_1 complex Crofts and Wraight (1983). The re-reduction of $[BChl]_2^+$ by cytochrome $(c_2 + c_1)$ is biphasic: there is a rapid phase with a half-time between $3\,\mu s$ and $30\,\mu s$, followed by a slower phase with a half-time of several hundred microseconds (Dutton *et al.*, 1975; Overfield *et al.*, 1979). It is very likely that the fast phase is contributed by cytochrome c_2 and that the slower phase is due in part to cytochrome c_1 oxidation, although this important point needs clarification.

The emerging dogma for *Rhodobacter* sp., that photooxidized $[BChl]_2^+$ is reduced by cytochrome c_2 which can shuttle along the membrane interface between reaction centre and b/c_1 complex, has recently been shaken. It was found that genetically engineered mutants of *Rb. capsulatus* deficient in cytochrome c_2 were capable of growth under phototrophic conditions. In these mutants, cytochrome c_1 in the b/c_1 complex (see page 179) appeared to be the immediate electron donor to P^+ (Daldel *et al.*, 1986; Prince *et al.*, 1986). However, that the position is still not yet clear is demonstrated by other work on cytochrome c_2 deficient mutants of *Rb. sphaeroides* which are not photosynthetically competent (Donohue *et al.*, 1987).

It will be evident from foregoing discussions on reaction centre structure that there is a slightly different interaction between cytochrome c and $[BChl]_2^+$ in organisms like *Rps. viridis*. Here the cytochrome c subunit, containing 4 haems, is tightly associated with the reaction centre subunit. The haem closest to the special pair, recently described as cytochrome c_{559}, can reduce the photooxidized $[BChl]_2^+$ in about 0.3 µs. This haem is then re-reduced by the next haem in line, cytochrome c_{556}, in about 2.5 µs (Drachev *et al.*, 1986).

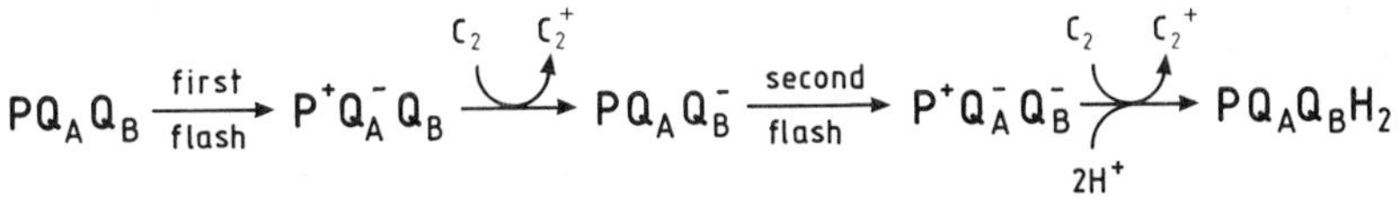

$$PQ_AQ_B \xrightarrow[\text{flash}]{\text{first}} P^+Q_A^-Q_B \xrightarrow{\overset{c_2 \quad c_2^+}{\curvearrowright}} PQ_AQ_B^- \xrightarrow[\text{flash}]{\text{second}} P^+Q_A^-Q_B^- \xrightarrow[2H^+]{\overset{c_2 \quad c_2^+}{\curvearrowright}} PQ_AQ_BH_2$$

Fig. 7.7   The two-electron gate of photosynthetic reaction centres. P represents the special pair, $[BChl]_2$; c_2 represents reduced cytochrome c_2 (c_2^+ is the oxidized form). The cytochrome c_2 binds weakly to the reaction centres presumably at some kind of 'docking site'. This figure is summarized from the results described in Bowyer *et al.* (1979), Wraight (1977) and Vermeglio (1977).

7.2.4.2   The acceptor side of the reaction centre

The electron acceptor side of the photosynthetic reaction centre has been described as a 'two-electron gate': electrons are fired through the reaction centre, one at a time, but emerge in two electron equivalents as fully reduced ubiquinol (Crofts and Wraight, 1983). The mechanism of this process has been elucidated with trains of single-turnover flashes, each of a few microseconds (or less) in duration, which are short enough to drive only single electrons through the reaction centre (Fig. 7.7). Q_A, reduced to the semiquinone during the first flash (see above, pages 339 and 344), transfers its electron to the second of the two bound quinones, Q_B, in the reaction centre in less than 50 µs in *Rb. capsulatus* chromatophores (Bowyer *et al.*, 1979). Provided the 'back reaction' (the return of the electron to $[BChl]_2^+$) is prevented by rapidly reducing $[BChl]_2^+$ with electrons from cytochrome c_2, semiquinone Q_B^- is stable for many minutes (Wraight, 1977; Vermeglio, 1977). A second flash drives another turnover of the primary photochemistry and therefore results in the transfer of a second electron to Q_B. Provided the proton-binding requirement ($Q + 2e^- + 2H^+ \rightarrow QH_2$) is satisfied (see below) this results in the formation of fully reduced and protonated quinol (UQH_2) at the Q_B site. This species dissociates from the reaction centre and enters the ubiquinone 'pool'. The Q_B site then re-binds another molecule of ubiquinone and becomes available for further electron transport. Because the semiquinone has a characteristic absorbance spectrum, 'oscillations' between Q and Q^- at the Q_B site in reaction centres on alternate flashes can be observed: the semiquinone is generated after odd numbers of flashes and disappears after even numbers of flashes. Proton binding during these processes can be followed in unbuffered suspensions of reaction centres or chromatophores by monitoring the absorbance changes of added pH indicators. At low pH in reaction centres, the proton-binding reaction ($2H^+$ per reaction centre) only takes place after even numbers of flashes as shown by the model in Figure 7.7 (Wraight, 1979; Barouch and Clayton, 1977). At higher pH values the oscillations in proton binding by reaction centres become less distinct, probably indicating the importance of acid/base groups in the reaction centre protein in mediating proton-binding to Q_B (Wraight, 1979).

Because Q_B is lost during the preparation and crystallization of *Rps. viridis* reaction centres, its position has not yet been directly observed. Analysis of crystals soaked in inhibitors of Q_B binding suggest that it is located approximately symmetrically with Q_A about the iron atom within the LM subunit diad. The recent work with crystals of *Rb. sphaeroides* reaction centres which do not lose their Q_B confirms this location. This would suggest a possible role for the iron atom in electron transfer between Q_A and Q_B. However, this has not been substantiated in experiments with iron-extracted reaction centres and the function of the metal remains an enigma (Debus *et al.*, 1985).

7.2.5   Electrogenic processes in the photosynthetic reaction centre revealed by electrochromism

A summary of events catalysed by the photosynthetic reaction centre is shown in Figure 7.8; and the energetics of electron transport through the bacterial reaction centre are summarized in Figure 7.9, in which the pigments and other redox centres are arranged according to their redox potentials. It is useful at this stage to anticipate the next sections of this chapter and to reflect on what has been achieved during electron transport through the reaction centre:

(a) it has resulted in the generation of ubiquinol and oxidized cytochrome c_2 which serve as electron donor and acceptor respectively for the cytochrome b/c_1 complex (see page 322);

(b) in that electron transport has taken place across the cytoplasmic membrane, it has resulted in the generation of an electric potential gradient ($\Delta\psi$) across that membrane.

Because $\Delta\psi$ is the major component of the proton electrochemical gradient that probably provides the driving force for ATP synthesis, the electrogenic activity of the reaction centre is of central significance. The methods used for monitoring this activity are described below.

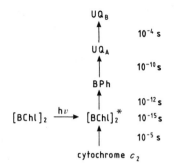

Fig. 7.8   Kinetics of component reactions of the cytochrome c/ubiquinone oxidoreductase in photosynthetic reaction centres. The bacteriochlorophyll special pair is represented as [BChl]$_2$ (this is P in Fig. 7.7). BPh is bacteriophaeophytin.

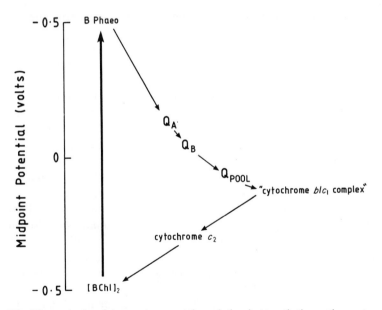

Fig. 7.9   The energetics of electron transport through the photosynthetic reaction centre of *Rb. sphaeroides*. The heavy line indicates the primary photochemical reaction and the fine lines show the subsequent chemical processes. The components are positioned with respect to the left-hand scale according to their approximate midpoint redox potential at pH 7.0. The cytochrome bc_1 complex cannot be displayed as a simple linear progression of redox centres, as discussed on page 353. Similar diagrams to these for other phototrophs are given in Fig. 1.4 (page 11) and Fig. 1.25 (page 61).

7.2.5.1   Electrochromism as a probe of membrane potential

Optical absorption spectra can be distorted by electrical fields. The effect, known as 'electrochromism', is usually small. It can become significant when the transition dipole of a molecule (the difference in dipole moment between the ground state and the excited state) is large and at high electrical field strengths. For example, chlorophylls and carotenoids extracted from the photosynthetic membranes of plants and bacteria and embedded in arachidonate films between the plates of a large electrical capacitor, undergo small shifts to longer wavelengths in their absorption maxima upon application of fields in the region of 10^6 volts cm^{-1} (Schmidt *et al.*, 1970).

It will be appreciated that a potential in the region of 200 mV across a biological membrane about 4 nm thick would give rise to an electrical field strength of approximately 5×10^5 volts cm^{-1} within the membrane. Significant electrochromic effects on the membrane pigments are therefore to be expected. Depending on their local environment in the light-harvesting or reaction centre protein complexes, individual chlorophyll or carotenoid molecules may be subject to electrical fields arising from many sources: local electrical fields within membranes and proteins, as well as transmembrane fields could influence the absorption spectrum. Consequently, the combination of local electrical field changes and transmembrane electrical field changes during photosynthetic activity might be expected to produce complicated spectral effects and such complex changes are indeed observed (Cogdell *et al.*, 1977). Fortunately, in many photosynthetic membranes spectral changes can be identified which originate only from transmembrane electric fields (Wraight *et al.*, 1978; Junge and Jackson, 1982). In fact, in *Rb. sphaeroides* and *Rb. capsulatas* the membrane potential-indicating electrochromic change is dominant and, moreover, the electrochromic effect linear with the applied membrane potential (Jackson and Crofts, 1969). It arises from a subpopulation of carotenoids which are located in light-harvesting complex II (otherwise known as the B800–850 antenna complex). The optical absorption spectrum of the subpopulation is unusually red-shifted (Symonds *et al.*, 1977; Matsura *et al.*, 1980; Scolnick *et al.*, 1980; Webster *et al.*, 1980). It has been suggested that local electrical fields due to fixed charges adjacent to the carotenoid pigments in LHII raise the sensitivity of the response to transmembrane electric fields into a region of linear dependence (Sewe and Reich, 1977).

The electrochromic response of the carotenoids in membranes of *Rb. sphaeroides*, like that observed *in vitro*, is a shift to longer wavelengths of the absorbance maximum (Fig. 7.10a). However, it is convenient to measure the effect, not as a wavelength shift but as an absorbance change at a single wavelength (or sometimes at a pair of wavelengths)—see Figure 7.10b. The wavelengths can be chosen to maximize the response and minimize interference from other absorbance changes. In this way, the kinetics of the electrochromic absorbance change provide a linear indicator of membrane potential. The feature which gives electrochromism an advantage over other methods of membrane potential determination is its broad frequency response: with appropriate instrumentation it can be monitored on a time-scale of $> 10^2$ to 10^{-6} s (limited by the formation of carotenoid triplet states at $< 10^{-6}$ s) (Junge and Jackson, 1982).

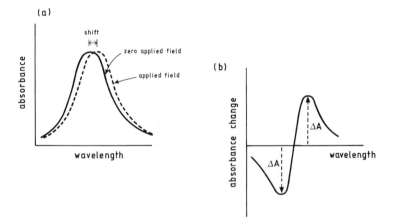

Fig. 7.10 Electrochromism. (*a*) This shows how an electric field can shift an absorption spectrum to longer wavelengths; (*b*) this shows the amplified difference spectrum in the presence and absence of the electric field. The wavelength of maximum absorbance change is indicated.

Electrochromic absorbance changes attributable to transmembrane electrical fields can be observed in appropriate conditions in chromatophore vesicles or intact cells of *Rb. sphaeroides, Rb. capsulatus* or *Rhs. rubrum* during the following: photosynthetic electron transport or respiration in the dark with oxygen (Clark *et al.*, 1983); respiration in the dark with N_2O, NO_3^- or trimethylamine-N-oxide (McEwan *et al.*, 1987); ATP or pyrophosphate hydrolysis in the dark (Baltscheffsky, 1969); and the imposition of diffusion potentials with ion gradients (Jackson and Crofts, 1969). In a general way this provides useful evidence for an important role of the membrane potential during energy conservation in these bacteria (for reviews see Wraight *et al.*, 1978; Junge and Jackson, 1982). Although most of these processes have been studied in depth, we shall concentrate here on the importance of electrochromic absorbance changes to our understanding of the photosynthetic reaction centre as a generator of $\Delta\psi$. How other techniques reinforce and extend the conclusions made from experiments on electrochromism will then be described. On page 357 there is an equivalent discussion of how the cytochrome b/c_1 complex operates as a generator of $\Delta\psi$ and on the manner in which electrochromic absorbance changes yield information on $\Delta\psi$ consumption during ATP synthesis. Because of their dominant electrochromic carotenoid absorbance bands, *Rb. sphaeroides* and *Rb. capsulatus* have been the most commonly used subjects for this work.

Unfortunately, *Rps. viridis* has insignificant electrochromic carotenoid absorbance bands. In view of the new information on its reaction centre structure, an analysis of electrochromic chlorophyll absorbance changes in this organism would be worthwhile but, as yet, little information is available. Although there is overwhelming evidence that appropriately selected carotenoid absorbance changes on a rapid time-scale accurately reflect $\Delta\psi$ (see below), on a slow time-scale there are both quantitative and qualitative discrepancies with measurements of $\Delta\psi$ by redistribution of phosphonium cation (Clark and Jackson, 1981; Crieland *et al.*, 1987). While this might indicate interference by other absorbance changes in the electrochromic determinations, complexities in phosphonium translocation seem more likely to distort that response (e.g. Midgley *et al.*, 1986).

7.2.5.2   Electrochromic absorbance changes attributable to transmembrane electric potential generation by the photosynthetic reaction centre

When chromatophores or intact cells of *Rb. sphaeroides* or *Rb. capsulatus*, treated with an inhibitor to block electron transport through the cytochrome b/c_1 complex, are exposed to short flashes of photosynthetic light, they undergo an electrochromic absorbance change in the carotenoid region of the spectrum (Jackson and Crofts, 1971; see Fig. 7.11). The absorbance change takes place very rapidly (see below) and decays slowly, on a time-scale of several seconds. The decay can be accelerated with ionophores such as valinomycin and protonophorous uncoupling agents; this shows that the electrochromic absorbance change is in response to a *trans*membrane electrical potential gradient. With chromatophore suspensions, only one molecule of ionophore per membrane vesicle is sufficient to produce a significantly accelerated decay, suggesting that the potential gradient to which the carotenoids respond is 'smeared' across the entire membrane surface; i.e. it is a delocalized membrane potential ($\Delta\psi$) (Saphon *et al.*, 1975; Packham *et al.*, 1980). Data such as those described in Figure 7.11 can therefore be viewed as a linear read-out of $\Delta\psi$ across the chromatophore, a 'membrane voltmeter' in the terminology of Junge and Witt (1968) who discovered this phenomenon in chloroplasts.

7.2.5.3   Two-step generation of $\Delta\psi$ by photosynthetic reaction centres

By recording the absorbance changes on a faster time-scale than that shown

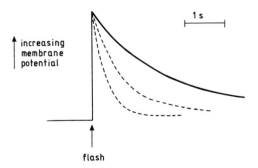

increasing
membrane
potential

1 s

flash

Fig. 7.11 The formation and decay of membrane potential indicated by electrochromic absorbance changes after short light flashes. The solid line shows the rapid build-up and slow decay of membrane potential after an excitation flash. The dashed lines show the effect of adding valinomycin/K^+ or protonophore.

in Figure 7.11 it has been possible to identify and characterize components in this electrogenic reaction, this characterization being aided by the 'redox poise and flash' technique which has been valuable in many aspects of photosynthesis research (Jackson and Dutton, 1973; Dutton *et al.*, 1975; Dutton, 1978). The principle of the technique as applied to the electrochromic absorbance changes which result from electron transport through the reaction centre of *Rb. sphaeroides* is shown in Figure 7.12.

The chromatophore suspension is incubated in the dark with low concentrations of redox 'mediators' which slowly but completely catalyse redox equilibrium among all the electron transport components in the membrane. The equilibrium redox potential (E_h), measured with electrodes, is 'poised' at a required value by the occasional addition of small quantities of oxidant or reductant. At a chosen redox poise, the suspension is exposed to a light flash and subsequent absorbance changes are recorded. It is assumed that on a rapid time-scale (milliseconds or less) the slowly reacting mediators do not donate or accept electrons to or from the chromatophore electron transport components at a significant rate. Two situations are shown in Figure 7.12 for chromatophores whose cytochrome b/c_1 complex is disabled with antimycin. At an E_h poise of $+245\,mV$, cytochrome $c_1 + c_2$ ($E_{m7} = 260\,mV$ and $340\,mV$ respectively) and the special pair, $[BChl]_2$ (E_{m7}) $= 450\,mV$) are almost completely reduced (Fig. 7.12A). A flash of light results in the generation of $[BChl]_2{}^+ Q_A{}^-$. This is followed by electron transport from Q_A to Q_B and by subsequent electron transport from the c-type cytochromes to re-reduce the special pair. However, in chromatophore suspensions held at a redox poise of $+385\,mV$ before the flash, the $[BChl]_2$ is reduced but the c-type cytochromes are largely oxidized (Fig. 7.12B). Consequently, although $[BChl]_2^+ Q_A^-$ is again generated by a flash, this is not followed by electron transfer from cytochrome c to the reaction centre. The electrochromic absorbance changes show differences in amplitude and in kinetics at the two E_h values. When cytochrome c oxidation is prevented (at $E_h = +385$ mV), only about one half of the total electrochromic absorbance change is observed. This is generated very rapidly ($<10^{-6}$ s) and arises from the reaction $[BChl]_2 \rightarrow Q_A$. The other half of the signal becomes evident when cytochrome c is already reduced before the flash is fired. It is slower and, significantly, has similar kinetics to those of cytochrome c oxidation and $[BChl]_2^+$ re-reduction after a flash when measured in parallel experiments (Dutton *et al.*, 1975): it must therefore arise from the reaction cytochrome $c \rightarrow [BChl]_2^+$.

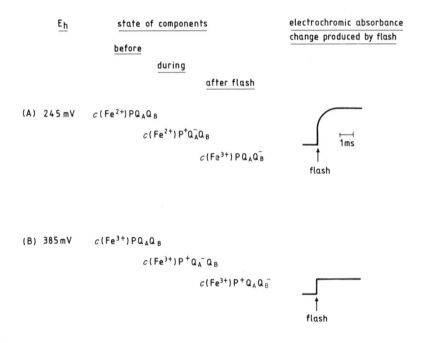

E_h	state of components	electrochromic absorbance change produced by flash

(A) 245 mV $c(Fe^{2+})PQ_AQ_B$
             $c(Fe^{2+})P^+Q_A^-Q_B$
             $c(Fe^{3+})PQ_AQ_B^-$

(B) 385 mV $c(Fe^{3+})PQ_AQ_B$
             $c(Fe^{3+})P^+Q_A^-Q_B$
             $c(Fe^{3+})P^+Q_AQ_B^-$

Fig. 7.12 The formation on a rapid time-scale of membrane potential indicated by electrochromic absorbance changes after short light flashes. On the left, the redox potential poise of the experiments is listed. The electron transfer reactions induced by short flashes are described in the centre, and the electrochromic absorbance changes on a rapid time-scale are shown on the right. The reduced and oxidized forms of cytochrome c_2 are represented by $c(Fe^{2+})$ and $c(Fe^{3+})$ respectively, and P represents the bacteriochlorophyll special pair, $[BChl]_2$. Taken from Dutton *et al.* (1975).

Although these electrochromic absorbance changes can be seen to arise from discrete electron transport reactions, it is unlikely that they are a consequence of local fields set up between the adjacent carriers, because they persist when further electron transport restores the starting redox state. It seems that *both* of these two absorbance changes are in response to transmembrane electric fields. First they have identical spectra indicating that the same pigments are responding; secondly, when allowance is made for a competing back reaction (from $Q_A^- \rightarrow [BChl]_2^+$), it can be shown that they are equally sensitive to ionophores (Packham *et al.*, 1980).

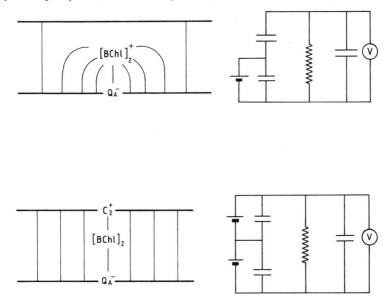

Fig. 7.13   The two-stage generation of membrane potential by photosynthetic reaction centres. On the left the membrane is depicted as a medium of homogenous dielectric constant separating two highly conducting aqueous phases. The bacteriochlorophyll special pair is placed at the centre of the membrane and the cytochrome c and Q_A are assumed to be equipotential with the two aqueous phases. The thin lines represent 'lines of force' arising from charge separations between $[BChl]_2$ and Q_A (top) and cytochrome c and Q_B (bottom). Analogue electrical circuits are shown on the right. Note that only an instantaneous voltage (V) will be generated across the membrane for the partial charge separation from $[BChl]_2$ to Q_A and that this will subsequently leak away across the membrane resistance. The two batteries and their associated capacitors represent the two electrogenic reactions $[BChl]_2 \rightarrow Q_A$ and cytochrome $c_2 \rightarrow [BChl]_2$. A single-membrane resistance and capacitance is shown.

These observations are explained in Figure 7.13. It is supposed that cytochrome c is on one side of the membrane, Q_A on the other and the bacteriochlorophyll special pair somewhere in between. This of course is the arrangement that is seen from the crystal structure of the *Rps. viridis* reaction centre (Fig. 7.5), although strict comparison with *Rb. sphaeroides* is somewhat precluded by the different arrangement of the cytochromes *c*. Because it is perpendicular to the plane of the membrane, the very rapid electron transfer from $[BChl]_2$ to Q_A creates a transmembrane potential, even though it does not completely span the membrane: the positive charge on the $[BChl]_2$ is 'capacitatively coupled' to the external bulk phase on the cytochrome c_2-side of the membrane through polarization of the dielectric medium of the membrane (Zimanyi and Garab 1982; see Fig. 7.13). The generation of the electric field lines in Figure 7.13, and therefore of this transmembrane field, follows the generation of $[BChl]_2{}^+ Q_A{}^-$ with the speed of propagation of electromagnetic radiation. The subsequent transfer from cytochrome c_2 to $[BChl]_2{}^+$ completes the transmembrane movement of the electron: the electric field is fully formed and is almost completely delocalized between the bulk aqueous phases on either side of the membrane (Fig. 7.13). Because the electrochromic absorbance changes generated under these circumstances are $> 90\%$ sensitive to valinomycin/K^+, the responsive carotenoids in the B800–850 antenna complex must be situated sufficiently far from the reaction centre that only the transmembrane field is sensed (Packham *et al.*, 1980).

In principle, the relative contributions to the total electrochromic absorbance change from the components arising from $[BChl]_2 \rightarrow Q_A$ and cytochrome $c \rightarrow [BChl]_A{}^+$ can be used to predict the position of $[BChl]_2$ within the membrane if a homogenous dielectric constant is assumed (Dutton *et al.*, 1975; Jackson and Dutton, 1973). Since the contributions are approximately equal, the special pair would be predicted to lie about half way across the membrane dielectric. This should be done cautiously, however, with existing data because the antimycin used in the original experiments is now recognized to inhibit only partly the electrogenic activity of the cytochrome b/c_1 complex (Glaser and Crofts, 1984).

The minimal conclusion from these experiments, assuming that the electrochromic measurements are being properly interpreted, is that the $\Delta\psi$-generating activity of the photosynthetic reaction centre arises from at least two components of vectorial electron transport across the membrane; between cytochrome $c_2 \rightarrow [BChl]_2$ and between $[BChl]_2 \rightarrow Q_A$.

7.2.6   Membrane potential generation in reaction centres reconstituted into artificial membranes

Scientists at the Moscow State University have been instrumental in the development of procedures for measuring gradients of electric potential across biological membranes. As part of a series of experiments in the mid-1970s, they fused lipid vesicles incorporated with reaction centres from *Rhs.*

rubrum to one side of a planar lipid bilayer (Drachev *et al.*, 1976). Illumination in the presence of artificial electron donors resulted in the generation of a potential difference across the bilayer and was taken as evidence for the electrogenic nature of the reaction centres. More recently this group has used an alternative technique which, although employing a less well-defined membrane structure gives them a much faster response: either chromatophores or reconstituted reaction centre vesicles are added to one side of a collodion film impregnated with asolectin in n-decane (Drachev *et al.*, 1981, 1986a,b). After several hours incubation in the presence of $CaCl_2$, the membranes associate with the collodion-phospholipid film. Illumination of the film with short light flashes then gives rise to potential differences which can be recorded with macroscopic electrodes.

A more direct demonstration of the electrogenic activity of photosynthetic reaction centres has been provided by two independent groups working in the United States and by a group in Switzerland (Apell *et al.*, 1983). They have made electrical measurements across lipid bilayers incorporated with reaction centres from *Rb. sphaeroides*. The bilayers were produced, either by blowing a solution of reaction centres in octanephospholipid into a hole in a teflon septum and allowing bilayer to develop spontaneously (Packham *et al.*, 1980, 1982), or by building the bilayer from two monolayers of reaction centre phospholipid spread at an air/water interface (Schonfield *et al.*, 1979; Gopher *et al.*, 1985). In each case the outcome is similar and is shown diagrammatically in Figure 7.14.

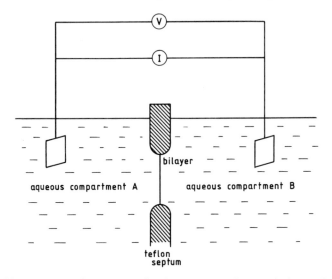

Fig. 7.14 Measurements of currents and voltages across photosynthetic reaction centres incorporated into artificial bilayers.

The artificial membrane, containing reaction centres, separates two aqueous compartments. Electrodes in these compartments can be used to measure either voltages developed across the membrane or current flowing through the membrane. The reaction centres incorporated into the bilayer take on two opposite orientations with equal probability (see Fig. 7.15). Since electrical contributions from the two orientations cancel, illumination of the membrane fails to give rise to any response. Anisotropy has to be imposed on the system. In the experiments of Schonfeld *et al.* (1979) this was achieved by adding reduced mammalian cytochrome *c* to one side of the membrane and ubiquinone (UQ_0) to the other (Fig. 7.15a). Under these circumstances prolonged illumination resulted either in the development of a membrane potential or a membrane current, depending on the conditions of measurement. It was reasoned that these responses arise from light-driven electron transfer from cytochrome *c* to UQ_0 through the photosynthetic reaction centre. Packham *et al.* (1980, 1982), who included ubiquinone (UQ_{10}) within their reconstituted membranes, imposed anisotropy by adding ferricyanide to one side of the membrane and reduced cytochrome *c* to the other side (Fig. 7.15b). The membrane-impermeant ferricyanide oxidized the $[BChl]_2$ in one population of reaction centres and thus rendered them photochemically inactive. Light-induced electrical currents under voltage-clamp could then be observed after single turnover flashes of light or during prolonged periods of illumination.

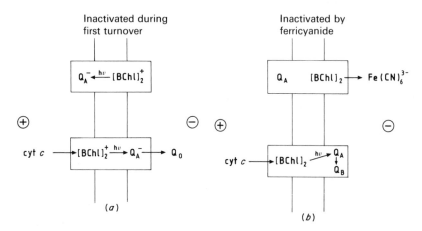

Fig. 7.15 Generation of currents and voltages across artificial bilayers containing photosynthetic reaction centres. In both (*a*) and (*b*) two orientations of the reaction centre in the membrane are depicted. These orientations are adopted with equal probability. In (*a*) the system was made anisotropic by adding cytochrome *c* to one side of the membrane and ubiquinone (UQ_0) to the other. In (*b*) anisotropy in reaction centres containing excess ubiquinone was imposed by adding ferricyanide to one side and cytochrome *c* to the other.

7.2.7   Electrogenic components in the photosynthetic reaction centre revealed by experiments with artificial membranes

Elaboration of the above procedures has led to experiments which show how components in the photosynthetic reaction centre can separately contribute to the development of the overall membrane potential. The conclusion from electrochromic measurements (see above) that the reactions cytochrome $c \rightarrow [BChl]_2^+$ and $[BChl]_2 \rightarrow Q_A$ are both electrogenic was supported by electrical measurements with reaction centre phospholipid bilayer membranes in two ways (Packham *et al.*, 1982).

First, excitation with short flashes, of reaction centres made assymmetric only by the addition of reduced cytochrome c to one side, gave rise to a transient transmembrane electrical current. Under these conditions it was shown that the electrical response from single turnovers of opposed populations of $[BChl]_2 \rightarrow Q_A$ completely cancelled one another during the flash and only cytochrome $c \rightarrow [BChl]_2^+$ was left to make a net contribution. Secondly, experiments were performed with ascorbate added to one side and ferricyanide added to the other to reduce and oxidize respectively the $[BChl]_2$ in the two populations. The electrical response after single short flashes was recorded before and after adding reduced cytochrome c to the ascorbate side of the membrane: the addition of cytochrome c almost doubled the transmembrane photocurrent, a result in close accord with the experiments following the electrochromic response in chromatophore membranes (Fig. 7.12).

The electrogenic nature of the reaction between cytochrome c and the bacteriochlorophyll special pair was also evident in experiments with *Rhs. rubrum* reaction centres incorporated into phospholipid-impregnated collodion film (Drachev *et al.*, 1986a). In the absence of cytochrome c, only a fast phase of charge separation was observed (less than 200 ns, the instrument response time). The provision of mammalian cytochrome c led to an additional slower component, comprising about 25% of the total photoelectric response. Whether the relative differential contributions (25% in *Rhs. rubrum* and 50% in *Rb. sphaeroides*) reflect real structural differences in the two types of reaction centre or whether they arise from the different experimental techniques remains to be seen.

Gopher *et al.* (1985) studied the reverse current in reaction centre bilayers from the back reaction (Q_A^- to $[BChl]_2^+$) at the instant the photosynthetic light was switched off. They discovered that when the natural ubiquinone in the Q_A site was replaced with anthraquinone, then the rate of the charge recombination reaction could be influenced profoundly by the magnitude of an electric potential applied across the membrane with an external voltage source. When the externally applied potential was made negative on the Q_A

side of the membrane the rate of charge recombination was stimulated and *vice versa*. Now, anthraquinone bound within the Q_A site is known to have a significantly lower redox potential than ubiquinone (Gunner *et al.*, 1982); and consequently charge recombination from the former is more likely to proceed by way of the bacteriophaeophytin intermediate. This provides an explanation for the effect of membrane potential on the rate of charge recombination: the back reaction ($Q_A^- \rightarrow$ bacteriophaeophytin) evidently lies partly across the membrane so that its rate can be stimulated or retarded by the existence there of an electric field. The conclusion therefore is that the forward reaction (bacteriophaeophytin $\rightarrow Q_A$) must be electrogenic. The precise dependence of the rate of charge recombination upon the size of the externally applied potential suggests that electron transport from bacteriophaeophytin $\rightarrow Q_A$ contributes at least one-seventh of the potential generated by the charge separation between $[BChl]_2$ and Q_A. Clearly, this is a refinement of the conclusion made from electrochromic experiments that the *overall* reaction $[BChl]_2 \rightarrow Q_A$ (*via* bacteriophaeophytin) is electrogenic. Again it is evident from the structure of *Rps. viridis* reaction centres (Fig. 7.5) that electron transport from bacteriophaeophytin b to Q_A is expected to be electrogenic since it lies in a direction perpendicular to that expected of the membrane plane.

The X-ray structure of reaction centres of *Rps. viridis* shows that the Q_A site and the probable Q_B site lie at the same depth in the membrane (Deisenhofer, 1984; see page 328). This shows that electron transport from Q_A to Q_B is parallel to the plane of the membrane and is therefore unlikely to be electrogenic. Indeed, this view is confirmed by observations by electrochromism in chromatophore membranes and on the photoelectric response of reaction centre bilayers with the inhibitor *o*-phenanthroline, which blocks electron transfer from Q_A to Q_B (Junge and Witt, 1968; Packham *et al.*, 1982). The original interpretation from experiments with the collodion membrane system that $Q_A^- \rightarrow Q_B$ is electrogenic (Drachev *et al.*, 1981) now appears to have been discounted (Drachev *et al.*, 1986a,b). However, there is a recent interesting suggestion which could be consistent with the crystal structure that the movement of protons from the aqueous phase to Q_B^- within the membrane makes a small ($\sim 10\%$) contribution to the net electrogenic reaction (Kaminskaya *et al.*, 1986).

7.2.8   Electrical activity of reaction centres embedded in monolayers on solid supports

An interesting system for demonstrating the electrogenic nature of photosynthetic reaction centres is the Langmuir-Blodgett film. In one such arrangement, monolayers of *Rb. sphaeroides* reaction centre in detergent are spread at an air/water interface and then picked up on an indium tin oxide-coated glass slide which acts as one of the electrodes (Popovic *et al.* 1986). The monolayer is coated with a polymer 'blocking' layer and the second, indium electrode (see Fig. 7.16). Photovoltages accompanying the formation of $[BChl]_2^+Q_A^-$ can be recorded. The size of the photovoltage can be increased by building up the number of monolayers in the film. For a five-coating, 'multilayer' light-induced transients in the region of 300 mV are observed.

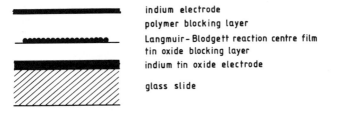

indium electrode
polymer blocking layer
Langmuir-Blodgett reaction centre film
tin oxide blocking layer
indium tin oxide electrode

glass slide

Fig. 7.16 Electrical activity of reaction centres in solid-state devices.

As well as providing an interesting solid-state photochemical device, the advantage of the Langmuir-Blodgett film is that it can withstand much larger externally applied electric fields than can a phospholipid bilayer. Consequently, the photoresponse with ubiquinone-containing reaction centres can be either enhanced or restricted by oppositely polarized electric fields applied across the film: electron transport from $[BChl]_2$ to Q_A is either favoured or retarded by the external electric field (Popovic *et al.*, 1986).

7.2.9 Direct recording of the electrical activity of photosynthetic reaction centres *in situ*

It has not yet been possible to insert microelectrodes into an intact cell or a chromatophore to measure directly the electric potential difference generated across the membrane during photosynthesis. However, a technique has been developed for photosynthetic membranes whereby electric potential gradients can be measured with two electrodes placed in the external aqueous phase (Witt and Zickler, 1973; Fowler and Kok, 1974; Fig. 7.17). The suspension of photosynthetic vesicles (chloroplasts in the original experiments) is excited by *weak* flashes of light. Because of the strong light absorption in the photosynthetic membranes, the side of each vesicle closer to the actinic flash receives more excitations than the side further away. Consequently, the side closest to the actinic light undergoes more photosynthetic charge separations. Within the suspension, an array of oriented dipoles is therefore generated and a photovoltage should be measurable by macroscopic electrodes placed in a direction parallel with the flash. The photovoltage will decay as charges recombine through migration in the plane of the membrane. For vesicles which catalyse outward electrogenic electron transport, the electrode further from the flash should become more negative. This indeed was found to be the case in the original experiments with chloroplasts. However, the method is fraught with technical difficulties. For example, in some chloroplast experiments, electrical responses of the 'wrong' polarity have been observed (Graber and Trissl, 1981; Trissl *et al.*, 1982). In equivalent experiments, photovoltages were measured in intact cells and chromatophores from photosynthetic bacteria and were tentatively assigned to events in the reaction centre on the basis of their kinetics, but interpretation was necessarily cautious (Trissl *et al.*, 1982).

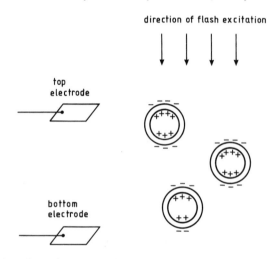

Fig. 7.17   Detection of membrane potentials with macroscopic electrodes during excitation with weak flashes.

There has recently been a considerable improvement in the technique and it is beginning to look very much more promising. The crucial factor seems to have been the construction of a smaller measuring cell with planar electrodes, having a consequently larger electrical capacitance. In this way the 'stray' capacitance of the external measuring circuitry is relatively insignificant and artefact is minimized. The response time of the method is very good and has permitted some valuable subnanosecond measurements on the electrogenic activity of reaction centres in intact cells of *Rps. viridis* (Trissl and Kunze, 1985; Deprez *et al.*, 1986). Following a 30 ps laser flash the development of the voltage was biphasic. The fast phase (<40 ps) was equated with charge separation between [BChl]$_2$ and BPh$^-$ and was possibly limited by the rate at which excitation energy in the antenna system was transmitted to the reaction centre. The slower phase in the generation of the photovoltage (125 ps) was blocked in 'poise-and-pulse' experiments (see page 339) when the ambient redox potential was such that Q$_A$ was already reduced before the laser pulse. Hence it was suggested that the slower phase is a consequence of electron transfer from BPh to Q$_A$. Because the extents of the two phases were similar it seems that [BChl]$_2 \rightarrow$ BPh and [BPh]$^- \rightarrow$ Q$_A$ make equal contributions to the generation of the total photovoltages (Deprez *et al.*, 1986). This is just what is expected on the basis of the reaction centre crystal structure (see Fig. 7.6). A much slower contribution to the photovoltage arising from electron donation to [BChl]$_2{}^+$ from the cyto-chrome *c*-haems (0.3 to 2.5 µs; see Dracher *et al.*, 1986c), analogous to that displayed by *Rb. sphaeroides* reaction centres (see pages 332 and 339), which would also be predicted from the *Rps. viridis* crystal structure, was not reported. It might be *too slow* to be detected by this electrode procedure: the dipoles generated by the flash (Fig. 7.17) can recombine in several micro-seconds through the process of ionic diffusion in the plane of the mem-brane (Trissl, 1985).

7.2.10   General comments on the electrogenic activity of the photosynthetic reaction centre

From experiments with electrochromism, membrane reconstitution and light gradients, using macroscopic electrodes, a consistent story emerges on the electrogenic nature of the photosynthetic reaction centre. This story neatly fits the picture from the X-ray data of the reaction centre crystals. It is clear that as an electron moves stepwise across the membrane from cytochrome *c* to [BChl]$_2$ to BPh to Q$_A$ each step makes an electrical contribution to the overall $\Delta\psi$. Before moving on, two general points remain to be considered.

First, it has been illustrated how the membrane potential is set up by the reaction centre without any proton translocation having taken place. Ele-mentary chemiosmotic schemes are often formally represented as though H$^+$ translocation is responsible for the generation of $\Delta\psi$, but in some circumstances this can be misleading. The proton-binding reactions associ-ated with the reduction of Q$_B$ in isolated reaction centres and in chromato-phores are *slower* than the electrochromic absorbance changes discussed above. These proton-binding reactions and proton-release reactions on the opposite side of the membrane are essential for the *sustained* electrogenic activity of the reaction centre in subsequent turnovers through the electri-cally neutral electron transport reactions catalysed by ubiquinone, but in the main they do not contribute directly to membrane potential development. As mentioned above, there may be a small (10%) contribution to $\Delta\psi$ generation by the reaction centre due to electrogenic H^+ transfer to the Q_B site (Kaminskaya *et al.*, 1986). A summary of the electrogenic processes which contribute to the formation of membrane potential in photosynthetic reaction centres is shown in Figure 7.18.

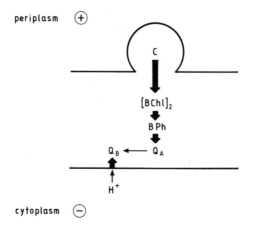

Fig. 7.18   Electrogenic reactions in the photosynthetic reaction centre. Electrogenic processes (either electron or proton translocation) are shown as thick arrows. Other (non-electrogenic) electron transfers are shown as thin arrows.

A second point worth making here is that, since the *Rps. viridis* reaction centre structure has revealed that the redox components are arranged across the membrane (Figs 7.4, 7.5), it is now clear that electron transport through the complex *inevitably* gives rise to a change in membrane potential. The magnitude of this potential for a single turnover of the reaction centre could be calculated if the membrane surface area and capacitance and the density of reaction centres in the membrane were known. In *Rb. sphaeroides* chromatophore membranes of 36 nm diameter, measurements by electron microscopy and spectroscopy show that there are approximately 11 reaction centres per vesicle (Packham *et al.*, 1978). Assuming a membrane capacitance of $1 \, \mu F \, cm^{-2}$ (a typical value for biological membranes), then a single turnover flash is expected to generate a membrane potential of about 45 mV. Although this calculation was originally done in reverse (to estimate membrane capacitance from membrane potentials *measured* by electrochromism; Packham *et al.*, 1978), its value here is that it shows how the crystal structure plus some rather basic measurements lead to the inescapable conclusion that a substantial $\Delta\psi$ is generated during even a single turnover. This is interesting in the context that, in some models of energy transduction, the electron transport complexes are resistively shielded from the aqueous phases on either side of the membrane and are prevented from contributing to the development of membrane potential (e.g. Westerhoff *et al.*, 1984).

7.3 Electrogenic reactions in the cytochrome b/c_1 complex

The photosynthetic reaction centre provides both substrates, the reductant (ubiquinol) and the oxidant (oxidized cytochrome c_2), for the cytochrome b/c_1 complex. The complex bears a strong resemblance to the equivalent protein from mitochondria but in many ways it is easier to study and is often used as a model. The special techniques available with photosynthetic membranes—short flashes to generate single equivalents of oxidant and reductant, redox poise-and-pulse experiments, electrochromism—all favour mechanistic studies of the chromatophore enzyme. A useful hybrid system of bacterial photosynthetic reaction centre and mitochondrial cytochrome b/c_1 complex in detergent has been developed which displays all the characteristics of chromatophore electron transport and serves to highlight the analogy (Packham *et al.*, 1980). Studies on the genetics of *Rb. capsulatus* and *Rb. sphaeroides* are developing rapidly and support from this direction to biophysical investigations of the chromatophore cytochrome b/c_1 complex also seems imminent.

7.3.1 Structure of the cytochrome b/c_1 complex

Unfortunately the three-dimensional analysis of the cytochrome b/c_1 complex from chromatophores and from other sources remains at a rather low resolution. Purified preparations from *Rb. sphaeroides* possess four identifiable redox centres: cytochrome c_1, low potential and high potential cytochrome b (cytochrome b_{566} and b_{561} respectively) and a high potential Rieske-type iron sulphur centre (Gabellini *et al.*, 1982). When run on sodium dodecylsulphate polyacrylamide gel electrophoresis, the complex stains for 4 protein bands (Hauska *et al.*, 1983). However, only three structural genes have been identified in the operon of the cytochrome b/c_1 complex from *Rb. sphaeroides** and *Rb. capsulatus* (Gabellini *et al.*, 1985; Daldal *et al.*, 1986; Davidson *et al.*, 1986). They are the genes for the FeS protein, the cytochrome b protein and the cytochrome c_1 protein. The complete sequences of the three genes are now available (Davidson *et al.*,

---

* On the basis of the DNA sequences it has been suggested by F. Daldal and colleagues (Daldal *et al.*, 1986; Davidson *et al.*, 1986) that the organism described as *Rb. sphaeroides* (Gabellini *et al.*, 1985) was *Rb. capsulatus*.

1986). They show strong homology with those of the cytochrome b/c_1 complex from mitochondria from various sources (although these have a more complex subunit structure) and with the cytochrome b_6/f complex of chloroplasts from higher plants. On this basis, it is anticipated that the low-resolution models which have been proposed for the related complexes might also be relevant to the b/c_1 from *Rb. capsulatus*.

Figure 7.19 shows the strongly hydrophobic cytochrome *b* as a membrane-spanning protein with little structure protruding into the aqueous phases; the haems of the cytochrome *b* may lie transversely across the membrane, bound by histidine residues in adjacent transmembrane helices (Widger *et al.*, 1984). For reasons which will become apparent in the next section, the low potential cytochrome *b* haem is positioned on the periplasmic side of the membrane. The FeS protein (Harnisch *et al.*, 1986) and cytochrome c_1 (Wakabayashi *et al.*, 1981) from mitochondria have large domains carrying the redox centres, which extend into the intercristal space and which are anchored to the membrane by way of a stretch of hydrophobic residues. A similar arrangement is adopted in Figure 7.19 for the *Rb. capsulatus* enzyme: the FeS and cytochrome c_1 haems are positioned on the cytochrome c_2 side of the membrane to account for the recognized electron transport sequence from FeS $\rightarrow c_1 \rightarrow c_2$ (see below).

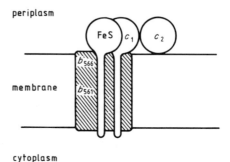

Fig. 7.19 Low-resolution model of the cytochrome bc_1 complex of *Rhodobacter*, including likely approximate positions of the redox centres. The cytochrome *b* subunit is shown as the shaded area.

7.3.2   The mechanism of electron transport through the cytochrome b/c_1 complex

Studies on the mechanism of electron transport through the cytochrome b/c_1 complex of mitochondria and phototrophic bacteria and the b_6/f complex of thylakoids have been the preoccupation of many laboratories in the last decade. The reader is referred to detailed reviews for further information (Cramer and Crofts, 1982; Crofts and Wraight, 1983; Rich, 1984; Dutton, 1986). It now seems unlikely that electron transport through the cytochrome b/c_1 complex proceeds by way of a simple, linear chain of redox centres but, as yet, there is no consensus on the precise mechanism. Here we shall side-step the controversy and discuss a pathway favoured by many working in this field, the 'protonmotive Q-cycle' (Mitchell, 1976). It will serve to illustrate some of the key experimental observations, to highlight the complexity of the problem, and it will provide a useful vehicle for the discussion of the electrogenic reactions in the cytochrome b/c_1 complex. A 'modified Q-cycle' (Crofts *et al.*, 1983) will be outlined because, although agreement in detail has not been reached, in its minimal form it seems to satisfy many of the criteria laid down in a number of laboratories for electron transfer through the complex.

7.3.2.1 Redox poise and electron transfer through the cytochrome b/c_1 complex

The redox poise and pulse technique described on page 339 has been nowhere more valuable than in the study of electron transfer through the cytochrome b/c_1 complex of chromatophores. The redox properties of electron carriers in the complex are known from equilibrium titrations: Table 7.2 shows the midpoint potentials of the essential components at pH 7.0. Consequently, for any redox potential imposed by the experimenter, the redox state of each couple can be assessed and, upon flashing the sample with photosynthetic light, the starting concentrations of individual components can be known. The rate of individual steps in the complex after short light flashes is profoundly influenced by the redox poise and this is important experimentally.

Table 7.2 Midpoint potentials at pH 7.0 of components in the cytochrome b/c_1 complex of *Rhodobacter* sp. (taken from Crofts and Wraight, 1983)

Cytochrome b_{566}	$- 80 \, \text{mV}$
Cytochrome b_{561}	$+ 50$
Cytochrome c_1	$+265$
Rieske FeS centre	$+285$
Ubiquinone pool	$+ 90$
Cytochrome c_2	$+345$

In isolated chromatophores at neutral pH, the rate of electron transport through the cytochrome b/c_1 complex is fastest when the redox potential is held at about $+ 100 \, \text{mV}$ and, in intact bacteria under anaerobic conditions, the redox potential is homeostatically maintained close to this optimal value (Dutton and Jackson, 1972; Cotton and Jackson, 1982). The mechanisms responsible for controlling the entry and exit of reducing equivalents to and from the electron transport chain are not known but in some circumstances a newly discovered family of enzymes in the periplasmic compartment of the cell appears to be involved (McEwan *et al.*, 1985). They are the auxiliary terminal oxido-reductases to nitrate, nitrous oxide and trimethylamine-N-oxide (for review see McEwan *et al.*, 1987).

7.3.2.2   A modified protonmotive Q-cycle for electron transfer through the cytochrome b/c_1 complex

In a dark suspension of chromatophores poised at a redox potential of $+100$ mV, the Rieske FeS centre and cytochrome c_1 are reduced and both the b-cytochromes are predominantly oxidized. Cytochrome c_2 at the membrane interface is reduced. These carriers are shown at their postulated positions in the b/c_1 complex within the membrane in Figure 7.20.

The essence of the protonmotive Q-cycle is that there are two sites within the b/c_1 complex at which quinone can interact (Mitchell, 1976). These are the Q_z site (also called Q_o or the quinol oxidase site) and the Q_c site (also called Q_i or the quinone reductase site). Experimentally, the two sites can be distinguished on the basis of their inhibitor sensitivity: antimycin A is a potent inhibitor of Q_c (O'Keefe and Dutton, 1971; Van den Berg *et al.*, 1979; Robertson *et al.*, 1984), and myxothiazol of Q_z (Meinhardt and Crofts, 1982). In Figure 7.20, the two sites are placed on opposite sides of the membrane. There is no direct evidence for this location but, as will be seen below, it provides a convenient framework within which to describe the electrogenic reactions of the complex.

There is a large molar excess of ubiquinone over other redox components in the chromatophore membrane and it is supposed that this quinone behaves as a 'pool' $(Q + QH_2)$ which can diffuse freely within the membrane and mediate electron transport between the reaction centre and the b/c_1 complex (Takamiya and Dutton, 1979; Meinhardt and Crofts, 1982; Crofts *et al.*, 1983; Snozzi and Crofts, 1984; Venturoli *et al.*, 1986). At a poise of $+100$ mV and a pH of 7.0 the quinone pool will be approximately 30% reduced. There is some discussion as to whether the quinone sites (Q_z and Q_c) on the b/c_1 complex are stoichiometrically *occupied* by quinone/quinol (Prince and Dutton, 1977) or whether the sites merely reflect catalytic locations at which the randomly diffusing quinone/quinol undergo reduction and oxidation (Crofts *et al.*, 1983; Robertson *et al.*, 1986). Although the former view cannot be ruled out, the latter is adopted here: the Q_z and Q_c in Figure 7.20 are shown as sites at which electron transfer with the Q and QH_2 can take place upon collision.

(A)

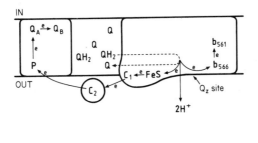

(B)

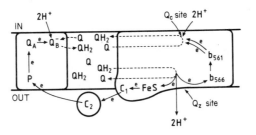

Fig. 7.20 A protonmotive Q-cycle for chromatophores of *Rhodobacter* (taken from Crofts *et al.* 1983). Thick lines delineate the membrane protein complexes (reaction centre on left and cytochrome b/c_1 complex on right). Electron and proton transfers are shown as thin lines. Dashed lines indicate the binding and release of quinone and quinol from their sites on the protein complexes. P represents the bacteriochlorophyll special pair, $[BChl]_2$. During experimental measurement, the first turnover (A) of the bc_1 complex is driven by half of the reaction centres of a preparation, and the second turnover (B) is driven by the other half (see page 356). For convenience, in (A) a reaction centre in which Q_B is initially in the fully oxidized form is shown, whereas in (B) a reaction centre in which Q_B is initially in the semiquinone form is shown. In practice, in chromatophores poised at a redox potential of $+100\,mV$, about 50% of the total Q_B is semiquinone (for details of the processes occurring in the reaction centre see Fig. 7.7 and page 332). Q_A is the quinone that is tightly bound in the reaction centre (menaquinone in Fig. 7.5, page 328). Q_B is more like Q_z and Q_c of the b/c_1 complex; it is the site at which quinone from the pool binds and becomes reduced (by way of a bound semiquinone) to quinol. Q_z is the site on the cytochrome b/c_1 complex at which pool quinol is oxidized (also called Q_o or the quinol oxidase site); Q_c is the site at which pool quinol is reduced (also called Q_i or quinol reductase site).

The net result of the Q-cycle is that for every two turnovers of the b/c_1 complex (2 photons) the reaction centre catalyses the reduction by two molecules of cytochrome c_2 of a quinone from the pool to quinol; the cytochrome b/c_1 complex catalyses the oxidation of one molecule of quinol in the pool to one quinone with the reduction of two molecules of cytochrome c_2; and four protons are transferred from the cytoplasm to the periplasm. A summary of this cycle is given in Fig. 1.31 (page 70).

We are now in a position to follow what might happen upon excitation of the photosynthetic reaction centres with a single turnover flash of light (the arrows in Fig. 7.20). It is assumed that in half the reaction centres Q_B exists initially as the fully oxidized quinone (Crofts *et al.*, 1983) which is reduced upon flashing to Q_B^- (see Fig. 7.20A). In the other half of the reaction centres Q_B exists initially as the semiquinone (Fig. 7.20B) so that, upon flashing, QH_2 is generated in those reaction centres and released into the pool. At about the same time as the Q_B reduction reactions ($< 100 \, \mu s$), cytochrome c_2 is oxidized by the reaction centre. The oxidized cytochrome c_2 initiates electron transfer in the "high potential" segment of the cytochrome b/c_1 complex (Fig. 7.20A):

$$*(i) \quad \text{cyt } c_2(\text{ox}) + \text{cyt } c_1(\text{red}) \underset{}{\overset{150 \, \mu s}{\rightleftharpoons}} \text{cyt } c_2(\text{red}) + \text{cyt } c_1(\text{ox})$$

$$(ii) \quad \text{cyt } c_1(\text{ox}) + \text{FeS(red)} \underset{}{\overset{< 200 \, \mu s}{\rightleftharpoons}} \text{cyt } c_1(\text{red}) + \text{FeS(ox)}$$

Then follows an interesting disproportionation reaction at the Q_z site (Fig. 7.20A); the FeS centre, oxidized after the flash, and the cytochrome b_{566} which was oxidized even before the flash, serve as separate electron acceptors for the two reducing equivalents from a ubiquinol from the pool:

$$(iii) \quad UQH_2 + \text{FeS(ox)} + \text{cyt } b_{566}(\text{ox}) \underset{}{\overset{300 \, \mu s}{\rightleftharpoons}} UQ + \text{FeS(red)} + \\ \text{cyt } b_{566}(\text{red}) + 2H^+$$

The ubiquinone returns to the pool and the reduced, low potential cytochrome b_{566} donates its electron to the high potential cytochrome b_{561}:

$$(iv) \quad \text{cyt } b_{566}(\text{red}) + \text{cyt } b_{561}(\text{ox}) \underset{}{\overset{< 300 \, \mu s}{\rightleftharpoons}} \text{cyt } b_{566}(\text{ox}) + \text{cyt } b_{561}(\text{red})$$

All of these reactions are summarized in Figure 7.20A. It is evident that thus far a single ubiquinol has been oxidized at the Q_z site. One reducing equivalent from the quinol has been transferred back to the reaction centre *via* cytochrome c_2 and the 'high potential' chain. The other reducing equivalent now resides on cytochrome b_{561}.

To complete the process, a second turnover of the Q_z site is required (Fig. 7.20B). Crofts *et al.* (1983) have estimated that the photosynthetic reaction centre content of their chromatophores is approximately twice that of the cytochrome b/c_1 complex. Consequently, a flash of light which drives a single turnover of *all* the reaction centres in a sample can also initiate the second turnover of the Q_z site: the 'high potential' segment of the b/c_1

---

* The figure shown above the arrows is the half-time of the reaction measured at a redox potential of $+100 \, mV$ (Crofts *et al.*, 1983).

complex is again oxidized by the reaction centres *via* cytochrome c_2 (i and ii) and a second ubiquinol from the pool reacts at centre Q_z according to (iii). Both *b*-type cytochromes are now reduced and reduction of a quinone at Q_c can ensue:

$$\text{(v) cyt } b_{566}(\text{red}) + \text{cyt } b_{561}(\text{red}) + \text{UQ} + 2\text{H}^+ \underset{}{\overset{1.5\,\text{ms}}{\rightleftharpoons}} \text{cyt } b_{566}(\text{ox}) + \text{cyt } b_{561}(\text{ox}) + \text{UQH}_2$$

Ubiquinol is released into the Q pool from Q_c. Note that the reduction of quinone at Q_c can take place only when both b_{561} and b_{566} are reduced. Hence this reaction is a 'two electron gate' analogous to the reduction of Q_B by the reaction centre (see page 333 and Fig. 7.7).

The final outcome of this sequence of events is that each b/c_1 complex has reduced one ubiquinone (at the Q_c site) plus two cytochromes c_2; and each b/c_1 complex has oxidized two ubiquinols (at the Q_z site). At starting redox potentials other than $+100\,\text{mV}$ at pH 7.0, the sequence and timing of these events can be altered. For example, at greater values of the starting redox potential, the quinone pool becomes increasingly oxidized. Consequently, after a single turnover flash, quinol is not available in the pool to provide promptly the necessary reducing equivalents at Q_z. In this case the overall rate is retarded by the time taken for ubiquinol produced at the reaction centre to diffuse to the Q_z site on the b/c_1 complex (Crofts *et al.*, 1983; Venturoli *et al.*, 1986).

7.3.3   Electrogenic activity in the cytochrome b/c_1 complex

Earlier, it was explained how studies with three different techniques are beginning to give rise to a consistent description of the electrogenic properties of the photosynthetic reaction centre from bacteria (pages 349–351). For the cytochrome b/c_1 complex, work with electrochromic absorbance changes in chromatophore membranes must be more heavily relied upon.

The purified cytochrome b/c_1 complex from *Rb. sphaeroides* has been reconstituted into artificial liposomes and its ubiquinol–cytochrome oxidoreductase activity has been shown to be electrogenic by three criteria:

(a) the rate of the reaction was stimulated by ionophores expected to relieve the back pressure from the protonmotive force;

(b) the rate of proton uptake accompanying the reaction was increased in the presence of valinomycin and K^+;

(c) the reaction led to absorbance changes of added cyanine dye which are characteristic of the formation of $\Delta\psi$ (Hurt *et al.*, 1983).

These experiments were important in helping to establish the electrogenic nature of the complex but the recording techniques were too slow to be of value in the determination of reaction mechanism. Direct electrical measurements on planar bilayers containing cytochrome b/c_1 complex together with photosynthetic reaction centres (cf. page 343) have not been performed and, unfortunately, recordings with macroscopic electrodes in suspensions exposed to light gradients (page 347) are not likely to be successful for observing electrogenic events on the timescale relevant to the cytochrome b/c_1 complex (10^{-4}–10^{-2} s). The only fast electrode measurements which have successfully demonstrated the electrogenic activity of the cytochrome b/c_1 complex have been performed with chromatophores fused to thick, lipid-impregnated collodion films (see page 345). In the presence of ascorbate and tetramethylphenylenediamine, an antimycin-sensitive electrical response was observed in the region of 700 μs after short light flashes (Drachev *et al.*, 1987). The most thorough analysis of the electrogenic activity of the b/c_1 complex has come, however, from measurements of electrochromic absorbance changes.

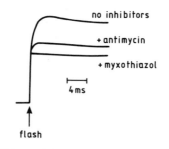

Fig. 7.21   The effect of inhibitors of the cytochrome bc_1 complex on electrochromic absorbance changes following short flash excitation of *Rb. sphaeroides* (taken from Glaser and Crofts, 1984). As in earlier figures (7.11, 7.12), an upward deflection indicates an increase in membrane potential. Note the difference in time-scale.

Figure 7.21 shows a typical electrochromic absorbance change resulting from single turnover flash excitation of *Rb. sphaeroides* chromatophores poised at the 'optimal' ambient redox potential of +100 mV. About one-half of the total flash-induced absorbance change is abolished after treatment of the chromatophores with inhibitors of the cytochrome b/c_1 complex such as myxothiazol (Jackson and Dutton, 1973; Dutton *et al.*, 1975; Bashford *et al.*, 1979; Glaser and Crofts, 1984). The residual absorbance change in the presence of myxothiazol is attributed to the electrogenic activity of the photosynthetic reaction centre and was discussed at length in Section 7.2.5.3. The myxothiazol-sensitive absorbance change is considerably slower ($t_{\frac{1}{2}} = 1.5$ ms) and is believed to result from the electrogenic activity of the cytochrome b/c_1 complex. On the basis of the relative extent of these components of the electrochromic absorbance change it appears that the photosynthetic reaction centre and the cytochrome b/c_1 complex make equal contributions to the generation of $\Delta\psi$ after a flash. It is known that the $\Delta\psi$ generated by the reaction centre arises from the transfer of a single electron across the membrane after a single turnover flash (page 333) so a similar stoichiometry is expected for the b/c_1 complex, i.e. one elementary charge translocated across the membrane for every reducing equivalent leaving the complex for cytochrome c_2. Although there have been some suggestions that the translocated charge might be the proton, i.e. that the b/c_1 complex is a 'proton pump' (Matsura *et al.*, 1983), most models envisage the electrogenic process as being electron transfer itself (Cramer and Crofts, 1982; Crofts and Wraight, 1983; Rich, 1984).

Recently, Glaser and Crofts have carried out a thorough re-evaluation of the slow (10^{-4}–10^{-2} s) electrochromic absorbance changes following flash excitation of chromatophores of *Rb. sphaeroides* (Glaser and Croft, 1984). The electrochromic absorbance change attributable to the electrogenic activity of the b/c_1 complex can be resolved into two subcomponents; one is inhibited by myxothiazol; the other by myxothiazol and by antimycin (this is evident from Fig. 7.21). The first subcomponent matches kinetically with the reduction of cytochrome b_{561} and the second with the oxidation of this cytochrome. The conclusion is therefore that the two electron transfer reactions, cytochrome $b_{566} \rightarrow$ cytochrome b_{561}, and cytochrome $b_{561} \rightarrow Q_c$ *span* the membrane and each contributes to the generation of $\Delta\psi$ (see Fig. 7.20). The Q_z site and cytochrome b_{566} are proposed to be at the same electrical potential as the periplasmic phase, and the Q_c site at the same potential as the cytoplasmic phase. On the basis of the relative extents of the two electrogenic components, and bearing in mind the fact that the complex turns over twice in the uninhibited state (Fig. 7.20) but only once when antimycin is present, the redox centre of cytochrome b_{561} is probably 35–50% of the way across the membrane from the periplasmic face. The location of the cytochrome b_{561} haem towards the centre of the membrane has been supported by other experiments. By raising the suspension pH or by partial extraction of the chromatophore ubiquinone, the Q_c site can be made to operate in reverse and so reduce cytochrome b_{561} after a flash of light (Glaser *et al.*, 1984; Robertson *et al.*, 1984). According to the model described by Figure 7.20, it will be clear that, in this process, electrons are driven back across the membrane in the 'wrong' direction (for ATP synthesis) and this would constitute an 'anti-electrogenic' reaction. It is therefore consistent with the model that an electrochromic absorbance change of opposite polarity to that normally found (i.e. characteristic of a blue shift rather than a red shift of carotenoid absorbance) is observed under these conditions (Robertson *et al.*, 1986). Quite clearly, the predictions of the positions of the prosthetic group locations within the b/c_1 complex made on the basis of biophysical studies present a challenge for further structural investigations.

7.3.4   Protolytic reactions associated with the cytochrome b/c_1 complex

The model in Figure 7.20 requires that two protons should be taken up on the cytoplasmic side of the membrane per quinone reduced at the Q_c site, and that two protons should be released on the periplasmic side of the membrane per quinol oxidized at the Q_z site. At a redox potential of $+100\,mV$ and at low pH, the available data are consistent with these expectations (Petty et al., 1977; Jones et al., 1987). Measurements with pH indicator dyes which report from the external aqueous phase of chromatophore suspensions show that $2\,mol\,H^+$ per mol of reaction centre are picked up on the cytoplasmic membrane face after a single turnover flash (Petty et al., 1977). One proton-binding reaction is insensitive, and one is sensitive to inhibition by antimycin A; the former is ascribed to H^+ binding during the reduction of Q_B in the reaction centre and the latter to H^+-binding during the reduction of quinone at the Q_c site in the b/c_1 complex: since two turnovers of the reaction centre are required for the reduction of one Q_c (see page 354), the stoichiometry is as predicted. At an ambient redox poise of $+100\,mV$ the rate of H^+ binding to Q_c ($t_\frac{1}{2} \simeq 1.5\,ms$) is also in good agreement with the model shown in Figure 7.20 (Petty et al., 1977).

In anaerobic suspensions of intact cells measurements with pH indicators after single turnover flash excitation show that $1.1\,mol\,H^+$ are released into the external medium per mol of reaction centre (Jones et al., 1987). The H^+-releasing reaction is completely inhibited by myxothiazol which suggests that it occurs, as predicted by Figure 7.20, at the Q_z site. The rate of H^+ appearance into the external medium of the bacterial suspension ($t_\frac{1}{2} \simeq 35\,ms$) is considerably slower than the rate of oxidation of quinol at Q_z. This might be due in part to slow proton diffusion through the immobile buffering groups of the periplasmic space, although measurements on spheroplast-derived vesicles have yielded similar discrepencies (Arata et al., 1987; Junge and McLaughlin, 1987).

There are a number of properties of the H^+-binding and H^+-releasing reactions which quite clearly point to a more complicated operation of the Q_c and Q_z sites than is suggested by Figure 7.20. In particular, the anomalously fast kinetics of the antimycin-sensitive proton-binding reaction at high ambient redox potential and the unexpectedly low pH at which H^+-binding become attenuated, merit further consideration but these problems are outside the scope of this chapter (see Matsura et al., 1983; Petty et al., 1977, 1979). The descriptions of the electron transfer reactions in the chromatophore b/c_1 complex and the charge translocation processes revealed by electrochromism are satisfyingly convergent. The answer to the puzzling anomalies in the proton binding and proton release reactions might be found in a better understanding of the protein domains which constitute the Q_c and Q_z sites and the way in which these domains channel protons to and from the aqueous medium on either side of the membrane.

7.4   Charge recombination during ATP synthesis

During rapid phototrophic growth the F_0F_1–ATP synthase is probably the major consumer of the Gibbs Free Energy made available from the electron transfer reactions in the cytoplasmic membrane (Clark *et al.*, 1983). The enzyme isolated from the photosynthetic bacteria resembles that from other bacterial sources and from mitochondria and chloroplasts. Its inhibitor sensitivity more closely resembles that of the mitochondrial enzyme than that of chloroplasts or of *Escherichia coli*. The genes coding for the ATP synthase from *Rps. blastica* and *Rhs. rubrum* have been cloned and sequenced and the amino acid sequences of the protein subunits have been deduced (Falk *et al.*, 1985; Walker *et al.*, 1985). The subunit structure and stoichiometry of the membrane-spanning F_0 sector and the peripheral, water-soluble F_1 component of related enzymes are described in Falk *et al.* (1985) and Senior (1985). In general, studies on the structure of the ATP synthase from photosynthetic bacteria have gone in parallel with those on the enzyme isolated from other sources and there have been few surprises. Although outside the scope of this chapter, special mention should be made, however, of recent structural studies on the ATP synthase from *Rbs. rubrum*. A remarkable finding was that the β subunit of F_1 could be removed simply by washing chromatophores of *Rhs. rubrum* with concentrated LiCl (Philosoph *et al.*, 1981). This has led to the identification of adenine nucleotide binding sites on this subunit, confirmation that it possesses the active site for ATP synthesis and that it is capable of limited rates of ATP hydrolysis (Gromet-Elhanan and Khananshvili, 1984; Harris *et al.*, 1985). These discoveries are important in a general context to our understanding of this complex enzyme.

In the years following the inception of the chemiosmotic hypothesis, evidence accumulated that ATP synthesis is dependent on the development of a protonmotive force across chromatophore membranes from photosynthetic bacteria, as well as the membranes isolated from other bacteria, and mitochondria and chloroplasts. In accord with observations on other membranes, ATP synthesis was found to be driven by 'artificial' proton electrochemical potential gradients generated by acid-base transitions and K^+-diffusion potentials in darkened suspensions of chromatophores (Leiser and

Gromet-Elhanan, 1974). Evidence from experiments with photosynthetic bacteria which show that ATP synthesis is accompanied by the translocation of charge across the cytoplasmic membrane through the ATP synthase is discussed below. In keeping with the theme of this chapter, it is shown how the special properties of photosynthetic organisms enable experiments to be carried out with a time resolution that is not yet possible with chemotrophic bacteria.

7.4.1   Charge recombination through the ATP synthase as revealed by electrochromic absorbance changes

It will be recalled that after single turnover flash excitation of *Rb. sphaeroides* or *Rb. capsulatus* chromatophores the electrochromic absorbance change is generated in two stages (Fig. 7.22). The first, which is complete in about 100 μs, is a consequence of the electrogenic activity of the photosynthetic reaction centre; the second, complete in about 10 ms (at a redox poise of $+100$ mV), is due to electrogenic activity in the cytochrome b/c_1 complex. At the end of this 10 ms period, cyclic electron transport is complete and the electrochromic absorbance change begins to decay slowly, over a period of seconds, to the pre-flash level. The slow decay can be accelerated by membrane-permeant ions, by ionophores such as valinomycin in the presence of K^+ and by protonophorous uncouplers (Fig. 7.11). These observations gave rise to the view that the membrane potential, generated during cyclic electron transport and sensed by the electrochromic pigments, is discharged by the electrophoretic movement of ions across the membrane. In the absence of added ionophores (etc.) the decay is presumably limited by the intrinsic permeability of the membrane to ions in the system. In simple ionic media, measurements of the absorbance change of added pH indicator dyes show that the re-release of H^+ accumulated immediately after a flash has the same kinetics as the decay of the electrochromic change, i.e. the membrane potential is collapsed by the outward movement of protons from the chromatophore lumen (Saphon *et al.*, 1975).

When chromatophore suspensions are supplemented with the substrates for phosphorylation, MgADP and Pi, at a suitable pH the decay of the electrochromic absorbance change after a single turnover flash is accelerated (Saphon *et al.*, 1975) (Fig. 7.22). The period of acceleration begins within a few milliseconds of flash excitation. The ATP synthesized after a single turnover flash (red) to drive photosynthesis can be measured *in situ* from the light (green) emitted from firefly luciferin and luciferase added to the chromatophore suspension (Lundin *et al.*, 1977; Petty and Jackson, 1979b). It has been established that the accelerated decay of the electrochromic absorbance change in the presence of ADP and Pi has an identical substrate concentration dependence and inhibitor sensitivity to the single flash yield of ATP synthesis (Petty and Jackson, 1979a). This therefore represents rather direct evidence that the process of ATP synthesis is accompanied by the translocation of charge across the chromatophore membrane, that is, by a transmembrane ionic current. It does not establish that protons are the translocated species but that is the expectation of the chemiosmotic hypothesis. When an additional pathway of ionic current is introduced into the chromatophore membrane, for example by adding valinomycin in the presence of K^+, then the yield of ATP after a flash is decreased in parallel with the decrease in the accelerated decay of the electrochromic absorbance change in the presence of ADP and Pi (Petty and Jackson, 1979a). This means that K^+-translocation by way of valinomycin, and charge transduction accompanying ATP synthesis, are driven by the same force and that this force, presumably $\Delta\psi$, is reported by the electrochromic absorbance change.

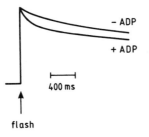

Fig. 7.22 Acceleration of the decay of the electrochromic absorbance change during photo-phosphorylation. Antimycin and Pi were present in both experiments.

A characteristic feature which is clearly shown in Figure 7.22 is that the accelerative effect of ADP and Pi on the decay of the electrochromic absorbance change is only partial. Semilogarithmic plots of the decay show that the accelerated period lasts for only a few hundred milliseconds. One reason for this is that the rate of ATP synthesis in chromatophores has a 'threshold' dependence on the value of the protonmotive force (Baccarini-Melandri *et al.*, 1977; Casadio *et al.*, 1978; Clark *et al.*, 1983). The ATP synthesis rate is extremely slow below the threshold Δp and it increases disproportionately through the threshold region. This interesting phenomenon has been thoroughly investigated in thylakoids from higher plants where similar considerations probably apply (Junesch and Gräber, 1985). The existence of the threshold might arise mainly for thermodynamic reasons: the free enthalpy of the protonmotive force ($-nF\Delta p$, where n is the H^+/ATP stoichiometry and F is Faraday's constant) must be sufficient to displace the phosphorylation reaction (whose free enthalpy is given by $\Delta G = \Delta G^{0'} + RT\ln[\text{ADP}][\text{Pi}]/[\text{ATP}]$ where $\Delta G^{0'}$ is the Standard Gibbs Free Energy for ATP hydrolysis). However, the ATP synthase activity might also be controlled kinetically by the protonmotive force. For example a kind of allosteric 'gating' of the enzyme by Δp can be envisaged. Following the flash of light, the accelerated membrane discharge (Fig. 7.22) is presumed to accompany the rapid rate of ATP synthesis during the period in which the absolute value of $\Delta\psi$ lies above threshold. The subsequent slow component of the decay which has a $t_{\frac{1}{2}}$ of about 10 seconds after single turnover flashes must represent outward proton diffusion either through other energy-conserving consumers of the protonmotive force or through energy-dissipating leaks. This rather simple explanation of the decay kinetics of the electrochromic absorbance change does not, however, provide a complete description. There are indications that other controlling parameters affect the activity of the ATP synthase (see Jackson *et al.*, 1975; page 365).

Charge translocation accompanying ATP synthesis can also be demonstrated by monitoring electrochromic absorbance changes in intact cells of the photosynthetic bacteria (Cotton and Jackson, 1982). Of course in this case the concentration of the phosphorylation substrates cannot be manipulated; the cytoplasmic concentrations of ADP, ATP and Pi are in the region of 10^{-3}, 10^{-3} and 10^{-2} M respectively in darkened, anaerobic suspensions of intact cells. ATP synthesis, however, can be blocked in control experiments by treating the cells with venturicidin, an oligomycin-like inhibitor of the F_0 segment of the ATP synthase. In untreated cells a rapid component of the decay of the electrochromic absorbance change immediately after the flash can be distinguished. The rapid decay component, presumably accompanying ATP synthesis, is inhibited by venturicidin but the subsequent slower decay is unaffected. The threshold effect is also evident in intact cells although at a larger flash-induced change in $\Delta\psi$ than is required in chromatophore experiments. The threshold is exceeded when the cells are excited either by single flashes in the absence of antimycin or by three closely spaced flashes in the presence of antimycin. The increased threshold for ATP synthesis in intact cells is probably a reflection of their large internal phosphorylation potential.

7.4.2   Are there interactions between the components of the photosynthetic electron transport chain and the ATP synthase other than those mediated by the protonmotive force?

According to the chemiosmotic hypothesis, a protonmotive force is an obligate 'intermediate' between the processes of electron transport and ATP synthesis. However, there might be other interactions between the electron transport chain and the ATP synthase which serve to control the process of energy coupling. For example, it could be imagined that the redox state of an electron transport carrier could directly (by oxidation or reduction of diffusible intermediate) or indirectly (e.g. by activating a protein kinase) alter the catalytic activity of the ATP synthase. There are indeed some indications that such interactions might operate. The important experiments of Melandri and colleagues showed that the dependence of the rate of ATP synthesis on Δp in chromatophores is not unique—it depends on the way in which Δp is varied (Melandri *et al.*, 1972; Casadio *et al.*, 1978). When Δp was decreased with electron transport inhibitors (in steady-state experiments) the rate of ATP synthesis was depressed more steeply than when Δp was decreased by uncoupling agents. Hence for the same value of Δp, the rate of ATP synthesis was higher in the presence of uncoupler than in the presence of electron transport inhibitor. Of course, in the simplest formulation of the chemiosmotic hypothesis, the ATP synthase would not 'know' whether Δp has been depressed by inhibiting electron transport or by dissipation through a protonophore. The redox state of (some of) the electron transport carriers, however, would be differently affected by inhibitor and uncoupler and this led Melandri and colleagues to predict co-operative effects on the ATP synthase.

Experiments of this sort have been repeated in mitochondria and chloroplasts but the results, and the interpretation of the results, have been controversial. Some investigators have felt that data such as those described by Melandri and colleagues detract from the appealing simplicity of the chemiosmotic hypothesis and call for an undesirable *ad hoc* explanation to rescue it. For this reason there have been some radical formulations of the coupling mechanism (e.g. Westerhoff *et al.*, 1984). Another consideration is that the experiments are technically complex. It has been argued and counterargued that systematic experimental errors might distort the true relationship (see Cotton *et al.*, 1987 and references therein). For another consumer of the protonmotive force in chromatophores, the 'energy-linked' pyridine nucleotide transhydrogenase (which is much easier to measure than ATP synthesis), it has been shown that the dependence of rate upon Δp is indeed unique and is not affected by the manner in which Δp is varied (Cotton *et al.*, 1987).

The problem of the precise mechanism of the energy coupling reaction is a complex one and the reader is referred to an article by Ferguson (1985) for a critical review of the experimental data in conflict with the chemiosmotic hypothesis. The controversy extends across a range of experimental biological material, and the photosynthetic bacteria continue to play their part. The consensus view is probably that the chemiosmotic hypothesis (or one of its derivatives) currently provides the best available model on which to base further research but, unfortunately, the largest gap in our understanding is the manner in which the energy of the protonmotive force is used to drive the ATP synthase (see Chapter 9 for further discussion of this controversy).

7.4.3 The mechanism of ATP synthesis

What should be, in view of its importance, one of the longest sections in this chapter, will in fact be one of the shortest. It is not known how the energy of the protonmotive force is transmitted to the catalytic site of the ATP synthase. Some favour a direct mechanism in which the translocated protons have a catalytic role at the enzyme active site; others favour the idea that proton translocation through F_0 gives rise to an energized conformation which is then transmitted to the catalytic site in the F_1. There is no evidence to discriminate between these possibilities in the chromatophore enzyme or in any other F_0F_1 ATP synthase.

What is likely to be influential in the pursuance of this matter is the recent finding that the equilibrium constant of the phosphorylation reaction *on the enzyme* is close to unity (Boyer, 1981). Penefsky (1985) has analysed the individual rate constants for ATP hydrolysis under conditions which he describes as 'single site catalysis' for the F_1 on submitochondrial particles according to the following scheme:

$$\text{ADP} + F_1 \underset{K_1}{\rightleftharpoons} \text{ADP}.F_1 \overset{\text{Pi}}{\underset{K_2}{\rightleftharpoons}} \text{ADP.Pi}.F_1 \underset{K_3}{\rightleftharpoons} \text{ATP}.F_1 \underset{K_4}{\rightleftharpoons} \text{ATP} + F_1$$

It turns out that the equilibrium described by K_3 is fairly centrally poised, whereas in aqueous solution at neutral pH the equilibrium constant for ATP synthesis is in the region of 10^{-5} M^{-1}. The reason for this is that ATP binds very tightly to the enzyme relative to ADP and Pi. This suggests that energy input during ATP synthesis is not, as was originally thought, at the phosphorylation step but rather at the level of ATP release from the enzyme: it shifts the emphasis of the problem towards an understanding of how proton conduction through the F_0F_1 can promote the release of tightly-bound ATP.

Another important aspect of ATP synthesis to emerge recently is the recognition that the ATP synthase might have multiple catalytic sites (many favour three sites per F_1) and that the sites 'co-operate' during enzyme turnover: it is supposed that catalysis at one site is linked to events at another site on the same enzyme via long range conformational changes. The sites would then operate in unison but out of phase (Gresser *et al.*, 1982). How proton conduction through the F_0 channel would operate to drive the three sites for synthesis is another interesting matter for conjecture and for further research. There is further discussion of this topic in Chapter 1, page 76).

7.5 Concluding remarks: the protonmotive current and bacterial growth

This chapter has concentrated on experiments in which bacterial electron transport and ATP synthesis are driven by short, intense pulses of photosynthetic light. Although valuable in scientific investigation, these are not the conditions which normally prevail in nature. There the situation is of course likely to be complex but we can probably come closer to natural conditions by performing experiments in steady state. Many experiments in bioenergetics on photosynthetic and non-photosynthetic organisms and organelles are, in fact, carried out in situations which approximate to steady state: conditions are established such that electron transport, ATP synthesis, solute translocation (etc.) remain at a constant rate during the period of measurement. Experiments with photosynthetic organisms in which key reactions can be followed on a rapid time-scale by spectroscopy show that a new steady state (or at least, conditions which approximate to steady state) can be established within about 100 ms of a stepwise change in light intensity but that subsequent changes then take place during the following seconds (e.g. activation of ATP synthase; Cotton and Jackson, 1984) and even following hours (induction of pigment synthesis; Schmidt *et al.*, 1970).

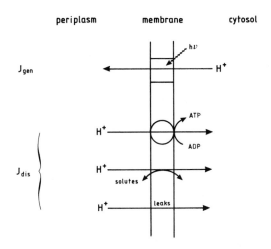

Fig. 7.23   The chemiosmotic proton circuit.

During steady-state, electron-transport-driven ATP synthesis (e.g. oxidative and photophosphorylation) the protonmotive force becomes constant: the rate at which Δp is generated is equal to the rate at which it is dissipated. The outward flux of H^+, driven by electron transport across the cytoplasmic membrane of a bacterium (J_{gen}) will be exactly counter-balanced by an ionic current (J_{dis}) driven by Δp and responsible for ATP synthesis and other energy-requiring reactions or leaks (see Fig. 7.23). If the activity of any one component in the proton circuit is modified then the flux through other components might change.

The dependence of J_{dis} on $\Delta\psi$, revealed by electrochromic experiments with chromatophores of a photosynthetic bacterium is shown in Figure 7.24. This is virtually the current/voltage profile of the bacterial membrane (Jackson, 1982). It turns out that the membrane does not behave like a simple (Ohmic) conductor; it behaves like a diode. As $\Delta\psi$ is increased there is a disproportionate increase in the ionic current flowing across the membrane. To put this another way, the ionic conductance of the membrane (the slope of the $J_{dis}/\Delta\psi$ relation) is not constant: it increases with $\Delta\psi$.

In large measure, the diodic nature of the bacterial membrane arises from the threshold dependence of the ATP synthesis reaction discussed earlier, see p. 363 (Clark *et al.*, 1983). Evidently, the increase in ATP synthesis above the threshold region in $\Delta\psi$ results in an increased proton current through the ATP synthase and this is reflected in the disproportionate increase in J_{dis}. Interestingly, other consumers of the protonmotive force also appear to have threshold dependences, for example the NADH/NADP$^+$ transhydrogenase (Cotton *et al.*, 1987).

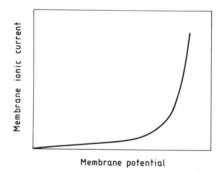

Fig. 7.24 The current/voltage profile of the chromatophore membrane.

The diodic nature of the bacterial cytoplasmic membrane has an important physiological consequence: it facilitates simple homeostatic control of the protonmotive force (see Jackson, 1982; Ferguson and Sorgato, 1982).

This can be illustrated by considering Figure 7.24. It was pointed out (above) that in the steady state, $J_{gen} = J_{dis}$. In fact, the data in Figure 7.24 were obtained by progressively reducing J_{gen} in a series of experiments with either myxothiazol or with antimycin A. It can be seen, therefore, that a considerable decrease in J_{gen} leads to only a small drop in $\Delta\psi$: the effect of the decrease in J_{gen} on the value of $\Delta\psi$ is partly offset by a decrease in the membrane ionic conductance (there is a *disproportionate* decrease in the dissipative ionic current flowing across the membrane). The important result is that $\Delta\psi$ is maintained at a moderately high value.

Now, because of the threshold relationship between ATP synthesis and $\Delta\psi$, even a small drop in the value of $\Delta\psi$ can result in a considerable fall in the rate of ATP synthesis. To some extent, of course, the properties of the ATP synthase itself govern the current/voltage properties of the membrane (Clark *et al.*, 1983). The outcome is that a large decrease in the rate of electron transport is accompanied by a large decrease in the ATP synthesis rate, even though $\Delta\psi$ is changed but little. The process of electron transport-driven ATP synthesis is apparently confined to large values of $\Delta\psi$.

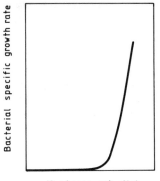

Fig. 7.25   The dependence of specific growth rate of a bacterial culture on membrane potential.

It seems likely that a moderately large value of $\Delta\psi$ is maintained by very low rates of electron transport so that other consumers, perhaps with a higher priority for cell viability, such as the ion pumps needed to maintain osmotic stability, can continue to idle during cell maintenance in the absence of growth (Taylor and Jackson, 1985). This chapter concludes with the results of an experiment on the growth rate of the photosynthetic bacteria (Fig. 7.25; Taylor and Jackson, 1985). It shows that the rate of growth has a threshold dependence on $\Delta\psi$ (equal to protonmotive force in these conditions) which correlates closely with the dependence of J_{dis} on $\Delta\psi$ (Fig. 7.25). This correlation in a general way highlights the importance of the membrane ionic current and the protonmotive force in cellular physiology. More specifically, it illustrates that when the growth rate is reduced to very low values, for example by decreasing the photosynthetic light intensity, then the ionic current flowing across the membrane is also extremely low, but $\Delta\psi$ is still in the region of 60% of its maximum value. Perhaps significant further loss of $\Delta\psi$ would be lethal.

References

Aargaard, J. and Sistrom, W. R. (1972). *Photochem. Photobiol.* **15**, 209–225.

Apell, H.-J., Snozzi, M. and Bachofen, R. (1983). *Biochim. Biophys. Acta* **724**, 258–277.

Arata, H., Takenaka, I. and Nishimura, N. (1987). *J. Biochem.* (in press).

Baccarini-Melandri, A., Casadio, R. and Melandri, B. A. (1977). *Eur. J. Biochem.* **78**, 389–402.

Baltscheffsky, M. (1969). *Arch. Biochem. Biophys.* **130**, 646–652.

Barouch, Y. and Clayton, R. K. (1977). *Biochim. Biophys. Acta* **462**, 785–788.

Bashford, C. L., Prince, R. C., Takamiya, K. I. and Dutton, P. L. (1979). *Biochim. Biophys. Acta* **545**, 223–235.

Bowyer, J. R., Tierney, G. V. and Crofts, A. R. (1979a). *FEBS Lett.* **101**, 201–206.

Bowyer, J. R., Tierney, G. V. and Crofts, A. R. (1979b). *FEBS Lett.* **101**, 207–212.

Boyer, P. D. (1981). *In* 'Chemiosmotic proton circuits in biological membranes' (Eds. V. P. Skulachev and P. C. Hinkel), pp. 395–406. Addison-Wesley, Reading, Massachusetts.

Casadio, R., Baccarini-Melandri, A. and Melandri, B. A. (1978). *FEBS Lett.* **87**, 323–328.

Chang, C.-H., Tiede, D., Tang, J., Smith, U., Norris, J. and Schiffer, M. (1986). *FEBS Lett.* **205**, 82–86.

Clark, A. J. and Jackson, J. B. (1981). *Biochem. J.* **200**, 389–397.

Clark, A. J., Cotton, N. P. J. and Jackson, J. B. (1983). *Eur. J. Biochem.* **130**, 575–580.

Clark, A. J., Cotton, N. P. J. and Jackson, J. B. (1983). *Biochim. Biophys. Acta* **723**, 440–453.

Clayton, R. K. (1980). 'Photosynthesis: physical mechanisms and chemical patterns'. IUPAB Biophysics Series. Cambridge University Press, Cambridge.

Clayton, R. K. and Straley, S. C. (1970). *Biochem. Biophys. Res. Commun.* **39**, 1114–1119.

Clayton, R. K. and Wang, R. T. (1971). *Methods Enzymol.* **23**, 696–704.

Cogdell, R. J. (1986). *FEBS Lett.* **87**, 252–254.

Cogdell, R. J., Celis, S., Celis, H. and Crofts, A. R. (1977). *FEBS Lett.* **80**, 190–194.

Cotton, N. P. J. and Jackson, J. B. (1982). *Biochim. Biophys. Acta* **679**, 138–145.

Cotton, N. P. J. and Jackson, J. B. (1984). *Biochim. Biophys. Acta* **767**, 618–626.

Cotton, N. P. J., Myatt, J. F. and Jackson, J. B. (1987). *FEBS Lett.* **219**, 88–92.

Cramer, W. A. and Crofts, A. R. (1982). *In* 'Photosynthesis I. Energy conversion by plants and bacteria' (Ed. Govindjee), pp. 387–468. Academic Press, New York.

Crielaard, W., Cotton, N. P. J., Jackson, J. B., Hellingwerf, K. J. and Konings, W. (1987). In press.

Crofts, A. R. and Wraight, C. A. (1983). *Biochim. Biophys. Acta* **726**, 149–185.

Crofts, A. R., Meinhardt, S. W., Jones, K. R. and Snozzi, M. (1983). *Biochim. Biophys. Acta* **723**, 202–218.

Daldal, F., Cheng, S., Applebaum, J., Davidson, E. and Prince, R. C. (1986). *Proc. Nat. Acad. Aci. USA*. **83**, 2012–2016.

Daldal, F., Davidson, E., Cheng, S., Naiman, B. and Rook, S. (1986). *In* 'Current Communications in Molecular Biology', 'Microbial energy transduction' (Eds. D. C. Youvan and F. Daldal), pp. 113–119. Cold Spring Harbor Laboratory.

Davidson, E., Rook, S. and Daldal, F. (1986). *In* 'Abstracts VII International Congress on Photosynthesis', pp. 307–400. Brown University, Providence, USA.

Debus, R. J., Okamura, M. Y. and Feher, G. (1985). *Biophys. J.* **47**, 3a.

Deisenhofer, J., Epp, O., Miki, K., Huber, R. and Michel, H. (1984). *J. Mol. Biol.* **180**, 385–398.

Deisenhofer, J., Epp, O., Miki, K., Huber, R. and Michel, H. (1985a). *Nature* **318**, 618–624.

Deisenhofer, J., Michel, H. and Huber, R. (1985b). *TIBS* **10**, 243–248.

Deprez, J., Trissl, H.-W. and Breton, J. (1986). *Proc. Natl. Acad. Sci. USA* **83**, 1699–1703.

Dickerson, R. E., Timkovich, R. and Almassy, R. J. (1976). *J. Mol. Biol.* **100**, 473–491.

Donohue, T. J., McEwan, A. G., van Doren, S., Crofts, A. R. and Kaplan, S. (1987). *Proc. Nat. Acad. Sci. USA* (in press).

Drachev, L. A., Frolov, V. N., Kaulen, A. D., Kondrashin, A. A., Samuilov, V. D., Semenov, A. Yu. and Skulachev, V. P. (1976). *Biochim. Biophys. Acta* **440**, 637–660.

Drachev, L. A., Demenov, A. Y., Skulachev, V. P., Smirnova, I. A., Chamorovsky, S. K., Kokonenko, A. A., Rubin, A. B. and Uspenskaya, N. Y. (1981). *Eur. J. Biochem.* **117**, 483–489.

Drachev, L. A., Kaminskaya, O. P., Konstantinov, A. A., Kotova, E. A., Mamedov, M. D., Samuilov, V. D., Semenov, A. Y. and Skulachev, V. P. (1986a). *Biochim. Biophys. Acta* **848**, 137–146.

Drachev, L. A., Kaminskaya, O. P., Konstantinov, A. A., Mamedov, M. D., Samuilov, V. D., Semenov, A. Y. and Skulachev, V. P. (1986b). *Biochim. Biophys. Acta* **850**, 1–9.

Drachev, S. M., Drachev, L. A., Zaberezhnaya, S. M., Konstantinov, A. A., Semenov, A. Yu. and Skulachev, V. P. (1986c). *FEBS Lett* **205**, 41–46.

Drachev, L. A., Mamedov, M. D. and Semenov, A. Y. (1987). *FEBS Lett.* **123**, 128–132.

Drews, G. (1985). *Microbiol. Revs.* **49**, 59–70.

Dutton, P. L. (1978). *In* 'Methods in Enzymology' (Eds. S. P. Colowick and N. O. Kaplan), LIIV, pp. 411–435. Academic Press.

Dutton, P. L. (1986). *In* 'Encyclopaedia of Plant Physiology. New Series. Photosynthesis' (Eds. L. A. Staehelin and C. J. Arntzen), Vol. 19, pp. 197–237. Springer-Verlag, Berlin, Heidelberg, New York, Tokyo.

Dutton, P. L. and Jackson, J. B. (1972). *Eur. J. Biochem.* **30**, 4957–5010.

Dutton, P. L. and Prince, R. C. (1978). *In* 'The Photosynthetic Bacteria' (Eds. R. K. Clayton and W. R. Sistrom), pp. 525–570. Plenum Press, New York and London.

Dutton, P. L., Petty, K. M., Bonner, H. S. and Morse, S. D. (1975). *Biochim. Biophys. Acta* **387**, 536–556.

Edwards, P. A. and Jackson, J. B. (1976). *Eur. J. Biochem.* **62**, 7–14.
Fajer, J., Brune, D. C., Davis, M. S., Forman, A. and Spaulding, L. D. (1975). *Proc. Nat. Acad. Sci. USA* **72**, 4956–4960.
Falk, G., Hampe, A. and Walker, J. E. (1985). *Biochem. J.* **228**, 391–407.
Feher, G. (1971). *Photochem. Photobiol.* **14**, 373–387.
Feher, G. and Okamura, M. Y. (1978). *In* 'The Photosynthetic Bacteria' (Eds. R. K. Clayton and W. R. Sistrom), pp. 349–386. Plenum Press, New York and London.
Ferguson, S. J. (1985). *Biochim. Biophys. Acta* **811**, 47–95.
Ferguson, S. J. and Sorgato, M. C. (1982). *Ann. Rev. Biochem.* **51**, 185–217.
Fowler, C. F. and Kok, B. (1974). *Biochim. Biophys. Acta* **357**, 308–318.
Gabellini, N., Bowyer, J. R., Melandri, B. A. and Hauska, G. (1982). *Eur. J. Biochem.* **126**, 105–111.
Gabellini, N., Harnisch, U., McCarthy, J. E. G., Hauska, G. and Sebald, W. (1985). *EMBO J.* **4**, 549–553.
Glaser, E. G. and Crofts, A. R. (1984). *Biochim. Biophys. Acta* **766**, 322–333.
Glaser, E. G., Mainhardt, S. W. and Crofts, A. R. (1984). *FEBS Lett.* **178**, 336–342.
Gopher, A., Blatt, Y., Schonfeld, M., Okamura, M. Y. and Feher, G. (1985). *Biophys. J.* **48**, 311–320.
Govindjee (Ed.) (1982). 'Photosynthesis', Vol. I, 'Energy Conservation by Plants and Bacteria'. Academic Press, New York.
Graber, P. and Trissl, H.-W. (1981). *FEBS Lett.* **123**, 95–99.
Gresser, M. J., Myers, J. A. and Boyer, P. D. (1982). *J. Biol. Chem.* **257**, 12030–12038.
Gromet-Elhanan, Z. and Khananshvili, D. (1984). *Biochemistry* **23**, 1022–1028.
Gunner, M. R., Tiede, D. M., Prince, R. C. and Dutton, P. L. (1982). *In* 'Functions of Quinones in Energy Conserving Systems' (Ed. B. L. Trumpower), pp. 265–269. Academic Press, New York.
Harnisch, U., Weiss, H. and Sebald, W. (1986). *Eur. J. Biochem.* **149**, 95–99.
Harris, D. A., Boork, J. and Baltscheffsky, M. (1985). *Biochemistry* **24**, 3876–3883.
Hauska, G., Hurt, E., Gabellini, N. and Lockau, W. (1983). *Biochem. Biophys. Acta* **726**, 97–133.
Hurt, E. C., Gabellini, N., Shahak, Y., Lockau, W. and Hauska, G. (1983). *Arch. Biochem. Biophys.* **225**, 879–885.
Imhoff, J. F., Truper, H. G. and Pfennig, N. (1984). *Int. J. Syst. Bact.* **34**, 340–343.
Jackson, J. B. (1982). *FEBS Lett.* **139**, 139–143.
Jackson, J. B. and Crofts, A. R. (1969). *FEBS Lett.* **4**, 185–188.
Jackson, J. B. and Crofts, A. R. (1971). *Eur. J. Biochem.* **18**, 120–130.
Jackson, J. B. and Dutton, P. L. (1973). *Biochim. Biophys. Acta* **325**, 102–113.
Jackson, J. B., Saphon, S. and Witt, H. T. (1975). *Biochim. Biophys. Acta* **408**, 83–92.
Jones, M. R., Taylor, M. A. and Jackson, J. B. (1987). Unpublished observations.
Junesch, U. and Graber, P. (1985). *Biochim. Biophys. Acta* **809**, 429–434.
Junge, W. and Jackson, J. B. (1982). *In* 'Photosynthesis', Vol. I, 'Energy Conservation by Plants and Bacteria', (Ed. Govindjee), pp. 589–646. Academic Press, New York and London.
Junge, W. and McLaughlin, S. (1987). *Biochim. Biophys. Acta* **390**, 1–5.
Junge, W. and Witt, H. T. (1968). *Z. Naturforsch.* **23b**, 244–254.
Kaminskaya, D. P., Drachev, L. A., Konstantinov, A. A., Semenov, A. Y. and Skulachev, V. P. (1986). *FEBS Lett.* **202**, 224–228.
Kaufman, K. J., Dutton, P. L., Netzel, T. L., Leigh, J. S. and Rentzepis, P. M. (1975). *Science* **188**, 1301–1304.
Kaufman, N., Reidle, H. H., Golecki, J. R., Garcia, A. F. and Drews, G. (1982). *Arch. Microbiol.* **131**, 313–322.

Leiser, M. and Gromet-Elhanan, Z. (1974). *FEBS Lett.* **43**, 267–270.
Lommen, M. A. J. and Takemoto, J. (1978). *J. Bacteriol.* **136**, 730–741.
Lundin, A., Thore, A. and Baltscheffsky, M. (1977). *FEBS Lett.* **79**, 73–76.
Madigan, M. T. and Gest, H. (1979). *J. Bacteriol.* **137**, 524–530.
Matsura, K., Ishikawa, T. and Nishimura, M. (1980). *Biochim. Biophys. Acta* **590**, 339–344.
Matsura, K., O'Keefe, D. P. and Dutton, P. L. (1983). *Biochim. Biophys. Acta* **722**, 12–22.
McEwan, A. G., Cotton, N. P. J., Ferguson, S. J. and Jackson, J. B. (1985). *Biochim. Biophys. Acta* **810**, 140–147.
McEwan, A. G., Greenfield, A. J., Wetzstein, H. G., Jackson, J. B. and Ferguson, S. J. (1985). *J. Bacteriol.* **164**, 823–830.
McEwan, A. G., Wetzstein, H. G., Ferguson, S. J. and Jackson, J. B. (1985). *Biochim. Biophys. Acta.* **806**, 410–417.
McEwan, A. G., Jackson, J. B. and Ferguson, S. J. (1987). *FEMS. Microbiol. Revs* (in press).
Meinhardt, S. W. and Crofts, A. R. (1982). *FEBS Lett.* **149**, 217–222.
Melandri, B. A., Baccarini-Melandri, A. and Fabbri, E. (1972). *Biochim. Biophys. Acta* **275**, 383–395.
Michel, H. (1982). *J. Mol. Biol.* **158**, 567–572.
Michel, H. (1983). *TIBS* **8**, 56–59.
Michel, H., Weyer, A., Greenberg, H. and Lottspeich, F. (1985). *EMBO J.* **4**, 1667–1672.
Midgley, M., Iscander, N. S. and Dawes, E. A. (1986). *Biochim. Biophys. Acta* **865**, 45–49.
Mitchell, P. (1976). *J. Theor. Biol.* **62**, 327–367.
Myatt, J. F., Cotton, N. P. J. and Jackson, J. B. (1987). *Biochim. Biophys. Acta* **890**, 251–259.
Norris, J. R., Uphaus, R. A., Crespi, H. L. and Katz, J. J. (1971). *Proc. Nat. Acad. Sci. USA* **68**, 625–629.
Okamura. M. Y., Steiner, L. A. and Feher, G. (1974). *Biochemistry* **13**, 1394–1403.
Okamura. M. Y., Isaacson, R. A. and Feher, G. (1975). *Proc. Nat. Acad. Sci. USA* **72**, 3491–3495.
Okamura. M. Y., Feher, G. and Nelson, N. (1982). *Ibid.* pp. 195–272.
O'Keefe, D. P. and Dutton, P. L. (1981). *Biochim. Biophys. Acta* **635**, 149–166.
Overfield, R. E., Wraight, C. A. and DeVault, D. (1979). *FEBS Lett.* **105**, 137–142.
Overfield, R. E. and Wraight, C. A. (1980a). *Biochemistry* **19**, 3322–3327.
Overfield, R. E. and Wraight, C. A. (1980b). *Biochemistry* **19**, 3328–3334.
Packham, N. K., Berriman, J. A. and Jackson, J. B. (1978). *FEBS Lett.* **89**, 205–210.
Packham, N. K., Greenrod, J. A. and Jackson, J. B. (1980). *Biochim. Biophys. Acta* **592**, 130–142.
Packham, N. K., Packham, C., Mueller, P., Tiede, D. M. and Dutton, P. L. (1980). *FEBS Lett.* **110**, 101–106.
Packham, N. K., Tiede, D. M., Mueller, P. and Dutton, P. L. (1980). *Proc. Nat. Acad. Sci. USA* **77**, 6339–6343.
Packham, N. K., Dutton, P. L. and Mueller, P. (1982). *Biophys. J.* **37**, 465–473.
Parson, W. W. and Ke, B. (1982). *In* 'Photosynthesis', Vol. I, 'Energy Conservation by Plants and Bacteria' (Ed. Govindjee), pp. 331–385. Academic Press, New York, London.
Penefsky, H. S. (1955). *J. Biol. Chem.* **260**, 13728–13734.
Petty, K. M. and Jackson, J. B. (1979a). *Biochim. Biophys. Acta* **547**, 463–473.
Petty, K. M. and Jackson, J. B. (1979b). *Biochim. Biophys. Acta* **547**, 474–483.

Petty, K. M., Jackson, J. B. and Dutton, P. L. (1977). *FEBS Lett.* **84**, 294–303.
Petty, K. M., Jackson, J. B. and Dutton, P. L. (1979). *Biochim. Biophys. Acta.* **546**, 17–42.
Philosoph, S., Khananshvili, D. and Gromet-Elhanan, Z. (1981). *Biochem. Biophys. Res. Commun.* **101**, 384–389.
Popovic, Z. D., Kovacs, G. J., Vincett, P. S., Alegria, G. and Dutton, P. L. (1986). *Biochim. Biophys. Acta* **851**, 38–48.
Prince, R. C. and Dutton, P. L. (1977). *Biochim. Biophys. Acta* **462**, 731–747.
Prince, R. C., Davidson, E., Haith, C. E. and Daldal, F. (1986). *Biochemistry* **25**, 5208–5214.
Prince, R. C. and Youvan, D. C. (1987). *Biochim. Biophys. Acta* **890**, 286–291.
Prince, R. L., Baccarini-Melandri, A., Hauska, G. A., Melandri, B. A. and Crofts, A. R. (1975). *Biochim. Biophys. Acta* **387**, 212–227.
Prince, R. C., Leigh, J. S. and Dutton, P. L. (1976). *Biochim. Biophys. Acta* **440**, 622–636.
Remsen, C. C. (1978). *In* 'The Photosynthetic Bacteria' (Eds. R. K. Clayton and W. R. Sistrom), pp. 31–60. Plenum Press, New York and London.
Rich, P. R. (1984). *Biochim. Biophys. Acta* **768**, 53–79.
Robertson, D. E., Prince, R. C., Bowyer, J. R., Matsuura, K., Dutton, P. L. and Ohnishi, T. (1984). *J. Biol. Chem.* **259**, 1758–1763.
Robertson, D. E., Giangiacomo, K. M., de Vries, S., Moser, C. M. and Dutton, P. L. (1984). *FEBS Lett.* **178**, 343–350.
Robertson, D. E., Prince, R. C., Davidson, E., Marrs, B. L. and Dutton, P. L. (1986). *In* 'Current Communications in Molecular Biology: Microbial Energy Transduction' (Eds. D. C. Youvan and F. Daldal), pp. 93–98. Cold Spring Harbor Laboratory.
Robertson, D. E., Davidson, E., Prince, R. C., van den Berg, W. H., Marrs, B. L. and Dutton, P. L. (1986). *J. Biol. Chem.* **261**, 584–591.
Rockley, M. G., Windsor, M. W., Cogdell, R. J. and Purson, W. W. (1975). *Proc. Nat. Acad. Sci. USA* **72**, 2251–2255.
Saphon, S., Jackson, J. B., Lerbs, V. and Witt, H. T. (1975). *Biochim. Biophys. Acta* **408**, 58–66.
Saphon, S., Jackson, J. B. and Witt, H. T. (1975). *Biochim. Biophys. Acta* **408**, 67–82.
Schonfeld, M., Montal, M. and Feher, G. (1979). *Proc. Nat. Acad. Sci. USA* **76**, 6351–6355.
Schmidt, S., Reich, R. and Witt, H. T. (1970). *Naturwissenschaften* **58**, 414–415.
Schultz, J. E. and Weaver, P. F. (1982). *J. Bacteriol.* **149**, 181–190.
Scolnick, P. A., Zannoni, D. and Marrs, B. L. (1980). *Biochim. Biophys. Acta* **593**, 230–240.
Senior, A. E. (1985). *Curr. Top. Membranes Transp.* **23**, 137–157.
Sewe, K. U. and Reich, R. (1977). *Z. Naturforsch* **32C**, 161–171.
Shuvalov, V. A. and Duysens, L. N. M. (1986). *Proc. Nat. Acad. Sci. USA* **83**, 1690–1694.
Snozzi, M. and Crofts, A. R. (1984). *Biochim. Biophys. Acta* **766**, 451–463.
Sprague, S. G. and Varga, A. R. (1986). *In* 'Encylopaedia of Plant Physiology. New Series Vol. 19. Photosynthesis III' (Eds. L. A. Staehelin and C. J. Arntzen), pp. 603–619. Springer-Verlag, Berlin, Heidelberg, New York, Tokyo.
Strayley, S. C., Parsons, W. W., Mauzerall, D. C. and Clayton, R. K. (1973). *Biochim. Biophys. Acta* **305**, 597–609.
Symons, M., Swysen, C. and Sybesma, C. (1977). *Biochim. Biophys. Acta* **462**, 706–717.

Takamiya, K. I. and Dutton, P. L. (1979). *Biochim. Biophys. Acta* **546**, 1–16.

Taylor, M. A. and Jackson, J. B. (1985). *FEBS Lett.* **192**, 199–203.

Thornber, J. P. (1986). *In* 'Encyclopaedia of Plant Physiology. New Series. Vol. 19. Photosynthesis III' (Eds. L. A. Staehelin and C. J. Arntzen), pp. 98–142. Springer-Verlag, Berlin, Heidelberg, New York, Tokyo.

Thornber, J. P., Cogdell, R. J., Seftor, R. E. B. and Webster, G. D. (1980). *Biochim. Biophys. Acta* **593**, 60–75.

Tiede, D. M., Prince, R. C. and Dutton, P. L. (1976). *Biochim. Biophys. Acta* **449**, 447–469.

Trissl, H.-W. (1985). *Biochim. Biophys. Acta* **806**, 124–135.

Trissl, H.-W. and Kunze, U. (1985). *Biochim. Biophys. Acta* **806**, 136–144.

Trissl, H.-W., Kunze, U. and Junge, W. (1982). *Biochim. Biophys. Acta* **682**, 364–377.

Van den Berg, W. H., Prince, R. C., Bashford, C. L., Takamiya, K. I., Bonner, W. D. and Dutton, P. L. (1979). *J. Biol. Chem.* **254**, 8594–8604.

Van der Rest, M. and Gingras, G. (1974). *J. Biol. Chem.* **249**, 6446–6453.

Venturoli, G., Fernandez-Velasco, J. G., Crofts, A. R. and Melandri, B. A. (1986). *Biochim. Biophys. Acta* **851**, 340–352.

Vermeglio, A. (1977). *Biochim. Biophys. Acta* **459**, 516–524.

Vignais, P. M., Colbeau, A., Willison, J. C. and Jouanreau, Y. (1985). *Adv. in Microbial Physiol.* **26**, 155–234.

Wakabayashi, S., Matsubara, H., Kim, C. H. and King, T. E. (1981). *J. Biol. Chem.* **257**, 9335–9344.

Walker, J. E., Fearnley, I. M., Gay, N. J., Gibson, B. W., Northrop, F. D., Powell, S. J., Runswick, S. J., Sarastee, M. and Tybulewicz, V. L. J. (1985). *J. Mol. Biol.* **184**, 677–701.

Wasielewski, M. R. and Tiede, D. M. (1986). *FEBS Lett.* **204**, 368–372.

Webster, G. D., Cogdell, R. J. and Lindsay, J. G. (1980). *Biochim. Biophys. Acta* **591**, 321–330.

Westerhoff, H. V., Melandri, B. A., Venturoli, G., Azzone, G. F. and Kell, D. B. (1984). *Biochim. Biophys. Acta* **768**, 257–292.

Widger, W. R., Cramer, W. A., Herrmann, R. and Trebst, A. (1984). *Proc. Nat. Acad. Sci. USA* **81**, 674–678.

Williams, J. C., Steiner, L. A., Feher, G. and Simon, M. I. (1984). *Proc. Nat. Acad. Sci. USA* **81**, 7303–7307.

Witt, H. T. and Zickler, A. (1973). *FEBS Lett.* **37**, 307–310.

Wraight, C. A. (1977). *Biochim. Biophys. Acta* **459**, 525–531.

Wraight, C. A. (1979). *Biochim. Biophys. Acta* **548**, 309–327.

Wraight, C. A., Cogdell, R. J. and Chance, B. (1978). *In* 'The Photosynthetic Bacteria' (Eds. R. K. Clayton and W. R. Sistrom), pp. 471–511. Plenum Press, New York and London.

Youvan, D. C., Bylina, E. J., Alberti, M., Begusch, H. and Hearst, J. E. (1984). *Cell* **37**, 949–957.

Youvan, D. C., Ismail, S. and Bylina, E. J. (1985). *Gene* **38**, 19–31.

Zimanyi, L. and Garob, G. (1982). *J. Theor. Biol.* **95**, 811–821.

Zinth, W., Kaiser, W. and Michel, H. (1983). *Biochim. Biophys. Acta* **723**, 128–131.

Zuber, H. (1986a). *In* 'Encyclopaedia of Plant Physiology. New Series. Vol. 19. Photosynthesis III' (Eds. L. A. Staehelin and C. J. Arntzen), pp. 238–251. Springer-Verlag, Berlin, Heidelberg, New York, Tokyo.

Zuber, H. (1986b). *TIBS* **11**, 414–419.

8 Bacterial transport: energetics and mechanisms

I. R. Booth

8.1	**Introduction**	**378**
8.1.1	Classes of transport systems	378
8.1.2	Criteria for assignment of energy coupling	380
8.2	**Group translocation**	**381**
8.2.1	Introduction	381
8.2.2	Organization of the PTS	382
8.2.3	The reactions of the PTS	382
8.2.4	Molecular basis of transport	384
8.2.5	The PTS and catabolite repression	384
8.3	**Periplasmic transport systems**	**386**
8.3.1	Discovery	387
8.3.2	Structure of the periplasmic systems	387
8.3.3	Energy coupling to periplasmic transport systems	389
8.3.4	Molecular evidence for ATP as an energy source	391
8.3.5	Mechanism of periplasmic systems	393
8.3.6	The solute-translocating ATPases	394
8.4	**Ion-linked transport systems**	**396**
8.4.1	Introduction	396
	8.4.1.1 Symport	397
	8.4.1.2 Antiport	397
	8.4.1.3 Uniport	397
8.4.2	Energy coupling	398
	8.4.2.1 Introduction	398
	8.4.2.2 Uncouplers and ionophores	398
8.4.3	Experimental systems	403
	8.4.3.1 Introduction	403
	8.4.3.2 Cells *vs* vesicles: some pros and cons	404
	8.4.3.3 On the value of energy-depleted cells	405
8.4.4	Fundamental tenets of ion-linked transport	406
8.4.5	The protonmotive force and transport kinetics	410
8.4.6	Ion-linked transport: thermodynamic considerations	414
	8.4.6.1 Leaks and slips	415
	8.4.6.2 Variable coupling	416

8.4.7   Molecular mechanisms of ion-linked transport 417
    8.4.7.1   Purification and reconstitution 417
    8.4.7.2   Organization of the carrier in the membrane 418
8.5   Transport systems and homeostasis 420
8.5.1   Introduction 420
8.5.2   Control of potassium accumulation in enteric bacteria 422
8.5.3   Coping with extremes of pH: control over internal pH 424
    8.5.3.1   Acidophiles 424
    8.5.3.2   Alkalophiles 425
8.6   Conclusion 426
References 426

8.1   Introduction

All biological membranes share the property of being selectively permeable to hydrophilic solutes and, in consequence, sustaining gradients of solutes across them. This derives from the lipid bilayer being a barrier to hydrophilic molecules and from the presence of transmembrane protein complexes that can discriminate between structurally related molecules such that one is accumulated and another is excluded. Thus, in addition to selectivity the membrane proteins must also be capable of coupling metabolic energy to solute translocation in order to sustain solute gradients. Recently it has become clear that the activity of some transport systems can be regulated so that they can participate in the control of the constitution of the cytoplasm. These systems are only active when cytoplasmic homeostasis is perturbed and are otherwise largely insensitive to changes in the magnitude of energy or solute gradients.

This chapter addresses the basic features of transport systems and considers our understanding of their molecular mechanisms and their role in homeostasis. The principal citations are of reviews which cover in more depth the main points made. Individual papers are only listed when they have made a particularly important contribution to understanding of transport.

8.1.1   Classes of transport systems

Transport systems can be broadly classified as active or passive, depending on whether or not energy is expended at the translocation step (Table 8.1). The passive transport systems do not involve energy transduction and simply facilitate the equilibration of the solute across the membrane by diffusion through a protein pore. The best-studied examples of these passive systems are the porins of the Gram-negative outer membrane which facilitate the penetration of small molecular weight ions and molecules across the membrane. The discrimination of these systems is generally poor, although some such as the LamB pore have greater specificity.

Active transport systems have been subclassified as either primary or secondary, according to whether the translocation step is associated with the direct hydrolysis of chemical bonds or is coupled to the transfer of an ionized species down its electrochemical gradient (Mitchell, 1973). An alternative classification is that of group and substrate translocation (Hamilton, 1977) which distinguishes between transport processes on the basis of the state of the translocated solute. In group translocation, the solute is chemically modified as part of the transport step and consequently no transmembrane gradient of solute is established. In essence this is vectorial metabolism with an enzyme system (or components of an enzyme system) arranged across the membrane such that the substrate approaches one face of the membrane and the product is released from the other. In substrate translocation, the transported species is not chemically transformed during the transport step and energy must be expended continuously to sustain the solute gradient. The different classes of transport systems are summarized in Table 8.1 and an excellent introduction to energetic aspects of transport is presented in Harold (1986).

Table 8.1 Classes of bacterial transport systems

(a)	*Passive* (energy-independent)	
	Passive diffusion:	permeation of lipid-soluble cations and acids.
	Facilitated diffusion:	outer membrane porins (e.g. OmpF, OmpC, LamB)
(b)	*Active* (energy-dependent)	
	Primary systems:	1. group translocation (e.g. phosphotransferase system, PTS)
		2. periplasmic, binding protein-dependent, systems (osmotic shock-sensitive or OSS system)
		3. solute-translocating ATPases (e.g. Kdp system)
	Secondary systems:	Ion-linked symports and antiports, and ion uniports

Note: An alternative classification of active transport systems divides them into the group translocation systems and the substrate translocation systems (i.e. all the others).

As with many other aspects of microbiology the enteric bacteria have provided the most rewarding framework for the development of an understanding of transport at the energetic and molecular levels. Three types of transport systems have merited intensive scrutiny in the last ten years:

(a) PEP-dependent group translocation systems (PTS) (or phosphotransferase systems);

(b) periplasmic binding protein-dependent systems;

(c) ion-coupled solute transport systems (symports, antiports, uniports).

At the molecular level, these transport systems exhibit different levels of complexity. For example, the translocator proteins of the PEP-dependent phosphotransferase system are believed to be single proteins, although other proteins are involved in the transfer of the phosphate group of PEP to the sugar. On the other hand, it is now well established that the periplasmic, binding protein-dependent systems consist of at least one periplasmic protein and up to three membrane-located proteins. Finally, by contrast, the lactose, melibiose, arabinose and xylose transport systems which are coupled to the electrochemical gradient of protons, are believed to comprise a single membrane-spanning polypeptide. In Mitchell's (1973) terms the first two are both primary systems, while the third is a secondary system; the first carried out group translocation and the second and third substrate translocation. From this it is clear that these classifications themselves do not offer clues as to the complexity of the system.

8.1.2 Criteria for assignment of energy coupling

Details of the specific experiments that have been undertaken to classify a transport system in terms of its energy coupling are described under the appropriate section. However, one point needs to be made about the use of uncouplers in transport investigations. The commonest criterion used to assign energy coupling to the protonmotive force is the sensitivity of the transport process to uncouplers. The uncoupled proton influx dissipates the protonmotive force and thus inhibits transport and causes the pre-accumulated solute to leak from the cell. This criterion is only useful if backed up by other tests (described below) and can only be rigorously applied if the transported solute is non-metabolizable. If the substrate can be metabolized then the effect of uncouplers may be ambiguous. Thus the fate of the accumulated solute depends upon the nature of the compound to which it has been converted. For example, *E. coli* cells convert transported glucose to acetate, pyruvate and carbon dioxide. Addition of uncoupler to cells transporting radioactive glucose leads to the rapid exit of the radiolabelled end-products of metabolism, such as acetate, giving the appearance of inhibition of transport and efflux of the accumulated material. In this case, the actual route of transport was the well-characterized PEP-dependent phosphotransferase system and apparent uncoupler-sensitivity was due to end-products that could rapidly pass out of the cell when the protonmotive force was dissipated.

Such errors are most easily avoided by the use of non-metabolizable analogues or, in the absence of a suitable analogue, an equivalent mutant which is unable to metabolize the compound.

8.2   Group translocation

8.2.1   Introduction

Of the group translocation systems the phosphotransferase or phosphoenol-pyruvate (PEP)-dependent transport system (PTS), of bacteria is one of the best characterized (Postma and Lengeler, 1985). The system catalyses the translocation of many sugars, disaccharides and sugar alcohols into the cell concomitant with their phosphorylation at the expense of PEP:

$$sugar_o + PEP_i \rightarrow sugar\text{-}phosphate_i + pyruvate_i$$

As expected, PEP-dependent transport systems usually only occur in bacteria in which PEP is an important intermediate in the pathway for catabolism (usually fermentation) of the sugar. This pathway is usually glycolysis; bacteria with alternative routes for carbohydrate metabolism (e.g. the aerobic pseudomonads) do not usually have a PTS.

Since its discovery in extracts of *Escherichia coli* in 1964 (Kundig *et al.*, 1964) the PTS has been studied in many organisms and, despite small species-dependent variations, there is considerable uniformity in the structural components of the system. Sugar translocation requires the presence of at least two general 'soluble' proteins, Enzyme I and HPr, and a membrane-located, sugar-specific Enzyme II protein. HPr is a small heat-stable, protein (hence HPr) common to all the systems. The sugar-specific components are indicated by a superscript, thus enzyme II for glucose is written II^{glc}. Additionally, another 'soluble' protein, enzyme III, that is specific for a particular enzyme II, is involved in phosphate transfer in some systems. It has recently been determined that some systems, such as the β-glucoside PTS system, which appear not to require enzyme III proteins, have the enzyme II and enzyme III proteins fused to form a single membrane-bound entity (Bramley and Kornberg, 1987). Thus, the pathway of phosphate transfer is well characterized (Fig. 8.1).

All the phosphoryl links to the proteins of the PTS are via histidine residues. Indeed current data suggest that the position of the essential histidine in the protein may be conserved and that in some instances the amino acid sequence around the essential histidine residues is also strongly conserved. Each of the phosphorylation steps is reversible until the final transfer to the sugar is effected; thus the equilibrium constant for the transfer of phosphate from PEP to HPr via enzyme I has been estimated to be

11 ± 7.7 and consequently phospho-HPr can be used to phosphorylate pyruvate.

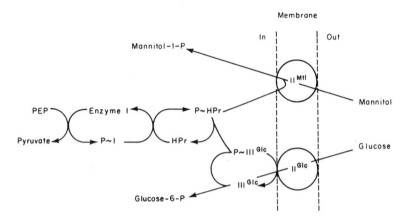

Fig. 8.1 The PEP-dependent transport system (PTS) of enteric bacteria. Enzyme I and HPr are the general proteins and the enzyme II and III are the sugar-specific proteins. Only two enzyme II proteins are shown, enzyme II^{mtl} does not require an enzyme III, but enzyme II^{glc} is dependent upon enzyme III^{glc} and does not interact with HPr. $P \sim I$, $P \sim HPr$ and $P \sim III$ represent the phosphorylated forms of these proteins.

8.2.2　Organization of the PTS

The definitions of components as 'soluble' relates only to their distribution in cell-free extracts and it is probable that in intact cells all the components interact at the inner face of the membrane. Indeed, enzymes III tend to be quite hydrophobic in character and thus partition between membrane and soluble fractions. Membrane-attached complexes that transfer phosphate from PEP through to sugar have been isolated (Saier *et al.*, 1982) and this makes the membrane the likely location for this set of reactions *in vivo*. However, from a consideration of current views of catabolite repression (see below) it is evident that enzyme III^{glc} (at least) must be free to diffuse through the cell, since it can interact with both soluble enzymes and membrane-located transport systems (Nelson and Postma, 1984).

8.2.3　The reactions of the PTS

The PTS is unique among transport systems in that its activity can be assayed in a variety of experimental systems: cell-free extracts; membrane vesicles; purified components; permeabilized cells; and whole cells. Of course, the transport reaction can only be studied in intact cells and in membrane vesicles as long as PEP is supplied intravesicularly. The essential link between transport and phosphorylation was established early in the development of the PTS system by elegant experiments with membrane vesicles (Kaback, 1968). Vesicles were pre-loaded under non-phosphorylating conditions with [^{14}C] glucose and then [^{3}H]-glucose and PEP were added at the outside of the vesicle. Only [^{3}H]-glucose 6-phosphate was formed, indicating that internal sugars could not be phosphorylated by the PTS and that translocation and phosphorylation were coupled. Although these experiments have been criticized (see Postma and Lengeler, 1985), subsequent repetition has largely confirmed the original conclusion.

In addition to the net phosphorylation of sugars at the expense of PEP, the enzyme II protein can carry out an exchange phosphorylation which is independent of both PEP and the soluble components:

$$\text{sugar}^*_o + \text{sugar-phosphate}_i \leftrightarrow \text{sugar}^*\text{-phosphate}_i + \text{sugar}_o$$

This reaction, which requires only the specific enzyme II and high concentrations of sugar-phosphate on the inside of the cell, may be of little physiological significance. Despite this limited capability for exchange there is no well-documented evidence to suggest that the enzyme II normally carries out facilitated diffusion. Thus mutants lacking enzyme I and HPr do not transport sugars into the cell via enzyme II. Furthermore, when an excess of sugar is provided externally during uptake of the radiolabelled sugar there is no exchange with the pool of sugar phosphates already accumulated. On the other hand, there is evidence to suggest that in *E. coli* enzyme II^{mtl} will allow the facilitated diffusion of galactose into the cells lacking any other transport system for this sugar. No PEP-dependent phosphorylation of galactose was demonstrated and galactokinase was essential for growth. Enzyme II^{mtl} has also been proposed to act as an exit pathway for mannitol accumulated intracellularly after hydrolysis of galactosyl-mannitol by β-galactosidase. Efflux of non-phosphorylated PTS sugars via their specific enzymes II has been implicated in the regulation of sugar utilization in the *Streptococci* and may play a role in detoxification of phosphorylated non-metabolizable analogues (see below).

The structural gene for enzyme II^{glc} can undergo mutation to allow facilitated diffusion; in *S. typhimurium* such a mutant can transport glucose through enzyme II^{glc} in the absence of a phosphorylation step. Furthermore, the mutant also lost the ability to phosphorylate glucose in the presence of PEP, HPr, enzyme I and factor III. Glucose transport in the mutant is characterized by a low affinity (10 mM compared with 5 µM) but the same overall maximum velocity. On this basis it has been suggested that the enzyme II is a sugar-specific pore which can be opened by phosphorylation at the inner face, and that the above mutation results in a changed conformation of the pore such that translocation occurs in the absence of phosphorylation.

8.2.4 Molecular basis of transport

In terms of transport the most interesting components of the system are the enzyme II molecules which catalyse the translocation of the sugar across the membrane as well as the final phosphorylation. Some insight into the structure of this class of proteins has come from a biochemical and genetic analysis. The enzyme II^{mtl} of *S. typhimurium* has been purified to homogeneity by hydrophobic chromatography and the structural gene has been sequenced. This large protein (M_r, 67 893) appears to be composed of two domains, the N-terminal 336 amino acids are largely hydrophobic, and the last 301 amino acids are more hydrophilic (Lee and Saier, 1983). Unlike the C-terminal domain of enzyme II^{bgl} there appears to be little homology of this region of enzyme II^{mtl} to enzyme III^{glc} (Bramley and Kornberg, 1987). However, it is likely that this domain plays the same role as that predicted for the analogous domain in enzyme II^{bgl}, namely the transfer of phosphate from HPr to the active site of the translocator section of the enzyme II protein.

An analysis of the topology of enzyme II^{mtl} in spheroplasts and inverted membrane vesicles using labelling reagents, proteases and polyclonal antibodies, has revealed significant asymmetry (Jacobsen *et al.*, 1983). Little of the protein is accessible from the periplasmic face of the membrane while extensive sections appear to protrude from the inner face of the membrane. It has been proposed that the protein consists of a membrane-located channel, possibly made up of membrane-spanning helices and a globular component which is exposed to the cytoplasm and may carry out the terminal phosphorylation reaction. A similar structure has been suggested for enzyme II^{bgl}, and structure predictions indicate that the histidine groups are located at the inner surface of the membrane (Bramley and Kornberg, 1987).

8.2.5 The PTS and catabolite repression

The current understanding of the PTS derives from both biochemical and genetic analysis of the system. Around the time that the PTS was discovered mutants of *Staphylococcus aureus* were isolated, some of which were unable to grow on a single sugar and others of which were pleiotropically affected in sugar utilization. Subsequently, it was determined that the former class possessed mutations affecting the gene for an enzyme II, while the latter group was found to possess inactive HPr. Genetic studies have shown that enzyme I mutants are also pleiotropically affected in the utilization of sugars. The analysis of such mutants was rendered somewhat complex by discovery that in *E. coli* and *Salmonella typhimurium* many mutants affected in the genes for enzyme I (*ptsI*) and HPr (*ptsH*) were unable to grow on a wide range of carbon sources, some of which were known *not* to be transported by the PTS (reviewed by Postma and Lengeler, 1985). The resolution of this apparent contradiction has come through the demonstration that the enzyme III^{glc} plays a central role in catabolite repression phenomena in these organisms (Fig. 8.2).

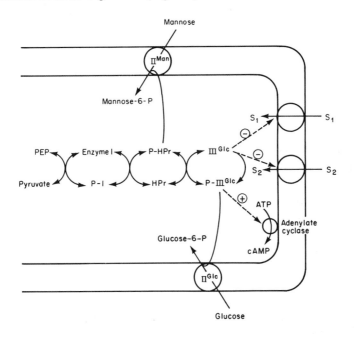

Fig. 8.2 Catabolite repression and the PTS. Enzyme III^{glc}–P acts as an activator of adenyl cyclase, and enzyme III^{glc} as an inhibitor of membrane transport systems (and at least one enzyme system) which allow the entry of inducer.

As with all the PTS components, enzyme III^{glc} exists in two states, namely phosphorylated and dephosphorylated. Addition of glucose to cells changes the balance of the two forms in favour of the dephosphorylated form of III^{glc}. It is believed that in its phosphorylated form III^{glc} acts as an activator of the enzyme adenyl cyclase, increasing its activity several-fold (but note also that many other factors also regulate adenyl cyclase activity; reviewed by Postma and Lengeler, 1985). The dephosphorylated form acts as an inhibitor of inducer entry; for example, the lactose permease of *E. coli* has been shown to form a 1:1 complex with enzyme III^{glc}. This interaction only takes place in the presence of a substrate for the lactose permease and leads to the inhibition of both active transport and facilitated diffusion via the lactose permease (Nelson *et al.*, 1983). Thus the glucose effect has two components: inhibition of the synthesis of cAMP, an essential positive effector of the expression of the genes of catabolic systems; and inhibition of the transport of the inducer.

The pleiotropy of enzyme I and HPr mutants is now readily explained by the fact that, in such mutants, the capacity to phosphorylate enzyme III^{glc} is severely reduced; thus inducer exclusion is maximal and cAMP synthesis is impaired. Indeed, it has been found that mutants resistant to catabolite repression can be isolated from strains carrying a mutation affecting either enzyme I or HPr. Such *crr* mutations arise in the gene for enzyme III^{glc} and result in loss of synthesis of this protein. Since the regulation of adenyl cyclase by enzyme III^{glc} is incomplete, sufficient cAMP is synthesized to allow expression of many operons that are normally subject to catabolite repression.

In some Gram-positive bacteria glucose is used in preference to other PTS sugars. The mechanism involves both the impairment of inducer uptake and the expulsion of pre-accumulated inducer. The system has three principal features:

(a) the ATP-dependent phosphorylation of the HPr molecule at a serine residue which specifically impairs its activity in catalysing the transfer of phosphate from PEP to the enzyme III molecules of inducible systems. This 'inactivation' of HPr blocks the further entry of the inducer of the system.

(b) the activation of a hydrolase which cleaves the accumulated sugar phosphates; and

(c) the expulsion of the de-phosphorylated sugar molecule via its enzyme II (Reizer and Peterkofsky, 1987).

Thus, in a wide variety of organisms the PTS system plays a central regulatory role in cell physiology in addition to its role as a mechanism of sugar transport and metabolism.

8.3   Periplasmic transport systems

The periplasmic transport systems derive their generic name from their possession of solute-binding proteins, which are an essential element of the transport system and are located in the periplasm. The possession of a periplasmic binding protein does place an extra complexity on the molecular mechanism of these systems; however, in terms of their bioenergetics, they may not differ significantly from other ATP-dependent systems which lack a periplasmic component (such as the Kdp potassium transport system). However, data derived from molecular studies suggest that the type of ATPase involved in the periplasmic systems is distinct from the Kdp-type of system. In general, ATP-dependent systems are often high-affinity scavenging systems capable of sustaining very high concentration gradients across the membrane.

8.3.1 Discovery

In 1965, Neu and Heppel demonstrated that the application of an osmotic shock to the cells of Gram-negative bacteria results in the release of a fraction of cellular protein which is distinct from the cytoplasm. In this procedure cells are incubated in high sucrose concentrations in the presence of Tris buffer and EDTA. The tris/EDTA loosens the outer cell membrane and the sucrose causes the cytoplasm to shrink, When such cells are diluted into low osmolarity buffer, the proteins that lie between the inner and outer membrane are released into the external medium. The space between the membranes is termed the periplasm and the proteins released by osmotic shock are the periplasmic proteins (see Chapter 3). Among these proteins has been found a large number of solute-binding proteins many of which have now been purified and characterized (reviewed by Ames, 1986). It was subsequently demonstrated that osmotic shock leads to the inactivation of many transport systems and that these systems exhibit substrate specificities identical to the periplasmic solute-binding proteins. They thus were called the 'osmotic shock-sensitive systems' (OSS systems). Finally, as vesicle systems became widely established for transport studies it became clear that the OSS systems were absent from vesicles. Those transport systems that are not affected by osmotic shock and which are present in membrane vesicles are termed the 'osmotic shock-resistant' (OSR) systems (the ion-linked transport systems described in Section 8.4, page 396).

8.3.2 Structure of the periplasmic systems

Among bacterial transport systems, the periplasmic systems are unique in that the greater part of their analysis has been carried out with the tools of molecular biology rather than with those of bioenergetics. The emerging picture of the structure of such systems is the result of the analysis of several transport systems in *E. coli* and *S. typhimurium*. Genetic analysis of the periplasmic histidine transport system of *S. typhimurium* indicates that more than one gene product is involved in the transport process since mutants could be isolated that were transport-deficient but had an intact histidine-binding protein. Thus models for the transport system which required only the binding protein could be discarded. Similarly, the binding protein could be demonstrated to be essential by the isolation of transport-defective mutants which had lost only this component. The absolute requirement for the binding protein has recently been confirmed by reconstitution of the active periplasmic system through shocking the binding protein into the periplasm of cells deprived of this protein either by mutation or by osmotic shock (see Ames, 1986).

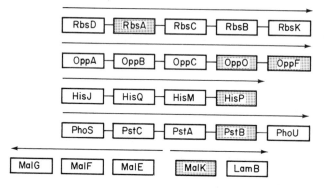

Fig. 8.3   The organization of the genes for the components of Periplasmic (OSS) systems. The genes for the ribose (Rbs), oligopeptide (Opp), histidine (His), phosphate (Pst) and maltose (Mal) OSS transport systems. The putative energy-coupling subunit (hatched) genes are indicated. The other gene products are involved with the translocation of the solute across the cytoplasmic membrane with the exception of LamB which encodes a maltose-specific outer membrane pore.

The periplasmic (OSS) systems appear to possess at least three membrane components. Thus the DNA encoding all seven systems analysed to date (histidine, maltose, branched-chain amino acids, oligopeptides, ribose, β-methylgalactoside and phosphate) has been found to contain three transcription units in addition to that for the periplasmic binding protein (Fig. 8.3). A general model of these systems may be emerging from the new DNA sequence information. Thus it is suggested that two of the membrane components show significant sequence homology and that they may form a heterodimer in the membrane. The third membrane-associated component has a predicted protein sequence which is much more hydrophilic than the other two membrane components and, in addition, shows extensive sequence homology with the analogous component of other periplasmic transport systems. This sequence homology extends to a large family of unrelated proteins, several of which are known to bind ATP; this subunit is considered, therefore, a possible ATP-binding unit (Higgins *et al.*, 1986, 1988; see below). The structures of the other subunits show little constancy of size but do show some commonality in the predicted organization of their membrane components (Hiles *et al.*, 1987).

It is well established that the membrane components may interact with more than one binding protein but usually the solutes transported are structurally related. Several of the periplasmic binding proteins not only interact with the membrane components of the transport system but also are components of the chemotactic system (Ames, 1986).

8.3.3 Energy coupling to periplasmic transport systems

The mechanism of energy coupling to the periplasmic systems has not been fully resolved but elegant physiological experiments by Berger in the early 1970s indicated that ATP or a similar metabolite is the energy source. Subsequent experiments have not significantly changed this view and bio-chemical studies are beginning to provide additional support for the idea. Berger used *unc* mutants of *E. coli* to investigate energy coupling to periplasmic (OSS) systems and ion-linked (OSR) systems (Berger, 1973; Berger and Heppel, 1974). Cells with mutations in the *unc* genes are unable to either synthesize ATP by oxidative phosphorylation or generate a protonmotive force by ATP hydrolysis when respiration is prevented. A combination of studies with these mutants, and with uncouplers and inhibitors of metabolism allowed the demonstration that the OSS and OSR systems use different energy sources (Fig. 8.4).

(a) Energy-depleted cells of the *unc* mutant were incubated with either glucose or PMS/ascorbate as energy source and the activity of OSS and OSR systems assessed. Oxidation of PMS/ascorbate can generate a protonmotive force but not ATP, whereas glucose metabolism leads to the generation of both. It was observed that OSR systems could be energized by either energy source but that the OSS systems were dependent upon active glycolysis. These experiments suggested that the protonmotive force alone was insufficient for activity of the periplasmic (OSS) systems

(b) Arsenate is a potent inhibitor of ATP synthesis and it was observed that this compound was inhibitory to OSS systems but less so to the OSR systems. From such experiments it was concluded that ATP or a similar metabolite was essential for the activity of the periplasmic (OSS) systems.

(c) When cells of the *unc* mutant were incubated anaerobically with glucose, OSR systems were inactive whereas OSS systems retained 50–

100% of the activity expressed in aerobic incubations. Under anaerobic conditions the *unc* cell is able to make ATP by substrate-level phosphorylation but cannot generate a protonmotive force because the ATPase is defective. Similarly, when the ability of aerobic cells to sustain a protonmotive force was impaired by the addition of an uncoupler, the OSR systems were more seriously impaired in their activity than were the OSS systems. Again, the absence of the ATPase prevents the diminution of the ATP pool when the pmf is reduced. Furthermore, it is clear that the inability to generate a protonmotive force did not seriously impair transport via the periplasmic (OSS) systems.

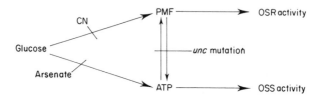

Fig. 8.4   Summary of energy-flow in *E. coli* cells. Metabolism of glucose leads to the synthesis of ATP by both substrate level and oxidative phosphorylation both of which can be prevented by arsenate. Cyanide inhibits respiration but not the synthesis of ATP by substrate level phosphorylation. The presence of an *unc* mutation prevents ATP synthesis by oxidative phosphorylation and prevents the generation of a protonmotive force by ATP hydrolysis under anaerobic conditions. The use of ascorbate/ PMS as experimental substrate leads to the generation of a protonmotive force and hence of ATP in a normal strain but not in an *unc* mutant.

On balance these experiments support the conclusion that ATP or a similar metabolite is necessary for the activity of the periplasmic (OSS) systems. Subsequent extensions of these studies in a number of laboratories have confirmed Berger's original findings for a wide range of periplasmic systems. On the other hand, there have been several reports which have suggested variously that ATP is not the energy source (Hong *et al.*, 1979), that the systems require both ATP and the protonmotive force (Plate, 1979) and that lipoic acid or some component of oxo-acid dehydrogenases (Richarme, 1985) is required for the activity of the periplasmic systems. In each of these cases the supporting evidence is equivocal.

Hong has proposed that acetyl phosphate or a metabolite derived from it is the energy donor rather than ATP (Hong *et al.*, 1979). Mutants lacking the capacity to make acetyl phosphate were reported to be deficient in glutamine transport via a periplasmic system. Subsequently, the glutamine-binding protein was reconstituted into membrane vesicles and active glutamine transport was elicited by the addition of pyruvate (Hunt and Hong,

1983). However, these vesicles exhibited a range of cytoplasmic metabolic abilities and in consequence no conclusion could be drawn about the identity of the energy donor. At present, no unequivocal evidence has been presented which would place acetyl phosphate at the centre of models for periplasmic transport systems.

The requirement for the protonmotive force for the activity of periplasmic systems has been suggested several times despite the earlier experiments of Berger which indicated that it was not required. Recently it has been shown that the activity of at least one ATP-dependent system in *Streptococcus* is very sensitive to reduction of the cytoplasmic pH (Poolman *et al*, 1987). Similarly, Ames has observed that the histidine transport system of *Salmonella typhimurium* is inhibited by lowering of the cytoplasmic pH (G. F-L. Ames, pers. comm.). Since many of the treatments used to lower the protonmotive force also result in acidification of the cytoplasm, this may explain the apparent pmf-dependence. The inhibition of periplasmic systems by 5-methoxyindole-2-carboxylic acid, a known inhibitor of oxo-acid dehydrogenases, may also fall into this same category of artefacts. The high concentrations of the weak acid that are required for inhibition may collapse the pH gradient rather than cause inhibition of the dehydrogenase.

In summary, Berger's experiments (Berger, 1973; Berger and Heppel, 1974) clearly indicate that there are two classes of transport system while at the same time remaining vague about the actual nature of the energy source for the periplasmic (OSS) systems. The experiments suffer from the rather indirect approach that can be made to the problem. It is likely that side-effects of some of the inhibitors could complicate the interpretation of the data. One certain complication is that the accumulation of any amino acid can occur via several transport systems, each of which might be of the OSS or the OSR class. Where two transport systems operate in parallel for a single solute and are differently energized, the result of any single manipulation may not be clearcut. Notwithstanding these caveats the experiments of Berger find general acceptance, with the broad conclusion that the protonmotive force is not essential to the periplasmic (OSS) systems and that ATP or a similar metabolite is the energy source.

8.3.4   Molecular evidence for ATP as an energy source

Support for the role of ATP as the energy source has recently emerged from analysis of the predicted amino acid sequences of the genes for periplasmic systems and from biochemical studies of the membrane components of the transport system. It has been established that one class of nucleotide-binding protein shares a small stretch of sequence homology which is believed to reflect the conservation of the amino acid side chains that are essential components of the nucleotide-binding site (Walker *et al.*, 1982). A search of the amino acid sequence of some of the membrane components of the periplasmic systems has revealed a similar sequence in one component of each of the systems examined (Fig. 8.5; the sole exception is the oligopeptide permease which appears to have two such 'energy-coupling' proteins). Such a finding is consistent with the binding of ATP or a related nucleotide to these proteins.

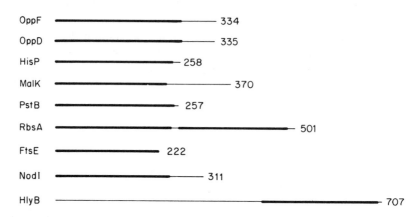

Fig. 8.5   Homology between different bacterial ATP-binding proteins and components of the periplasmic (OSS) systems. The homologous regions are shown as a thick black line superimposed upon a thinner line depicting the total sequence of the protein. The genes shown are: OppF, OppD, MalK, RbsA, PstB and HisP which are components of the periplasmic systems depicted in Fig. 8.3; HlyB is a component of a protein secretory system, NodI is a gene product of unknown function that is essential for nodulation in *Rhizobium* and FtsE is an *E. coli* gene involved in cell division. The number at the end of each line indicates the number of amino acids in each protein.

The sequence homology is not restricted to just those few residues which make up the nucleotide-binding motif (Fig. 8.5). Comparison of the amino acid sequences of a number of proteins known to hydrolyse ATP with those of the periplasmic systems has revealed extensive homology within the domain responsible for ATP binding (Higgins *et al.*, 1988). Particularly important is the finding that the functions of the other proteins studied is quite diverse, extending from DNA repair (UvrA) to nodulation (NodI) and cell division (FtsE). Furthermore, this sequence conservation is also evident in the P-glycoprotein, a mammalian transport system involved in drug-resistance in cancer cells. This extensive conservation of sequence in a protein common to diverse systems has been proposed to reflect the recruitment of an ancestral ATP hydrolytic subunit to the periplasmic transport systems (and other energy-dependent) systems at some time in the distant past.

Additional support for the role of ATP in the periplasmic systems comes from direct labelling studies with ATP and its analogues. Membrane components of the systems for oligopeptide (OppD), histidine (HisP) and maltose (MalK) transport have been shown to bind ATP or its analogue. Thus the OppD protein has been specifically labelled with 5'-p-fluorosulphonylbenzoyl adenosine (Higgins *et al.*, 1985) and the HisP and MalK proteins have been labelled with 8-azido-ATP (Hobson *et al.*, 1984). In each case the subunit which binds ATP is also that which has extensive sequence homology to other ATP-binding proteins.

In conclusion there can be little doubt that all the evidence emerging favours the involvement of ATP (or a related nucleotide) as the energy source for the periplasmic binding protein systems.

8.3.5 Mechanism of periplasmic systems

Most acceptable models for the periplasmic transport systems envisage an interaction between the periplasmic binding protein, its substrate and several membrane components which effect the translocation process. The precise mechanism of energy input is unclear. The binding protein itself is usually essential for transport but mutants have been obtained which possess altered membrane components such that solute translocation is now independent of the periplasmic component.

The periplasmic binding proteins possess a high affinity for their transport solute (usually in the range 0.1–1 μM) and, on binding their substrate, the proteins undergo a conformational change. Direct evidence that this conformational change triggers interaction of the binding proteins with the membrane components is lacking. However, there are several lines of evidence which support a model which envisages such an interaction (reviewed by Ames, 1986).

(a) Structural and genetic analyses have revealed that the binding proteins have two functionally separate domains. First, a mutant form of the histidine-binding protein of *S. typhimurium* has been isolated which retains the capacity to bind histidine but is unable to operate in transport. Analysis of this mutant by NMR demonstrates that, although histidine was bound, no conformational change occurred.

(b) Suppressor mutations which restore histidine transport to this mutant arise in the *hisP* gene which encodes a membrane component of the system. This suggests that a conformational change in the membrane components allows the mutant-binding protein to become functional and thus that the binding protein can interact with at least one of the membrane components of the transport system.

(c) Comparison of the protein sequences of two binding proteins (histidine binding protein and the related lysine–arginine–ornithine-binding protein) that share a common set of the membrane components revealed considerable sequence conservation. The greatest sequence homology was in the domain which had been implicated in the interaction of the histidine-binding protein with the membrane components (Ames, 1986). Figure 8.6 shows one model which is consistent with the known data (Ames and Higgins, 1983).

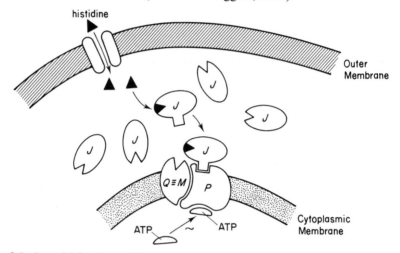

Fig. 8.6 A model for the histidine periplasmic (OSS) system. Histidine interacts with the histidine binding protein (HisJ) which changes conformation. The complexed protein interacts with the membrane components (His P, Q and M) and on hydrolysis of ATP the histidine is translocated across the membrane. The squiggle (~) suggests an involvement of ATP in energy coupling in an unknown way.

In summary, current models of this class of transport system would encompass the following features. In the periplasm the binding protein binds its solute and, after undergoing a conformational change, interacts at the membrane with one or more membrane components. As a result of this interaction, and possibly coupled to the hydrolysis of ATP, the solute is translocated across the membrane.

8.3.6   The solute-translocating ATPases

A separate group of primary transport systems are the solute-translocating ATPases, so-named because they can be responsible for activities as diverse as the accumulation of potassium and amino acids and the expulsion of calcium and arsenate. These systems have been found in both Gram-negative and Gram-positive organisms and the lack of a periplasmic binding protein may be only one of several mechanistic distinctions from the periplasmic (OSS) systems. The best-studied example of this class of system is the *E. coli* potassium ATPase encoded by the *KdpABC* genes (reviewed by Epstein, 1985).

Genetic analysis of *E. coli* mutants impaired in growth at low external potassium concentrations has been used to define two sets of genes (*kdp* and *trk*) involved in potassium uptake (Epstein, 1986). The *kdp* genes define the components of an inducible high affinity potassium uptake system (*kdpABC*) and its regulatory elements (*kdpDE*) which ensure that the transport system is induced whenever turgor is reduced. Induction of the Kdp transport system leads to the appearance of a high affinity potassium-stimulated ATPase in the cell membrane. Mutations which affect the affinity of the Kdp system affect the threshold potassium concentration required for ATPase activity in an identical manner. The products of the three structural genes are located in the inner membrane, and genetic and biochemical approaches have been used to assign functions to the *kdpA* and *kdpB* gene products. The KdpA protein is believed to be a membrane-spanning protein of approximately 47 000 Daltons. Mutations which affect the affinity of the Kdp system for potassium have been mapped to the *kdpA* gene and thus, to a first approximation, the product of this gene may play a role in binding the cation. Genetic analysis of such mutants suggests that the protein may form oligomers (Epstein, 1985).

The KdpB protein is 90 000 Daltons, carries the ATP-binding site and has been shown to be phosphorylated by ATP. The addition of potassium to the incubation mixture reverses the phosphorylation. Sequencing of the genes for several ATPases has revealed that the KdpB protein has extensive sequence homology with several other cation-transporting ATPases, notably the Ca^{++}-ATPase from muscle (Hesse *et al.*, 1984) and other fungal and mammalian ATPases. The region of homology surrounds the phosphorylation site and shows no homology with the conserved sequences found in the ATP-binding site of the periplasmic transport systems. No function has yet been ascribed to the KdpC protein, although its activity is essential for Kdp transport and ATPase activity (Epstein, 1985).

Study of this class of system is still in its infancy and it is evident that there may be more than one biochemical mechanism among the different ATP-dependent transport systems. From the standpoint of bioenergetics it is fair to say that the ATP-dependent systems have received little detailed attention. For example, the relationship between the phosphorylation potential (ΔGp) and the gradient of the transported solute has not been systematically investigated. The reversibility and the capacity of the periplasmic systems for exchange reactions has also never been established. Thus, while the organization and molecular biology of these systems is being established there is a gap in our knowledge of the fundamental properties of these systems.

8.4 Ion-linked transport systems

8.4.1 Introduction

The ion-linked transport systems are usually low-affinity, high-velocity transport systems which share several unique features:

(a) they are readily reversible and indeed they are the only active transport systems which can operate by facilitated diffusion when the cell is de-energized;

(b) in the absence of metabolically derived energy they can be driven by artificially generated ion and charge gradients;

(c) in general, this is the only class of active transport system which can be found in membrane vesicles (the PTS may also be present but this carries out group translocation rather than active transport);

(d) each of the ion-linked transport systems is the product of a single gene.

These properties have been so invaluable in the purification of the membrane components of transport systems that members of this class are the only membrane components of transport systems to have been purified to homogeneity. The proteins have been reconstituted in artificial membrane vesicles with reasonable success and have been studied in isolation from other protein components of the membrane. In this respect, the study of the ion-linked systems is more advanced than that of the other classes discussed. However, in molecular terms, the transport proteins are still a long way from being really understood.

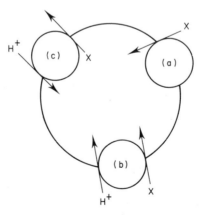

Fig. 8.7 Classes of secondary transporters: (a) uniporter; (b) symporter and (c) antiporter. The solute (X) may be charged or neutral except in the case of uniports when X must be charged.

In essence, the ion-linked systems are very simple: they are proteins which couple the movement of one ion (usually H^+ or Na^+) down its electrochemical gradient to the movement of another ion or solute up its chemical gradient. Mitchell (1973) designated three classes of ion-linked system; symport, antiport and uniport (Fig. 8.7 and Fig. 1.8, page 24).

8.4.1.1   Symport

The solute and the coupling ion traverse the membrane in the same direction. Predominantly these systems function for the accumulation of molecules or ions in the cytoplasm, although some are utilized as routes for excretion of fermentation end-products.

8.4.1.2   Antiport

The solute and the coupling ion move in opposite directions across the membrane. The function of these systems depends on the solute being considered and, occasionally, on the precise circumstances. For example, a Na^+/H^+ antiport can function with sodium as the solute leading to the expulsion of Na^+ from the cell and the establishment of a transmembrane Na^+ gradient which can be utilized to drive the uptake of other solutes by Na^+-linked symports. On the other hand, in some alkalophiles the antiport may function with Na^+ as the driving ion and the protons as the solute leading to acidification of the cytoplasm (see below). Also falling into this category are systems that can serve simply to exchange solutes, e.g. the glucose 6-phosphate uptake system can catalyse phosphate/glucose 6-phosphate exchange (reviewed by Maloney *et al.*, 1987).

8.4.1.3   Uniport

A uniport is a system in which there is no coupling ion, and the driving force for uptake is the charge on the molecule itself under the influence of the membrane potential (Fig. 8.7). Uniports often have the apparent property of irreversibility. However, in reality this often arises from their substrates, generally cations, becoming complexed with cellular components (Kashigawa *et al.*, 1986). Only in membrane vesicles do these systems display their true reversibility.

Of these three, the symport is the best documented, largely as a result of many years' work on the lactose permease but more recently because of an expanding array of similar systems which has become more accessible through gene cloning and the general ease with which genetic studies can be accomplished. Antiports are also well documented but their properties have been less intensively studied because their reaction is generally that of solute expulsion from the cell, which makes them intrinsically less easy to study.

Evidence for the existence of uniports is equivocal since the critical experiments that would eliminate the translocation of any other ion from the mechanism have not been carried out. The transport systems for lysine in *Staphylococcus aureus*, for polycations in Gram-negative bacteria, and possibly for the ammonium ion and some antibiotics may be via uniports.

8.4.2   Energy coupling

8.4.2.1   Introduction
Energy coupling for the ion-linked systems ultimately stems from the transmembrane protonmotive force (pmf) which is established across all bacterial cytoplasmic membranes by the action of reversible proton pumps (Chapter 1). The protonmotive force generated by the primary pumps is made up of a transmembrane proton gradient (ΔpH) and a membrane potential ($\Delta\psi$) (see Table 1.2, page 12). The contribution of the pH gradient and the membrane potential to the pmf varies in reciprocal fashion with the external pH. This reflects the tendency of the cell to maintain its cytoplasmic pH constant despite variations in the external pH (Fig. 8.8). Thus, bacteria living in extremely acidic conditions possess a large pH gradient, alkaline inside, and a membrane potential which may be positive inside. Bacteria existing in the neutral and mildly acidic pH range generally have a pmf which, in simple numerical terms, consists almost equally of a pH gradient, alkaline inside, and a membrane potential, negative inside. Above pH 8.0 these organisms and the obligate alkalophiles have a pH gradient, acidic inside, and a membrane potential, negative inside. The changing distribution of the components of the pmf with external pH is illustrated in Figure 8.8.

Investigation of ion-linked transport systems has frequently involved the use of uncouplers and ionophores and some discussion of their properties is in order.

8.4.2.2   Uncouplers and ionophores
Ionophores are compounds that facilitate the movement of ions across the cytoplasmic membrane and thus increase the intrinsic membrane permeability to that ion. The action of ionophores and uncouplers is independent of any membrane protein. These compounds are generally very lipid-soluble and dissolve in the lipid bilayer and either form channels or shuttle between the two faces of the membrane where they load or unload ions. When the ion translocated is a proton, the compound mediating transfer of the proton is referred to as an uncoupler; when the ion translocated is another cation (e.g. K^+, Na^+, Ca^+) the compound is referred to as an ionophore, even though movement of the cation may be via exchange with

a proton. This is not simply a semantic point but relates to the effect of the compound on the pmf. Uncouplers bring about the almost complete dissipation of the pmf, whereas most ionophores selectively dissipate either $\Delta\psi$ or ΔpH. This definition is used throughout this text. Further, it should be realized that dissipation of one component of the pmf often leads to enhancement of the other with consequently small effects on the total pmf. By contrast, small changes in the composition and magnitude of the pmf can lead to major changes in the activity of a transport system (see Table 8.2, page 402).

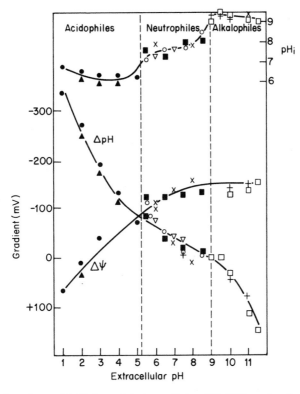

Fig. 8.8 pH-dependence of ΔpH, $\Delta\psi$ and the intracellular pH (pHi) in various bacteria able to grow across part of the pH range pH 1–12. Symbols: ($\times$) *E. cloi*; ($\bigcirc$) *Micrococcus lysodeikticus*, ($\blacksquare$) *Halobacterium halobium*, ($\triangle$) *Streptococcus faecalis*, ($\blacktriangle$) *Bacillus subtilis*, ($\blacktriangle$) *Bacillus acidocaldarius*, ($\bullet$) *Thiobacillus ferrooxidans*, ($\square$) *Bacillus alcalophilus*, and ($+$) *Bacillus firmus*.

(a) Uncouplers
These are weakly acidic compounds in which both the undissociated weak acid and the anion exhibit a high lipid permeability. On addition to cells or

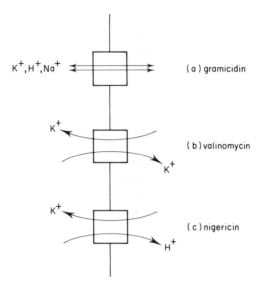

Fig. 8.9   The mode of action of ionophores. Ionophores either form channels through the membrane (e.g. gramicidin) or shuttle ions from one face of the membrane to the other (e.g. valinomycin, trichlorocarbanilide and nigericin). See text for further details.

vesicles the undissociated acid permeates across the membrane and dissociates on the inside; the liberated anion then exits from the cell under the influence of the membrane potential, and the net effect is that the anion cycles across the membrane, leading to proton influx. The uncoupled proton influx dissipates the pmf. One generally has a free choice of which uncoupler to use; however, the following should be borne in mind:

(i) *Side-effects*: both FCCP and CCCP have the capacity to act as sulphydryl reagents and this can inhibit respiration by direct interaction with the components of the respiratory chain. DNP elicits potassium efflux from *E. coli* cells when used at high concentrations, whereas other uncouplers have no effect, suggesting that this compound may interact with some membrane proteins to change their activity. Consequently, the concentration range of these compounds should be carefully controlled.

(ii) *Association with membranes*: as stressed above, the uncouplers are all lipid-soluble and thus in Gram-negative bacteria they tend to associate with both the outer and the inner membrane. Consequently the concentrations that must be used in whole cells tend to be rather higher than for membrane vesicle preparations. In general, it is likely that any organism with a high lipid content will appear to be resistant to uncouplers due to the preferential partitioning of these compounds into the lipid layer. Indeed, mutations to resistance to uncouplers frequently involves the production of greater amounts of lipid.

(iii) *Changes in lipid composition*: changes in the lipid composition may also affect the resistance to uncouplers. Thus it has been reported that the ether lipids of the archaebacteria may lead to high resistance to uncouplers. In passing, it is also worth noting that there is some evidence to suggest that the cytoplasmic membranes of eukaryotic organisms are more resistant to uncouplers than are those of procaryotes. The membranes of the subcellular organelles do not show any significant resistance to uncouplers and it may be the presence of sterols in the cytoplasmic membrane of the higher organisms that causes resistance.

(b) Ionophores

Of the ionophores that have been discovered, only valinomycin, gramicidin, nigericin, monensin and A23186 have become generally accepted in the study of transport processes. The mode of action of each of these compounds is given in Fig. 8.9 and their effect of the protonmotive force is discussed below (see also Bakker, 1979). With the exception of gramicidin, the ionophores shuttle back and forth across the membrane as a complex with an ion. In the case of gramicidin, a transmembranous channel is formed which allows the free passage of K^+, Na^+, H^+ and probably NH_4^+. Addition of gramicidin will completely dissipate the pmf, rather than achieve the selective dissipation of one component. One further general point: the Gram-negative bacteria are relatively impermeant to ionophores due to the presence of the outer membrane. Isolation of rough mutants which lack the lipopolysaccharide layer, or treatment with EDTA (Leive, 1968), increases the sensitivity of the cell to most ionophores.

(i) *Valinomycin.* This antibiotic has proved to be one of the most useful in the investigation of energy coupling to transport. It is a cyclic depsipeptide produced by certain strains of *Streptomyces*. As with other ionophores (and uncouplers) it is insoluble in water and must be dissolved in ethanol. Valinomycin greatly increases the permeability of the membrane to cations and exhibits the preference $H^+ > Rb^+ > K^+ > Cs^+ > Na^+ > Li^+$. At physiological pH and in the absence of rubidium, the primary effect is an increase in the permeability of the membrane to potassium, and it is in this mode that it has found greatest utility. The valinomycin molecule has been shown to encapsulate the potassium ion in a cage-like structure which has a hydrophobic exterior, thus facilitating the transfer of the ion across the lipid bilayer.

Table 8.2   Relationship of the solute gradient to ΔpH and $\Delta\psi$

System	m	n	Electrochemical gradient (volts)
Uniport	1	0	$\Delta\psi$
	>1	0	$> \Delta\psi$
Symport	0	1	$\Delta\psi - Z\Delta\text{pH}$
	-1	1	$- Z\Delta\text{pH}$
	-1	2	$\Delta\psi - 2Z\Delta\text{pH}$
	$+1$	1	$2\Delta\psi - Z\Delta\text{pH}$
Antiport	1	-1	$+ Z\Delta\text{pH}$
	1	-2	$-\Delta\psi + 2Z\Delta\text{pH}$

The transmembrane electrochemical gradient of solute ($\Delta\bar{\mu}S/F$) is calculated from the formula:

$$\Delta\bar{\mu}S/F = (m + n) \Delta\psi - nZ\Delta\text{pH}$$

where m is the charge on the solute and n is the number of protons cotranslocated with the carrier; see also Table 1.2, page 12).

Valinomycin can be used in two principal ways: it can be used either to dissipate a membrane potential preformed by the action of one of the transmembrane proton pumps or it can be used to generate a membrane potential in de-energized cells or vesicles. In the former case the presence of valinomycin allows potassium to equilibrate across the membrane in response to the existing membrane potential. The membrane potential represents the imbalance between net outward and net inward movement of charged species. As the potassium concentration is raised, more potassium crosses the membrane into the cell and the differential between outward and inward fluxes is eroded, leading to a fall in the membrane potential. This occurs despite an increase in the rate of proton extrusion occasioned by the release of the proton pump from the feedback control by the membrane potential. However, one consequence of increased proton pumping is that the loss of the membrane potential is partially compensated for by an increase in the pH gradient. A drawback to this use of valinomycin is that swelling of the cell can take place even in bacteria which are usually thought to be tightly constrained by the peptidoglycan layer. Ultimately, the incubation of cells in high potassium concentrations (plus valinomycin) leads to a generalized increase in membrane permeability and osmotically sensitive cells may burst.

The use of valinomycin to generate a membrane potential in energy-depleted cells or vesicles is quite simple. The cells or vesicles are incubated such that they become replete in potassium and then they are diluted into potassium-free medium in the presence of valinomycin. In the presence of the ionophore, the potassium flows down its concentration gradient and the movement of charge causes the build-up of a membrane potential. The use of this technique is discussed below (see pages 408–410).

(ii) *Nigericin, monensin and A23187.* Each of these ionophores catalyses the electroneutral exchange of a cation for a proton; their specificities are nigericin, K^+, monensin, Na^+ and A23187, Ca^{2+}. In each case, when there is a significant cation gradient across the membrane the addition of the ionophore will collapse the gradient and lower or raise the pH gradient, depending on the orientation of the cation gradient. Since the potassium gradient is oriented outward the addition of nigericin will cause loss of potassium and lowering of the cytoplasmic pH due to the influx of protons. Both sodium and calcium are usually expelled by cells and thus the addition of the respective ionophore causes cation influx thus dissipating the ion gradient and causing proton efflux.

Each of these three ionophores has its specific uses but, given the ubiquity of proton-coupled systems, it is nigericin which has been most frequently used. Dissipation of the pH gradient with nigericin leads to a small enhancement of the membrane potential for reasons similar to those for valinomycin. As with valinomycin, the Gram-negative envelope acts as a barrier to the penetration of nigericin; this can be overcome by the use of lipopolysaccharide (LPS) mutants and less successfully with the use of EDTA treatments. Nigericin has the further disadvantage that it forms dimers when present at high concentrations. The dimeric form can facilitate electrogenic exchange of potassium and protons, which leads to the dissipation of both the pH gradient and the membrane potential. Trichlorocarbanilide (TCC) (an alternative to nigericin) facilitates the electroneutral exchange of chloride for hydroxyl ions and thus similarly collapses the pH gradient. The advantage of this compound is that it can be used without pretreatment of the cells. (The trialkyltin compounds have a similar action.)

8.4.3 Experimental systems

8.4.3.1 Introduction

Two basic experimental systems have been useful in the study of ion-linked transport systems: whole cells and membrane vesicles. Whole cells are the starting-point for most investigations because of their ease of preparation, but they are often found to have shortcomings which can be overcome by the preparation of membrane vesicles. Membrane vesicles are made by lysing cells in such a way that the membrane reseals to give predominantly closed vesicles in which the cytoplasm is replaced by buffer. Great care needs to be taken in the preparation of the vesicles in order to ensure that the majority have the same orientation. By osmotic lysis of protoplasts it is possible to prepare populations of vesicles which consist of more than 95% of right side out orientation (Kaback, 1971). Lysis of cells at low pressure (approx. 8000 psi) in a French pressure cell leads to the inversion of the vesicle membrane (Futai *et al.*, 1974). Inverted membrane vesicles generally have a smaller internal volume than those prepared by osmotic lysis and this can limit their utility in transport experiments. However, inverted vesicles have been invaluable in the analysis of extrusion systems and were used to demonstrate that the asymmetry of function of ion-linked transport systems is due to imposition of a pmf and is not inherent to the structure of the permeases (see below). Membrane vesicles can be stored in the deep-freeze ($-70°C$), so that identical membrane preparations can be studied over an extended period.

8.4.3.2   Cells *vs* vesicles: some pros and cons

In cells it is difficult to control the generation of the pmf; the preferential use of fermentative bacteria in the early studies of ion-driven transport systems arose from the ease with which their pmf could be reduced to close to zero simply by removing the energy source. Removal of an external carbon source from respiring organisms will often result in only a modest drop in the total pmf due to their ability to respire endogenous reserves. Membrane vesicles on the other hand are essentially totally de-energized until a respiratory substrate is added. Such preparations can thus readily be used for studies of facilitated diffusion, transport driven by respiration, and uptake energized by artificial gradients.

It used to be the case that vesicles could not be readily energized by ATP, although by incorporating pyruvate kinase, PEP and ADP in the lysis buffer during vesicle preparation, ATP can be supplied intravesicularly. This has recently been simplified by the cloning of the *glfP*-encoded transport system of *Salmonella* which will transport PEP (Hugenholz *et al.*, 1981). Thus any strain into which the plasmid-borne gene has been introduced may now be used to prepare vesicles with ADP and pyruvate kinase entrapped in which the supply of ATP can be controlled by the addition of PEP.

In membrane vesicles the composition of the pmf can be controlled by the addition of ionophores with much greater ease than can be effected in whole cells. Whole cells of Gram-negative bacteria are quite resistant to the effects of ionophores such as gramicidin, nigericin and valinomycin. The inherent susceptibility of Gram-positive organisms to ionophores was again a factor which favoured their exploitation for bioenergetic studies.

The preparation of vesicles allows transport of a substrate to be studied without its concomitant metabolism or incorporation into cellular components. This can also be achieved in whole cells by the isolation of mutants unable to metabolize the compound, by the synthesis of appropriate analogues or by the use of inhibitors. The preparation of vesicles is sometimes a simpler way to separate the transport step from non-essential components of the cell metabolism and yet retain the essential characteristic of concentrative transport.

Finally, many of the manipulations undertaken will cause secondary changes in the composition of the cytoplasm. Whether these are significant will depend on the system under study. For example, washing cells with Na^+-containing buffers leads to an exchange of internal K^+ for external Na^+. The accumulated Na^+ is often sufficient to inhibit the activity of Na^+-driven transport systems (Stewart and Booth, 1984). Addition of potassium in the presence of an energy source leads to exchange of internal Na^+ for external potassium and restores the activity of the transport system. Consequently, what the experimentalist sees is a transport system stimulated by potassium ions which is, in fact, coupled to Na^+ ions (indeed K^+ stimulation of transport is often a sign of sodium coupling!). In membrane vesicles, the lumen of the vesicle has a very precise composition which can be dictated by the experimentalist. However, even here it is essential to use very pure reagents in order to avoid misleading conclusions.

The major disadvantages of vesicles is that they are inherently leaky to ions and in consequence $\Delta\psi$ is much lower than routinely observed in cells. On the other hand, vesicles display the same pH-dependence of the pH gradient as that shown by cells. Vesicles also lack cytoplasmic and periplasmic components which are essential for some transport activities. However, there have been several recent reports of the re-addition of periplasmic proteins to membrane vesicles with restoration of transport.

8.4.3.3 On the value of energy-depleted cells
Notwithstanding the value of vesicle systems, many of the early important observations on transport were made using energy-depleted cells and have subsequently been confirmed with vesicles. Energy-depleted cells may be made in one of two principal ways. First, cells can simply be incubated for lengthy periods in the absence of an energy source under aerobic conditions until the endogenous reserves are depleted. This incubation period can be extensive, periods of greater than 24 h are not unusual. The alternative approach is to cause the dissipation of the protonmotive force by anaerobic incubation of the cells in the presence of iodoacetic acid, a powerful inhibitor of glycolysis. One danger of this last approach is that some transport systems are sensitive to direct inhibition by iodoacetate which is a

sulphydryl reagent. However, after 30–40 min incubation cells treated in this way lack a significant pmf (one note of caution, the continued absence of oxygen is essential since once oxygen is reintroduced a pmf will be regenerated). In general, the two types of cells have been used for the investigation of transport in response to artificial imposed ion gradients, and ion movements associated with solute transport, respectively.

8.4.4   Fundamental tenets of ion-linked transport

There are two fundamental tenets of the chemiosmotic view of transport. First, the imposition of a solute gradient on de-energized cells will lead to the transfer of protons across the membrane in the direction governed by the mechanism of the carrier, i.e. inwards for a symport and outwards for an antiport. Secondly, the imposition of a transmembrane pmf will result in translocation of a solute against its gradient.

West demonstrated lactose-induced proton uptake in de-energized cells of *E. coli* possessing a functional lactose permease (West, 1970; West and Mitchell, 1973). The addition of lactose to cells generates a transmembrane gradient of solute, high outside, low inside. In the absence of a pmf, lactose enters the cell by facilitated diffusion driven by the concentration gradient and transfers a coupling proton with it (Fig. 8.10). The transfer of the proton generates a membrane potential, inside positive, which must be dissipated in order that further influx will not be impeded. This can be achieved either by the presence of thiocyanate (SCN^-) in the incubation buffer or by treatment of cells with valinomycin and incubation with high potassium concentrations. In either case, an ion is then present which can diffuse freely across the membrane and neutralize the charge on the proton. Thus thiocyanate (or valinomycin/potassium) stimulates the rate of lactose-induced proton uptake. Proton uptake can be prevented either by preincubation with uncouplers which recycle the transported proton or by addition of inhibitors of the lactose permease.

A logical development of these experiments is that solute efflux down its gradient should lead to the generation of a pmf. This has been demonstrated both as a phenomenon and as a mechanism of energy sparing. De-energized *E. coli* cells pre-loaded with either β-galactoside or gluconate, and incubated in a sugar-free buffer were found to be able to drive the accumulation of proline (Bentaboulet *et al.*, 1979). Of more practical value is the proposal that fermentative organisms gain energy-sparing through the generation of a pmf by the efflux of lactate via an electrogenic transport system (reviewed by Konings, 1985). In the homofermentative *Streptococci*, lactate is the major fermentation product and so the gradient of lactate is generated and maintained by metabolism. Efflux via an electrogenic carrier leads to the removal of protons from the cytoplasm, thus generating a membrane potential and a pH gradient. Such efflux has been demonstrated to be sufficient for the accumulation of other amino acids by ion-linked transport systems.

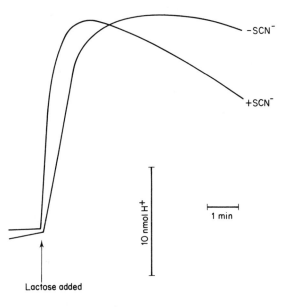

Fig. 8.10   Proton uptake driven by lactose transport. *E. coli* ML308-225 cells were incubated in weakly buffered medium at pH 7.0 in the presence of iodoacetate to inhibit glycolysis, and under a stream of nitrogen to exclude air (West and Mitchell, 1973). Lactose (4 mM, final concentration) was added and the resulting proton uptake measured (I. R. Booth, unpublished data).

Similar experiments have since been used to demonstrate proton uptake with a variety of sugars and amino acids, and the counterflux of protons with Na^+ via the Na^+/H^+ antiport. Furthermore, sodium ion electrodes have been used to demonstrate the movement of sodium ions into de-energized *E. coli* cells during the transport of melibiose and proline which are accumulated by Na^+-symports.

Following on from the experiments of West, other groups sought to establish that the pmf could drive transport. The demonstration that an artificially generated pmf could drive transport was critical to establishing the validity of the concept of ion-linked transport systems. An artificial pH gradient was established by preincubation of cells or vesicles at high pH and dilution into a lower pH buffer; and a membrane potential was generated by the dilution of potassium-loaded cells/vesicles into a potassium-free buffer in the presence of valinomycin. In the former experimental regime, the intrinsic low permeability of the membrane to protons prevents the rapid equilibration of the pH of the two compartments, and so a pH gradient is established across the membrane. A pH gradient can also be generated by pre-loading cells/vesicles with a weak acid and then diluting the cells/vesicles into an acid-free medium. The weak acid leaves the cell down its concentration gradient and in so doing removes protons from the inside of the cell/vesicle (Fig. 8.11). The generation of a membrane potential across the membrane relies on the increased permeability of the membrane to potassium ions in the presence of valinomycin. The movement of the cation down its concentration gradient establishes a membrane potential which dissipates gradually as protons or other ions move down the charge gradient (Fig. 8.11).

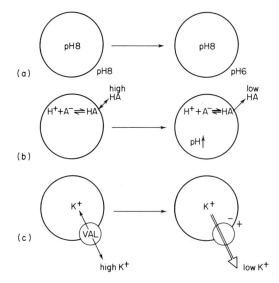

Fig. 8.11 Generation of an artificial pH gradient or membrane potential. (*a*) Generation of a pH gradient by transfer of cells/vesicles to a lower external pH; (*b*) generation of a pH gradient by weak acid efflux down its concentration gradient; (*c*) generation of a membrane potential by potassium efflux down its concentration gradient in cells/vesicles in which the membrane permeability to potassium ions has been increased by the presence of valinomycin.

These experimental regimes were used to demonstrate the transport of a number of amino acids in *E. coli* membrane vesicles and of thiomethylgalactoside (TMG) in de-energized cells of *Streptococcus lactis* (Kashket and Wilson, 1973; Hirata *et al.*, 1974; Figs 8.12 and 8.13). The transport of the solute in response to the imposed gradient was just as sensitive to uncouplers as that energized by a respiratory substrate in cells or in membrane vesicles. However, accumulation of the solute was transient, i.e. uptake was followed by efflux of the accumulated material. At the steady state, a membrane potential is balanced by a pH gradient of opposite polarity and there is no net driving force across the membrane. For a neutral solute, such as TMG, the transport system is coupled to both components of the pmf and, as this force is dissipated, the solute gradient exceeds the driving force and reverses the direction of net flux. For a charged solute, such as lysine, the driving force is the membrane potential, and as this is maintained at the steady state, lysine does not readily leave the vesicle.

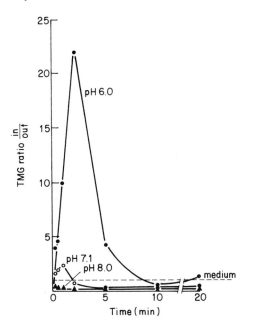

Fig. 8.12 Uptake of TMG in response to an artificial pH gradient in *S. lactis* cells. Energy-depleted cells of *S. lactis* were diluted into buffers of different pH values and radioactive TMG added to the cells. The pH gradient formed by dilution of the cells into buffer at low pH is sufficient to cause the transient accumulation of TMG by a proton symport. At higher pH values, the pH gradient is either zero (pH 7.0) or inverted (pH 8.0).

One further refinement of these techniques was used to demonstrate that the lactose transporter is functionally symmetrical (Lancaster and Hinkle, 1977). Membrane vesicles with opposed orientations were prepared from *E. coli* cells induced for the lactose transport system. A pmf was imposed upon the vesicles using valinomycin/K to generate a membrane potential and an acetate gradient to generate the pH gradient. Despite the different orientation of the membranes, both vesicle preparations accumulated lactose against a concentration gradient. The carrier is thus functionally symmetrical and the direction of movement of lactose and proteins via the carrier is determined by the magnitude and orientation of the lactose gradient and of the pmf.

Subsequent studies have concentrated on the thermodynamics of transport and on the elucidation of the molecular mechanism of the translocation step. At present, neither has been totally resolved but the former attracts a greater consensus of opinion.

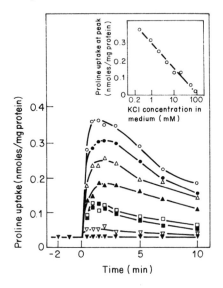

Fig. 8.13 Proline uptake energized by an artificial membrane potential. *E. coli* membrane vesicles were potassium-loaded and then diluted into sodium-phosphate buffer containing different concentrations of potassium. As the external concentration of potassium is raised the membrane potential generated is reduced and the maximum accumulation of proline is consequently lower.

8.4.5   The protonmotive force and transport kinetics

It is generally accepted that the pmf exerts its influence over ion-linked carriers through the modulation of the kinetic parameters of the translocation process. Indeed, the initial proposal that energy coupling to the lactose carrier was mediated through changes in the kinetics of efflux was first made over 20 years ago (Winkler and Wilson, 1966). In the intervening years there has been less agreement on which steps in the translocation cycle are affected and, indeed, these may differ from one carrier to the next. At present there is inadequate information available to allow discrimination between alternative models of the mechanism of energy coupling.

It has been proposed that a simplified view of energy coupling can be obtained by dividing the effects of the pmf into mass action effects, affinity effects and kinetic effects (Overath and Wright, 1983). Consider the reaction cycle depicted in Figure 8.14. The carrier protein C can exist in four states— C, CH, CHS and CS—each of which can be found exposed at either face of the membrane. It is assumed that there is only one binding site and that this site is alternately exposed at each face of the membrane. It has generally been assumed that the binding reactions are rapid relative to the rate of translocation itself, and thus that the rate-determining step is the reorientation of the carrier (but see Sanders *et al.*, 1984, for a detailed discussion of alternatives).

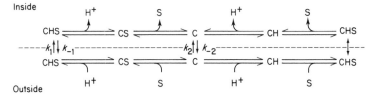

Fig. 8.14 The reaction cycle for a proton symport. The figure depicts the binding of solute (S) and proton (H) to the carrier (C) and that the order of binding is random. Only the unloaded (C) and the loaded (CSH) forms of the carrier are depicted crossing the membrane.

(a) Mass action effects

From Figure 8.14 it is clear that the formation of the protonated forms of the carrier will depend upon the local proton concentration and the pK of the groups involved in proton binding. Thus, changes in the pH gradient could affect the distribution of the carrier at the two faces of the membrane through the trapping of the carrier at the low pH side of the membrane. The rate of exit of lactose from both *E. coli* cells and vesicles is reduced three- to four-fold by lowering the external pH from pH 8.0 to pH 5.5. Exchange of lactose under the same conditions is pH-insensitive suggesting that translocation itself is not affected by pH. In terms of Figure 8.14, trapping of the carrier at the external face of the membrane can explain the pH-dependence of efflux, since it reduces the effective concentration of the carrier available to participate in solute efflux from the lumen of the cell or vesicle. Other rate constants for lactose transport are largely unaffected by pH and it has been proposed that energy coupling to the pH gradient is mediated through mass action effects. However, given that the intracellular pH is relatively constant it follows that an increase of ΔpH affects transport primarily by relocating the carrier to the outer face.

(b) Affinity effects

Changes in the affinity of the lactose carrier have been noted when membrane vesicles are energized. This observation is reproducible but its significance is controversial. Facilitated diffusion exhibits a Kd of 20 mM for lactose compared with a Kt of 0.2 mM for active transport (Kaback, 1986). However, this change in kinetics has been reported to be an all or none change and to occur at low Δψ and not to occur when ΔpH is the driving force for transport.

(c) Translocation effects

Let us make the initial assumption that only the C and CHS forms of the carrier can relocate across the membrane (although see below for a caveat to this generalization). It follows that the balance of the transport cycle is dictated by the numerical values of four translocation constants, k_1, k_{-1}, k_2, k_{-2} and the relative concentrations of the C and CHS forms of the carrier at the two faces of the membrane. Accumulation of solute could be achieved if these rate constants were influenced by the magnitude of the pmf. In the case of the lactose carrier it has been suggested that it behaves as though negatively charged such that imposition of a membrane potential relocates the unloaded carrier to the outer face of the membrane.

These effects are summarized in Figure 8.15.

Much of this discussion of the kinetics of transport systems is based upon studies of the lactose permease. The generality of the conclusions has yet to be tested. Two further aspects of kinetics deserve comment. First, it has long been established that at the steady state the rate of exchange of the solute via the carrier is equal to the initial rate of net uptake, i.e. in general there is no feedback control over carrier activity. Secondly, increases in transport rate are non-linear with respect to protonmotive force. This property has been termed 'non-ohmic conductance' (see Chapters 7 and 9) and appears to be a universal property of ion-linked energy transducers. From the model described in Figures 8.14 and 8.15 one expects a tenfold increase in the rate of transport for every 59 mV increase in the protonmotive force; and the non-linear relationship observed is confirmation of this prediction. For example, the initial rate of lactose uptake increases approximately sixfold for a 58 mV increase in the magnitude of the membrane potential (Ghazi and Shechter, 1981; Fig. 8.16; similar observations have been made for proton conductance of membranes and the activity of the ATP synthase). Consequently, addition of a respiratory substrate to

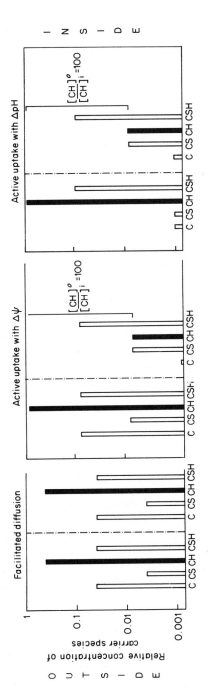

Fig. 8.15 A simple model scheme for the action of the pH gradient and the membrane potential on the β-galactoside carrier of *E. coli*. The figure depicts the relative concentrations of the different forms of the carrier at the outside (left half of each panel) and the inside (right half of each panel). The depicted effects assume perfect coupling and a driving force of either 118 mV membrane potential (negative inside) or two pH units (also 118 mV).

endogenously respiring cells leads to a five- to tenfold increase in the rate of uptake (depending on the ratio of protons to solute transported and the charge on the solute, see equation below) for a change in the pmf of only 20–30 mV.

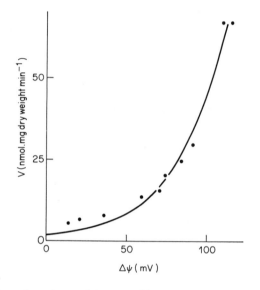

Fig. 8.16   Non-linear dependence of the rate of lactose transport on the magnitude of the membrane potential (negative inside) (see text for details).

8.4.6   Ion-linked transport: thermodynamic considerations

The maximum possible gradient of a solute generated via an ion-linked system is quite simply described by the following equation:

(i) $\Delta\mu S/F = (m + n)\Delta\psi - nZ\Delta pH$

where, $\Delta\mu S$, $\Delta\psi$ and ΔpH are the transmembrane gradients of solute, membrane potential and pH, respectively; m is the charge on the solute and n is the number of protons cotranslocated by the carrier; Z is the term which converts gradients into mV and equals $2.303\ RT/F$ where R is the gas constant, T is temperature in degrees absolute and F is the Faraday constant. Using this equation and a given set of values of $\Delta\psi$ and ΔpH one can set the upper limit of accumulation for any solute through the value of n (Table 8.2, page 402).

8.4.6.1 Leaks and slips

It has come to be realized that the maximum gradient is rarely if ever reached for a number of reasons.

(a) Leaks

It is rarely the case that there is a single pathway for a solute to enter the cell and, consequently, the gradient is dependent upon the relative activities of the different pathways and upon their coupling to energy. An elegant treatment of this problem has been given for the lipid-permeable lactose analogue thiomethyl-β-galactoside (TMG) in *E. coli* cells induced for various levels of the lactose permease (Maloney and Wilson, 1974). In this instance, as the degree of induction of the *lac* operon is increased, the influence of the leak pathway is diminished and the ability of the cell to accumulate the analogue increases. However, even as the permease pumps the analogue inwards the outward leak of the accumulated material becomes significant and activity of the leak pathway sets a limit on the accumulation that can be achieved. More complex examples also exist. For example, it is clear that the degree of tetracycline-resistance expressed by *E. coli* cells carrying a Tet resistance determinant is affected by the number and activity of the tetracycline entry routes as well as by the activity of the antiport encoded by the resistance gene.

(b) Slips

Simply defined, a slip is a pathway which is intrinsic to the carrier protein and which diminishes the accumulation of solute. For many solutes there is no evidence for significant leak pathways yet the accumulation of the solute via an ion-linked transport system does not achieve equilibrium with the observed driving force. Careful quantitative studies on a number of systems leads to the conclusion that the failure to reach thermodynamic equilibrium is not simply the result of poor quantitation of the parameters in equation (i) (page 414) (Ahmed and Booth, 1981). Careful studies in a number of laboratories have shown that the internal concentration of solute is a saturable function of the external concentration and thus that the gradient of solute (and hence $\Delta\mu S$) diminishes as the external concentration of solute is raised. Similarly, it can be shown that lowering the pmf does not result in a quantitatively equivalent reduction of the solute gradient (Fig. 8.17). Furthermore, it has been evident for some time that the steady-state accumulation of any two solutes via the same carrier is different despite the same driving force being available. Clearly the solute gradient cannot be at equilibrium with the driving force.

How does one explain this paradox? There is no clear-cut answer—only several equally plausible suggestions, of which only two will be presented.

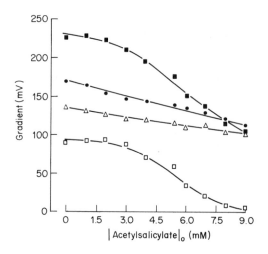

Fig. 8.17   Lactose transport and thermodynamic equilibrium. Cells of *E. coli* ML308-225 were incubated at pH 5.5 in the presence of different concentrations of acetylsalicylic acid and the magnitude of the lactose gradient, membrane potential and pH gradient determined in parallel incubations. The experiment shows that the lactose accumulation (●) is not at equilibrium with the total protonmotive force (■) and that equilibrium is attained once the magnitude of the protonmotive force has been reduced with a weak acid (data from Ahmed and Booth, 1981, with permission). Other symbols: (□) ΔpH; (△) Δψ.

The original definition of slip was the transport of solute by the carrier without the coupling ion (Eddy, 1980). This reaction is clearly unfavourable for efficiency and was suggested only to occur when the carrier is exposed to high solute concentrations. Since the carrier is functionally symmetrical this means that, whenever the internal concentration of solute is high, the carrier itself acts as a pathway allowing movement of solute without the coupling ion. The second possibility is that the solute carrier becomes saturated with solute at both sides of the membrane and thus is incapable of dissociating solute. Under such conditions net accumulation ceases despite the availability of a driving force favouring accumulation. This has been termed the carrier–solute interaction since the affinity of any solute for the carrier is specific to that solute; and thus the solute limits its own accumulation (Booth and Hamilton, 1984). It has been suggested that the poor accumulation of thiodigalactoside via the *lac* permease can be explained in this way (Overath and Wright, 1983).

8.4.6.2   Variable coupling
One obvious way by which the cell could manipulate its capacity to accumulate a solute would be by varying *n* (the number of protons

cotranslocated with the solute; see equation (i), page 414 and Table 8.2). The most significant variations in n have been proposed to occur in response to changes in the external pH. This phenomenon was first noted by Kaback during investigations of anion uptake at acid and at alkaline pH (Ramos and Kaback, 1977; reviewed by Konings and Booth, 1981). Subsequent investigations have shown that indeed many transport carriers show pH-dependency of the value of n but that the critical controlling parameter is the internal pH. Thus carriers for anions appear to be electrogenic at acidic pH as long as the internal pH of the cell/vesicle is held above pH 6.8–7.0. When the internal pH is reduced below that value electroneutral solute translocation occurs (Bassilana *et al.*, 1984). The degree to which variable stoichiometries are a physiologically significant control mechanism on the accumulation of solute is still unclear. However, there is no doubt that the electrogenic functioning of many systems which operate electroneutrally (e.g. anion/proton symports and cation/proton antiports) greatly benefits the transport capacity of the cell by coupling the membrane potential to the accumulation (or exclusion) of the solute (Table 8.2).

8.4.7   Molecular mechanisms of ion-linked transport

In some respects the ion-linked transport systems pose the greatest challenge of all in understanding the molecular dynamics of translocation. One requires the protein to transfer two solutes either simultaneously or in a controlled sequential and coupled manner. Progress towards an understanding of transport at the molecular level has increased recently through the introduction of methods for purification of carrier proteins, for peptide synthesis and for the generation of specific antibodies; and through the advent of gene cloning and *in vitro* mutagenesis. This panoply of techniques is beginning to supply insights not available from standard bioenergetic approaches, although they are most valuable when combined with earlier techniques. The only major deficiency at present is the failure to make crystals of a transport protein to allow X-ray crystallography.

8.4.7.1   Purification and reconstitution

Since the discovery of the *lac* operon it has been proposed that the *lac Y* gene product is the only protein required to effect pmf-dependent β-galactoside transport. Since then there have been suggestions that a common energy-coupling unit might be required for active transport. The purification and reconstitution of the lactose carrier into purified liposomes has since confirmed that only a single protein is required for the lactose-driven proton uptake and for pmf-driven lactose uptake. The purified carrier exhibits similar kinetic and bioenergetic properties to those found in more complex vesicle systems (reviewed in Kaback, 1986). Thus only a single protein is required for transport activity.

8.4.7.2 Organization of the carrier in the membrane

At present, understanding of the organization of transport proteins is based on structure predictions made from the amino acid sequences of transport proteins. Such analyses indicate a common structure for the proton symports that have been examined consisting of between 12 and 14 membrane-spanning α-helices linked by short hydrophilic sequences. In common with many other membrane proteins, the majority of the charged residues are found at the surface of the membrane. However, there are charged residues that are predicted to be buried in the membrane-spanning helices and which may play a central role in the transfer of protons across the membrane.

Site-directed polyclonal antibodies and monoclonal antibodies raised against specific segments of the lactose permease have been used to test the validity of the structural predictions (Kaback, 1986). In addition, such antibodies have been used to probe the structure of the lactose permease in mutants specifically defective in energy coupling to lactose influx. The carboxyl terminus and segments 5 and 7 (Fig. 8.18) have been located at the inner surface by this technique. These studies have confirmed the structure predictions and additionally have revealed significant changes in conformation in the mutants (Kaback, 1986). Clearly such antibodies are powerful probes for the analysis of mutant transport proteins since they facilitate the identification of those changes that cause major alterations in conformation.

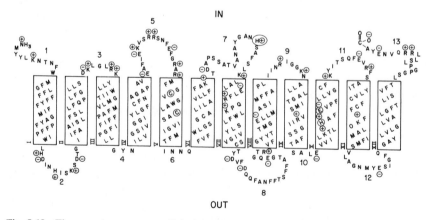

Fig. 8.18   The secondary structure of the lactose permease based upon the *hydropathy* profile of the protein. Hydrophobic segments are shown in boxes as transmembrane, α-helical domains connected by hydrophilic segments. The carboxyl terminus and hydrophilic segments 5 and 7 are shown located on the cytoplasmic face of the membrane and cys-148, cys-154, his-205, his-322 and glu-325 are circled.

Genetic analysis of ion-linked transport systems proceeds on two levels. The comparison of the primary sequence of transport proteins can reveal common features and conserved elements, as was described earlier for the putative energy coupling unit of the periplasmic transport systems. Extensive homology has been found to exist between the xylose and arabinose proton symports of *E. coli* and the mammalian glucose transporters (Maiden *et al.*, 1987). Further homologies have been found between ion-linked carriers with substrates as diverse as arabinose and tetracycline. On the other hand, the sequences of the lactose and melibiose permeases of *E. coli* show no significant homology with each other or with any other transporter sequenced to date. Thus there are likely to be families of transporters with common evolutionary roots.

The recent application of *in vitro* mutagenesis to the lactose transport system by Kaback's group has allowed systematic evaluation of the importance of different amino acid residues (Kaback, 1987). The interpretation of the data generated from this technique is limited by the lack of a crystal structure which would reveal the possible interactions of different side chains. However, some insights have been gained of which two examples will serve to illustrate the power of the technique.

(a) cys-148

It has long been established that N-ethylmaleimide (NEM) inactivates the *lac* carrier and that this can be prevented by inclusion of thiodigalactoside (TDG) in the incubation buffer. The residue labelled by NEM and protected by TDG was found to be cys-148 and it was proposed that this residue might participate in substrate binding or translocation. When the cys-148 was replaced with either a serine or a glycine residue the carrier retained activity but was now relatively insensitive to NEM; the slow NEM-labelling of protein that did occur was not prevented by TDG. Thus the cys-148 residue is not essential for carrier function. Intriguingly another cysteine residue, cys-154, was identified by mutagenenic replacement as essential for translocation but not for binding of the β-galactosides

(b) his-322 and his-205

Chemical modification studies of the lactose permease had revealed that histidine reactive agents such as Rose Bengal and diethylpyrocarbonate inactivate lactose transport. Each of the four histidine residues was systematically replaced by arginine, glutamine or asparagine. Replacement of his-35 or his-32 with arginine had no effect on the activity of the carrier. His-205 is also essential for activity but can be replaced by either asparagine or glutamine which suggests that the role of this residue is involved in hydrogen bonding. Replacement of his-322 with arginine converts the carrier to a facilitated diffusion system without concomitant proton translocation. Replacement of glu-325 with alanine also leads to loss of active transport but not of facilitated diffusion. Thus the effect of loss of either his-322 or glu-325 is similar and this has led to speculation that they are either involved in ion-pairing which stabilizes a particular conformation of the carrier or that they are directly involved in a proton relay (Kaback, 1987).

In the absence of a three-dimensional model of the transport system there is an obvious danger of over-interpreting these examples, but the results obtained to date do indicate that it is possible to separate ion translocation from solute movement by specific alterations of the primary sequence of the carrier.

8.5   Transport systems and homeostasis

8.5.1   Introduction

The two major homeostases of the cell are cytoplasmic pH and cell turgor (Booth, 1985; Epstein, 1986; Booth *et al*, 1988). Transport systems that are subject to feedback control are major components of the mechanisms that serve to maintain cytoplasmic pH and turgor relatively constant. The need for such feedback can be seen from an examination of properties of transport systems.

Cells maintain large transmembrane solute gradients by virtue of the selectivity of their transport systems and the coupling of energy to solute translocation. Externally supplied solutes can be concentrated in the cytoplasm and cytoplasmic pools of intracellularly generated metabolites can be sustained either by preventing leakage or by facilitating recapture of lost molecules. For example, *E. coli* cells sustain high potassium gradients without appreciable loss, even when the external concentration is low. In mutants lacking potassium uptake systems, the leakage of potassium becomes evident due to the inability of the cell to recapture the lost cation. Similar findings have been made for amino acids and other molecules which occur at high concentrations in the cytoplasm. The maintenance of solute gradients is thus a dynamic process in which uptake competes with leakage to maintain a constant pool. In this simplest of contexts, the transport system plays a direct role in homeostasis since the cytoplasmic pool is to a significant extent maintained by solute transport across the cell membrane.

Although the role of transport systems in maintaining metabolite pools is homeostatic, regulation of their activity is not essential and only rare cases of such feedback control have been reported. By contrast, any transport system which has the primary function of restoring homeostasis must be subject to significant feedback control. The differences between 'normal' transport systems and regulated systems are illustrated in Figure 8.19. The activity of a 'normal' transport system will respond passively to changes both in the external concentration of the transported solute X and in the driving force. For example, an increase in the concentration of X will normally elicit an increase both in the velocity of the transport system and in the internal concentration of solute (Fig. 8.19). Similarly, when the energy for transport is diminished the gradient of solute and the velocity of transport will be reduced. As was emphasized above, the precise magnitude of the changes in velocity and accumulation will depend upon how tightly coupled the system is, and to what extent other transport activities influence the magnitude of the gradient.

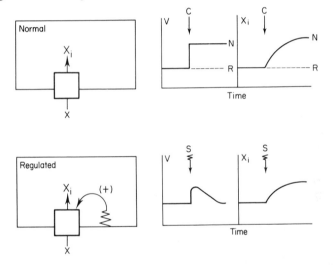

Fig. 8.19   The properties of normal and regulated transport systems. The diagram illustrates the response of the two types of system to an increase (at C) in the external concentration of solute (X) (top) or to a perturbation of homeostasis (S) (bottom). N and R indicate the responses of the normal system and the regulated system, respectively; V is the initial rate of solute transport and X_i is the internal concentration of the solute X. It is assumed that the regulated system is activated by the signal S but inhibition of some systems may occur. Note that the transport activity (V) of the regulated system returns to the basal level once homeostasis is restored but that this can involve a change in X_i. An example of such a series of responses is that of the Trk potassium transport system to a reduction of cell turgor (see Epstein, 1986).

The activity of a regulated system will be dictated primarily by the control system(s) rather than by the external solute concentration or by changes in the magnitude of the driving force. Consequently the system will often be quiescent and thus relatively insensitive to changes in the concentration of the transported solute and indeed to alterations of the energy supply. Whenever the homeostatic state is perturbed a signal (S) is generated (Fig. 8.19) which may either activate or inhibit the transport system according to the contribution the system makes to restoring homeostasis. In the activated state, the capacity of the transport system to restore homeostasis will be limited by the concentration of the solute to be transported and the energy available to drive the system. In this respect these systems do not transgress against any major thermodynamic principle. However, unlike normal systems, the gradient of the solute sustained by the transport system will usually be dictated by the control system, and will be held far from equilibrium by regulation of the transport protein. The unique feature of such systems is their capacity to respond to perturbations of the cell such that their activity restores the cell to a condition which is closer to the optimum for cell growth.

The best understood examples of such feedback control of transport are to be found in the studies of osmoregulation in the enteric bacteria (Epstein, 1986; Booth *et al*, 1988). However, there is also evidence for the regulation of proton-translocating complexes that are involved in pH homeostasis.

8.5.2   Control of potassium accumulation in enteric bacteria

Potassium ions are the major cytoplasmic osmolyte of all Gram-negative bacteria. The transport systems that are active in the regulation of potassium accumulation have been best studied in *Escherichia coli* (Epstein, 1986). There are two transport systems for the accumulation of potassium and at least three potassium efflux systems. The two uptake systems are Kdp and Trk and their genetics and kinetic properties are summarized in Table 8.3. Understanding of these systems is incomplete but accurate predictions of their response to either positive or negative changes in turgor can be made from the existing studies.

Both the Kdp and the Trk systems for potassium uptake show properties consistent with feedback control of their activity. Net potassium uptake into depleted cells takes place at a high velocity but exchange occurs at only 10–25% of the rate of net influx and, in the case of the Kdp system, this is inhibited by high external potassium concentrations (greater than 0.1 mM). A sudden reduction in turgor stimulates the activity approximately fourfold. While the mechanism of feedback control has not yet been determined in the case of the Trk system, an examination of the energy requirements for transport has yielded some insight.

Using the methods developed by Berger, the energy-coupling to the Trk system was examined. It was demonstrated that net potassium uptake has a requirement for both a pmf and for ATP, while exchange requires only ATP (Rhoads and Epstein, 1977). It has recently been proposed that the feed-back control of Trk activity can be accommodated within this dual energy requirement. The activity of the Trk system may be controlled by phos-phorylation of one of the Trk components and the net translocation of potassium alone requires the pmf. Specifically it has been suggested that the Trk carrier is only active when phosphorylated and that accumulation of potassium stimulates dephosphorylation such that exchange can only occur at much lower rates than for net uptake (Stewart *et al.*, 1985). Such a cycle of phosphorylation/dephosphorylation could be under the control of turgor leading to renewed phosphorylation when turgor falls.

Table 8.3   The Kdp, Trk and Kef potassium transport systems of *E. coli*

System	Affinity	Structural genes	Regulation of Activity
Uptake			
Kdp	High (2 μM)	*kdpA,B,C*	Turgor[a]
Trk	Low (1 mM)	*trkA,D,E,G,H*	Turgor
Efflux			
KefC	N.D.	*kefC*	Glutathione
KefB	N.D.	*kefB*	Glutathione
KefX[b]	N.D.	N.D.	Turgor and cytoplasmic pH

[a] The Kdp system is also induced by conditions of low turgor (see Epstein, 1986).
[b] The presence of the KefX system can be deduced from studies of mutants lacking all the other transport systems (Bakker *et al.*, 1987).

It is generally held that the Trk transport system is probably a potassium-proton symport. As descibed above (Table 8.2) the driving force for accumulation would be $2\Delta\psi–Z\Delta pH$. Within the physiological pH range this driving force ($\Delta\psi = -100$ to $-180\,mV$; $Z\Delta pH = 0$ to $+120\,mV$) for a potassium–proton symport could attain gradients of $10^3–10^5$. At an external potassium concentration of $1\,mM$ (close to the Km for Trk) internal concentrations of $1–100\,M$ are theoretically possible. In media of moderate osmolarity, the internal potassium is closer to $200–300\,mM$ (Epstein, 1986) which demonstrates the effectiveness of the control mechanisms.

Relatively little is known of the potassium efflux systems in the enteric bacteria. Recent genetic studies have shown that there are at least three efflux systems (Table 8.3). Two of these systems, KefB and KefC, are subject to feedback control by reduced glutathione. Lowering of the cytoplasmic glutathione pool, either by mutation or by chemical treatment, leads to rapid potassium efflux via these systems. Neither the physiological significance nor the molecular mechanism of this regulation has been elucidated.

Potassium efflux is also controlled by turgor and by cytoplasmic pH. When the turgor pressure of the cell is raised by the addition of a solute which can be accumulated in the cytoplasm, the cell responds by a controlled exodus of potassium. Similarly, when the internal pH is raised, potassium efflux occurs by exchange for protons until the internal pH is restored to the normal value (Booth, 1985).

Regulation of the potassium pool is thus effected by feedback control of the uptake and efflux systems. The magnitude of the pool is primarily determined by the osmolarity of the medium. However, when a compatible solute, betaine, is present in the environment, exchange of betaine for the internal potassium takes place. This exchange is effected in a controlled manner such that the transmembrane gradient of betaine is determined by the osmolarity of the medium. The two betaine transport systems ProU and ProP are regulated at the level of gene expression and their activity *in situ* is regulated by the osmotic pressure of the environment. Thus, both the osmolarity of the cytoplasm and its constitution are regulated via integrated control of several solute transport systems (reviewed by Booth *et al.*, 1988).

8.5.3   Coping with extremes of pH: control over internal pH

As stressed earlier, most bacterial cells regulate the internal pH at a value close to neutrality (Padan *et al.*, 1981; Booth, 1985). A more detailed appraisal suggests that internal pH values span a range of two pH units from pH 6.5 to pH 8.5, with acidophilic bacteria exhibiting the lower values and alkalophiles the higher (Fig. 8.7; Padan *et al.*, 1981). The alkalophilic bacteria are thus unique since the environmental pH is often more alkaline than that of the cytoplasm and these organisms consequently have an inverted pH gradient. The acidophiles and alkalophiles represent the extremes in terms of composition of the pmf. The former have a very large pH gradient which at very low pH is counterbalanced by a membrane potential, positive inside. In the alkalophiles the pH gradient will oppose the entry of solutes by proton symport and the synthesis of ATP by the ATP synthase. Such cells must therefore possess a large membrane potential, negative inside, in order to overcome the effect of the inverted pH gradient such that the pmf still favours ATP synthesis and solute translocation. In addition, the alkalophile faces the problem of generating and maintaining an internal pH more acidic than that of the environment (see Krulwich, 1987).

8.5.3.1   Acidophiles
Acidophilic bacteria exhibit considerable metabolic diversity, as to be

expected from a group of organism linked only by their capacity to grow at low pH. The bioenergetics and general *modus vivendi* of these organisms has been extensively reviewed (Ingledew 1982; Cobley and Cox, 1983; Krulwich and Guffanti, 1983; Matin, 1984).

The problem facing these organisms is primarily that they maintain their internal pH between pH_i 6.0–7.0 while that of the environment is in the range pH_o 1.0–2.0. It has recently been established that these organisms are no less sensitive to perturbation of internal pH than is *E. coli* or *Streptococcus faecalis* (Leach and Ingledew, 1987). Thus the cells must limit proton entry and it seems likely that control over cation permeability plays a major role in preventing acidification of the cytoplasm. For example, initial studies on a range of acidophiles suggested that uncouplers and respiratory inhibitors did not cause the complete dissipation of the pH gradient. However, when K^+–H^+ exchange was stimulated by the addition of nigericin at acid pH, complete acidification of the cytoplasm occurred and death followed (Guffanti *et al.*, 1979). Clearly, the efficacy of the uncoupler may be reduced by low permeability of the cell membrane to cations such as potassium. Similarly, in *Bacillus acidocaldarius*, transport via H^+-symports is more sensitive to ion-exchange ionophores (nigericin) and channel-forming antibiotics (gramicidin) than to protonophores (Krulwich *et al.*, 1979). This again suggests that an increase in cation conductivity of the membrane is essential for good uncoupling.

8.5.3.2 Alkalophiles

The transport of sodium ions plays a major role in the generation of an acidic interior in alkalophiles (McLaggan *et al.*, 1984; Krulwich *et al.*, 1985; reviewed by Booth, 1985). Acidification of the cytoplasm is achieved by the exchange of internal Na^+ for external H^+ via the sodium/proton antiport. However, the sodium/proton antiport also serves to maintain the transmembrane sodium gradient. These two functions will frequently be at odds with the needs of the cell. Thus when an influx of sodium occurs via a sodium symport, acidification of the cytoplasm will occur without regard to the requirement for a lowering of cytoplasmic pH. On the other hand, a specific sodium entry mechanism is necessary to allow the cell to cope with alkalinization of the cytoplasm (Booth, 1985). Since both a sodium gradient and a low internal pH are maintained by alkalophiles there must be either tight regulation of sodium ion flux or tight regulation of the cytoplasmic pH by another system which functions independently of sodium ions. In one alkalophile, *Exiguobacterium aurantiacum*, the latter appears to be the case (Booth, 1985).

The bioenergetic problems faced by the alkalophiles have been explored by Krulwich and Guffanti (1983). The principal problem is the magnitude of the membrane potential which has been reported to be too low to offset the inverted pH gradient. Furthermore, it has been reported that while vesicles do support oxidation-dependent ATP synthesis and transport, an artificially imposed membrane potential of equal magnitude is inoperative as a driving force (Guffanti *et al.*, 1984). The resolution of these questions poses some major problems for conventional Chemiosmotic thinking (see Chapter 9, this volume). At least one reported solution is that of an energy-transducing system based on Na^+, as found in *Vibrio alginolyticus*, in which it is envisaged that the respiratory chain and the ATPase translocate sodium rather than protons (reviewed by Dimroth, 1987).

Without doubt the bioenergetics of the alkalophiles represents one of the major challenges for the future.

8.6 Conclusion

From the above it can be appreciated that bacteria have a considerable diversity of transport mechanisms. This chapter has attempted to provide an overview of what are regarded to be the major classes of transport systems. Given the wealth of information available and the range of systems that have been examined it is not surprising that our appreciation of transport systems now ranges from bioenergetics through regulation to molecular structure. Over the next ten years it is the aim of most workers in this field to move descriptions further into the molecular realm so that one day the translocation process may be described in fine detail.

References

Ahmed, S. and Booth, I. R. (1981). *Biochem. J.* **200**, 583–589.

Ames, G.-F. L. (1986). *Ann. Rev. Biochem.* **55**, 397–425.

Ames, G.-F. L. and Higgins, C F. (1983). *Trends Biochem. Sci.* **8**, 97–100.

Bakker, E. P. (1979). *In* 'Antibiotics' (Ed. F. E. Hahn), Vol. 1, pp. 67–97. Springer-Verlag, Berlin, Heidelberg, New York.

Bakker, E. P., Booth, I. R., Dinnbier, U., Epstein, W. and Gajewska, A. (1987). *J. Bacteriol.* **169**, 3743–3749.

Bassilana, M., Damiano, E. and Leblanc, G. (1984). *Biochmeistry* **23**, 5288–5294.

Bentaboulet, M., Robin, A. and Kepes, A. (1979). *Biochem. J.* **178**, 103–107.

Berger, E. A. (1973). *Proc. Natl. Acad. Sci. USA* **70**, 1514–1518.

Berger, E. A. and Heppel, L. A. (1974). *J. Biol. Chem.* **249**, 7747–7755.

Booth, I. R. (1985). *Microbiol. Rev.* **49**, 359–378.

Booth, I. R. and Hamilton, W. A. (1984). *In* 'Membranes and Transport' **2** (Ed. A. N. Martonosi), pp. 41–46. Plenum, New York.

Booth, I. R., Cairney, J., Sutherland, L. and Higgins, C. F. (1988). *J. Appl. Bact.* (in press).

Bramley, H. F. and Kornberg, H. L. (1987). *Proc. Natl. Acad. Sci. USA* **84**, 4777–4780.

Cobley, J. G. and Cox, J. C. (1983). *Microbiol. Rev.* **47**, 579–595.

Dimroth, P. (1987). *Microbiol. Rev.* **51**, 320–340.

Eddy, A. A. (1980). *Biochem. Soc. Trans.* **8**, 271–273.

Epstein, W. (1985). *Curr. Top. Memb. Transp.* **23**, 153–175.

Epstein, W. (1986). *FEMS Microbiol. Rev.* **39**, 73–78.

Futai, M., Sternweis, P. C. and Heppel, L. A. (1974). *Proc. Natl. Acad. Sci. USA* **71**, 2725–2729.

Ghazi, A. and Shechter, E. (1981). *Biochim. Biophys. Acta* **645**, 305–315.

Guffanti, A. A., Davidson, L. F., Mann, T. B. and Krulwich, T. A. (1979). *J. Gen. Microbiol.* **114**, 201–206.

Guffanti, A. A., Fuchs, R. T., Schneier, M., Chui, E. and Krulwich, T. A. (1984). *J. Biol. Chem.* **259**, 2971–2975.

Hamilton, W. A. (1977). *Symp. Soc. Gen. Microbiol.* **27**, 185–216.

Harold, F. M. (1986). 'The Vital Force: a Study of Bioenergetics'. W. H. Freeman, New York.

Hesse, J. E., Wieczorek, L., Altendorf, K., Reicin, A. S., Dorus, E. and Epstein, W. (1984). *Proc. Natl. Acad. Sci. USA* **81**, 4746–4750.

Higgins, C. F., Hiles, I. D., Whalley, K. and Jamieson, D. J. (1985). *EMBO J.* **4**, 1033–1040.

Higgins, C. F., Hiles, I. D., Salmond, G. P. C., Gill, D. R., Downie, J. A., Evans, I. J., Holland, I. B., Gray, L., Buckel, S. D., Bell, A. W. and Hermodson, M. A. (1986). *Nature* **323**, 448–450.

Higgins, C. F., Gallagher, M. P., Mimmack, M. L. and Pearce, S. R. (1988). *Bioessays* (in press).

Hiles, I. D., Gallagher, M. P., Jamieson, D. J. and Higgins, C. F. (1987). *J. Mol. Biol.* **195**, 125–142.

Hirata, H., Altendorf, K. and Harold, F. M. (1974). *J. Biol. Chem.* **249**, 2939–2945.

Hobson, A. C., Weatherwax, R. and Ames, G. F-L. (1984). *Proc. Natl. Acad. Sci. USA* **81**, 7333–7337.

Hong, J-S., Hunt, A. C., Masters, P. S. and Lieberman, M. A. (1979). *Proc. Natl. Acad. Sci. USA* **76**, 1213–1217.

Hugenholz, J., Hong, J-S. and Kaback, H. R. (1981). *Proc. Natl. Acad. Sci. USA* **78**, 3446–3449.

Hunt, A. G. and Hong, J-S. (1983). *J. Biol. Chem.* **256**, 11988–11991.

Ingledew, W. J. (1982). *Biochimica et Biophysica Acta* **683**, 89–117.

Jacobsen, G. R., Kelly, D. M. and Findlay, D. R. (1983). *J. Biol. Chem.* **285**, 2955–2959.

Kaback, H. R. (1968). *J. Biol. Chem.* **243**, 3711–3724.

Kaback, H. R. (1971). *Meth. Enzymol.* **22**, 99–120 (Ed. W. B. Jakoby). Academic Press, London.

Kaback, H. R. (1986). *In* 'Physiology of Membrane Disorders' (Eds. E. Thomas, M. D. Andreoli, J. F. Hoffman, D. D. Fenestil and S. G. Schultz), pp. 387–407. Plenum Press, New York.

Kaback, H. R. (1987). *Biochemistry* **26**, 2071–2076.

Kashigawa, K., Kobayashi, H. and Igarashi, K. (1986). *J. Bacteriol.* **165**, 972–977.

Kashket, E. R. and Wilson, T. H. (1973). *Proc. Natl. Acad. Sci. USA.* **70**, 2866–2869.

Konings, W. N. (1985). *Trends Biochem. Sci.* **10**, 317–319.

Konings, W. N. and Booth, I. R. (1981). *Trends Biochem. Sci.* **7**, 257–262.

Krulwich, T. A. (1987). *In* 'Sugar Transport and Metabolism in Gram Positive Bacteria' (Eds. J. Reizer and A. Peterkofsky), pp. 333–364. Ellis-Horwood, Chichester.

Krulwich, T. A. and Guffanti, A. A. (1983). *Adv. Microbiol. Physiol.* **24**, 173–214.

Krulwich, T. A., Davidson, L. F., Filip, S. J., Zukerman, R. S. and Guffanti, A. A. (1979). *J. Biol. Chem.* **253**, 4599–4603.

Krulwich, T. A. Federbush, J. G. and Guffanti, A. A. (1985). *J. Biol. Chem.* **260**, 4055–4058.

Kundig, W., Ghosh, S. and Roseman, S. (1964). *Proc. Natl. Acad. Sci. USA* **52**, 1067–1074.

Lancaster, J. R. and Hinkle, P. C. (1977). *J. Biol. Chem.* **252**, 7657–7661.

Leach, A. S. and Ingledew, W. J. (1987). *J. Gen. Microbiol.* **133**, 1171–1179.

Lee, C. A. and Saier, M. H. (1983). *J. Biol. Chem.* **258**, 10761–10767.

Leive, L. (1968). *J. Biol. Chem.* **243**, 2373–2380.

Maiden, M. C. J., Davis, E. O., Baldwin, S. A., Moore, D. C. M. and Henderson, P. J. F. (1987). *Nature* **325**, 641–643.

Maloney, P. C. and Wilson, T. H. (1974). *Biochem. Biophys. Acta* **330**, 196–204.

Maloney, P. C., Ambudkar, S. V. and Sonna, L. A. (1987). *In* 'Sugar Transport and Metabolism in Gram Positive Bacteria' (Eds. J. Reizer and A. Peterkofsky), pp. 134–149. Ellis-Horwood, Chichester.

Matin, A. (1984). *In* 'Microbial Growth on C1 Compounds' (Ed. R. L. Crawford), pp. 62–71. John Wiley, New York.

McLaggan, D., Selwyn, M. J. and Dawson, A. P. (1984). *FEBS Lett.* **165**, 254–258.

Mitchell, P. (1973). *J. Bioenergetics* **4**, 63–91.

Nelson, S. O. and Postma, P. W. (1984). *Eur. J. Biochem.* **139**, 29–34.

Nelson, S. O., Wright, J. K. and Postma, P. W. (1983). *EMBO J.* **2**, 715–720.

Neu, H. C. and Heppel, L. A. (1965). *J. Biol. Chem.* **240**, 3685–3692.

Overath, P. and Wright, J. K. (1983). *Trends Biochem. Sci.* **8**, 404–408.

Padan, E., Zilberstein, D. and Schuldiner, S. (1981). *Biochim. Biophys. Acta* **650**, 151–166.

Plate, C. A. (1979). *J. Bacteriol.* **137**, 221–225.

Poolman, B., Hellingwerf, K. J. and Konings, W. N. (1987). *J. Bacteriol.* **169**, 2272–2276.

Postma, P. W. and Lengeler, J. W. (1985). *Microbiol. Rev.* **49**, 232–269.

Ramos, S. and Kaback, H. R. (1977). *Biochemistry* **16**, 4271–4275.

Reizer, J. and Peterkofsky, A. (1987). *In* 'Sugar Transport and Metabolism in Gram Positive Bacteria' (Eds. J. Reizer and A. Peterkofsky), pp. 333–364. Ellis-Horwood, Chichester.

Rhoads, D. B. and Epstein, W. (1977). *J. Biol. Chem.* **252**, 1394–1401.

Richarme, G. (1985). *Biochem. Biophys. Acta* **815**, 37–43.

Saier, M. H., Cox, D. F., Feucht, B. V. and Novotny, M. J. (1982). *J. Cell. Biochem.* **18**, 231–238.

Sanders, D., Hansen, U.-P., Gradmann, D. and Slayman, C. L. (1984). *J. Membr. Biol.* **77**, 123–152.

Stewart, L. M. D. and Booth, I. R. (1983). *FEMS Microbiol. Lett.* **19**, 161–164.

Stewart, L. M. D., Bakker, E. P. and Booth, I. R. (1985). *J. Gen. Microbiol.* **131**, 77–85.

Walker, J. E., Saraste, M., Runswick, M. J. and Gay, N. (1982). *EMBO J.* **1**, 945–951.

West, I. C. (1970). *Biochem. Biophys. Res. Commun.* **41**, 655–661.

West, I. C. and Mitchell, P. (1973). *Biochem. J.* **312**, 587–592.

Winkler, H. H. and Wilson, T. H. (1966). *J. Biol. Chem.* **214**, 2200–2211.

9 Protonmotive energy-transducing systems: some physical principles and experimental approaches

D. B. KELL

9.1 Introduction and scope 430
9.2 Elementary chemical thermodynamics: chemical and electrical
    potentials 433
9.2.1 Chemical potentials 433
9.2.2 Electrochemical potentials 436
9.3 Energy coupling of reactions and phenomenological non-
    equilibrium thermodynamics 439
9.4 Statistical thermodynamics and the relationship between
    microscopic and macroscopic descriptions of bioenergetic
    systems 442
9.5 Transition state theory 447
9.6 Non-thermally activated processes 450
9.7 Metabolic control theory 455
9.8 Hill diagrams and the 6-state proton pump 456
9.9 Is the protonmotive force an energetically-significant
    intermediate in electron transport-driven phosphorylation? ... 459
9.9.1 Introduction 459
9.9.2 Phosphorylation driven by an artificially applied pmf 461
9.9.3 Direct estimation of the pmf using microelectrodes 462
9.9.4 Ion- and acid/base distribution methods for estimating
      the pmf 466
9.9.5 Spectroscopic methods for estimating $\Delta\psi$ in bacteria 470
9.9.6 Relationships between the apparent pmf and rates and
      extents of phosphorylation 471
      9.9.6.1 Introduction 471
      9.9.6.2 Respiration-driven proton translocation 472
9.9.7 Double inhibitor titrations 478
9.9.8 Co-reconstituted systems: the emperor is not yet clothed ... 480
9.10 Concluding remarks 482
References 482

9.1   Introduction and scope

If the metabolites are extracted from a cell, or observed *in situ* using a technique such as nuclear magnetic resonance, a morass of different compounds will, of course, be seen. Questions which one might wish to ask about groups of these compounds (Fig. 9.1) include:

    (a) are they direct precursors or products of each other;

    (b) if so what are the pathways by which they are interconverted; and

    (c) does their intracellular activity differ substantially from their concentration (as estimated on the basis of (n)mol compound per unit membrane-enclosed volume)?

If they *do* bear a direct precursor–product relationship to each other, we might then enquire:

    (d) as to the nature and mechanism of the enzymes catalysing their interconversion; and

    (e) whether the metabolites are organized as a diffusible pool or whether there is a direct transfer of the protein-associated product of one reaction to the enzyme catalysing the 'next' reaction.

To give an answer to these types of question, the *criteria* which are involved in answering these questions must be considered, since a proper understanding of such criteria underpins much of what follows. The first part of this chapter, then, will address these points at a relatively fundamental level, seeking to indicate the difficulties which attach even to the simple, textbook types of question above. Since the present work constitutes an overview of bacterial bionenergetics, there will be special cause to ponder the fact that, whilst similar schemes to those of Figure 9.1 may be written for *bioenergetic* systems, there is an important additional difficulty. In contrast to metabolic substructures such as carbon skeletons, *free energy* is *not* perfectly conserved, and the degree of its conservation is not independent of time (Welch and Kell, 1986; Kamp and Westerhoff, 1987; Westerhoff and Kamp, 1987). Thus, even while clarifying the type of description sought for metabolic systems of the type given in Figure 9.1, it needs to be known to what extent such a description is suitable, even in principle, for bioenergetic systems. In other words, a distinction must be sought between *fundamental* and *phenomenological* descriptions of bioenergetic systems.

In the present context, this chapter enquires to what extent a minimal chemiosmotic scheme of the type shown in Figure 9.2 provides an adequate description of energy coupling in protonmotive systems. As the main example will be taken the ideas and data that pertain to the overall process of electron transport phosphorylation, not only in bacteria, but in other,

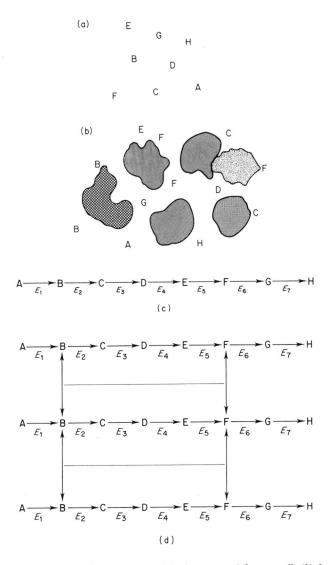

Fig. 9.1 (*a*) An ensemble of molecules as might be extracted from a cell. (*b*) *In vivo* some molecules are bound and some are free. Both their activity coefficients and standard chemical potentials may differ from those of the dilute aqueous extract in (*a*). (*c*) *In vivo* the molecules may form part of a metabolic pathway which operates in a particular order and a particular direction. The order is determined by the thermodynamics of the reaction A ... → ... H, the catalysts being enzymes of particular specificities. (*d*) Although one may observe a metabolic flux from A to H, as in (*c*), this does not show whether metabolites are free, diffusible, 'pool' intermediates (such as B and F here) or whether their transfer from enzyme to enzyme is direct, and not via a pool with a macroscopically definable activity or concentration.

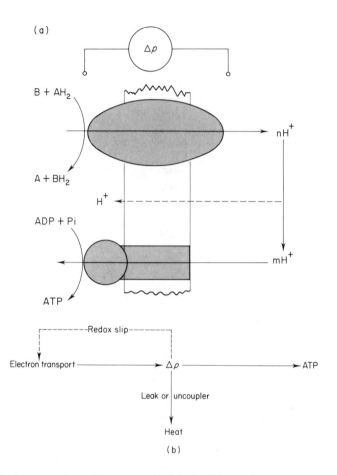

(a)

Δp

B + AH$_2$

A + BH$_2$

ADP + Pi

ATP

nH$^+$

H$^+$

mH$^+$

------Redox slip-------

Electron transport ⟶ Δp ⟶ ATP

Leak or uncoupler

Heat

(b)

Fig. 9.2 A chemiosmotic coupling scheme. (*a*) A (redox-linked) primary proton pump creates a protonmotive force (pmf, Δp) across a (relatively) ion-impermeable coupling membrane. Protons may leak back across the membrane, a process stimulated by so-called uncoupler molecules, or may drive an otherwise endergonic reaction catalysed by a secondary proton pump, in this case an H$^+$-ATP synthase. No topological relationships between the primary and secondary proton pumps are specified since coupling is via a pmf. (*b*) The coupling between the chemical reactions and proton transfer catalysed by the H$^+$ pumps themselves may not be perfect, a phenomenon known as 'redox slip'. 'Leak' refers to any H$^+$ pumped across the membrane by the primary pump but not coupled to the chemical reaction catalysed by the secondary H$^+$ pump, so that this definition includes slip within the secondary H$^+$ pump. Δp is by definition a delocalized intermediate, in equilibrium with the pmf between the two bulk aqueous phases that the coupling membrane serves to separate. It is thus freely available to all enzymes in the membrane vesicle. In the stationary state, the relation zF$\Delta p \geqslant \Delta G_P$ holds, where ΔG_P is the phosphorylation potential in the phase to which the ATP synthase is adjacent. F, Faraday's constant; z, the number of protons passing through the ATP synthase during the synthesis of one molecule of ATP (see also Table 1.2, p. 12).

evolutionarily related, free-energy-conserving membrane systems such as mitochondria and chloroplast thylakoids. Since the subject and its relevant literature are broad, if not limitless (Frank Harold's recent and beautifully structured overview required some 600 pages; Harold, 1986), and the shades of opinion, arguments and counterarguments ever-changing, it is necessary to be selective about the topics covered in any detail, notwithstanding a substantial list of references.

The major themes of this chapter are:

(a) that there are reasons strongly to doubt that the simplest purely chemiosmotic scheme for energy coupling is an adequate description of reality as reflected in the existing data;

(b) that something approximating the most extreme alternative viewpoint is as useful and defensible a hypothesis as is the scheme of Figure 9.2; i.e. That *electron transport-linked* phosphorylation *in vivo* is in *no* case *coupled* via a delocalized protonmotive force;

(c) that there are experiments and predictions which follow from some of the more plausible alternative schemes that are not simply consequential upon the chemiosmotic description embodied in the scheme of Figure 9.2.

At the more fundamental level it will be argued that it is unlikely *in principle* that the types of question that dominate the current bioenergetic literature are, in fact, altogether pertinent to the types of answer genuinely required. This is a much more difficult area, not for dwelling on here were it not that it is believed that the present book may foster the thinking and experimentation of the bioenergetics and bioenergeticists of the future. This said, it is time to look a little more closely at the metabolic system of Figure 9.1 and the energy coupling system of Figure 9.2, beginning with a discussion of some relevant chemical thermodynamics. A brief introduction to this topic is given in Chapter 1 (pages 9–11).

9.2 Elementary chemical thermodynamics; chemical and electrochemical potentials

9.2.1 Chemical potentials

The chemical potential of a substance i, usually denoted μ_i, is an intensive property of a system which describes the force which causes that system to relax to equilibrium. It may be thought of as the increase in the free energy of the system when one mole of compound i is added to an infinitely large quantity of the mixture so that the mixture does not significantly change its composition. The chemical potential is, in fact, the partial derivative corresponding to the change in Gibbs free energy per unit change in the number of molecules of type i under conditions in which all other parameters such as temperature, pressure and the number of molecules of substances other than i remain constant (see e.g. Smith, 1982). The change in Gibbs free energy dG associated with a particular chemical reaction occurring in a single thermodynamic phase is given by:

$$dG = VdP - SdT + \Sigma\mu_i dn_i \tag{9.1}$$

where V, P, S and T are respectively the volume, pressure, entropy and temperature of the (closed) system and dn_i the change in the number of molecules of i. For almost all the biochemical reactions needed to be considered, temperature and pressure are controlled. Thus the first two terms on the right-hand side of equation (9.1) may be ignored (although remember that this is not so for special cases such as those involving changes in osmotic pressure in one or more phases), giving:

$$dG = \Sigma\mu_i dn_i \tag{9.2}$$

For a pure substance dissolved in water at a mole fraction x_i and forming an ideal solution, the chemical potential is:

$$\mu_i = \mu°_i + RT \ln x_i \tag{9.3}$$

$\mu°_i$ is the *standard chemical potential* of the substance in the phase to which the substance i is being added. This standard chemical potential may be defined in various ways, but the most usual is to define it for a solution of unit molality. For a *real* solution, equation (9.3) does not hold, due mainly to interactions between solute and solvent molecules (Bockris and Reddy, 1970). Thus the *activity* a_i of substance i is defined in terms of its chemical potential in the solution and in its standard state:

$$\mu_i = \mu°_i + RT \ln a_i \tag{9.4}$$

The activity may be considered to represent the *effective concentration* per unit of component i relative to that in its standard state. They are related to each other by the activity coefficient γ_i:

$$a_i = \gamma_i x_i \tag{9.5}$$

For an ionic solution, such as a KCl solution in which the KCl molecules are fully ionized, there is a *different* standard state for each of the components, so that the chemical potential of a KCl solution is:

$$\mu_{KCl} = \mu°_{K+} + \mu°_{Cl-} + RT \ln a_{KCl} \tag{9.6}$$

It is particularly important to note that, in a mixture, the *standard chemical potential* of a substance depends upon the chemical potentials of *all* other substances in the thermodynamic phase under consideration. Particular difficulties arise here in knowing what the actual activities of different compounds in the aqueous cytoplasm and in other intracellular or intra-organellar phases are, and what proportion of the molecules of a given type are bound or complexed. An introduction to these topics may be found in Clegg (1984), Ling (1984) and Welch and Clegg (1987). Parenthetically, it should also be mentioned that although one may speak about the activity coefficients of single ions, e.g. a_{K+}, such a thing cannot ever be measured since single ions cannot be added to a solution without also adding their counterions. Thus the well-known definition of pH, viz.:

$$pH = -\log_{10} a_{H+} \qquad (9.7)$$

is not a thermodynamic definition but merely a convention.

For a chemical reaction such as $A + B \leftrightarrow C + D$, the *extent of reaction* χ, which may have a value between 0 and 1, may be defined. If the position of the reaction is such that the only molecules present are those on the left-hand side of the reaction $\chi = 0$, whilst if the position of the reaction is such that the only molecules present are those on the right-hand side of the reaction then $\chi = 1$. For an actual chemical *change* such as $A + B \rightarrow C + D$, we may refer to the change in Gibbs free energy per extent of reaction $dG/d\chi$, a variable that is often called ΔG (see, however, Welch, 1985). At equilibrium, $\Delta G = 0$ and the equilibrium constant of the reaction K_{eq} is given by the ratio of the activities of the products to those of the reactants when $\Delta G = 0$. The *standard* free energy change for the reaction, $\Delta G^{0'}$, is thus:

$$\Delta G^{0'} = -RT \ln K_{eq} \qquad (9.8)$$

The prime indicates that the standard state is defined in a way that differs from that usual in chemistry, where the standard states are all of unit activity. Since this would mean considering for protons a pH $\simeq 0$, a more convenient standard state is used for biochemical work, i.e. pH = 7.0. For reactions in which the mass action ratio (i.e. the product of the activities of the products divided by the product of the activities of the reactants) is not equal to the equilibrium constant, the *actual* free energy change ΔG associated with the transformation of a certain amount of reactants to the appropriate (stoichiometric) amount of products differs from the standard free energy change. In general:

$$\Delta G = \Delta G^{0'} + RT \ln \{(\Pi a_{PRODUCTS})/(\Pi a_{REACTANTS})\} \qquad (9.9)$$

Therefore, as is well known, the direction in which a reversible reaction such as $A + B \leftrightarrow C + D$ will go spontaneously depends upon the mass-action ratio of the different compounds involved and the equilibrium constant for the reaction. The direction of a particular reaction may thus be changed simply by changing the standard chemical potential or the activity (coefficient) of one of the reactants. Thus the *primary criterion* by which the possible metabolic interconversions of the molecules in the system of Figure 9.1 is determined is whether or not such an interconversion is thermodynamically favourable.

9.2.2   Electrochemical potentials

If the reaction in which one is interested involves separate thermodynamic phases, the difference in pressure between those thermodynamic phases (if applicable) must be included and, for reactions involving charged compounds, an electrical term added. So far, the contribution of electrical factors to thermodynamic potentials has been ignored since thus far only isolated, macroscopically homogeneous thermodynamic phases have been considered in which significant differences in electrical potential may be assumed not to exist. Strictly, however, the *electro*chemical potential (denoted $\tilde{\mu}$) of an ion is what is of interest, and this differs from the chemical potential (equation 9.4) by the electrical potential (ψ) times the charge on the ion (z) times Faraday's constant (F):

$$\tilde{\mu}_i = \mu_i + z_i F \psi = \mu^{\circ}_i + RT \ln a_i + z_i F \psi \qquad (9.10)$$

The splitting up of the electrochemical potential into a standard chemical potential and an electrical part is not strictly necessary; the electrical term could just as easily be incorporated into the standard chemical potential. However, whilst this separation is considered controversial in some quarters, it has become common in bioenergetics to adopt this experimentally convenient convention.

Since *absolute* values of the electrical potential are inaccessible, it might be wondered whether the ψ term has any physical meaning; this point is discussed further by Walz (1979), who gives a helpful description of much of the relevant chemical thermodynamics. What is especially important for our purposes is that *differences* in electrical potential between two thermodynamic phases *may* be measured, using reversible reference electrodes (such as the Ag/AgCl electrode); differences in the *electrochemical* potential of an ion between such phases may be measured using an electrode reversible to that ion (such as a potentiometric glass pH electrode for protons). If there is no difference in the standard chemical potential of the ion in the two phases, differences in *chemical* potential may thus be obtained by difference, by means of equation 9.9. Similarly, since the standard chemical potential of an ion in a particular thermodynamic phase is only separated from its measured activity coefficient by convention (it cannot be separated experimentally), a further simplification is acceptable: a system of interest may be allowed to come to equilibrium (so that there is no difference in electrochemical potential of the ion of interest between the two thermodynamic phases); the differences in *concentration* of the ion in the two phases can be measured; the standard chemical potential of the ion in the two phases be *assumed* to be the same; and all differences in concentration be *ascribed* to differences in the activity coefficients for the ion in the two phases. These points are illustrated, from both a theoretical and experimental standpoint, in Figure 9.3 and Table 1.2 (page 12).

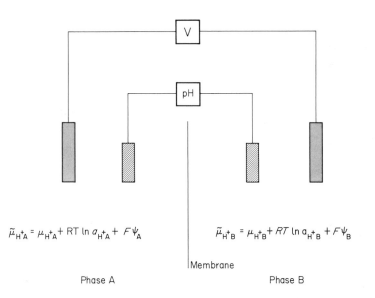

$$\tilde{\mu}_{H^+_A} = \mu_{H^+_A} + RT \ln a_{H^+_A} + F\psi_A \qquad \tilde{\mu}_{H^+_B} = \mu_{H^+_B} + RT \ln a_{H^+_B} + F\psi_B$$

Membrane

Phase A   Phase B

Fig. 9.3 The electrochemical potential of the proton in two adjacent thermodynamic phases separated by a membrane. The electrochemical potential of the proton in each phase is given (Eq. 9.9) by the expressions in the figure. However, since $pH = -\log_{10} a_{H^+}$, $RT \ln a_{H^+} = -2.303 \ RT$ pH (since $2.303 = \ln 10$). Thus the difference in the electrochemical potential of the proton in the two phases is:

$$\Delta \tilde{\mu}_{H^+ \ (A-B)} = \Delta\mu^{\circ}_{H^+ \ (A-B)} + F\Delta\psi_{(A-B)} - 2.303 \ RT \ \Delta pH_{(A-B)}.$$

If the standard chemical potentials are assumed to be the same on either side of the membrane:

$$\Delta \tilde{\mu}_{H^+ \ (A-B)} = F\Delta\psi - 2.303 \ RT \ \Delta pH_{(A-B)}, \text{ with the units in kJ.}$$

If the membrane is permeable to ions other than protons, to allow the electrical potentials to be defined, $\Delta\psi$ may be measured with a voltmeter connected to reversible electrodes (such as Ag/AgCl), $\Delta \tilde{\mu}_{H^+}$ with glass pH electrodes (which themselves contain reversible Ag/AgCl electrodes) and ΔpH by difference. For further details, see text, and also Table 1.2 (page 12).

Thus (Fig. 9.3) there is a definition of the difference in proton electro-chemical potential between two phases which is not only thermodynamically correct but experimentally useful. Similar expressions may be written for the difference in (electro)chemical potential of any other ion between two such phases.

The 'membrane' in Figure 9.3 was used just as a permeability barrier, to stop the equilibration of the electrochemical potential difference of the protons in the two phases, and no other properties were ascribed to it. Similarly, the individual compartments, which the membrane served to separate, were considered as themselves to be homogeneous and at electro-chemical equilibrium. Whereas the second description may be acceptable for present purposes, the first one is not sufficient, since there is always an *interfacial region* in which strict electroneutrality is not maintained. Especially in the case of a membrane which possessses fixed charges, and thus a surface potential (Fig. 9.4), the adjacent molecular layers will be excessively populated by ions of opposite charge, to form a so-called electrical double layer. The structure and thickness of this electrical double layer depend in a complex fashion upon the surface charge density, surface potential and number and valency of the ions of the bulk phase (e.g. Bockris and Reddy, 1970; McLaughlin, 1977; Barber, 1980, 1982; Pethig, 1986).

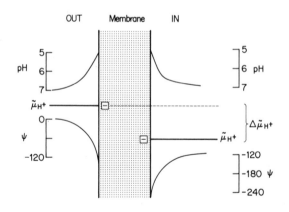

Fig. 9.4   The existence of fixed (surface) charges at the interface between a biomembrane and solution implies the existence of a surface potential. Under equilibrium conditions, this surface potential will change the extent to which Δp is distributed between $\Delta\psi$ and ΔpH but not the magnitude of Δp.

It is convenient initially to consider this interfacial region as one or more separate thermodynamic phases, and to think about it in the following way. As a negative charge is approached, the local electrical potential becomes more negative; however, the negative charge is also increasingly attractive to protons (for electrostatic reasons) so that the *chemical* potential of the proton is increased. Thus, provided the system on one side of the membrane is *in equilibrium* ($\Delta G = 0$), the *electrochemical* potential of the proton will not be changed from that in the bulk phase as the surface is approached, whilst its local distribution between ψ and pH may be changed (see e.g. Junge, 1977; Nicholls, 1982). This is experimentally useful (provided the equilibrium assumption holds), since measurements in 'bulk' phases are generally more convenient and reliable than those in complex and hetero-geneous interphases. However, and this is most important, spectroscopic (or other) probes *cannot* properly be used to measure the electrochemical potential difference between two bulk phases if the probes are at the same time bound to membranes or interfaces. *Only* if it is *known*, (a) that the system *is* in electrochemical equilibrium throughout the phases on a given side of the membrane, *and* (b) that the probes of the electrical and chemical parts of the proton electro-chemical potential difference bind in *identical* places, can success be hoped for. It is unlikely that either (let alone both) of these conditions is met in practical situations of interest to the bio-energeticist.

Acquaintance has now been made with the most important equations of macroscopic, equilibrium thermodynamics that are relevant to the questions to be discussed here, in particular the criterion to be used to decide whether it is *possible* that a given (named) reaction is taking place: is it associated, as written, with a negative change in (electro)chemical potential (or Gibbs free energy) of the system under the conditions prevailing? Of course, as is well known, the thermodynamics of initial and final states cannot tell us anything about reaction *rates*, but at least there is a criterion for deciding whether a particular rate may be expected to exceed zero! It now needs to be considered how (electro)chemical reactions may be coupled to each other.

9.3 Energy coupling of reactions and phenomenological non-equilibrium thermodynamics

The important reaction $ADP + P_i \rightarrow ATP + H_2O$ has a typical standard free energy change *in vivo* of some $+31.8$ kJ/mol. The actual value depends, for reasons given above, upon the pH, Mg^{++} concentration and other factors (Thauer *et al.*, 1977). In other words, ATP synthesis is a highly endergonic process, and *in vivo* is poised so that its free energy change is positive by some 43.9 kJ/mol; Thauer *et al.*, 1977). One might wonder, then, how this is possible. The answer is, of course, that ATP synthesis is coupled to the performance of reactions whose free energy change is yet more *ex*ergonic. Consider a reaction such as $S \rightarrow P$, with a ΔG value under a particular set of conditions equal to $-53\,kJ/mol$. Allowing these reactions to be coupled to each other means that the overall reaction:

$$S + ADP + P_i \rightarrow P + ATP + H_2O$$

has a free energy change that is $-9.1\,kJ/mol$ and thus favourable.

If the coupling is strictly stoichiometric, the equilibrium distribution of S, P, ADP, P_i and ATP may be calculated from a knowledge of the starting mixture and the free energy change of the constituent reactions. In real systems, however, energy coupling is rarely strict, since there are likely to be side reactions and uncoupled partial reactions (e.g. ATP hydrolase activity which leads only to heat production). To maintain a particular series of reactions at a particular poise (mass action ratio), then, a continuing input of free energy is necessary, and the system cannot attain a true (global) equilibrium. To begin to describe such reactions, the formalism of non-equilibrium thermodynamics (NET) must be used. A full introduction to this and related topics is given in the authoritative monograph of Westerhoff and van Dam (1987).

Fig. 9.5   A phenomenological, 'black box', non-equilibrium thermodynamic view of energy coupling. Here only input and output forces and fluxes are considered. For further details, see text.

A 'black box' may be considered (Fig. 9.5) catalysing an input reaction with a flux (i.e. at a rate) J_1 and held (by unspecified means) so that it has a free energy change ΔG_1. When the system reaches a stationary (i.e. steady) state, an output reaction may also be observed occurring at a rate J_2 and associated with a ΔG_2. (Obviously the ΔG of the output reaction is constantly changing, and times for which ΔG_2 has a 'constant' value are therefore considered short.) In the formalism of NET, an affinity X is defined, which is the negative of the free energy change (per extent of reaction). The efficiency of energy coupling in the system of Figure 9.5 is then defined as $-J_2X_2/J_1X_1$ (see e.g. Westerhoff *et al.*, 1982, 1983). Such a definition, based upon the observations of the effects upon the outside world of the 'black box' energy converter, is purely phenomenological, and cannot therefore of itself provide mechanistic answers to questions such as *why* a particular energy converter is more or less efficient than any other one. It is, however, interesting to note that, in contrast to what may be proved rigorously or expected from first principles (Kubo, 1969, 1986; Nicolis and Prigogine, 1977), it is often found that the 'domain of linearity' between macroscopic forces (affinities) and their concomitant fluxes in bioenergetic systems is very broad (e.g. Rottenberg, 1973; Caplan and Essig, 1983; Stucki *et al.*, 1983). The fundamental (and mechanistic) reasons for this are not understood, but one might speculate that they are perhaps related to the ability of such proteinaceous coupling devices to exhibit many more degrees of conformational freedom than can a small molecule.

Phenomenological NET certainly allows a more accurate description of living processes than does classical thermodynamics, but there is neither space here to do the subject justice nor in fact does it seem to have greatly aided the resolution of the central problems of membrane bioenergetics addressed in this chapter. Apart from an NET approach to the description of proton pumps (see later), it must suffice to cite several of the more relevant articles and books which must serve as an introduction to the literature for the interested reader (Stucki, 1978, 1980, 1982; Rottenberg, 1979a,b; Westerhoff and van Dam, 1979, 1987; Caplan and Essig, 1983; Stucki *et al.*, 1983). However, the following ideas from these mainly phenomenological descriptions are particularly important:

(a) that the efficiency of coupling in non-equilibrium systems that behave according to these NET principles is always less than 100%;

(b) that this efficiency depends upon each of the terms J_1, X_1, J_2 and X_2 (and therefore upon the load on the energy converter); and

(c) that the optimal degree of coupling for any type of coupling system depends upon whether it is J_2, X_2 or their product that it is desired to maximize.

The available evidence indicates, interestingly enough, that microorganisms have in general evolved to permit a maximum *flux* at the expense of yield (Westerhoff *et al.*, 1983; Kell, 1987a,b).

While, as shall be seen, there are reasons to doubt the genuinely rigorous basis even of the simplest NET treatments when applied to proton pumps, a number of authors have derived detailed specific models to describe the energy coupling of protonmotive systems of the type displayed in Figure 9.2. These models form some of the more rigorous descriptions of how one should *actually* expect a purely chemiosmotic system to behave. For this reason, they shall be treated in a separate section. However, it is first worthwhile considering the relationship between macroscopic and microscopic systems generally.

9.4   Statistical thermodynamics and the relationship between microscopic and macroscopic descriptions of bioenergetic systems

Thus far only macroscopic systems have been considered. What this means is that it has been assumed that it is suitable to treat an ensemble of molecules of a given type as though they were behaving identically. Of course, experimentally, huge ensembles of particles are of necessity what are observed. In fact, the term 'ensemble' has a special meaning in this context; it is used specifically to describe 'a random collection of systems, each of which corresponds to the "same" macroscopic thermodynamic state but which has a different microstructure'. This, however, leads us to problems of great difficulty, since modern thermodynamics is still built upon the foundations laid by pioneers such as Boltzmann in the nineteenth century (Brush, 1976, 1983). What is known from the work of Maxwell, Boltzmann and others is that an ensemble of gas molecules of a given type (at temperatures above absolute zero) is constituted such that individual molecules do not possess the same instantaneous energy since, even at thermodynamic equilibrium, they are exchanging heat energy quanta between themselves and the walls of the isothermal heat bath in which they are contained (of a magnitude equal to $\frac{1}{2}k_B T$ per degree of freedom, where k_B = Boltzmann's constant = 1.38×10^{-23} J/K). The factor $k_B T$ thus has the units of energy, and is taken as a yardstick by which free energy changes of interest are compared with the randomizing thermal forces pervading a system; at 25°C it is equal to some 2.48 kJ/mol or, in electrical units, 26 mV.

Imagine such a heat bath, containing a mixture of molecules of H_2, Cl_2 and HCl whose composition is such that it is at chemical equilibrium and further assume that the chemistry involves interconversions written as:

$$H_2 + Cl_2 \leftrightarrow 2\,HCl \qquad (9.11)$$

If it was observed that, despite the direction in which the reaction was started, the system evolved spontaneously and monotonically to give an identical mixture of molecules one might be confident that a true chemical equilibrium had been attained (Denbigh, 1981). This would, of course, be a *dynamic* equilibrium, for reversible chemical changes are known to be constantly taking place as thermally activated molecules cross and re-cross the 'energy barrier' separating them (see later). Since the system is (macroscopically) unchanging it would be assumed correctly (at equilibrium) that for every molecule participating in the reaction from left to right a stoichiometrically equal number is doing the reverse.

If one asks what is happening as a function of time to any *individual* molecule, the realms of statistical mechanics and statistical thermodynamics are entered (see e.g. Hill, 1960; Finkelstein, 1969; Gassier and Richards, 1974; Gopal, 1974; Knox, 1978; Brush, 1983). Their fundamental importance to the whole of thermodynamics and hence bioenergetics means that one must dwell here briefly, to develop one particular set of arguments. Given the present state of knowledge, however, it is not necessary to be acquainted with the mathematical foundation of the subject to acquire an understanding of the relevant physical behaviour and points at issue.

Statistical mechanics and thermodynamics consider the motion of particles through *phase space*, a mathematical construct in which the (generalized) three-dimensional position of a particle is plotted against the generalized momentum of the particle. According to Boltzmann's *ergodic hypothesis*, any particle in a closed ensemble of particles will sooner or later have the opportunity of passing through (or arbitrarily close to) all regions of phase space. It will have the opportunity of meeting and reacting with all other particles in the system and thus of experiencing (populating) any of the 'states' possible to the system and that determine *in toto* the system's physical and chemical constitution. Because the classical equations of motion are unchanged by time-reversal, Tolman (1938) was able to argue that direct interconversions between the various individual cells (points) making up phase space satisfy Boltzmann's conditions of the ergodic hypothesis and, from a statistical point of view, the equilibrium condition is the most probable one because sooner or later it was inevitably attained, and maintained (in the absence of outside influences) (Tolman, 1938).

Because the macroscopic state of the system is unchanging, under equilibrium conditions, *individual* quantal processes and their reverses must occur at equal rates and, since they are quantal, they must occur by the same pathway in each direction; this principle is known as *microscopic reversibility*. An extension of this principle states that equilibrium can therefore be maintained by balancing the number of particles moving into or out of a particular *region* of phase space; a principle known as the principle of *detailed balance* (Tolman, 1938). Whilst it is usually considered that microscopic reversibility holds for equilibrium and non-equilibrium systems, detailed balance holds only at or very close to equilibrium (Morrissey, 1975; Haken, 1977) (though fundamental problems remain even here, Zukav, 1980; Primas, 1981; Landsberg, 1982; Davies and Brown, 1986). Why is this important?

It is possible to calculate *from first principles* the dynamics of small gas molecules in a closed thermodynamic phase: to show that at equilibrium these dynamics display both microscopic reversibility and detailed balance; to derive the statistical distribution of these states; and to show both that this distribution is Boltzmannian and that it is the state of lowest energy available to the system throughout the entirety of phase space. In other words, the equilibrium observed represents a *global* free energy minimum and the above ideas have a rigorous foundation for small molecules forming an ideal gas at equilibrium.

Unfortunately the same cannot be said even about an ensemble of 'identical' aqueous globular protein molecules isolated in a heat bath *and which is believed to be at equilibrium*. The reasoning runs as follows (e.g. Jaenicke, 1984; Kell, 1987a, 1988). A protein of molecular mass 20 kD can, in principle, possess some 10^{80} conformational states (i.e. positions in phase space), but since the Universe is 'only' some 10^{17} seconds old (Barrow and Silk, 1983), even if one allows the protein to explore these states at a rate of 10^{15} per second, it cannot conceivably explore all of them in passing from the unfolded to the folded state. Thus, given experimental realities, even a protein isolated in a heat bath of solvent molecules at 'equilibrium' is not an ergodic system. Furthermore, *it cannot be known* whether it is at a local or a global free energy minimum and it seems probable that the relevant *macroscopic* thermodynamic properties cannot even be *measured* properly (Lumry, 1986). What is known, assuming that it *is* at 'equilibrium', is that the protein is exploring many areas of phase space and fluctuating wildly between conformations whose free energy differences are roughly those to be expected from an ensemble of particles at equilibrium in a heat bath. In other words, any conformation or conformational 'state' of a protein is constituted by a large variety of microstates whose energy lies within $k_B T$ of the average energy of the system.

One consequence of the ergodic hypothesis is that one should be able to make statements about an *individual* molecule, which might itself be isolated (but at thermal equilibrium) in a *microscopic* thermodynamic phase. Since a particle at equilibrium explores all regions of phase space with equal probability, the time-average of the energy of such molecules will be equal to the instantaneous average of the energies of an ensemble of such molecules. This is the central dogma of equilibrium statistical thermodynamics, and allows macroscopic concepts (such as concentration) legitimately to be used to describe the different states of small equilibrium systems (Hill, 1963). For non-equilibrium systems, however (which nonetheless for systems *near* equilibrium exploit the assumption of local equilibrium by virtue of the fluctuation-dissipation theorem (Kubo, 1969, 1986; Kreuzer, 1983), the situation is entirely different: the ergodic hypothesis does not hold, and, if the system *works* microscopically (i.e. significant parts of the system do *not* interact during the relevant time of observation), then it is *not* appropriate to describe that system macroscopically (Welch and Kell, 1986). In fact, the Boltzmannian derivation of the ergodic theorem contains the assumption (usually referred to as the *Stosszahlansatz*; Chester, 1969) that the motions of the molecules constituting the system *are uncorrelated before each collision that takes place.* Whilst this is acceptable for a dilute gas, it is obviously incorrect for a protein molecule in which the conformational flexibilities of individual atoms are weakly or highly correlated with those of other atoms. Thus, the fact that protein molecules themselves are individually so complex gainsays any rigorous application of macroscopic equilibrium thermodynamic principles to non-equilibrium systems. Whether this matters from an *experimental* standpoint or not depends simply upon the extent to which individual but isoenergetic ($\pm k_B T$) areas of phase space are explored by any particular molecule of interest over the relevant time-scale. Consider the 'ensemble' of protein molecules. Let them be present at 1 mM (i.e. 6×10^{17} molecules in 1 ml). Let each one explore 6×10^{12} *overall* conformations per second. Even in 1 millisecond (a typical turnover time for an enzyme) only 1 in 100 molecules even has a chance to encounter (for a miniscule fraction of the time) the same conformational state as any other. Since the real numbers are vastly greater than this, it should be obvious that the forward pathway taken by an enzyme during a reaction is *most unlikely* to be identical to that taken by it during any reverse reaction. This is just another way of saying that the principle of detailed balance cannot be applied *a priori* to working protein molecules (although experimentally such deviations *may* not be manifest). This point applies both to proteins and to organisms. A simple example, relevant to bacterial bioenergetics and based loosely on the famous thought experiment of Schrödinger's (1935) cat (Primas, 1981), may be used to illustrate this.

Consider an individual, obligately aerobic respiratory, bacterium whose behaviour depends upon (a) whether it spends half of its life exposed to 100 μM O_2 and half of its life anaerobic, or (b) whether it spends its entire life exposed to 50 μM O_2. Notwithstanding the difficulty of defining 'life' in the case of a microorganism (see e.g. Harris and Kell, 1985a; Mason *et al*, 1986), it would be agreed that after a greater or lesser period of anaerobiosis, prior to any re-exposure to oxygen, the microorganism would have 'snuffed it'. The probability of this microorganism 'dying' depends upon the *length of time* for which it experiences anaerobiosis. Given experimental realities connected with the accuracy and response time of oxygen electrodes, however, the two sets of oxygen tensions might be made to appear identical *from a macroscopic point of view.* In other words, because an oxygen electrode is rather slowly responding, its *mean* signal after a long time will be the same in the two cases because in each case the 'mean' does correspond to 50 μM O_2. Only *after* exposing this bacterium to one of the two regimes of oxygenation would one be able to tell if it were 'alive' or 'dead'. Similarly, the state of a system consisting of a culture of bacteria in which one half of the bacteria are exposed to 100 μM O_2 and one half anaerobic will differ from that of a system in which all bacteria are uniformly exposed to 50 μM

O_2. Since oxygen reduction is irreversible, individual cells work 'in isolation' and cannot share the free energy made available by respiratory electron flow. Again, one would be able to have two macroscopically stationary states which were apparently identical but which led the system, because it works *microscopically*, to behave in two entirely different ways! The time average is not equal to the ensemble average (Welch and Kell, 1986).

This general point, which underlies the so-called 'problem of scale-up' in biotechnological systems (Kell, 1986a), means that macroscopic descriptions of far-from-equilibrium systems, even in apparently macroscopically stationary states, do not of themselves give an adequate picture of the systems they are seeking to describe. Indeed, an overzealous application of macroscopic considerations to microscopic coupling systems may lead to apparent violations of the second law of thermodynamics. An example of relevance to membrane energy coupling is that given by Westerhoff and Chen (1985), and relevant considerations for purely electrical fields are given by Kell *et al.* (1988).

The general conclusion to be drawn is that non-equilibrium systems, containing ensembles of molecules with many degrees of freedom, individually may explore different parts of conformation space when working in the forward and reverse direction, i.e. when crossing and re-crossing an energy barrier. Since there is by definition a loss in free energy in one of the directions, and there cannot be a gain in free energy in the 'other' direction, irreversibility is to be seen as a property of individual molecules, and not of ensembles. This explains in another way why the principle of detailed balancing does not, in general, apply to microscopic systems of macromolecules transducing free energy *via* states which are far from thermal equilibrium (Steinberg, 1986).

Parenthetically, it is worth mentioning that while there is much lively and current debate about the philosophical status and interpretation of the behaviour of quantum mechanical systems (e.g. Bohm, 1980; Zukav, 1980; Gal-Or, 1981; Primas, 1981; MacKinnon, 1982; Popper, 1982; Wheeler and Zurek, 1983; Bohm, 1984; Garden, 1984; Davies and Brown, 1986), at the time of writing little of this debate (e.g. McClare, 1971; Blumenfeld, 1983; Welch and Kell, 1986; Kamp and Westerhoff, 1987; Westerhoff and Kamp, 1987) has filtered through to the fields of chemical thermodynamics and bioenergetics. However, it seems implausible that this unsatisfacatory state of affairs be allowed to continue indefinitely, and it can be concluded from this section that, while thermodynamics is often presented as beyond debate or discussion, a proper formalism for describing the energy states of even equilibrium proteinaceous systems, let alone those in the act of catalysing free energy transduction, is not yet to hand. This point will be considered when we discuss the degree of coupling within individual proton pumps.

9.5 Transition state theory

The *rates* of (electro)chemical reactions which possess favourable thermodynamics must now be considered. It is first necessary to distinguish transition states from systems exhibiting non-thermally activated collective behaviour.

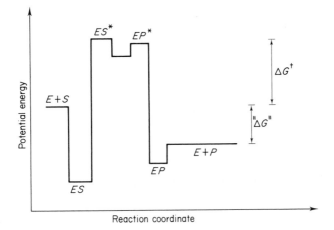

Fig. 9.6 The energetics of enzymes as assessed by transition-state theory. For further details, see text.

Whilst chemical kinetics may perhaps best be approached in terms of collision theory (Atkins, 1978; Knox, 1978), it is nowadays usual to treat the kinetics of enzyme-catalysed reactions in terms of absolute reaction rate theory (ART), commonly called transition-state theory (TST) (e.g. Glasstone *et al.*, 1941; Jencks, 1969; Laidler, 1969; Thornton and Thornton, 1978; Blumenfeld, 1981; Wharton and Eisenthal, 1981; Somogyi *et al.*, 1984; Fersht, 1985). The salient features of TST are summarized in Figure 9.6, which plots the potential energy of a system against the reaction coordinate (i.e. the physicochemical nature of the system at a given point in the reaction sequence). In Figure 9.6, the system chosen is an enzyme obeying Michaelis–Menten kinetics and catalysing the reaction $S \rightarrow P$ according to the reaction scheme:

$$E + S \leftrightarrow ES \leftrightarrow ES^* \rightarrow EP^* \rightarrow E + P \tag{9.12}$$

Starred states represent 'activated complexes' or 'transition states' and are at the peaks of such a diagram, while unstarred states represent more-or-less stable intermediates of the reaction sequence such as enzyme–substrate complexes. TST addresses the question of how *mechanistically* the overall reaction takes place and what factors underlie the *rate* of such a reaction. Theories such as TST therefore seek to bridge the gap between free energy terms (i.e. thermodynamics) and reaction rates (i.e. kinetics).

The central idea in transition-state theory is that reactions between molecules proceed via an activated complex whose rate of formation is a function of the absolute temperature and the so-called activation energy. Once formed, the activated complex, in this case *ES**, decays rapidly to give the products of the reaction. Neglecting second-order effects and assuming that every molecule of activated complex decays to give products (i.e. that the so-called transmission coefficient = 1), the rate of formation of the activated complex is given by the equation:

$$k = k_B T/h \exp(\Delta S_0^{\dagger}/R) \exp(-\Delta H_0^{\dagger}/RT) \qquad (9.13)$$

where h is Planck's constant ($= 6.63 \times 10^{-34}$ J s), R is the gas constant (8.31 J mol^{-1}K^{-1}) and $\Delta S_0^{\dagger}$ and $\Delta H_0^{\dagger}$ are respectively the entropy and enthalpy of activation, related to $\Delta G_0^{\dagger}$ (the free energy of activation) by the well-known general equation (applied at constant volume):

$$\Delta G = \Delta H - T\Delta S \qquad (9.14)$$

The existence of T in the exponent of equation 9.13 (which is derived from statistical mechanics) indicates that the production of this activated complex is caused by the absorption of thermal energy from the surroundings, and such a reaction is said to be *thermally activated*. Equation 9.13 resembles the Arrhenius rate law:

$$k = A \exp(-\Delta E_a/RT) \qquad (9.15)$$

in which ΔE_a is an *experimentally determined* 'activation energy' (and might better be referred to as ΔE_{expt}) whereas the entropy and enthalpy of activation in equation 9.14 are thermodynamic variables which may or may not be related to it, and which may or may not have some genuine meaning. It should be noted that the above assumes thermal and thermodynamic equilibrium between the states $E + S$ and $ES**$, so that their concentrations may be calculated from their differences in free energy (via Eq. 9.8) and that the exchange of heat quanta is so rapid that we may treat these states macroscopically as being in equilibrium with an ensemble of microstates of the energy $\pm k_B T$. Taking natural logarithms and differentiating equations 9.13 and 9.15 with respect to temperature:

$$d \ln k/dT = \Delta E_{expt}/RT^2 = 1/T + \Delta H_0^{\dagger}/RT^2 \qquad (9.16)$$

$$\text{i.e. } \Delta E_{expt} = \Delta H_0^{\dagger} + RT \qquad (9.17)$$

Thus, whilst the temperature-dependence of rate constants is often used experimentally to obtain the value of $\Delta H^{\ddagger}_0$ and $\Delta S^{\ddagger}_0$ from equations 9.13 and 9.14, the former equals the Arrhenius activation energy only at absolute zero and, as Blumenfeld (1981, p. 66) has put it, 'there is no physical meaning in substituting [Eq. 9.17] into [Eq. 9.13] with "T" as a variable.'

It has been seen that enzymes in a given 'state' (i.e. collection of conformational microstates with a free energy difference $\leqslant k_B T$) exhibit thermal fluctuations. In transition-state theory the idea is that thermal energy (derived mainly from the solvent heat bath) is used to drive the enzyme-substrate complex into a transition state whose energy may differ from that of the ground state by *many times* $k_B T$. Since the assumption of equilibrium between the ground state and the transition state means that the ordinate actually represents Gibbs Free Energy in a diagram such as Figure 9.6, it might be wondered how heat quanta, *under macroscopically isothermal conditions*, might be permitted to raise the free energy of a molecule, albeit transiently, without breaking the second law of thermodynamics. As Kemeny (1974) has put it, however, 'this does not mean that heat energy is turned into free energy but rather that the transduction of internal energy and heat exchange with the reservoir are part of the same mechanism.' This leads to the idea of *enthalpy–entropy* compensation (see Somogyi *et al.*, 1984; Lumry, 1986), i.e. the experimental finding that small changes in protein molecules (or their substrates) are often accompanied by equal and opposite changes in the values of ΔH_0 and ΔS_0 for a particular step such as ligand binding. The molecular meaning of this compensation behaviour remains unknown (Lumry, 1986), but it seems probable that it has the same fundamental basis as the linear free energy relationships (LFER) first observed for an enzymatic reaction by Fersht and his colleagues in engineered derivatives of the tyrosyl t-RNA synthetase of *Bacillus stearothermophilus* (Fersht *et al.*, 1986).

Linear free energy relationships (LFER) are well known in organic chemistry in the form of Hammett or Brønsted plots of the log of the rate constant for a particular reaction step versus the log of the equilibrium constant for that step (i.e. a ΔG term) (Jencks, 1985). The finding that LFER may be observed with (conservative) changes in the structure of an enzymatic active site (Fersht *et al.*, 1986) provides powerful (though not conclusive) evidence for the thermally based mode of energy transduction of at least some steps in an enzyme that simply catalyses the approach of a reaction to equilibrium. In particular, such a finding may be taken to bolster the case for a macroscopic treatment of enzyme kinetics (at least at room temperature), assuming microscopic reversibility and detailed balance for interconversions between macrostates (i.e. intermediates) and ignoring the microstates which indubitably contribute to any given 'state'. Against this, however, is the finding by Bechtold *et al.* (1986) that the pathway of electron transfer in a cytochrome *c* derivative is different in the forward and reverse directions. Whiist this may be just a clear-cut example of a general truth, Williams and Concar (1986) remark that 'if . . . [the data do not go away,] . . . we must re-examine a considerable part of our thinking about protein energy states and not just about electron transfer.' Similarly, in the carefully studied example of myoglobin (Ansari *et al.*, 1985), it has been shown that the transduction of free energy from initially non-equilibrium states does not pass through microstates that are in thermal equilibrium (Frauenfelder and Wolynes, 1985; cf. Bialek and Goldstein, 1985). Indeed, it is worth remarking that, while much is known about the dynamics of relatively simple synthetic polymers (Hedvig, 1977; Doi and Edwards, 1986) in which local, but not *large-scale* (i.e. cooperative) motions obey the Arrhenius activation law (Hedvig, 1977), much remains to be learned about the more complex, and above all *coupled* internal motions of proteins. In particular, formalizing the relationships between enzymatic fluctuations between microstates, the relevant thermodynamics and their functional consequences, remains an important goal (Somogyi *et al.*, 1984; Welch, 1986).

In summary, then, transition-state theory describes the behaviour of chemical reactions activated by the absorption of thermal energy and in internal thermal equilibrium. It should be noted, however, that linear free energy relationships are linear plots of *log k* versus ΔG (times a constant), whereas the linear non-equilibrium thermodynamic relationships are between forces (equivalent to ΔG) and rates (i.e. *k* rather than log *k*). The significance of this important difference remains unknown, but *might* be related to whether a particular step is thermally activated or not. What are non-thermally activated processes?

9.6   Non-thermally activated processes

Although many processes exhibit behaviour of the type concentrated on above, and which is consistent with the view that their activated states are in internal thermal equilibrium with the ground states from which they arise, this is by no means true of all physical systems. A well-known example is the laser (see e.g. Haken, 1977). Why then is a laser not like a light-bulb (i.e. a tungsten filament lamp)?

Figure 9.7a gives the relevant schematic diagram of a simple electrical circuit containing a light-bulb; the latter is simply an electrical resistor of magnitude *R* Ohms. Before the switch is turned on, the system is in thermal equilibrium with its heat bath. Since the electrical resistor consists of atomic particles, they exhibit Brownian motion, and, being electrically charged their

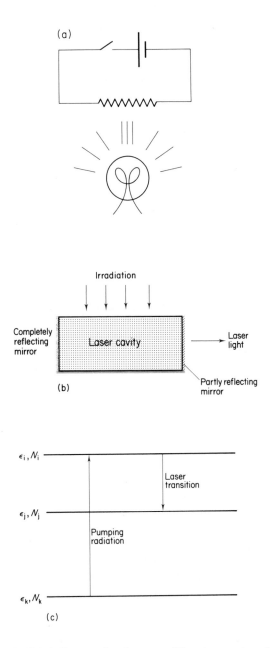

Fig. 9.7 A laser and a light bulb; examples of two very different energy-transducing systems in which the output is a flux of protons. (*a*) A light bulb is simply a resistor, transducing but dissipating the free energy input from the voltage source. (*b*) Block diagram of the action of a laser. (*c*) The three-level laser whose operation is outlined in the text.

motions give rise to a fluctuating displacement current. Since these currents are carried in an electrical resistor, they may be detected as a fluctuating voltage ('noise'), if the terminals of the resistor are connected to a voltmeter of high input impedance. Because of the equipartition of states, according to a Boltzmann distribution and consequent upon the stated thermal equilibrium, this noise is independent of frequency (i.e. 'white'). It was discovered by Johnson in 1927 and explained in these terms by Nyquist the following year (see Johnson, 1928; Nyquist, 1928; Beck, 1976; van der Ziel, 1976; De Felice, 1981).

If a steady current is allowed to flow through the resistor, an excess noise appears. This is not simply because the resistor becomes hotter (for such noise would still be white), since it has an approximately inverse dependence upon frequency. This so-called $1/f$ noise reflects the fact that, as energy flows through the system, some localized states with a higher-than-average energy appear with a probability that is increased the longer the period of observation. Nonetheless, the ratio of the noise voltages to the voltage across the resistor is miniscule, and little error (from a macroscopic standpoint) results in assuming that the current flowing is strictly proportional to the voltage drop produced across the resistor, i.e. that Ohm's Law holds. In other words, although the thermal equilibration (i.e. equipartition) of states is not *perfect* it is reasonable to treat such a system as if it were genuinely in internal thermal equilibrium.

In the laser, by contrast, the phenomenon of stimulated emission occurs (Fig. 9.7b and see e.g. Jones, 1969). In a solid-state laser, the active material can absorb photons, forming excited states. These can decay either back to the ground state or they may be stimulated to do so by absorption of radiation of frequency equivalent (via $E = h\nu$) to the difference in energy levels between the states. The simplest possible three-level laser consists of three energy levels (ε_i, ε_j and ε_k) populated with a number of atoms (N_i, N_j and N_k) (Fig. 9.7c). In this arrangement, the active part of which might be a crystal of ruby, the atoms in an energy state k may be induced to absorb radiation such that they are excited to a higher energy state i. The decay of atoms from this state to the states of lower energy may be by spontaneous or stimulated emission. In the case of spontaneous emission, the decay rate is simply proportional to the number of atoms in state k. In stimulated emission, by contrast, the decay is stimulated by absorption of radiation so that the decay is proportional both to the number of atoms in state i and to the intensity of the irradiation at this particular frequency (ν_{ij}). If spontaneous emission from state j to state k is much more rapid than that from state i to state k then an excess of population of atoms in state i over those in state j will occur in the steady state (known as a 'population inversion'). States of higher energy become more populated than those of lower energy and the system is evidently *not* in thermal equilibrium. (Application of the Boltzmann expression actually implies the existence of a negative temperature, which is simply misleading.) The build-up of this population inversion is what is responsible for the laser transition of frequency v_{ij}, since the absorption of radiation of this frequency stimulates suitably excited atoms to emit radiation at the same frequency. Ultimately, the build-up of photons of the relevant frequency inside the laser cavity leads to the amplification characteristic of laser action. Readers interested in exploring this topic further may find details in Smith and Sorokin (1966), Haken (1970, 1977) and Jenkins and White (1981).

What then distinguishes a laser from a light bulb is that in the laser the population of excited states exceeds that to be expected on the basis that they are in thermal equilibrium with the ground states of the system. This can lead to modes of free energy transduction that are *not* based upon the formation and decay of activated states that are in thermal equilibrium with the ground states from which they arise. Such effects are widespread in physics, and underlie many other important phenomena such as superconduction. The type of behaviour in which collections of molecules do not act independently from each other is usually referred to as collective behaviour (e.g. March and Parrinello, 1982). Evidently, since the distinction between 'thermal' and 'non-thermal' processes is so fundamental in physics, the question arises as to whether or not biological systems might possess non-thermal means of free energy transduction, especially in electron transport-linked phosphorylation, and introductory discussions of this distinction as applied to proteins and free energy-transducing molecular machines may be found in several recent articles (Blumenfeld, 1983; Somogyi *et al.*, 1984; Welch and Kell, 1986; Kell, 1987a, 1988; Westerhoff and Kamp, 1987).

Many of the collective states described above may arise even from linear processes, i.e. the rate of their production in response to a stimulus is a linear function of the extent to which that stimulus exceeds its starting value. There are other modes of free energy transductions which are highly non-linear, and two which are presently attracting widespread interest and discussion will be discussed below: solitary excitations (solitons) and Fröhlich processes.

As mentioned above, proteins, like other macromolecules, are highly dynamic entities, and contain a vast plethora of internal motions (e.g. Careri *et al.*, 1979; Gurd and Rothgeb, 1979; Welch *et al.*, 1982; Englander and Kallenbach, 1984; Somogyi *et al.*, 1984; Ringe and Petsko, 1985; Welch, 1986). Motions of individual groups or segments are in many cases highly correlated, and may be thought of as the motions of real or quasi-particles in a local potential well or a particular region of phase space (Welch and Smith, 1987). Such motions will generally contain both acoustic (mechanical) and electric modes. Acoustic modes are known as phonons. A soliton is a special kind of phonon or wave packet which, because of the type of wave equation to which it conforms, can carry energy over a 'long' distance in an essentially *dispersionless* fashion, i.e. without losing its energy by thermal exchange or viscous damping. It thus maintains its energy at a level greater than that expected on the basis of a Boltzmannian distribution of vibrational energies at the ambient temperature. Solitons are thus non-thermally excited modes. None of the workers who have considered solitons in a reasonably biological context has considered what happens when the motion of a soliton is tightly coupled to that of an 'energized' proton (e.g. Bilz *et al.*, 1981; Blumenfeld, 1983; Davydov, 1983; Jardetzky and King, 1983; Scott, 1983; Yomosa, 1983; Careri and Wyman, 1984; Carter, 1984; Chou, 1984; Del Giudice *et al.*, 1984; Lomdahl, 1984; Lomdahl *et al.*, 1984; Somogyi *et al*, 1984). However, as discussed elsewhere (Kell and Westerhoff, 1985; Kell, 1987a), if a proton is part of a solitary excitation, its free energy will be both inseparable from it and inadequately described by a macroscopic electrochemical potential. Since these are exactly the properties that it is necessary to invoke in order to solve many of the current problems of bacterial bioenergetics, we should not ignore the probability of this type of behaviour being of crucial importance.

Another type of collective behaviour or 'coherent excitation' which may be important in bioenergetic systems is that developed over a number of years by Fröhlich (see e.g. Fröhlich, 1968, 1969, 1980, 1986; Bilz *et al.*, 1981; Fröhlich and Kremer, 1983; Kell and Hitchens, 1983; Del Giudice *et al.*, 1984; Kell and Westerhoff, 1985; Kell, 1987a, 1988). In the most general terms, Fröhlich showed that *non-linear* coupling between the electric and acoustic modes of a system (such as an energy coupling membrane) with the thermal energy of a heat bath could lead to the condensation of many of the various possible modes into a single mode of a particular energy, a phenomenon equivalent to the Einstein condensation of a gas of bosons. This excited mode could then serve as a source of free energy which might be transduced in a *dispersionless* fashion, as with a soliton above. Fröhlich discusses in particular the idea that relevant frequencies of the exciting modes might be of the order of 10^{11} Hz, based upon the velocity of sound (phonons) in condensed media (ca. 10^3 ms^{-1}) and the thickness of biomembranes (ca. 10^{-8} m). However, many types of mode softening are possible through appropriate interactions, and the model may thus be generalized to take particular cases into account. Various types of experimental evidence, such as the effects of very weak microwaves on cellular growth (see Fröhlich and Kremer, 1983) and the laser Raman spectroscopic studies of Webb (1980; and see Del Giudice *et al.*, 1984), have been invoked in support of the Fröhlich model, and do not lend themselves to any other simple explanation. Suffice it to say that, given its generality, the Fröhlich model provides another good example of a *well-developed* physical model in which non-thermal behaviour has been seen as an important part of biological free energy transduction. Therefore, whereas most existing models of bio-energetic systems are thermally based, future workers might fruitfully bear in mind that such models represent only one subset (and perhaps the least interesting one) of the approaches that colleagues in the physical sciences have found necessary to describe the properties of condensed matter.

Despite the foregoing discussion, thermally based, macroscopic models are by far the most familiar to workers in the biological sciences, and the 'fundamental' or more theoretical part of this chapter is concluded by drawing attention to two particularly relevant and well-founded macro-scopic descriptions of relevant biochemical systems: metabolic control analysis and Hill diagrams.

9.7   Metabolic control theory

In a metabolic system, such as that of Figure 9.1, the question often arises as to which step in the pathway A → H is rate-limiting when one is measuring a steady-state flux through it. This is, in fact, an inappropriate question, since each enzyme (and other parameters) contributes to the control of a par-ticular flux. The last few years have seen a renaissance of interest in the metabolic control theory developed from the work of Kacser and Burns (1973) and Heinrich and Rapoport (1974). The formalism, which will be referred to as 'control analysis', considers only small departures from (asymptotically stable) stationary states, and is thus an *exact* theory. This formalism has been reviewed extensively several times recently so no more than its chief tenets need be mentioned here (Groen *et al.*, 1982; Westerhoff *et al.*, 1984a; Derr, 1985, 1986; Porteous, 1985; Kell and Westerhoff, 1986a,b; Kell, 1987; Westerhoff and van Dam, 1987).

Unfortunately, as with all theories, the relevant concepts come with names attached, i.e. there is a jargon to be learned. However, the minimal time necessary to acquire a working knowledge of the formalism is far outweighed by the improved understanding of metabolic systems that it brings. The treatment ascribes, by means of the flux-control coefficients, *quantitative measures* of the degree to which any step in a metabolic pathway is flux controlling. The elasticity coefficients provide a mechanistic basis for the contribution of a particular enzyme (or other external parameter) to flux control, and are related to the flux-control coefficients by means of the connectivity theorems. Other theorems relate to the relationship between metabolite concentrations and enzymic elasticities.

For present purposes, the main benefits of the control analysis are that it is exact (this is refreshing enough in itself) and that it provides an appropriate formalism for the treatment of stationary metabolic systems. Additionally, it serves to relate the properties of *systems* to those of the subsystems from which they are constructed. It implicitly contains one major assumption: that intermediates exhibit pool (i.e. macroscopic or delocalized) behaviour. Thus, any apparent *failure* of the system of interest to conform to the tenets of control analysis provides an excellent and *rigorous* set of criteria for the failure of the pool assumption in that particular case. Since chemiosmotic coupling systems are simply metabolic systems (or may be treated thus), and the protonmotive force is by definition a macroscopic, delocalized, pool intermediate, it is obvious that the application of control analysis to putatively chemiosmotic coupling systems may be particularly rewarding. A later section covers some relevant applications. What it means, however, is that such traditional ill-phrased questions as, 'is metabolite X "the" intermediate of process Y?' or 'is the rate of ATP synthesis controlled by the protonmotive force?', are quite inappropriate, and that special procedures and arguments must be applied for the proper description of stationary processes. One special procedure that has been applied to the description of proton pumping systems is the Hill diagram method, and especially since it gives one a useful intuitive feel for certain of the points at issue in bioenergetics, it is introduced here.

9.8   Hill diagrams and the 6-state proton pump

One way to consider biological free energy transduction is by means of the King-Altman (1956) method popularized by Hill (1977) and illustrated, for a typical proton pump, in Figure 9.8. This shows in schematic form the different macroscopic conformational states that may be adapted by a protein catalysing the transmembrane, electrogenic pumping of a proton coupled to the performance of a redox reaction. Each of these states, of course, comprises a multitude of micro- or sub-states, which are *assumed* (by definition) to be in internal equilibrium, and which therefore exhibit both microscopic reversibility and detailed balance. The population of the states may be calculated on the basis of a knowledge of the forward and reverse rate constants for each of the transitions between the intermediate 'states' (Hill, 1977). There is as yet little certainty regarding the strict validity of these assumptions for real proton pumps, and whether, in view of the discussion above, simple models containing small numbers of conformational states are applicable even in principle to large protein molecules. In any event, models such as that in Figure 9.8 have often been applied to the proteins thought to be involved in electron transport-linked phosphorylation.

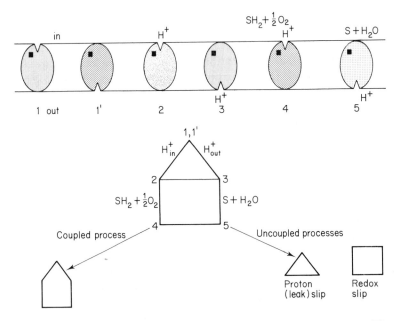

Fig. 9.8 A Hill diagram illustrating the operation of a redox-linked proton pump. The top part shows some of the plausible and identifiable conformational states of a membrane-located, redox-linked proton pump whilst the lower part shows a cycle diagram indicating the pathways of protein–ligand interactions. It is evident that there is one coupled cycle and two uncoupled ones.

Figure 9.8 also illustrates the fact that the protein may indulge in both coupled and uncoupled cycles between the macroscopic conformational states, in a fashion that may depend upon the prevailing protonmotive force. A similar diagram may be drawn for a proton pump coupled to ATP synthesis/hydrolysis. The idea is that in the uncoupled cycles, a proton pump may catalyse a redox (or ATP hydrolase) reaction without pumping a proton, or it may let a proton leak through without reversal of the scalar reaction. In the coupled cycle there is 'perfect' coupling with the so-called 'mechanistic' stoichiometry. The overall efficiency is varied by balancing the proportion of coupled and uncoupled cycles. An explicit version of this type of proton-pump model is given by Pietrobon and Caplan (1985b).

One important corollary of this 'balancing' between coupled and uncoupled cycles is that it permits a great deal of freedom in the modelling of chemiosmotic coupling systems. In other words, whatever the *mechanistic* stoichiometry of the coupled cycle of a redox-linked proton pump and an ATP synthase-linked proton pump embedded in the same coupling membrane, the observable stoichiometry will be modified in a fashion which depends upon the prevailing value of Δp (and *vice versa*). Indeed, it was in particular the unexpected properties of the relationship between the rate of electron transport and the apparent Δp (judged by ion-distribution techniques) which caused the invocation and development of the slip concept (Pietrobon *et al.*, 1981, 1982; Pietrobon *et al.*, 1986; Zoratti *et al.*, 1986). The general reasoning runs as follows.

In the stationary state, the net rate of formation of the pmf is zero; i.e. there is an exact balance between the generation and utilization of the pmf. The rate of generation of the pmf in electron transport-linked phosphorylation (J_{GEN}) is given by the rate of electron transport times the $\rightarrow H^+/e^-$ ratio at the prevailing value of Δp; and the rate of utilization of the pmf (J_{DIS}) is given by the pmf times the conductance of the membrane. If the conductance is itself a function of the pmf, the membrane is said to exhibit non-ohmic conductance. If this is accompanied by a variable extent of uncoupled transport of charge across the membrane, the phenomenon is referred to as non-ohmic leakage. At the time of writing, most of the evidence based upon measurement of the relation between the rate of phosphorylation (J_0) and the apparent Δp suggests that slip within redox pumps is of special importance in modulating the apparent stochiometries of oxidative phosphorylation (Pietrobon *et al.*, 1981, 1982, 1986; Zoratti *et al.*, 1986). Other evidence, particularly that based on spectroscopic measurements of J_{DIS} from the group of Jackson (Jackson, 1982; Clark *et al.*, 1983; Cotton *et al.*, 1984), would ascribe a more significant role to non-ohmic leaks. The data from these latter studies suggest that phosphorylation-coupled charge transport across the plane of the membrane may account for more than 90% of the utilization of the energized state set up by electron transport (Clark *et al.*, 1983).

The Hill diagram approach has the great merit of having an essentially rigorous basis (incorporating the assumptions of non-equilibrium thermodynamics described above), for the description of systems that actually work according to macroscopic chemiosmotic coupling principles. As with metabolic control analysis, any failure of these descriptions to exhibit self-consistency may be taken as good evidence against the significance of Δp in energy coupling. Thus another view (e.g. Welch and Kell, 1986; and see later) would have it that, whilst both slips and leaks in the sense of Figure 9.8 are of fundamental importance to the understanding and operation of protonmotive devices, their extents are not determined by Δp but by an *alternative* 'high energy' intermediate, which itself remains to be established (see Welch and Kell, 1986; and later). Whereas a 'chemiosmotic' analysis of Hill diagrams can account for much of the available data, there are other data for which the analysis cannot account. This author's view, particularly considering the difficulties in obtaining a correct value for the magnitude of the pmf (see later), is that these overall observations are best interpreted as yet further evidence that one of the axioms which they incorporate, (the primacy of the pmf as an energy coupling intermediate) is the simplest one to jettison in order most readily to accommodate the available data. Since this view leaves open the question of exactly what it is that many probes are measuring when they purport to measure an energetically-significant Δp, one must now enquire more closely into the methods which are generally used for the estimation of Δp, and the extent to which they are self-consistent and are likely to be accurate.

9.9   Is the protonmotive force an energetically significant intermediate in electron transport-driven phosphorylation?

9.9.1   Introduction

Although the above question is frequently asked, the answer obtained depends upon what one thinks the question actually means. In common with Ort and Melandri (1982), for instance, I shall take as a 'straw man' (model for destruction) the idea that the only free energy-transducing interactions between primary (redox-linked) and secondary (ATP synthase) proton pumps in energy coupling membranes are mediated via a delocalized protonmotive force (pmf). In this context 'delocalized' means macroscopic and implies that the 'energized state' set up by a given electron transport-linked (primary) proton pump may be used by all secondary proton pumps in the same membrane vesicle. Actually, this definition of 'delocalized' as 'macroscopic' already raises difficulties since, as seen above, the definition of a macroscopic thermodynamic force such as the pmf is based upon the acceptance of detailed balance and the attainment of a macroscopic stationary state. This restriction might appear unnecessarily limiting, because there is no doubt that energy-transducing membranes can make ATP under transient conditions; nor does it seem likely that they would do it differently under transient and steady-state conditions. One way round the problem is to consider a transient as a superposition of short-lived stationary states, so that one might compare the *instantaneous* value of the pmf (for example) with the instantaneous value of an output flux such as the rate of phosphorylation. It is implicitly assumed, on this basis, that one is averaging the pmf over all the vesicles in the system and for the time equivalent to the turnover time of an ATP synthase. It should be noted that, although it may be helpful in some cases, this has no rigorous quantitative basis, because the equivalence of a time average with an ensemble average follows from the ergodic hypothesis which requires an arbitrarily long time, and not a stated, limited time.

It was noted above that thermodynamic criteria are the most rigorous by which to decide whether or not a particular variable is an intermediate in a particular process. In order to gain an idea of what magnitude of pmf must be created by primary proton pumps in order to make credible the idea that energy coupling takes place via the intermediacy of a delocalized pmf some studies of phosphorylation driven by *artificially applied* pmf will be considered. This will then enable an assessment to be made of some of the methods actually used for estimating the pmf under steady-state conditions.

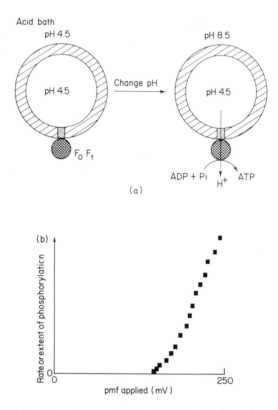

Fig. 9.9 Phosphorylation driven by an artificial pmf. (*a*) A typical experimental protocol, in which an 'inverted' vesicle (with the F_1 part of the ATP synthase which it contains facing outwards) is incubated in a medium of low pH and then immersed in a medium of high pH, possibly concomitant with the generation of a diffusion potential. It is thought that phosphorylation may be driven during the passage of H^+ through the H^+-ATP synthase, down their electrochemical potential. (*b*) The typical 'threshold' behaviour (of some 150 mV) found when the rate or extent of phosphorylation is varied by varying the magnitude of the pmf applied.

9.9.2   Phosphorylation driven by an artificially applied pmf

If a pmf couples electron transport to phosphorylation, it might be predicted that an artificially-applied pmf should also drive phosphorylation (Fig. 9.9). The key criterion is that it *should* do so, without a significant lag and at rates commensurate with those observed *in vivo*. Following the pioneering experiments of Jagendorf and Uribe (1966), many investigators have found that an artificially applied pmf, generated across the bulk aqueous phase that the coupling membrane serves to separate, can indeed drive phosphorylation catalysed by energy coupling membrane vesicles, and even by purified ATP synthases reconstituted in liposomes (e.g. Thayer and Hinkle, 1975a,b; Smith *et al.*, 1976; Wilson *et al.*, 1976; Sone *et al.*, 1977; Gräber, 1981; Hangarter and Good, 1982; Maloney, 1982; Mills and Mitchell, 1982; Gräber *et al.*, 1984; Horner and Moudrianakis, 1985; Schmidt and Gräber, 1985). The assumption in such experiments is that, whilst the incubation conditions may not be *identical* to those *in vivo*, there should be a combination of conditions (of pH, adenine nucleotide concentration, etc.) which should permit phosphorylation at approximately *in vivo* rates (or better). The following take-home messages may be distilled from the above references:

(a) rates of ATP synthesis driven by an artificially imposed pmf may indeed be made to approach or exceed those driven by electron transport, provided that the applied pmf is high enough;

(b) the initial rate of phosphorylation is very much more rapid than that measured after a few seconds;

(c) the relationship between the applied Δp and the rate of phosphorylation (J_P) or the phosphorylation yield (since ATP hydrolysis is slow under the conditions of a very low ATP:ADP ratio used) is highly non-linear;

(d) there is a 'threshold' value typically amounting to 150 mV or even higher, below which *no* phosphorylation takes place.

Finding (d) means that, since ΔG_P is less than $z \Delta p$ then z is not infinitely variable, and values greater than 3 may not be invoked (Kell, 1986b). (ΔG_P is the free energy change in the ATP synthase reaction and z the number of protons moving through the ATP synthase per ATP molecule synthesized.) Similarly, findings (b) and (c) mean that imperfect rapid-mixing and -quenching techniques used to initiate and terminate the reaction will tend to blur both the existence of the threshold, and the sharpness of the distinctions to be made between localized and delocalized coupling (Kell, 1986b). It is also worth pointing out that the 'threshold' behaviour observed may really reflect a more fundamental behaviour of a protonmotive system as a molecular energy machine (Welch and Kell, 1986). (For a very unusual and probably unimportant exception to the 150 mV threshold, see van Walraven *et al.*, 1985, 1986).

The general conclusion from this section is that we should expect the protonmotive force generated by electron transport to exceed at least 150 mV in order to obtain any phosphorylation at all, and that the steady-state values should be substantially higher to be consonant with reasonably rapid rates of phosphorylation. To what extent is this borne out in practice?

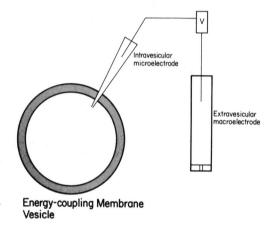

Fig. 9.10   Estimation of $\Delta\psi$ and/or ΔpH by means of microelectrodes. This type of approach is essentially that described in Fig. 9.3 by which the relative contributions of $\Delta\psi$ and ΔpH to Δp may be assessed. Care must be taken to ensure that a diffusion potential at the electrode tip does not contribute to the assessed potentials. Similarly, the necessity to puncture the energy coupling membrane means that care must also be taken to ensure that any low values measured are not due simply to a leak induced by the electrode puncture.

9.9.3   Direct estimation of the pmf using microelectrodes

As intimated in Figure 9.3, the most rigorous means by which one might aim to estimate the magnitude of the protonmotive force across an energy coupling membrane is to measure the electric membrane potential and the pH differential directly by means of (micro)electrode techniques (Fig. 9.10). This experiment is technically very difficult since the small size of mito-chondria (say) places great demands upon electrode placement. Generally, therefore, only the membrane potential is measured, but this type of experiment has been performed over a number of years in a variety of laboratories, particularly that of Tedeschi (Bulychev and Vredenberg, 1976; Vredenberg, 1976; Giulian and Diacumakos, 1977; Maloff *et al.*, 1978a,b; Tedeschi, 1980, 1981; Bulychev *et al.*, 1986; Tamponnet *et al.*, 1986). The results are quite clear and consistent in each case: the electron transport-linked, delocalized membrane potential across the mitochondrial or thylakoid membrane is energetically insignificant (< 50 mV), even under conditions in which a significant pH gradient either was not or could not have been formed. In the case of the single, procaryotic exception, the estimated membrane potential never extrapolated (after correction for the transmembrane conductance) to more than 140 mV (negative inside), which is still below the required threshold of 150 mV and even though the incubation conditions (of pH) were such that a pH gradient would have been absent (Felle *et al.*, 1980). Further, in this work it is by no means certain that the membrane potentials were generated metabolically, since the effect of uncoupler was not tested and the values measured were very sensitive to the external ionic incubation conditions (Felle *et al.*, 1978), strongly suggesting that an ionic diffusion potential was contributing to some extent to the values of $\Delta\psi$. This important experiment has not yet been reproduced.

A variety of explanations has been offered to account for the generally negligible values of $\Delta\psi$. Two of the more plausible were that the electrode is not inside the mitochondrion (or other vesicle), or that the mitochondrion is membrane-leaky (i.e. uncoupled); but these explanations have been discounted by experiments showing interior impalement of single mitochondria capable of making ATP. Many (but not all) of the experiments cited above required that the mitochondria be prepared from rats fed on cuprizone, which causes the formation of 'giant' mitochondria. Grinius (1986) has suggested that pinching off of the cristae in such mitochondria (Wakabayashi *et al.*, 1984) might allow net ATP synthesis by inverted, intramitochondrial vesicles when the parent mitochondria are incapable of forming a pmf. Obviously, this explanation can be invoked only for the case of mitochondria from cuprizone-fed rodents. Evidently, the simplest conclusion from all of these measurements is that it is most likely that the delocalized pmf is energetically insignificant. Astonishingly enough, however, this remains a minority viewpoint, largely since other methods purport to give values of the pmf at total variance with those observed with microelectrodes. As Ferguson (1985) has put it, 'some biochemists outside the immediate field of bioenergetics are puzzled why the apparent absence of a membrane potential according to microelectrodes does not create more concern amongst those working in the field of bioenergetics.' One can only agree, and trust that the future will bring further studies of this crucial topic from independent laboratories, perhaps using *really* giant mitochondria (1 mm diameter) produced by electrofusion (Zimmermann, 1982).

Patch clamping (Sakmann and Neher, 1983, 1984; Dwyer, 1985) is a related technique in which the electrical behaviour of (say) energy-transducing membranes may be measured by direct electrophysiological techniques, particularly since the conductivity of the membrane patch may be made exceedingly low. Hamamoto *et al.* (1985) did in fact make patch clamp measurements of the $\Delta\psi$ generated by cytochrome *o* from *Escherichia coli* incorporated into a black lipid membrane. The membrane potentials they measured were 2–4 mV, rather lower than the 200 mV expected and significantly lower than the threshold of 150 mV. Nonetheless, these authors (Hamamoto *et al.*, 1985) chose to argue that their low values were due to the low electrical resistance of the membranes as measured under *DC* conditions. As discussed in more detail elsewhere (Kell, 1986b) this is an entirely incorrect line of argument (since a resistance 1000 times greater than that measured would mean (by Ohm's law) a potential of 2–4 V, greater than the therymodynamic driving force for the reaction!). Future workers trying to perform patch clamp measurements on protonmotive systems should be aware that the potential generated by a putatively electrogenic source embedded in a membrane depends upon the impedance of the source *and* of the membrane, at the frequency of operation of the enzyme. Such impedances contain capacitive terms which it is not permitted to ignore (see below).

The existence and approximate biological constancy of the membrane capacitance per unit area (see Cole, 1972; Kell and Harris, 1985) provides one of the more clear-cut means for determining the veracity of the chemiosmotic coupling concept. An electrical potential between two electrodes causes the induction of an electrical potential *difference* ($\Delta\psi$) across the relatively ion-impermeable membrane of a spherical vesicle suspended between such extracellular electrodes (Fig. 9.11). For a DC field, this $\Delta\psi$ is given by 1.5 E_0 r cos θ, where θ is the angle between a particular portion of the bilayer and the field direction, E_0 is the electric field strength (equal to the potential difference between the electrodes divided by the distance between them) and r is the radius of the vesicle (e.g. Zimmermann, 1982; Kell and Harris, 1985; Tsong and Astumian, 1986). Below an apparent 'threshold' potential no damage results (but see Glaser, 1986). However, above this threshold there is a reversible or irreversible electrical breakdown (depending upon the length of time for which the electrical field is applied), caused by electromechanical forces and/or by the formation of conducting aqueous pores or 'electropores', a phenomenon for which the theoretical basis is reasonably well developed (see e.g. Zimmermann and Vienken, 1982; Chernomordik *et al.*, 1983; Tsong, 1983; Dimitrov and Jain, 1984; Sugar and Neumann, 1984; Glaser, 1986; Powell *et al.*, 1986). The cos θ relation between the field and the induced potential, coupled with the 'threshold'

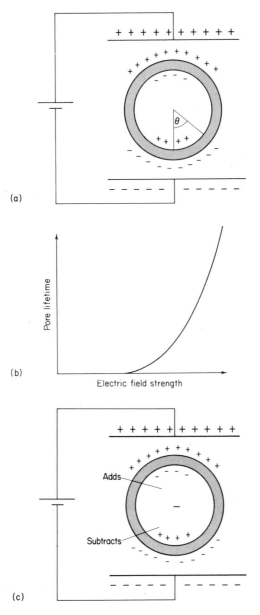

Fig. 9.11 (*a*) Induction of a transmembrane potential in a spherical vesicle by means of an electric field between two macroscopic electrodes. (*b*) The pore lifetime increases non-linearly with the potential induced. (*c*) If there is a pre-existing or metabolically induced electrical potential across the vesicle membrane of interest it will on one side *add to*, and on the other side *subtract from*, that induced by the extracellular electrodes, permitting an estimation of whether any such natural potential difference exists at a significant magnitude.

behaviour of the electrical breakdown voltage, means that (a) electrical breakdown just above the 'threshold' will be confined to regions of the vesicle facing the poles opposite the electrode faces and (b) *the existence of any electric membrane potential caused by other processes such as electron transport will have the effect of inducing an anisotropy in the electroporation process* (Fig. 9.11). Since the *true* field-induced potential may be observed optically (Gross *et al.*, 1986), and electroporation and its sidedness may be observed in the microscope by the release of appropriate fluorescently tagged molecules (Mehrle *et al.*, 1985; Sowers and Lieber, 1986), the presence or absence of a transbilayer electrical potential should severely modify, quantitatively and qualitatively, both the field/electroporation spectrum and in particular its anisotropy (Mehrle *et al.*, 1985). It should be noted that any negative intravesicular *surface* potential will also tend to induce a degree of anisotropy in the electroporation of individual vesicles, so that the *lack* of such an effect would constitute powerful evidence against the existence of an energetically significant, electron transport-linked bulk phase-to-bulk phase $\Delta\psi$. Obviously uncoupler-treated cells or vesicles provide appropriate controls, since the passive electrical properties of biological membranes observed with extracellular electrodes are only *very* weakly dependent upon the transmembrane conductance (see e.g. Harris and Kell, 1985b). This would seem to provide a potent and novel means of assessing the magnitude of any $\Delta\psi$ without having to attempt to measure it by the use of intracellular electrodes or, as shall now be seen, extrinsic probe molecules.

It has been seen that the threshold behaviour observed in 'artificial pmf' experiments means that it is of the first importance to have an accurate method for the estimation of the pmf under stationary conditions, so as to establish whether the signal purporting to reflect the pmf is consistent with the view that the pmf is a thermodynamically competent intermediate in processes such as electron transport phosphorylation. Readers will be aware that a great many methods have been used for this purpose, each with its problems; the different approaches tend to give different values under the 'same' conditions and are therefore possessed of a different degree of credibility depending upon one's standpoint, prejudices and the types of experiment with which one is familiar. Some of these methods and the type of results that they give are therefore discussed. Perhaps the primary or commonest method, often regarded as a benchmark method, is the ion- or acid/base-distribution method; it is therefore given special treatment.

9.9.4   Ion- and acid/base-distribution methods for estimating the pmf

First are considered ion-distribution methods that may be used to estimate the electrical potential difference between two thermodynamic phases.

Figure 9.11 shows a vesicle in which the inside lumen (bulk phase) has a *pre-existing* electrical potential that is maintained at a value positive with respect to the bulk phase outside the vesicle. Surface potentials of the type described in Figure 9.4 are not considered here. If the vesicle membrane is permeable to a particular charged 'probe' anion (X^- in Fig. 9.9), the ion will flow down its electrochemical potential until it is in equilibrium with the pre-existing or maintained electrical potential. (For this to occur, it must cross the membrane purely by uniport, and not be taken up or extruded by other means.) Under such equilibrium conditions, the transmembrane electrical potential $\Delta\psi$ is related to the internal and external concentrations (strictly activities, of course) of the ion by the equation:

$$\Delta\psi = -2.303\frac{RT}{F}\log([a_x\text{-}]_{in}/[a_x\text{-}]_{out}) \qquad (9.18)$$

This equation follows from equation 9.10 on the basis that the criterion for electrochemical equilibrium is that there is no difference in the electrochemical potential of the 'probe' ion in the internal and external phases when equilibrium is attained between its chemical potential difference across the membrane and the electrical potential difference across the membrane. Since the standard electrochemical potential of the probe ion is not known in either the inner or the outer phase it is assumed that it is the same for each phase and that all differences in thermodynamic factors *other* than the electrical potential are incorporated into the activity term. Evidently, then $\Delta\psi$ can be estimated by measuring the activity ratio of such probe ions inside and outside the vesicle of interest. (For reviews, see Rottenberg, 1975, 1979, 1985; Ferguson and Sorgato, 1982; Nicholls, 1982; Azzone *et al.*, 1984; Jackson and Nicholls, 1986).

It is generally assumed that the activity coefficients of probe ions are the same inside and outside the vesicle of interest. Whilst this assumption can never be checked, it probably does not introduce *enormous* errors; a twofold difference in the activity coefficients would introduce an inaccuracy into the estimation of $\Delta\psi$ (if calculated on the basis of free internal and external *concentrations*) of some 18 mV (equation 9.18). If one wishes to interpret certain data in terms of strictly delocalized chemiosmotic coupling, it is necessary to assume that a variation of Δp of this magnitude is sufficient to cover all rates of phosphorylation (from zero to maximal). In such cases it is obvious that an ignorance of the true values of $\Delta\psi$ of even this magnitude precludes any sensible interpretation of the relationship between, say, the rate of phosphorylation and the protonmotive force. Naturally, errors in the estimation of $\Delta\psi$ due to different causes are additive, and it is necessary to take all of them into account. On the (perhaps doubtful) assumption that the principle of the method described in Figure 9.11 actually holds for energy coupling membrane vesicles, the sources of such errors should be considered, so that one may be aware of the limited significance of reported values, which are often given, without error bars, to the nearest millivolt! (for those who like this sort of thing, a tabulation is given by Kashket, 1985).

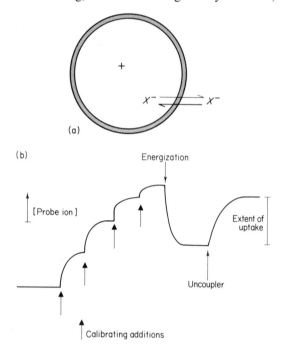

Fig. 9.12   (*a*) Principle of the estimation of $\Delta\psi$ by means of the ion-distribution technique, as described in the text. (*b*) A typical trace of the extracellular activity of the ion of interest, as measured with an electrode, showing how one may calculate the extent of energy-dependent ion uptake/binding.

With the exception of such techniques as nuclear magnetic or electron spin resonance, which have not been widely used in this context (Cafiso and Hubbell, 1981), most approaches are capable of measuring only the *amount* of probe ion taken up by the vesicle of interest. It is thus necessary to know the volume of the internal phase to convert amounts to concentration. Further, what matters is the activity of probe ions that are free in the lumen of the vesicle, so that any binding must be corrected for. The extent of *energy-dependent* binding cannot be corrected for, since one has no knowledge of the 'real' membrane potential which is what one is trying to measure. At least in bacterial cells, however, the energy-*in*dependent binding of probe ions can be very extensive; corrections for such binding cause the calculated values of $\Delta\psi$ to fall very sharply, to values well below the 150 mV 'threshold' (e.g. Lolkema *et al.*, 1982, 1983; Elferink *et al.*, 1985, 1986). In other vesicular systems, binding or energy-dependent extrusion (Midgley, 1986) of the probe may be sufficiently great to cause the estimated $\Delta\psi$ to be of the opposite *sign* to the 'real' one assessed (and in *these* cases generally accepted!) on the basis of microelectrode measurements (Ritchie, 1982; Barbier-Brygoo *et al.*, 1985). Controls which can help to alleviate these difficulties include: showing that variation of the concentration of vesicles or probe ion gives no change in apparent $\Delta\psi$; observing equal extents of uptake using probe ions having the same charge but which differ in their rate of uptake; and comparing the membrane potential estimated by probe methods with one of known magnitude generated artificially by means of a diffusion potential.

To avoid becoming enmeshed in the details of particular cases, only the following general findings are stated: sometimes different methods agree and sometimes they do not, and in no case known to the author have different methods, including all the controls listed above, been performed on the same system with a resulting $\Delta\psi$ exceeding 150 mV. Sometimes authors will pick one method in preference to another (say Rb^+ uptake in preference to the uptake of methyl triphenylphosphonium$^+$). Obviously there is no rational basis for such choices. Kaback (1986) continues to claim, apparently on the basis of unpublished observations, that $\Delta\psi$ values estimated from the steady-state accumulation of a variety of lipophilic cations are independent of the concentration of these cations 'over submicromolar to millimolar ranges'. For a $\Delta\psi$ of 120 mV, this implies internal concentrations of hundreds of millimolar lipophilic cation. Such values exceed the aqueous solubility of these cations and it is necessary to regard such a claim as a 'Churchillian' terminological inexactitude.

Since the imposition of an artificial pmf has the merit of more certainly representing what it claims to, it is particularly important to compare the ability of an artificial pmf to stimulate phosphorylation with that of what is claimed to be a real, *electron transport-generated* pmf of the same magnitude. This is rarely done in the same system at the same time. However, in alkalophilic bacteria in which an apparent pmf of significantly less than 50 mV is observed in spite of an unimpaired ability to synthesize ATP, the apparent $\Delta\psi$ measured by standard ion-distribution techniques is *not* equivalent to a diffusion potential of the same magnitude so far as its ability to drive phosphorylation is concerned (Krulwich and Guffanti, 1983; Guffanti *et al.*, 1984, 1985). One interpretation of this is that the ion-distribution method, whilst reliable in cases in which there *is* a pre-existing $\Delta\psi$, is *falsely* suggesting the existence of a significant, energy-dependent $\Delta\psi$ across the bacterial cytoplasmic membrane (and other membranes) for reasons yet to be explained (but see later). Since this is consistent with the general lack of a membrane potential estimated directly, as described above, and with other findings to be discussed later, this is an acceptable, simple interpretation of an otherwise inexplicable phenomenon.

The principle of the acid/base-distribution method for measuring ΔpH is similar to that of the ion-distribution method for estimating $\Delta\psi$, except that here the membrane of interest must be permeable only to the probe in its *uncharged* form (see e.g. Rottenberg, 1975a,b, 1979; Nicholls, 1982). At least in bacteria, the values of ΔpH are generally rather small (exceptions are acidophiles; Cobley and Cox, 1983), and are believed to be fairly reliable, not least since approaches such as ^{31}P-NMR give values similar to those estimated by acid/base-distribution (e.g. Nicolay *et al.*, 1981). What is generally wished to be known is the functional relationship between what is believed to represent the total pmf and the rate or extent of an output process such as ATP synthesis. To help sharpen the argument, and to decrease the number of variables that must be measured, it is often helpful to attempt to estimate $\Delta\psi$ under conditions (controlled by external pH or by the addition of an ionophore such as nigericin) in which ΔpH is zero, since it is well known that bacteria maintain a relatively constant internal pH (Padan *et al.*, 1981; Booth, 1985; Kaback, 1986). The fact that both output forces and fluxes are generally rather independent of the existence of a ΔpH (acidophiles are again an exception here; Cobley and Cox, 1983), might be taken as a further indication of the irrelevance of ΔpH as an energy coupling intermediate in electron transport phosphorylation in bacteria.

9.9.5   Spectroscopic methods for estimating $\Delta\psi$ in bacteria

Although most commentators regard probe-distribution as the method of choice for estimating Δp (despite the many technical difficulties raised above), this method suffers the grave disadvantage of being relatively slow to respond. Optical methods, such as the electrochromic response of carotenoids (see Jackson, Chapter 7), can be very rapidly responding and may give information unobtainable from stationary-state measurements. However, even here, for reasons which are not yet understood (Jackson and Nicholls, 1986), the steady-state value of $\Delta\psi$ estimated from ion-distribution methods is about half that estimated on the basis of the electrochromic carotenoid response (Ferguson *et al.*, 1979; Clark and Jackson, 1981; McCarthy and Ferguson, 1982). Evidently, any attempts to draw quantitative conclusions from the latter method concerning the relationship between $\Delta\psi$ and the rate of phosphorylation, while ignoring the fact that the $\Delta\psi$ given by ion distribution is below the 150 mV threshold, are doomed to failure. Until we can be more sure of what the various methods for estimating '$\Delta\psi$' are actually measuring there seems no hope of significant progress in solving the central problem of whether an energetically significant $\Delta\psi$ actually *exists* between the bulk aqueous phases that energy coupling membranes serve to separate.

Notwithstanding this gloomy prognosis, and the carotenoid/ion-distribution discrepancy is as marked as any, many workers have carried out measurements of the pmf under various conditions, titrating it with inhibitors of electron transport, with ionophorous uncouplers or with other molecules, and comparing these values with either the rate or extent of phosphorylation. These studies are briefly commented on below.

9.9.6 Relationships between the apparent pmf and rates and extents of phosphorylation

9.9.6.1 Introduction

On the highly questionable assumption that the Δp values measured by any given method are at least semi-quantitatively reliable, studies of the effects of variation in Δp on the rate or extent of ATP synthesis can give mechanistic information concerning the number of protons translocated across the membrane per ATP synthesized, and on the efficiency of coupling of Δp to phosphorylation.

It has been widely observed that the rate of phosphorylation (J_P) depends more upon the rate of electron transport than upon Δp. In some cases the relationship is entirely arbitrary. These findings have themselves led some reviewers to invoke 'parallel coupling' or alternatives to Δp as an intermediate in energy coupling (see e.g. reviews in Kell, 1979; Ferguson and Sorgato, 1982; Westerhoff *et al.*, 1984a; Ferguson, 1985; Rottenberg, 1985). However, since the rates of reactions may depend upon many factors besides the thermodynamic poise of an energy coupling intermediate (see above), and even on the relative magnitudes of $\Delta\psi$ and ΔpH at constant Δp (Boork and Wennerström, 1984; Sanders, 1987), such findings alone are probably not very informative as to the veracity of any coupling scheme. What *is* worth stressing, however, is that if one can see a significant rate of phosphorylation during a titration at a value of Δp that is well below the 'threshold', it is difficult to argue that this Δp is actually reflecting the Δp which is applied during an 'artificial pmf' experiment. Since this is exactly what has been done in many cases, albeit unknowingly, the titration studies do have the merit of pointing to some of the inadequacies of the methods for estimating Δp.

In a similar vein, there has been a recent controversy (Woelders *et al.*, 1985; Petronilli *et al.*, 1986) as to the relationship between Δp and ΔG_P

under static head conditions (actually in this case equilibrium is claimed) and whether or not this is a function of Δp as the latter is diminished with uncouplers. Whatever the reality, these studies show that continuing ATP synthesis can occur during the approach to static head under conditions in which the Δp (estimated by ion-distribution techniques) is well below 100 mV. Not only does this lead to unrealistic values of the H^+/ATP ratio (Petronilli *et al.*, 1986) but it is *quite inconsistent* with the view that the apparent values of Δp reflect the real values given the thresholds observed in artificial pmf experiments. Similar data are obtained with *Paracoccus denitrificans* (Kell *et al.*, 1978; McCarthy and Ferguson, 1983a,b). Noting the difficulties encountered above with the ergodic hypothesis of equilibrium in complex, heterogeneous structures, it might be prudent to realize how insecure is the idea that stationary extents of probe ion uptake correspond to an equilibrium with any putative $\Delta \psi$ across an energy-coupling membrane.

In view of all these difficulties, and the circularities inherent in attempting to *test* the role of Δp (i.e. delocalized chemiosmotic coupling) by trying to estimate Δp alone, many authors have tried to approach this problem using methods that do not require such an estimation. Some of these are considered next.

9.9.6.2   Respiration-driven proton translocation

The addition of a pulse of oxygen (as a small volume of air-saturated KCl) to a weakly buffered, anaerobic suspension of aerobic respiratory bacteria such as *Paracoccus denitrificans*, held in a reaction vessel whose pH is constantly monitored, might lead to the production of one of the pH traces of the type shown in Figure 9.13. In other words, there might be a slowish decrease in pH which then stays constant (trace a); or a rapid and larger decrease in pH which then turns into a pH increase towards the baseline (trace b); or no effect whatsoever (trace c). In fact, all three such traces have been observed by different workers studying suspensions of *Paracoccus denitrificans*, albeit with different interpretations of their underlying bases (Scholes and Mitchell, 1970; Kell and Hitchens, 1982; Hitchens and Kell, 1984). What are the specific conditions that determine which of these traces is observed?

The first trace is typically observed when cells are used 'as isolated' and when held in a medium containing KCl as osmotic support. The second trace is observed when a compound such as valinomycin is also present or (more commonly with bacteria) when a proportion of the KCl in the suspending medium is replaced by KSCN. The last trace is observed when an appropriate concentration of an uncoupler such as FCCP is present. Taken together, these results are interpreted within a chemiosmotic coupling

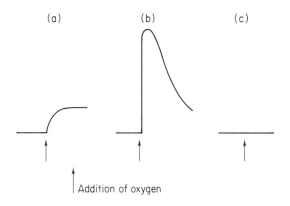

(a) (b) (c)

Addition of oxygen

Fig. 9.13 Typical traces observed in an O_2-pulse experiment. For further details, see text.

framework as follows (Fig. 9.14). The energy coupling, bilayer membrane may be represented as an electrical capacitor, whose magnitude may be calculated, and is observed, to be some 1 ± 0.5 $\mu F/cm^2$ (see e.g. Cole 1972; Harris and Kell, 1985b; Kell and Harris, 1985; Kell, 1986b; Pethig and Kell, 1987 and reference therein). If the primary, electron transport-linked proton pump is electrogenic in the sense that it pumps protons *electrogenically* from the inner bulk phase to the outer bulk phase of the suspension, the result of moving such electrically charged particles across the membrane capacitance is that the 'plates' of this capacitance (represented by the aqueous interfaces on either side of the membrane) become charged, the membrane potential V being given by $Q = CV$ where C is the total capacitance and Q the total number of charges moved. If the cells are spherical or of a known shape, and their number and hence their membrane area is known, the maximum number of charges necessary to obtain a given membrane potential may be estimated (Gould and Cramer, 1977; Kell and Hitchens, 1982; Kell, 1986b, 1987b).

In a typical arrangement such as that of Figure 9.13, the number of protons measured (if pumped *electrogenically*) is more than sufficient to charge fully the membrane capacitance, given that an excessive value of V leads simply to dielectric breakdown (see above). This, in turn, leads either to a redox slip or a non-ohmic leak such that the average, net number of protons observably pumped into the bulk aqueous phase is a function of the average membrane potential during the O_2 pulse. The presence of a substance such as the SCN^- ion, which can cross biological membranes in the charged form (e.g. Kell *et al.*, 1978), dissipates the membrane potential and permits the 'true', limiting stoichiometry of H^+ translocation to be observed. The presence of FCCP causes a rapid back-flow of protons before the pH electrode can respond, and the fact that the observable $\rightarrow H^+/O$ stoichiometry drops to zero is taken to mean that all H^+ are vectorially pumped, and not scalar protons accompanying unwanted chemical side-reactions.

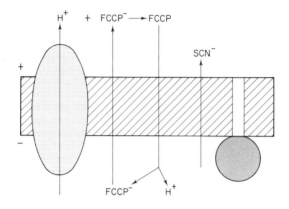

Fig. 9.14   Classical chemiosmotic explanation of the type of data observed in Fig. 9.13, as discussed in the text.

Such reasoning can explain the general features of respiration-linked proton translocation seen in a typical 'O₂-pulse' experiment of the type outlined. If asked to choose between 'slip' and 'non-ohmic leak' as the major explanation for the 'low' $\rightarrow H^+/O$ stoichiometry observed in the absence of SCN^- with a chemiosmotic framework one should plump for the former, since the observable rates of proton back-flux in the presence of SCN^- are very much greater than in its absence, a condition which should be accompanied by an *enhanced* pmf and which is already somewhat difficult to explain (Ferguson, 1985; Kell, 1986b). [It is expected that the electrical capacitance of the coupling membrane is very much smaller than the so-called differential buffering power (Mitchell, 1968).]

This explanation, however, whilst apparently plausible, is open to a more stringent test than that posed by the simple '$\pm SCN^-$' experiment just described (Gould and Cramer, 1977; Kell and Hitchens, 1982; Hitchens and Kell, 1984). If the translocation of a small number of protons causes the build-up of a membrane potential sufficient to induce a substantial slip or non-ohmic leak across the coupling membrane, so that the net observable $\rightarrow H^+/O$ stoichiometry is submaximal, then doubling the size of the O₂ pulse should not allow the appearance of further protons, since the membrane potential (or pmf) is supposedly already saturated. In practice, this prediction is not borne out (Gould and Cramer, 1977; Gould, 1979; Kell and

Hitchens, 1982; Hitchens and Kell, 1984); the $\rightarrow H^+/O$ stoichiometry is independent of the size of the O_2 pulse over a wide range, before secondary phenomena such as an unfavourable internal pH and an inadequacy of internal anions, come into play. One may take this experiment further, and seek to vary the rate of electron transport (Kell and Hitchens, 1983) or the extent of any putative non-ohmic leak (Hitchens and Kell, 1984) with the intention of comparing the $\rightarrow H^+/O$ stoichiometry under the various conditions using O_2 pulses of different sizes. When conditions are arranged such that this $\rightarrow H^+/O$ stoichiometry is independent of the size of the O_2 pulse, one may treat the system as though it has attained a macroscopic stationary state. In such a system, the pathways of electron transport are independent of the size of the O_2 pulse. What is found (Hitchens and Kell, 1984) is that the presence of venturicidin, a compound which inhibits ATP synthase (Ferguson and John, 1977) and which should also decrease any non-ohmic leak (Cotton *et al.*, 1981), *raises* the observable $\rightarrow H^+/O$ stoichiometry above that seen in its absence (and in the absence of SCN^-). Since this should raise the membrane potential (Kell *et al.*, 1978), it appears that slip is irrelevant to these observations (since a higher $\Delta\psi$ should exacerbate slip and *decrease* the observable $\rightarrow H^+/O$ stoichiometry). Thus (in spite of the argument in the previous section concerning the rate of H^+ backflux), non-ohmic leak is the only free variable to permit these observations to be explained within the framework of a delocalized, chemiosmotic coupling scheme. Unfortunately, as seen above, the stationary state conditions mean that the extent of non-ohmic leak should be a strictly inverse function of the observable $\rightarrow H^+/O$ stoichiometry. This is not the case, and so it must be construed that the lack of observable protons in the absence of SCN^- occurred not because they were not pumped (i.e. slip occurred), nor because they returned across the coupling membrane before the electrode could respond (non-ohmic leak), but simply because they were not pumped in to the bulk aqueous phase.

The calculations concerning the number of protons necessary to make a membrane potential of, say, 200 mV suggest that the great majority appearing in the bulk phase could not be appearing there electrogenically. They must therefore have got there electroneutrally, in symport or antiport with another ion. The question thus arises as to whether *any* of the observable protons were pumped electrogenically to the bulk phase, and were thus capable of creating an energetically significant, delocalized electric membrane potential. This question, which obviously relates to the question of the magnitude of the electron transport-linked $\Delta\psi$ estimated by ion-distribution methods, may be addressed by turning the above reasoning on its head (Gould and Cramer, 1977; Conover and Azzone, 1981; Kell and Hitchens, 1982; Kell, 1986b, 1987).

If an energetically significant $\Delta\psi$ *is* feeding back on the electron transport chain, then decreasing the number of electron transfer events per cell, and hence the number of H^+ pumped per cell, will decrease the maximum $\Delta\psi$ attainable (i.e. even if all observable protons are pumped electrogenically) to an insignificant value. If this is the case, then the net observable $\rightarrow H^+/O$ stoichiometry should be the same whether SCN^- is present or absent (since its presence can hardly dissipate an already negligible $\Delta\psi$). Again, when attempts are made to perform the experiment under conditions approximating a stationary state, the $\rightarrow H^+/O$ stoichiometry is independent of the size of the O_2 pulse over a wide range, and apparently strictly dependent upon the natural permeability of the membrane to ions other than protons (Gould and Cramer, 1977; Conover and Azzone, 1981; Kell and Hitchens, 1982; Kell, 1986b). There is only one conclusion it seems possible to draw from these types of observations (e.g. Kell and Westerhoff, 1985; Kell, 1986b; and see also Tedeschi, 1980, 1981): the apparent absence of an energetically-significant $\Delta\psi$ (as observed by microelectrodes) is due to the fact that in the stationary state all protons pumped by the respiratory chains of proton-motive systems do not appear in the bulk phase unless the 'primary macroerg' (Blumenfeld, 1983) or 'energized state' set up by electron transport is dissipated by the inclusion of a membrane-permeant ion which then allows each proton pumped across the membrane into an 'invisible' space to enter the bulk phase *electroneutrally*. It should be evident that such an explanation is not consistent with the generation, and hence intermediacy, of a chemi-osmotic membrane potential as a part of electron transport-linked phos-phorylation.

Experiments related to the above have recently been performed by Jackson and colleagues in photosynthetic bacteria, with similar results (in terms of the discrepancies between 'expected' and 'observed' $\rightarrow H^+/e^-$ ratios) but with, in part, a somewhat different interpretation (Taylor and Jackson, 1985a,b; Myatt and Jackson, 1986). In view of the discrepancies between the number of protons measured with a glass electrode and a pH indicator (approx. 43 and 144 in Fig. 2 of Taylor and Jackson, 1985b), and between the number of charges allegedly necessary to create the maximum membrane potential (approx. 5) judged by carotenoid spectroscopy, and these observed (approx. 2; Myatt and Jackson, 1986), one would retain the conclusion that the data obtained to date, using observations of respiration-linked translocation of protons, are consistent with the view that essentially *all* protons observed in the bulk phase are deposited there electroneutrally and thus cannot contribute to the generation of a delocalized membrane potential.

In this regard, it is worth drawing what I have found to be a useful analogy when discussing some of the points at issue in this chapter (Fig. 9.15). It is not disputed that the source of free energy both for charging a car battery via a dynamo and for turning the wheels is the engine (Fig. 9.15). Similarly, it is not conventional to draw a bidirectional arrow between the wheels and the dynamo to suggest that the latter may be an intermediary energy source for the former. Nevertheless, there is a remarkable tendency, faced with what amounts to the same logical structure, to assume that any ion or proton movements accompanying electron transport and entering the bulk phase might subsequently be coupled to otherwise endergonic processes such as ATP synthesis. It should be clear (Fig. 9.15) that this is not a logical procedure! The fact that bulk-phase proton movements accompany electron transport, and thus contribute to the creation of a small pH gradient, tells us nothing whatever about whether they may subsequently be used to drive phosphorylation. This rests on other criteria, which, when tested, are found wanting.

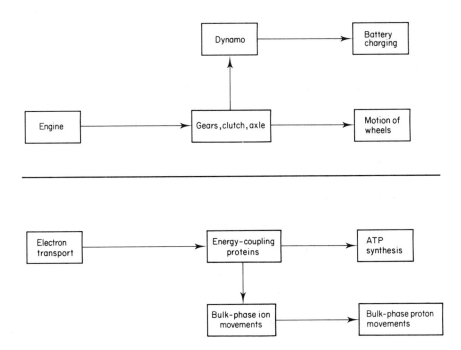

Fig. 9.15 Phenomena which may be observed concomitantly do not necessarily bear a given cause-and-effect relationship to each other. Turning on the engine of a car causes both dynamo and wheels to move, but it is not the dynamo which is driving the wheels. Similarly, electron transport causes both phosphorylation and observable movements of (bulk phase) protons and other ions to occur. This does not of course mean that the proton movements are driving phosphorylation!

Since everything seen so far might well have caused one to have the gravest suspicions about whether or not any signal purporting to represent Δψ actually leads one logically to the view that the latter is of sufficient magnitude to be an intermediate in electron transport-linked phosphorylation, other approaches which do not rely upon the measurement of Δ*p* but exploit its delocalized *nature* have been sought. One such approach is that known as 'double-inhibitor titrations'.

9.9.7   Double-inhibitor titrations

The general principle of the double-inhibitor titration (DIT) method was first expounded in a bioenergetic context by Kahn (1970) and by Baum (Baum *et al.*, 1971; Baum, 1978). Motivated by related experiments carried out by Venturoli and Melandri (1982), the approach was taken up and extended by Hitchens and Kell (1982a,b; 1983a,b) and by a number of other workers (Parsonage and Ferguson, 1982; Cotton and Jackson, 1983; Berden *et al.*, 1984; Krasinskaya *et al.*, 1984; Mills, 1984; Westerhoff *et al.*, 1984b; Davenport, 1985; Ferguson, 1985; Herweijer *et al.*, 1985, 1986; Pietrobon and Caplan, 1985a; van der Bend and Herweijer, 1985; van der Bend *et al.*, 1985; Chen, 1986; Kell, 1986b,c; Kell and Walter, 1986; Westerhoff *et al.*, 1986; Pietrobon and Caplan, 1986a,b; Kell, 1987b; Petronilli *et al.*, 1988; Slater, 1987; Westerhoff and Kell, 1988).

The idea behind these types of approach runs essentially as follows (Fig. 9.16).

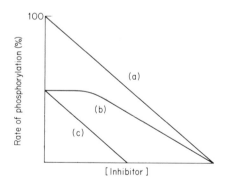

Fig. 9.16   Possible diagrammatic traces observed in a double-inhibitor titration using an inhibitor of electron transport in the absence (a) or in the presence (b,c) of a titre of ATP synthase inhibitor sufficient to inhibit the 'starting' rate of phosphorylation by 50%.

If a delocalized intermediate (such as the pmf) serves to link electron transport with phosphorylation, the addition of a titre of a tight-binding ATP synthase inhibitor will tend to raise the pmf and thus the titre of an inhibitor of electron transport or of an uncoupler necessary to inhibit phosphorylation, provided that these act to inhibit phosphorylation by decreasing the pmf, and that the rate of phosphorylation is controlled solely by the pmf. In practice, it is observed, under appropriate conditions, that such a treatment tends to leave unchanged the inhibitory titre of an electron transport inhibitor and actually *decreases* the titre of uncoupler necessary to inhibit phosphorylation by a given amount, results initially thought quite incompatible with a delocalized energy coupling scheme.

However, as seen above, when considering metabolic control analysis, metabolic fluxes are rarely, if ever, controlled by single parameters (and never by variables), and the later commentaries on this topic have invoked, or drawn attention to, conditions (of varying degrees of plausibility) in which the data observed *may* in fact be made to fit a delocalized energy coupling scheme. For instance, there may be redox-linked *activation* of the ATP synthases, or the elasticity of the primary proton pumps towards the pmf may be significantly different from that of the ATP synthase towards the pmf, so that uncouplers might (counterintuitively) be expected to work better when ATP synthases are partially inhibited by an appropriate titre of energy transfer (ATP synthase) inhibitor. Thus a largely quantitative analysis is required, although preferably one which does not of itself *rely* upon the accurate estimation of the values of the pmf.

The most recent and comprehensive analyses of this topic (Pietrobon and Caplan, 1987a,b; Petronilli *et al.*, 1988; Westerhoff and Kell, 1987), based on metabolic control analysis and permitting arbitrary (but *stated and self-consistent*) relations to exist between Δp and the partial reactions with which it is supposed to be coupled, lead to the following conclusions:

(a) no delocalized model can account for the finding that the titre of uncoupler necessary for *full* uncoupling is lower in the presence of a partially-inhibitory titre of ATP synthase inhibitor than in its absence;

(b) qualitatively, linear and non-linear chemiosmotic models can explain some but not all of the data; and

(c) when the P/2e ratio is taken into account, the degree of leakiness required to explain the data on a delocalized coupling basis is not consistent with that found.

It is worth remarking in this context that the above analysis is confined *strictly* to genuinely stationary states. It is of interest nonetheless that studies on the prestationary *initiations* of phosphorylation in thylakoids (Hangarter and Ort, 1985) show that this may, under some, but not all conditions

(Chiang and Dilley, 1987), be strongly controlled by cooperation between electron transfer chains.

Finally, the DIT approach opens up a further interesting means of testing whether a particular method such as ion uptake might plausibly be measuring a delocalized intermediate such as a membrane potential. The idea in such an experiment is that, given the relatively low electrical capacitance of coupling membranes discussed above, the pmf rapidly attains its 'steady-state' value. The uptake of 'probe' ions which supposedly monitor $\Delta\psi$ is then a slow response to this essentially 'clamped' $\Delta\psi$. Since in all cases, purely energy-transfer inhibitors tend to raise the $\Delta\psi$ estimated in the stationary state, it is to be expected that neither the rate nor extent of ion uptake should be decreased, during an uncoupler titration, by the inclusion of an energy transfer inhibitor in the reaction mixture.

A reductionist approach to biochemistry would have it that the best way to decide what is necessary for a given process, and thus what may be its mechanism, is to isolate, purify and reconstitute the enzymes believed to be involved in that process. Purportedly according to such reasoning, another means by which some authors have sought to establish the veracity of delocalized chemiosmotic coupling is by co-reconstitution experiments. However, since the theory and the data (as with several other aspects of the present topic) are utterly divergent (Kell and Westerhoff, 1985), it seems appropriate to draw attention to them here.

9.9.8 Co-reconstituted systems: the emperor is not yet clothed

The idea behind the co-reconstitution approach may be illustrated with reference to Figure 9.17; it is that the co-reconstitution, in a lipid vesicle, of a 'primary' proton pump, capable of creating a pmf, together with a 'secondary' proton pump, capable of using a pmf to drive a reaction, should permit the energy coupling of the chemical reactions catalysed by the two types of enzyme (complex). Following the initial experiments of Racker and Stoeckenius (1974) with bacteriorhodopsin and a very impure ATP synthase preparation, it is widely believed that such a successful co-reconstitution has been amply demonstrated, with the implication that energy coupling is, indeed, via a chemiosmotic type of mechanism. As stressed elsewhere (Kell and Morris, 1981; Kell and Westerhoff, 1985; Kell, 1986b) and as I shall briefly review once more, the presently available data actually justify rather the opposite conclusion.

In many of these co-reconstitution experiments, the assumption is made that because the reversible primary and secondary proton pumps do actually pump protons, then any coupled phosphorylative activity (for example) that is observed must also be taking place via a delocalized pmf. As seen above

(Fig. 9.15), this is an extraordinary piece of logic which may be expunged from our considerations. Any phosphorylative activity observed in a co-reconstituted system may be occurring via any type of mechanism, which may or may not be the same as that occurring *in vivo* and may or may not be via a pmf. (Enhanced and possibly free-energy-transducing collisional inter-actions between the proteins in the more fluid milieu of a liposome are one obvious possibility.) The question which should at the very least be asked in order to gain mechanistic information about the coupling in a co-reconstituted system is: what is the *turnover number* of the secondary proton pump (typically an ATP synthase preparation)? This, and not the absolute flux (in terms of nmol/minute/mg protein), is what can show that these systems are actually working according to the 'simple interpretation' (Kell and Wester-hoff, 1985) given in Figure 9.17.

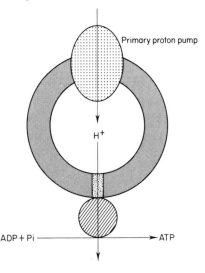

Fig. 9.17   The 'simple interpretation' of co-reconstitution experiments in which it is assumed that any phosphorylation observed is occurring via the generation of a pmf between *purified* and *spatially separate* proton pumps.

In one of the more thoughtful studies of that era (Hauska *et al.*, 1980), it was shown in a system co-reconstituted from a photosystem I preparation and a thylakoid CF_0F_1 that, whereas the *rate* of photophosphorylation was some 30% that of the rate *in vivo*, the *turnover number* of the ATP synthase molecules was *less than 1% of that in vivo*; i.e. a value that in other contexts is considered to be more or less zero. Since even in these relatively purified systems the components were certainly not 99% pure, one can as well argue that *purified* co-reconstituted systems can *not* in fact make ATP. As surveyed and discussed elsewhere (e.g. Kell and Morris, 1981; Casey, 1984; Kell and Westerhoff, 1985; Kell, 1986b), the data of Hauska *et al.* (1980) are quite typical for co-reconstituted systems using more-or-less *purified* components. Recent small improvements in turnover numbers (van der Bend *et al.*, 1984), and studies of the effects of alternating the ratio of primary to secondary proton pumps, have not qualitatively altered these conclusions; i.e. that energy coupling here is (i) extremely inefficient and (ii) not via a pmf of any credible magnitude.

9.10 Concluding remarks

Readers will be aware that the present overview differs from many in bioenergetics in seeking to argue a self-consistent case to the effect that the protonmotive force across energy coupling membranes catalysing electron transport-linked phosphorylation is energetically insignificant. Reasons of space have meant that many pertinent arguments (Kell and Westerhoff, 1985) could not be developed but hopefully, it has been conveyed, at the very least, that the subject of bacterial bioenergetics is alive, lively and offers plenty of scope for further research and study.

Finally, the Editor wishes me to add here that his aim in including this final chapter is to indicate that, although this is the close of the book, the book of bioenergetics is not closed, and also to emphasize that we should not believe all that we read (including this chapter).

References

Ansari, A., Berenden, J., Bowne, S. F., Frauenfelder, H., Iben, I. E. T., Shyam-sunder, E. and Young, R. D. (1985). *Proc. Natl. Acad. Sci. USA* **82**, 5000–5004.

Atkins, P. W. (1978). 'Physical Chemistry'. Oxford University Press, Oxford.

Azzone, G. F., Pietrobon, D. and Zoratti, M. (1984). *Curr. Top. Bioenerg.* **14**, 1–77.

Barber, J. (1980). *Biochim. Biophys. Acta* **594**, 253–308.

Barber, J. (1982). *Ann. Rev. Plant Physiol.* **33**, 261–295.

Barbier-Brygoo, H., Gibrat, R., Renaudin, J.-P., Brown, S., Pradier, J. M., Grignon, C. and Guern, J. (1985). *Biochim. Biophys. Acta* **819**, 215–224.

Barrow, J. D. and Silk, J. (1983). 'The Left Hand of Creation. The Origin and Evolution of the Expanding Universe'. Unwin, London.

Baum, H. (1978). 'The Molecular Biology of Membranes' (Eds. S. Fleishcer, Y. Hatefi, D. H. MacLennan and A. Tzagoloff). pp. 243–262. Plenum Press, New York.

Baum, H., Hall, G. S., Nalder, J. and Beechey, R. B. (1971). *In* 'Energy Transduction in Respiration and Photosynthesis' (Eds. E. Quagliariello, S. Papa and C. S. Rossi), pp. 747–755. Adriatica Editrice, Bari.

Bechtold, R., Kuehn, C., Lepre, C. and Isied, S. S. (1986). *Nature* **322**, 286–287.
Beck, A. H. W. (1976). 'Statistical Mechanics, Fluctuations and Noise'. Edward Arnold, London.
Berden, J. A., Herweijer, M. A. and Cornelissen, J. B. J. W. (1984). 'H$^+$-ATPase (ATP synthase); Structure, Function, Biogenesis of the F_0F_1 Complex of Coupling Membranes' (Eds. S. Papa, K. Altendorf, L. Ernster and L. Packer), pp. 339–348. Adriatica Editrice, Bari.
Bialek, W. and Goldstein, R. F. (1985). *Biophys. J.* **48**, 1027–1044.
Bilz, H., Buttner, H. and Frohlich, H. (1981). *Z. Naturforsch* **36B**, 208–212.
Blumenfeld, L. A. (1981). 'Problems of Biological Physics'. Springer, Heidelberg.
Blumenfeld, L. A. (1983). 'Physics of Bioenergetic Processes'. Springer, Heidelberg.
Bockris, J. O.'M. and Reddy, A. K. N. (1970). 'Modern Electrochemistry'. Vols. 1 and 2. Plenun Press, New York.
Bohm, D. (1980). 'Wholeness and the Implicate Order'. Routledge and Kegan Paul, London.
Bohm, D. (1984). 'Causality and Chance in Modern Physics'. Routledge and Kegan Paul, London.
Boork, J. and Wennerström, H. (1984). *Biochim. Biophys. Acta* **767**, 314–320.
Booth, I. R. (1985). *Microbiol. Rev.* **49**, 359–378.
Brush, S. G. (1976). 'The Kind of Motion we Call Heat'. North-Holland, Amsterdam.
Brush, S. G. (1983). 'Statistical Physics and the Atomic Theory of Matter, from Boyle and Newton to Landau and Onsagar'. Princeton University Press, Princeton, New Jersey.
Bulychev, A. A. and Vredenberg, W. J. (1976). *Biochim. Biophys. Acta* **423**, 548–556.
Bulychev, A. A. *et al.* (1986). *Biochim. Biophys. Acta* **852**, 68–73.
Cafiso, D. S. and Hubbell, W. L. (1981). *Ann. Rev. Biophys. Bioeng.* **10**, 217–244.
Campo, M. L., Bowman, C. L. Tedeschi, H. (1984). *Eur. J. Biochem.* **141**, 1–4.
Caplan, S. R. and Essig, A. (1983). 'Bioenergetics and Linear Nonequilibrium Thermodynamics'. The Steady State. Harvard University Press.
Careri, G. and Wyman, J. (1984). *Proc. Natl. Acad. Sci. USA* **81**, 4386–4388.
Careri, G., Fasella, P. and Gratton, E. (1979). *Ann. Rev. Biophys. Bioeng.* **8**, 69–97.
Carter, F. L. (1984). *Physica* **10D**, 175–194.
Casey, R. P. (1984). *Biochem. Biophys. Acta* **768**, 319–347.
Chen, Y. (1986). *Biochim. Biophys. Acta* **850**, 490–500.
Chernomordik, L. V., Sukharev, S. I., Abidor, I. G. and Chizmadzhev, Yu. A. (1983). *Biochim. Biophys. Acta* **736**, 203–213.
Chester, G. V. (1969). *In* 'Many-body Problems' (Eds. W. E. Parry, R. E. Turner, D. ter Haar, J. S. Rawlinson, G. V. Chester, N. M. Huygenholtz and R. Kubo), pp. 125–186. W. A. Benjamin, New York.
Chiang, G. and Dilley, R. A. (1987). *Biochemistry* **26**, 4911–4916.
Chou, K.-C. (1984). *Biophys. Chem.* **20**, 61–71.
Clark, A. J. and Jackson, J. B. (1981). *Biochem. J.* **200**, 389–397.
Clark, A. J., Cotton, N. P. J. and Jackson, J. B. (1983). *Eur. J. Biochem.* **130**, 575–580.
Clegg, J. S. (1984). *Amer. J. Physiol.* **246**, R133–R151.
Cobley, J. G. and Cox, J. C. (1983). *Microbiol. Rev.* **47**, 579–595.
Cole, K. S. (1972). 'Membranes, Ions and Impulses'. University of California Press, Berkeley.

Conover, T. E. and Azzone, G. F. (1981). *In* 'Mitochondria and Microsomes' (Eds. C. P. Lee, G. Schatz and G. Dallner), pp. 481–518. Addison-Wesley, New York.

Cotton, N. P. J. and Jackson, J. B. (1983). *FEBS Lett.* **161**, 93–99.

Cotton, N. P. J., Clark, A. J. and Jackson, J. B. (1981). *Arch. Microbiol.* **129**, 94–99.

Cotton, N. P. J., Clark, A. J. and Jackson, J. B. (1984). *Eur. J. Biochem.* **142**, 193–198.

Davenport, J. W. (1985). *Biochim. Biophys. Acta* **807**, 300–307.

Davies, P. C. W. and Brown, J. R. (Eds.) (1986). 'The Ghost in the Atom. A Discussion of the Mysteries of Quantum Physics'. Cambridge University Press, Cambridge.

Davydov, A. S. (1983). *In* 'Structure and Dynamics: Nucleic Acids and Proteins' (Eds. E. Clementi and R. H. Sarma), pp. 377–387. Adenine Press, New York.

De Felice, W. J. (1981). 'Introduction to Membrane Noise'. Plenum Press, New York.

Del Guidice, E., Doglia, S., Milani, M. and Fontana, M. P. (1984). *Cell Biophys.* **6**, 117–129.

Denbigh, K. (1981). 'The Principles of Chemical Equilibrium', 4th edn. Cambridge University Press, Cambridge.

Derr, R. F. (1985). *Biochem. Arch.* **1**, 239–247.

Derr, R. F. (1986). *Biochem. Arch.* **2**, 31–44.

Dimitrov, D. S. and Jain, R. K. (1984). *Biochim. Biophys. Acta* **779**, 437–468.

Doi, M. and Edwards, S. F. (1986). 'The Theory of Polymer Dynamics'. Clarendon Press, Oxford.

Dwyer, T. (1985). *J. Electrophysiol. Technol.* **12**, 15–29.

Elferink, M. G. L., Hellingwerf, K. J. and Konings, W. N. (1985). *Eur. J. Biochem.* **153**, 161–165.

Elferink, M. G. L., Hellingwerf, K. J. and Konings, W. N. (1986). *Biochim. Biophys. Acta.* **848**, 58–68.

Englander, S. W. and Kallenbach, N. R. (1984). *Q. Rev. Biophys.* **16**, 521–655.

Felle, H., Stetson, D. L., Long, W. S. and Slayman, C. L. (1978). *In* 'Frontiers of Biological Energetics' (Eds. P. L. Dutton, J. S. Leigh and A. Scarpa), Vol. 2, pp. 1399–1407. Academic Press, New York.

Felle, H., Porter, J. S., Slayman, C. C. and Kaback, H. R. (1980). *Biochemistry* **17**, 3585–3590.

Ferguson, S. J. (1985). *Biochim. Biophys. Acta* **811**, 47–95.

Ferguson, S. J. and John, P. (1977). *Biochem. Soc. Trans.* **5**, 1525–1527.

Ferguson, S. J. and Sorgato, M. C. (1982). *Ann. Rev. Biochem.* **51**, 185–217.

Ferguson, S. J., Jones, O. T. G., Kell, D. B. and Sorgato, M. C. (1979). *Biochem. J.* **180**, 75–85.

Fersht, A. R. (1985). 'Enzyme Structure and Mechanism', 2nd edn. W. H. Freeman, New York.

Fersht, A. R., Leatherbarrow, R. J. and Wells, T. N. C. (1986). *Nature* **322**, 284–286.

Finkelstein, R. J. (1969). 'Thermodynamics and Statistical Physics'. W. H. Freeman, San Francisco.

Frauenfelder, H. and Wolynes, P. G. (1985). *Science* **229**, 337–345.

Fröhlich, H. (1968). *Int. J. Quantum Chem.* **2**, 641–649.

Fröhlich, H. (1969). *In* 'Theoretical Physics and Biology' (Ed. M. Morris), pp. 13–22. North Holland, Amsterdam.

Fröhlich, H. (1980). *Adv. Electron. Electron. Phys.* **53**, 85–152.

Fröhlich, H. (1986). *In* 'The Fluctuating Enzyme' (Ed. G. R. Welch), pp. 421–449. Plenum Press, New York.

Fröhlich. H. and Kremer, F. (Eds.) (1983). 'Coherent Excitations in Biological Systems'. Springer, Heidelberg.

Gal-Or, B. (1981). 'Cosmology, Physics and Philosophy'. Springer, Heidelberg.

Garden, R. W. (1984). 'Modern Logic and Quantum Mechanics'. Adam Hilger, Bristol.

Gassier, R. P. H. and Richards, W. G. (1974). 'Entropy and Energy Levels'. Clarendon Press, Oxford.

Giulian, D. and Diacumakos, E. G. (1977). *J. Cell. Biol.* **72**, 86–103.

Glaser, R. W. (1986). *Studia Biophys.* **116**, 77–86.

Glasstone, S., Laider, K. J. and Eyring, H. (1941). 'The Theory of Rate Processes'. McGraw-Hill, New York.

Gopal, E. S, R. (1974). 'Statistical Mechanics and Properties of Matter'. Ellis Horwood, Chichester.

Gould, J. M. (1979). *J. Bacteriol.* **138**, 176–184.

Gould, J. M. and Cramer, W. A. (1977). *J. Biol. Chem.* **252**, 5875–5882.

Gräber, P. (1981). *Curr. Top. Membr. Trans.* **16**, 215–245.

Gräber, P., Junesch, U. and Schatz, G. H. (1984). *Ber. Bunsenges. Phys. Chem.* **88**, 599–608.

Grinius, L. L. (1986). 'Macromolecules on the Move: Energy Transduction and Gene Transfer in Chemotrophic Bacteria'. Gordon & Breach, London.

Groen, A. K., van der Meer, R., Westerhoff, H. V., Wanders, R. J. A., Akerboom, T. P. M. and Tager, J. M. (1982). *In* 'Metabolic Compartmentation' (Ed. H. Sies), pp. 9–37. Academic Press, New York.

Gross, D., Loew, L. M. and Webb, W. W. (1986). *Biophys. J.* **50**, 339–348.

Guffanti, A. A., Fuchs, R. T., Schneier, M., Chiu, E. and Krulwich, T. A. (1984). *J. Biol. Chem.* **259**, 2971–2975.

Guffanti, A. A., Chiu, E. and Krulwich, T. A. (1985). *Arch. Biochem. Biophys.* **239**, 327–333.

Gurd, F. R. N. and Rothgeb, T. M. (1979). *Adv. Prot. Chem.* **33**, 73–168.

Haken, H. (1970). 'Laser Theory'. Springer, Heidelberg.

Haken, H. (1977). 'Synergetics'. Springer, Heidelberg.

Hamamoto, T., Carrasco, N., Matsushita, K., Kaback, H. R. and Montal, M. (1985). *Proc. Natl. Acad. Sci. USA* **82**, 2570–2573.

Hangarter, R. P. and Good, N. E. (1982). *Biochim. Biophys. Acta* **681**, 397–404.

Hangarter, R. and Ort, D. R. (1985). *Eur. J. Biochem.* **149**, 503–510.

Harold, F. M. (1986). 'The Vital Force: A Study of Bioenergetics'. Freeman, Oxford.

Harris, C. M. and Kell, D. B. (1985a). *Biosensors J.* **1**, 17–84.

Harris, C. M. and Kell, D. B. (1985b). *Eur. Biophys. J.* **13**, 11–24.

Hauska, G., Samoray, D., Orlich, G. and Nelson, N. (1980). *Eur. J. Biochem.* **111**, 535–543.

Hedvig, P. (1977). 'Dielectric Spectroscopy of Polymers'. Adam Hilger, Bristol.

Heinrich, R. and Rapoport, T. A. (1974). *Eur. J. Biochem.* **42**, 89–95.

Herweijer, M. A., Berden, J. A., Kemp, A. and Slater, E. C. (1985). *Biochim. Biophys. Acta* **809**, 81–89.

Herweijer, M. A., Berden, J. A., and Slater, E. C. (1986). *Biochim. Biophys. Acta* **849**, 276–287

Hill, T. L. (1960). An Introduction to Statistical Thermodynamics'. Addison-Wesley, New York.

Hill, T. L. (1963). 'Thermodynamics of Small Systems'. Part 1. Benjamin, New York.

Hill, T. L. (1977). 'Free Energy Transduction in Biology'. Academic Press, New York.

Hitchens, G. D. and Kell, D. B. (1982a). *Biochem. J.* **206**, 351–357.

Hitchens, G. D. and Kell, D. B. (1982b). *Biosci. Rep.* **2**, 743–749.

Hitchens, G. D. and Kell, D. B. (1983a). *Biochem. J.* **212**, 25–30.

Hitchens, G. D. and Kell, D. B. (1983b). *Biochim. Biophys. Acta* **723**, 308–316.

Hitchens, G. D. and Kell, D. B. (1984). *Biochim. Biophys. Acta* **768**, 222–232.

Horner, R. D. and Moudrianakis, E. N. (1985). *J. Biol. Chem.* **260**, 6153–6159.

Jackson, J. B. (1982). *FEBS Lett.* **139**, 139–143.

Jackson, J. B. and Nicholls, D. G. (1986). *Meth. Enzymol.* **127**, 557–577.

Jaenicke, R. (1984). *Angew. Chem. Int. Ed. Engl.* **23**, 395–413.

Jagendorf, A. T. and Uribe, E. G. (1966). *Proc. Natl. Acad. Sci. USA* **55**, 170–177.

Jardetzky, O. and King, R. (1983). *Ciba Found. Symp.* **43**, 291–309.

Jencks, W. P. (1969). 'Catalysis in Chemistry and Enzymology'. McGraw-Hill, New York.

Jencks, W. P. (1985). *Chem. Rev.* **85**, 511–527.

Jenkins, F. A. and White, H. E. (1981). 'Fundamentals of Optics', 4th edn. McGraw-Hill, New York.

Johnson, J. B. (1928). *Phys. Rev.* **32**, 97–107.

Jones, W. J. (1969). *Lasers. Q. Rev. Chem. Soc.* **23**, 73–79.

Junge, W. (1977). *Ann. Rev. Plant. Physiol.* **28**, 503–526.

Kaback, H. R. (1986). *Ann. Rev. Biophys. Biophys. Chem.* **15**, 279–319.

Kacser, H. and Burns, J. A. (1973). *In* 'Rate Control of Biological Processes' (Ed. D. D. Davies) (*Symp. Soc. Exp. Biol. Vol.* **27**), pp. 65–104. Cambridge University Press, Cambridge.

Kacser, H. and Porteous, J. W. (1987). *Trends in Biochem. Sci.* **12**, 1–14.

Kahn, J. S. (1970). *Biochem. J.* **116**, 55–60.

Kamp, F. and Westerhoff, H. V. (1987). *In* 'The Organisation of Cell Metabolism' (Eds. G. R. Welch and J. S. Clegg), pp. 357–365. Plenum Press, New York.

Kashket, E. R. (1985). *Ann. Rev. Microbiol.* **39**, 219–242.

Kell, D. B. (1979). *Biochim. Biophys. Acta* **549**, 55–99.

Kell, D. B. (1986a). *In* 'Biosensors: Fundamentals and Applications' (Eds. A. P. F. Turner, I. Karube and G. S. Wilson), pp. 429–470. Oxford University Press, Oxford.

Kell, D. B. (1986b). *Meth. Enzymol.* **127**, 538–557.

Kell, D. B. (1986c). *Biochem. J.* **236**, 931–932.

Kell, D. B. (1987a). *In* 'Energy Transfer Dynamics' (Eds. T. W. Barrett and H. A. Pohl), pp. 237–246. Springer-Verlag, Heidelberg.

Kell, D. B. (1987b). *J. Gen. Microbiol.*, **133**, 1651–1665.

Kell, D. B. (1988). *In* 'Biological Coherence and Response to External Stimuli' (Ed. H. Fröhlich). Springer, Heidelberg pp. 233–241.

Kell, D. B. and Harris, C. M. (1985). *J. Bioelectricity* **4**, 317–348.

Kell, D. B. and Hitchens, G. D. (1982). *Faraday Discuss. Chem. Soc.* **74**, 377–388.

Kell, D. B. and Hitchens, G. D. (1983). *In* 'Coherent Excitations in Biological Systems' (Eds. H. Fröhlich and F. Kremer), pp. 178–198. Springer-Verlag, Heidelberg.

Kell, D. B. and Walter, R. P. (1986). *In* 'The Organisation of Cell Metabolism' (Eds. G. R. Welch and J. S. Clegg), pp. 215–232. Plenum Press, New York.

Kell, D. B. and Westerhoff, H. V. (1985). *In* 'Organised Multienzyme Systems; Catalytic Properties' (Ed. G. R. Welch), pp. 63–139. Academic Press, New York.

Kell, D. B. and Westerhoff, H. V. (1986a). *FEMS Microbiol. Rev.* **39**, 305–320.

Kell, D. B. and Westerhoff, H. V. (1986b). *Trends in Biotechnol.* **4**, 137–142.

Kell, D. B., John, P. and Ferguson, S. J. (1978). *Biochem. J.* **174**, 257–266.

Kell, D. B., Astumian, R. D. and Westerhoff, H. V. (1988). 'Ferroelectrics' (in press).

Kemeny, G. (1974). *Proc. Natl. Acad. Sci. USA* **71**, 2655–2657.

King, E. L. and Altman, C. (1956). *J. Phys. Chem.* **60**, 1375–1378.

Knox, J. H. (1978). 'Molecular Thermodynamics: An Introduction to Statistical Mechanics for Chemists'. John Wiley, Chichester.

Krasinkskaya, I. P., Marshansky, V. N., Dragunova, S. F. and Yaguzhinsky. L. S. (1984). *FEBS Lett.* **167**, 176–180.

Kreuzer, H. J. (1983). 'Nonequilibrium Thermodynamics and its Statistical Foundations'. Clarendon Press, Oxford.

Krulwich, T. A. and Guffanti, A. A. (1983). *Adv. Micr. Physiol.* **24**, 173–214.

Kubo, R. (1969). *In* 'Many-body Problems'. (Eds. W. E. Parry, R. E. Turner, D. ter Haar, J. S. Rowlinson, G. V. Chester, N. M. Hugenholtz and R. Kubo), pp. 235–265. Benjamin, New York.

Kubo, R. (1986). *Science* **233**, 330–334.

Laidler, K. J. (1969). 'Theories of Chemical Reaction Rates'. McGraw-Hill, New York.

Landsberg, P. T. (1982). 'The Enigma of Time'. Adam Hilger, Bristol.

Ling, G. N. (1984). 'In Search of the Physical Basis of Life'. Plenum Press, New York and London.

Lolkema, J. S., Hellingwerf, K. J. and Konings, W. N. (1982). *Biochim. Biophys. Acta* **681**, 85–94.

Lolkema, J. S., Abbing, A., Hellingwerf, K. J. and Konings, W. N. (1983). *Eur. J. Biochem.* **130**, 287–292.

Lomdahl, P. S. (1984). *In* 'Nonlinear Electrodynamics in Biological Systems' (Eds. W. R. Adey and A. F. Lawrence), pp. 143–154. Plenum, New York.

Lomdahl, P. S., Layne, S. P. and Bigio, I. J. (1984). *Los Alamos Science* **10**, 2–31.

Lumry, R. (1986). *In* 'The Fluctuating Enzyme' (Ed. G. R. Welch), pp. 1–190. John Wiley, New York.

McCammon, J. A. (1984). *Rep. Progr. Phys.* **47**, 1–46.

McCarthy, J. E. G. and Ferguson, S. J. (1982). *Biochem. Biophys. Res. Comm.* **107**, 1406–1411.

McCarthy, J. E. G. and Ferguson, S. J. (1983a). *Eur. J. Biochem.* **132**, 417–424.

McCarthy, J. E. G. and Ferguson, S. J. (1983b). *Eur. J. Biochem.* **132**, 425–431.

McClare, C. W. F. (1971). *J. Theoret. Biol.* **30**, 1–34.

MacKinnon, E. M. (1982). 'Scientific Explanation and Atomic Physics'. University of Chicago Press, Chicago.

McLaughlin, S. (1977). *Curr. Top. Membr. Trans.* **9**, 71–144.

Maloff, B. L., Scordilis, S. P. and Tedeschi, H. (1978a). *Science* **199**, 568–570.

Maloff, B. L., Scordilis, S. P. and Tedeschi, H. (1978b). *J. Cell Biol.* **78**, 214.

Maloney, P. C. (1982). *Curr. Top. Membr. Trans.* **16**, 175–193.

March, N. H. and Parrinello, M. (1982). 'Collective Effects in Solids and Liquids'. Adam Hilger, Bristol.

Mason, C. A., Hamer, G. and Bryers, J. D. (1986). *FEMS Microbiol. Revs.* **39**, 373–401.

Mehrle, W., Zimmermann, U. and Hampf, R. (1985). *FEBS Lett.* **185**, 89–94.

Midgley, M. (1986). *J. Gen. Microbiol.* **132**, 3187–3193.

Mills, J. D. (1984). *In* 'H$^+$-ATPase (ATP synthase): Structure, Function, Biogenesis of the F_0F_1 Complex of Coupling Membranes' (Eds. S. Papa, K. Altendorf, L. Ernster and L. Pocher), pp. 349–358. Adriatica Editrice, Bari.

Mills, J. D. and Mitchell, P. D. (1982). *FEBS Lett.* **144**, 63–67.

Mitchell, P. (1968). 'Chemiosmotic Coupling and Energy Transduction'. Glynn Research, Bodmin.

Morrissey, B. W. (1975). *J. Chem. Ed.* **52**, 296–298.

Myatt, J. F. and Jackson, J. B. (1986). *Biochim. Biophys. Acta* **848**, 212–228.

Nicholls, D. G. (1982). 'Bioenergetics'. Academic Press, London.

Nicolay, K., Lolkema, J. S., Hellingwerf, K. J., Kaptein, R. and Konings, W. N. (1981). *FEBS Lett.* **123**, 319–323.

Nicolis, G. and Prigogine, I. (1977). 'Self-organisation in Non-equilibrium Systems'. John Wiley, Chichester.

Nyquist, H. (1928). *Phys. Rev.* **32**, 110–113.

Ort, D. R. and Melandri, B. A. (1982). *In* 'Photosynthesis: Energy Conversion by Plants and Bacteria' (Ed. Govindjee), Vol. 1, pp. 537–587, Academic Press, New York.

Padan, E., Zilberstein, D. and Schuldiner, S. (1981). *Biochim. Biophys. Acta* **650**, 151–166.

Parsonage, D. and Ferguson, S. J. (1982). *Biochem. Soc. Trans.* **10**, 257–258.

Pethig, R. (1986). *In* 'Modern Bioelectrochemistry' (Eds. F. Gutmann and H. Keyzer), pp. 199–239. Plenum Press, New York.

Pethig, R. and Kell, D. B. (1987). *Phys. Med. Biol.* (in press).

Petronilli, V., Pietrobon, D., Zoratti, M. and Azzone, G. F. (1986). *Eur. J. Biochem.* **155**, 423–431.

Petronilli, V., Azzone, G. F. and Pietrobon, D. (1988). *Biochim. Biophys. Acta.* **932**, 306–324.

Pietrobon, D., Azzone, G. F. and Walz, D. (1981). *Eur. J. Biochem.* **117**, 389–394.

Pietrobon, D. and Caplan, S. R. (1985b). *FEBS Lett.* **192**, 119–122.

Pietrobon, D. and Caplan, S. R. (1985c). *Biochemistry* **24**, 5764–5776.

Pietrobon, D. and Caplan, S. R. (1986a). *Biochemistry* **25**, 7682–7690.

Pietrobon, D. and Caplan, S. R. (1986b). *Biochemistry* **25**, 7690–7696.

Pietrobon, D., Zoratti, M., Azzone, G. F. and Caplan, S. R. (1986). *Biochemistry* **25**, 767–775.

Pietrobon, D., Zoratti, M., Azzone, G. F., Stucki, J. W. and Walz, D. (1982). *Eur. J. Biochem.* **127**, 483–494.

Popper, K. R. (1982). 'Quantum Theory and the Schism in Physics'. Hutchinson, London.

Porteous, J. W. (1985). *In* 'Circulation, Respiration and Metabolism' (Ed. R. Gilles), pp. 263–277. Springer, Berlin.

Powell, K. T., Derrick, E. G. and Weaver, J. C. (1986). *Bioelectrochem. Bioenerg.* **15**, 243–255.

Primas, H. (1981). 'Chemistry, Quantum Mechanics and Reductionism'. Springer-Verlag, Heidelberg.

Racker, E. and Stoeckenius, W. (1974). *J. Biol. Chem.* **249**, 662–663.

Ringe, D. and Petsko, G. A. (1985). *Progr. Biophys. Mol. Biol.* **45**, 197–235.

Ritchie, R. J. (1982). *J. Membr. Biol.* **69**, 57–63.

Rottenberg, H. (1973). *Biophys. J.* **13**, 503–511.

Rottenberg, H. (1975). *J. Bioenerg* **7**, 63–76.

Rottenberg, H. (1979a). *Meth. Enzymol.* **55**, 547–569.

Rottenberg, H. (1979b). *Biochim. Biophys. Acta* **549**, 225–253.

Rottenberg, H. (1985). *Modern Cell Biol.* **4**, 47–83.

Sakmann, B. and Neher, E. (1983). 'Single-channel Recording'. Plenum, New York.

Sakmann, B. and Neher, E. (1984). *Annu. Rev. Physiol.* **46**, 455–472.

Sanders, D. (1987). *In* 'Physiological Models in Microbiology' (Eds. M. Bazin and J. I. Prosser), Vol. 1. CRC Press, Boca Raton, Florida (in press).

Schmidt, G. and Gräber, P. (1985). *Biochim. Biophys. Acta* **808**, 46–51.

Scholes, P. and Mitchell, P. (1970). *J. Bioenerg.* **1**, 309–323.

Schrödinger, E. (1935). *Naturwiss.* **23**, 807–812, 823–828, 844–849.

Scott, A. C. (1983). *In* 'Structure and Dynamics: Nucleic Acids and Proteins' (Eds. E. Clementi and R. H. Sarma), pp. 389–404. Adenine Press, New York.

Shiner, J. S. (1982). *J. Theoret. Biol.* **99**, 309–318.

Slater, E. C. (1987). *Eur. J. Biochem.* **166**, 489–504.

Smith, E. B. (1982). 'Basic Chemical Thermodynamics', 3rd edn. Oxford University Press, Oxford.

Smith, W. V. and Sorokin, P. P. (1966). 'The Laser'. McGraw-Hill, New York.

Smith, D. J., Stokes, B. O. and Boyer, P. D. (1976). *J. Biol. Chem.* **251**, 4165–4171.

Somogyi, B., Welch, G. R. and Damjanovich, S. (1984). *Biochim. Biophys. Acta* **768**, 81–112.

Sone, N., Yoshida, M., Hirata, H. and Kagana, Y. (1977). *J. Biol. Chem.* **252**, 2956–2960.

Sowers, A. E. and Lieber, M. R. (1986). *FEBS Lett.* **205**, 179–184.

Steinberg, I. Z. (1986). *Biophys. J.* **50**, 171–179.

Stucki, J. W. (1978). *In* 'Energy Conservation in Biological Membranes' (Eds. G. Schafer and M. Klingenberg), pp. 264–287. Springer, Heidelberg.

Stucki, J. W. (1980). *Eur. J. Biochem.* **109**, 269–283.

Stucki, J. W. (1982). *In* 'Metabolic Compartmentation' (Ed. H. Sies), pp. 39–69. Academic Press, London.

Stucki, J. W., Compiani, M. and Caplan, S. R. (1983). *Biophys. Chem.* **18**, 101–109.

Sugar, I. P. and Neumann, E. (1984). *Biophys. Chem.* **19**, 211–225.

Tamponnet, C., Rona, J. P., Barbotin, J. N. and Calvayrac, R. (1986). *Proc. 2nd Int. Conf. Plant Mitochondria* (Eds. A. L. Moore and R. B. Beechey), p. 46.

Taylor, M. A. and Jackson, J. B. (1985a). *FEBS Lett.* **180**, 145–149.

Taylor, M. A. and Jackson, J. B. (1985b). *Biochim. Biophys. Acta* **810**, 209–224.

Tedeschi, H. (1980). *Biol. Rev.* **55**, 171–200.

Tedeschi, H. (1981). *Biochim. Biophys. Acta* **639**, 157–196.

Thauer, R. K., Jungermann, K. and Decker, K. (1977). *Bacteriol Rev.* **41**, 100–180.

Thayer, W. S. and Hinkle, P. C. (1975a). *J. Biol. Chem.* **250**, 5330–5335.

Thayer, W. S. and Hinkle, P. C. (1975b). *J. Biol. Chem.* **250**, 5336–5342.

Thornton, E. K. and Thornton, E. R. (1978). *In* 'Transition States of Biochemical Processes' (Eds. R. D. Gandour and R. L. Schowen), pp. 3–76. Plenum Press, New York.

Tolman, R. C. (1938). 'The Principles of Statistical Mechanics', pp. 152–165. Oxford University Press, London.

Tsong, T. Y. (1983). *Biosci. Rep.* **3**, 487–505.

Tsong, R. Y. and Astumian, R. D. (1986). *Bioelectrochem. Bioenerg.* **15**, 457–476.

van der Bend, R. L. and Herweijer, M. A. (1985). *FEBS Lett.* **186**, 8–10.

van der Bend, R. L., Cornelissen, J. B. W., Berden, J. A. and van Dam, K. (1984). *Biochim. Biophys. Acta* **767**, 87–101.

van der Bend, R. L., Peterson, J., Berden, J. A., van Dam, K. and Westerhoff, H. V. (1985). *Biochem. J.* **230**, 543–549.

van der Ziel, A. (1976). 'Noise in Measurements'. John Wiley, New York.

van Walraven, H. S., Haak, N. P., Krab, K. and Kraayenhof, R. (1986). *FEBS Lett.* **208**, 138–142.

van Walraven, H. S., Hagendvorn, M. J. M., Krab, K., Haak, H. P. and Kraayenhof, R. (1985). *Biochim. Biophys. Acta* **809**, 236–244.

Venturoli, G. and Melandri, B. A. (1982). *Biochim. Biophys. Acta* **680**, 8–16.

Vredenberg, W. J. (1976). *In* 'The Intact Chloroplast' (Ed. J. Barber), p. 53. Elsevier, Amsterdam.

Wakabayashi, T., Horiuchi, M., Kawamoto, S. and Onda, H. (1984). *Acta Pathol. Jpn.* **34**, 481–488.

Walz, D. (1979). *Biochim. Biophys. Acta* **505**, 279–253.

Webb, S. J. (1980). *Phys. Rep.* **60**, 201–224.

Welch, G. R. (1985). *J. Theoret. Biol.* **114**, 433–446.

Welch, G. R. (Ed.) (1986). 'The Fluctuating Enzyme'. John Wiley, New York.

Welch, G. R. and Clegg, J. S. (Eds.) (1987). 'The Organisation of Cell Metabolism'. Plenum Press, New York.

Welch, G. R. and Kell, D. B. (1986). *In* 'The Fluctuating Enzyme' (Ed. G. R. Welch), pp. 451–492. John Wiley, New York.

Welch, G. R. and Smith, H. A. (1987). *In* 'Proc. 3rd Int. Seminar on the Living State' (Ed. R. K. Mishra), Plenum Press, New York (in press).

Welch, G. R., Somogyi, B. and Damjanovich, S. (1982). *Progr. Biophys. Mol. Biol.* **39**, 109–146.

Westerhoff, H. V. and Chen, Y. (1985). *Proc. Natl. Acad. Sci. USA* **82**, 3222–3226.

Westerhoff, H. V. and van Dam, K. (1979). *Curr. Top. Bioenerg.* **9**, 1–62.

Westerhoff, H. V. and Kell, D. B. (1988). *Comments Mol. Cell. Biophys*, (in press).

Westerhoff, H. V. and van Dam, K. (1988). 'Thermodynamics and Control of Biological Energy Transduction'. Elsevier/North Holland, Amsterdam.

Westerhoff, H. V. and Kamp, F. (1987). *In* 'The Organization of Cell Metabolism' (Eds. G. R. Welch and J. S. Clegg), pp. 339–356. Plenum Press, New York.

Westerhoff, H. V., Groen, A. K. and Wanders, R. J. A. (1984a). *Biosci. Rep.* **4**, 1–22.

Westerhoff, H. V., Melandri, B. V., Venturoli, G., Azzone, G. F. and Kell, D. B. (1984b). *Biochim. Biophys. Acta* **768**, 257–292.

Westerhoff, H. V., van der Bend, R. L., van Dam, K., Berden, J. A. and Petersen, J. (1986). *Biochem. J.* **236**, 932–933.

Westerhoff, H. V., Hellingwerf, K. J. and van Dam, K. (1983). *Proc. Natl. Acad. Sci. USA* **80**, 305–309.

Westerhoff, H. V., Lolkema, J. S., Otto, R. and Hellingwerf, K. (1982). *Biochim. Biophys. Acta* **683**, 181–220.

Wharton, C. W. and Eisenthal, R. (1981). 'Molecular Enzymology'. Blackie, London.

Wheeler, J. A. and Zurek, W. H. (1983). 'Quantum Theory and Measurement'. Princeton University Press, Princeton.

Williams, R. J. P. and Concar, D. (1986). *Nature* **322**, 213–214.

Wilson, G. M., Alderete, J. G., Maloney, P. C. and Wilson, T. H. (1976). *J. Bacteriol.* **126**, 327–337.

Woelders, H., van der Zande, W. J., Colen, A.-M. A. F., Wanders, R. J. and van Dam, K. (1985). *FEBS Lett.* **179**, 278–282.

Yomosa, S. (1983). *J. Phys. Soc. Japan.* **52**, 1866–1873.

Zimmermann, U. (1982). *Biochim. Biophys. Acta* **694**, 227–277.

Zimmermann, U. and Vienken, J. (1982). *J. Membr. Biol.* **67**, 165–182.

Zoratti, M., Favaron, M., Pietrobon, D. and Azzone, G. F. (1986). *Biochemistry* **25**, 760–767.

Zukav, G. (1980). 'The Dancing Wu Li Masters'. Fontana/Collins, London.

Index

Pages marked f denote a figure and those marked t denote a table.

Acetaldehyde dehydrogenase, in SLP reactions, 88t

Acetate,
oxidation by acetyl-CoA pathway, 134
substrate for methanogenesis, 113, 115, 120–123
substrate for sulphate-reducing bacteria, 124, 125t, 128, 133, 134

Acetate as fermentation product from lactate, 103f

Acetate fermentations, 105–111
the homoacetogens, 106–109
the hydrogen-producing acetogens, 109–111

Acetate kinase,
in SLP reactions, 88t
role in sulphate reducers, 127

Acetic acid bacteria,
aldehyde dehydrogenase of, 297t
see also Acetobacter, Gluconobacter

Acetobacter,
cytochrome a_1, 51t, 274, 275t, 277t, 278
cytochrome d, 271
cytochrome o, 258
glucose dehydrogenase of, 305, 306
methanol dehydrogenase and cytochrome c of, 300

Acetobacterium, 106–109

Acetogens, 105–111
in microbial consortia, 135

Acetoin as fermentation product, 95f, 101

Acetolactate, fermentation intermediate, 101

Acetone fermentation, 96–99

Acetyl-CoA pathway, 107f
in acetogens, 106–109
in methanogens, 115
in sulphate-reducers, 134

Acetylene, inhibitor of ammonia monooxygenase, 198

Acetyl-phosphate, energy donor for transport, 390

Acid/base distribution for pmf measurement, 466–470

Acid production and metal corrosion, 188

Acid production and pmf in fermentation, 105

Acidophilic bacteria,
iron oxidizers, 220–223
pH control in, 399f, 424
methylotrophs, 300

Acinetobacter,
alcohol dehydrogenase, 301
cytochrome a_1, 277t
cytochrome b, 311, 312f
glucose dehydrogenase, 297t, 298, 305, 306, 311, 312f
glucose oxidation in periplasm, 161t, 171, 312f
PQQ requirement for growth of some strains, 298

Active transport systems, 379

Activity coefficients of probe ions, 467

Acrylyl-CoA in propionate fermentation, 104

Active transport systems, 379,
see also group translocation, periplasmic transport systems, and ion-linked transport systems

Adenosine phosphosulphate (APS)
  redox potential, 211f
  reductase, 57–60, 126, 127, 212f, 214
Adenylate cyclase, role in regulation of
    transport, 385
Alanine transport, 312
Alcaligenes,
  hydrogenase of, 31
  nitrite reductase of, 145
  periplasmic cytochrome *c*, 157t
Alcohol dehydrogenases (quinoprotein),
    297t, 301, 311
  see also methanol dehydrogenase
Aldehyde dehydrogenase (quino-
    protein), 297t
Alkalophilic bacteria, 15
  ATP synthesis in, 76
  pH control in, 23, 398, 399f, 425
  sodium translocating enzymes in, 33,
    397, 425
Alteromonas, 146
  periplasmic cytochrome *c*, 157t
Amicyanin, 45, 46,
  periplasmic location, 157t, 170f, 174,
    175
  properties and distribution, 304t
  interaction with methylamine
    dehydrogenase, 174
Ammonia monooxygenase, 197–201
Ammonia oxidation to nitrite, 195–203
  coupling to generation of proton-
    motive force, 202
  coupling of specialized enzymes to
    respiratory chain, 200–202
  oxidation of ammonia to hydroxyl-
    amine, 196–199
  oxidation of hydroxylamine to
    nitrite, 199
Ammonia oxidation, by a mono-
    oxygenase, 197–199
Ammonium ion transport, 398, 401
Anaerobic bacteria, energy transduction
    in, 83
Anaerobic energy coupling, mechan-
    isms of, 84–89
Anaerobic microbial consortia, 112, 135
Anaerobic respiration, 6–9, 112–146
  definition, 87
  fumarate as terminal electron
    acceptor, 136–141

Anaerobic respiration (*cont.*)
  fumarate reductase, 139–141
  oxides of nitrogen as terminal elec-
    tron acceptor, 137, 141–146
  proton translocation in, 69f
  by sulphate-reducers, 126–128
    see also methanogenesis, sulphate
      reduction, fumarate reduction,
      nitrate reduction
Antenna chlorophyll, 53, 322
Anthraquinone, in study of reaction
    centres, 345
Antibiotic transport, 397
Antimony oxidation, 187
Antimycin,
  effect on sulphite oxidation, 215
  in study of photosynthesis, 339, 342,
    354, 358–360, 364
Antiports, and ion-linked solute trans-
    port, 22–24, 396–398, 402
APS reductase, *see* adenosine phospho-
    sulphate reductase
Arabinose transport, 419
Arthrobacter, methylamine oxidase
    (quinoprotein), 297t, 307
Archaebacteria, 113, 124–126, 133
Arsenate, inhibitor of transport, 389,
    390f
Artificial bilayers, with reaction centres,
    342–347
Artificial membrane potential driving
    transport of proline, 410f
Artificial pH gradient, in vesicles, 408
ATPase,
  solute translocating, 394
  and histidine transport, 392
  and potassium transport, 394, 422–424
  see also ATP synthase
ATP,
  energy source for bacterial transport,
    389–395
  synthesis, 11–18, 21, 76–78, 459–482,
    see also ATP synthase,
  coupled to artificial pH gradient,
    460f, 481
  in methanogenesis, 120–123
  photosynthetic, and charge separat-
    ation, 361–366
ATP-binding proteins, homology with
    transport proteins, 388, 392

ATP/2*e* quotients, 21
and growth yields, 71
ATP/O ratios, 21
ATP sulphurylase, 127, 214
ATP synthase, 73–78
chemiosmotic coupling, 12–15, 20f,
361–367, 457–461, 472–482
inhibitors, 361, 478–480
see aurovertin, DCCD, oligomycin,
venturicidin
inhibitor titrations, 478–480
of iron oxidizers, 223
major component of membranes, 169
mechanism, 76–78, 366, 457, 472,
478–482
photosynthetic, 361–367
regulation of, 75
and reversed electron transport, 24,
25t
sodium-translocating, 33, 426
ATP yield
in chemolithotrophs, 194
in fermentation, *see* individual fermentations
in fumarate reduction, 139
in sulphate reducers, 129
Aurovertin, inhibitor of ATP synthase,
76
Autoreduction of cytochrome c_L, 300
Autotrophy, 3, 193, 195
Azotobacter,
cytochrome a_1. 277t
cytochrome *d*, 265, 266–268, 273
o-type oxidase of, 251t, 256, 260
proton translocation in, 67f
respiratory chains of, 60f
transhydrogenase of, 28
Azurin, 46
periplasmic location and role, 157t,
170f, 174–176, 303, 304t, 310,
311f

Bacillus,
ATP synthase of, 73
protonmotive force and pH and ion-
linked transport, 399f, 425
cytochrome *a* in, 246, 247, 249
see also thermophile PS3

Bacterial photosynthesis, *see* photosynthesis, bacterial
Bacterial transport, energetics and
mechanisms, 377–426
Bacteriochlorophyll, 10, 11f, 53–56, 322
accessory, 328f
antenna, 53, 322
arrangement in reaction centres, 325–
328, 330
special pair [BCh]$_2$ (*P*), in reaction
centres, 56, 324, 328–335, 339–
350, 355f
structures, 54f, 328f, 331f,
Bacteriophaeophytin (BPh), 10, 11f,
53–56, 325–328
function in reaction centres, 329,
335f, 346, 349–350f
arrangement in reaction centres, 328f,
330
Bacteriorhodopsin, 10, 22, 480
Bacillus,
a-type oxidases of, 176, 178, 238, 241t
cytochrome *c* in, 176–178
incapable of denitrification, 178
lack of periplasm, 176–179
Bactopterin, *see* molybdopterin
Beggiatoa, 215–216
see also Vitreoscilla
Beneckea,
cytochrome a_1, 277t
CO-binding cytochrome *c*, 256
cytochrome *o*, 256
Betaine transport, 424
Bifidobacterium, 95
Bifidum pathway of glucose fermen-
tation, 94f–96
Binding proteins, 393, *see* periplasmic
transport systems
Bisulphite, 210, 211
see sulphite
Blue copper proteins, 45, 303
see cupredoxin, amicyanin, azurin
Blue-green bacteria, *see* Cyanobacteria
Bluff? *see* Kell
Boltzmannian distribution, 433, 452,
454
Butanol fermentation, 96–100
Butanediol, fermentation product, 101
Butyrate,
substrate for acetogens, 110

Butyrate (*cont.*)
 substrate for sulphate-reducers, 125t,
  131
Butyrate and butanol/acetone fermentations, 96–100
Butyryl-CoA oxidation coupled to pmf,
  111f

Calcium-ATPase, 395
Calcium ionophore, 403
Caldariella, novel quinones in, 43
Calvin cycle, 194, 196
Campylobacter,
 formate oxidation, 161t
 hydrogen oxidation, 161t
 periplasmic reactions in, 161t, 163t
 nitrite reduction, 163t
 TMAO reductase, 146
Car analogy for phosphorylation and
  proton movements, 477f
Carbamate kinase, in SLP reactions,
  88t
Carbon assimilation in methanogens,
  115
Carbon dioxide,
 assimilation in chemolithotrophs, 193
 substrate for methanogenesis, 119
Carbon dioxide reduction factor
  (CDR), 117–119
 see methanofuran
Carbon monoxide,
 difference spectra, 48, 50, 243f, 251t,
  253f, 257f, 270f, 276f
 reaction with *a*-type cytochromes,
  233–235, 237, 243f, 248, 274,
  277t, 278
 reaction with *b*-type cytochromes,
  262–266, 268–272, 263t, 270f
 reaction with cytochrome *cd*$_1$, 280
 reaction with *d*-type cytochromes,
  261 263t, 276f
 reaction with *o*-type cytochromes,
  249–260, 251t, 253f, 257f
 reaction with haemoprotein *b*-590,
  276f
 reaction with haemoprotein of
  Vitreoscilla, 279
 redox potential, 186

Carbon monoxide dehydrogenase, 36
 in acetyl-CoA pathway
  acetogens, 107f, 108
  methanogens 115,
  sulphate-reducers, 134
 and pmf in methanogens, 123
Carbon monoxide oxidoreductase, 36
 see carbon monoxide dehydrogenase
Carotenoids in photosynthesis, 53, 322,
  336–338
 see also electrochromism
Carrier, in ion-linked transport, 411–
  414, 418
Catabolite repression, and PEP-dependent transport systems, 384–386
Catalase, 280–283
CCCP, 400, *see also* uncouplers
CDR, *see* carbon dioxide reduction
  factor, methanofuran
Cell turgor, regulation, 420–424
Charge recombination (photosynthesis)
  during ATP synthesis, 361–367
Charge separation in photosynthesis,
  319, 341f
 see also electrogenic reactions
Chemical potential, 433, 436, *see also*
  thermodynamics
Chemiosmotic coupling of proton
  pumps to ATP synthesis, 11,
  14f, 322f, 362–370, 432, 459–
  462
Chemiosmotic explanation of oxygen-
  pulse experiments, 64–70
Chemiosmotic function of cytochrome
  oxidases,
 a-type oxidases, 248
 d-type oxidases, 272
 o-type oxidases, 258–260
Chemiosmotic mechanism of energy
  transduction, 11–27, 64–78,
  162–169, 432, 433, 459, 474
 in chemolithotrophs, 191, 196f, 202,
  205–207, 221–225
 in fumarate reduction, 139
 in methanogens, 117, 120–123
 in methylotrophs, 308–311
 in photosynthetic bacteria, 322f,
  355f, 360–370
Chemiosmotic proton circuit, 14f, 322f,
  367f

Chemiosmotic proton circuit (*cont.*)
see also electrogenic reactions, protonmotive force
Chemiosmotic view of solute transport,
406–410, 414–416
Chemoautotrophy, 185
Chemolithoautotroph, 185
Chemolithotrophic fermentation, 135
Chemolithotrophic metabolism in
methanogens, 114
Chemolithotrophs, 3t
Chemolithotrophs
acid production and chemical corrosion, 188
biochemical features common to,
189
chemical data relevant to, 185–188
energy utilization and efficiency in
(general), 194
enzyme location in (generalizations),
192
growth yields, 71
importance of carbon dioxide fixation, 193
oxidases of, 190t, 191
oxidation of metal ions, 220–225
oxidation of ammonia to nitrite,
195–203
oxidation of inorganic sulphur compounds, 208–220
oxidation of nitrite, 203–207
proton-translocation and respiration
in (introduction), 69f, 223
respiration in, 6, 7f, 189
reverse electron transport in, 25, 191,
196f, 207f, 221f
reverse proton flow for NADH synthesis, 25, 191
Chemolithotrophy, 184–225
definition, 185
restriction to Gram-negative bacteria?, 178
see also chemolithotrophs
Chemoorganotrophs (chemoheterotrophs), 3t, 320
Chemotaxis, 389
Chlorin, 47f, 50
see haem *d*
Chlorobium,
bacteriochlorophyll of, 55t

Chlorobium (*cont.*)
light harvesting complex in, 54
quinone of, 61f
see also green bacteria
Chloroflexaceae, bacteriochlorophyll of,
55t
Chlorophyll, 11f
see bacteriochlorophyll
Chromatium,
cytochrome *o* in, 252
reaction centres, 329
sulphite reductase in, 217
Chromatophores, 323f, 324, 355f, 368f
see photosynthetic cytochrome *bc*$_1$ and
photosynthetic reaction centre
Citrate,
anaerobic degradation, 104
fermentation to diacetyl, acetoin and
lactate, 95f, 96
Classification according to carbon and
energy sources, 3t
Clostridium,
acetyl-CoA pathway in, 106–109
anaerobic citrate degradation, 104
butyrate, butanol and acetone
fermentation in, 96–100
relationship to sulphate-reducers, 124
Cobamide cofactor in acetyl-CoA
pathway, 107f–109
Coenzyme M, 117–120
Coenzymes, unusual ones in methanogenesis, 117–119
Collisional interactions and ATP synthesis, 15
Commamonas, hydroperoxidase, 282t
Component B in methanogenesis, 118
see mercaptoheptanoylthreonine
phosphate
Conformational changes in ATP synthase, 77f
Conformational proton pump, 15, 16f,
19, 68f
Control, see homeostatis, metabolic
control theory, regulation
Copper,
in *a*-type oxidases, 235–238, 241t,
244–245, 255, 278
in *o*-type oxidases, 251t, 252
in amine oxidase (quinoprotein), 302,
307

Copper (*cont.*)
  in ammonia monooxygenase, 199
  in nitrite reductase, 145
  oxidation by thiobacilli, 220
  in superoxide dismutase, 283
  see also cupredoxins
Co-reconstituted systems with ATP
    synthase, 480
Corrinoid cofactors, in acetyl-CoA
    pathway, 107f, 108, 115
Corrosion of metals by chemolitho-
    trophs, 188
Corynebacterium, cytochrome a_1-like
    protein, 275t
p-Cresol oxidation in periplasm, 155t
Cupredoxin (blue copper protein), 45,
    303–305, 310, 311f
  prosthetic group, 45f
  redox potential, 5f, 45
  spectrum, 45f
  see also amicyanin, azurin, plasto-
    cyanin
Cuproteins, 45, 303
  see amicyanin, azurin, cupredoxin,
    plastocyanin
Cyanide, ligand to cytochromes, 243f,
    254, 264, 266, 271, 280
  -resistant respiration, 255, 283
Cyanobacteria (blue-green bacteria),
  photosynthesis in, 10, 11f, 56, 61f, 62
  quinones of, 45
Cyclic AMP, 385f, 386
Cyclic electron transfer in photosyn-
    thesis, 10, 11f, 61, 322f, 355f
Cytochrome a_1, 51, 274–278
  in chemolithotrophs, 190t, 204, 221f
  classification, 274
  component similar in d-type oxidase,
    261, 262, 264–268, 271–273
  cytochromes believed to be oxidases,
    274–278
  in iron oxidation, 221f, 222
  in nitrite oxidoreductase, 204, 275t
  Type I, 274, 278
  Types II and III, 274
  diversity of similar proteins, 275t
  see also cytochromes (introduction)
Cytochrome "a_1", 274–276
  hydroperoxidase role, 281
  spectrum in whole membranes, 49f

Cytochrome $a_1 c_1$, in nitrite oxidation,
    204
Cytochrome aa_3, caa_3 and $c_1 aa_3$, 51,
    238–249
  in chemolithotrophs, 190t, 203
  chemiosmotic function, 248, 308
  distribution, 238
  genetics and molecular biology, 249
  in methylotrophs, 308, 310
  properties, 241t
  purification and properties of sub-
    units, 240–242
  reaction with oxygen, 246
  redox centres (copper), 244
  redox centres (haems), 242
  redox properties, 245–248
  spectrum, 49f, 243f, 247f,
    see also cytochromes (introduction)
    and individual organisms
Cytochrome b,
  in glucose oxidation, 306, 311, 312f
  periplasmic in *Acinetobacter*, 171,
    306, 311
  of nitrate reductase, 143
  of succinate dehydrogenase, 34
  and fumarate reductase, 139, 140f
  spectrum in whole membranes, 49f
    see also cytochromes (introduction)
Cytochrome b_{561}, 351–359
Cytochrome b_{566}, 351–359
Cytochrome b_{558} in d-type oxidase,
    265, 267–271, 273
Cytochrome b_{590}, *see* haemoprotein
    b_{590}
Cytochrome b_{595}, 266–268, 274, 276
Cytochrome b_{562} (periplasmic), 162
Cytochrome bc_1, 50
  in chemolithotrophs, 189
Cytochrome bc_1 (photosynthetic), *see*
    photosynthetic cytochrome bc_1
Cytochrome bc_L, 309f
Cytochrome bd complex, 261, 276
  see cytochrome d
Cytochrome c,
  in chemolithotrophs, 189, 191–193,
    200–201, 204–207, 212, 214–
    216, 219–222
  classification into three types, 48, 189
  in iron oxidation, 221–223
  interaction with cupredoxins, 174, 222

Cytochrome *c* (*cont.*)
  interaction with methylamine dehydrogenase, 305, 310
  in nitrite oxidation, 204, 205
  in oxidation of sulphur compounds, 212f, 214–216, 219, 220
  periplasmic location and role, 47, 157t, 160, 170, 193
  post-translational modification of, 176
  prosthetic group of, 47f
  spectrum, 49f
Cytochrome c_2,
  diffusion, and doubt over accepted role, 179, 332
  in photosynthetic electron transport, 321–323, 327f, 332–335, 339–345, 351–357
  in photosynthetic Q-cycle, 354–357
  redox potential, 353t
Cytochrome c_3, 126, 130, 159, 160
Cytochrome c_4 and c_5, 260
Cytochrome c', and nitrous oxide reduction, 171
Cytochrome cd_1, 53, 280
  see also nitrite reductase
Cytochrome c_H, 174, 175, 299, 305, 308–311
Cytochrome c_L, and methanol dehydrogenase, 174–176, 298–301, 303, 307–311
Cytochrome c_{554}, acceptor for hydroxylamine oxidoreductase, 171, 196f, 200–202
Cytochrome c_{559} (photosynthetic), 333
Cytochrome *c* oxidase of mitochondria, 236–238f
Cytochrome *co*, 49f, 51, 250, 252, 308, 309f, 310
  regulation, 255, 308, 309f
Cytochrome *d*, 51, 261–273
  chemiosmotic function, 272
  definition, distribution and nomenclature, 261
  EPR properties, 266
  genetics and molecular biology, 272
  in glucose oxidation, 311, 312f
  optical properties, 264–266
  potentiometric studies, 267
  purification and properties of subunits, 262–264

Cytochrome *d* (*cont.*)
  reaction with oxygen and other ligands, 264, 268–272
  regulation, 262, 283
  spectrum, 49f, 270f
  in *Vireoscilla*, 279
Cytochrome d_{650}, 269–271
Cytochrome *f*, 61f
Cytochrome *o*, 249–261
  chemiosmotic aspects, 258–260, 311, 464
  definition and distribution, 249
  with GDH in vesicles, 311
  genetics, 255, 260
  in glucose oxidation, 311 312f
  purification and properties, 250–252
  reactions with oxygen and catalytic activity, 256–258
  regulation, 255
  spectral, potentiometric and ligand-binding properties of the haems, 252–256
  spectrum, 49f, 253f, 257f
  reaction with oxygen (spectrum), 256–258
  table of properties, 251t
  see also cytochrome *co*
Cytochrome *o*-like haemoprotein of *Vitreoscilla*, 278
Cytochrome oxidase,
  mitochondrial (a survey), 236
  proton pumping during nitrite oxidation, 206
  see also cytochromes *a, co, d, o*
Cytochrome oxidases, 50–53, 231–284
  historical perspective and terminology, 232–235
  references to major sources, 239t
  regulation, 52, 239, 255, 283, 262, 273, 308, 309f
  see also individual cytochromes
Cytochrome P-450, 198
Cytochrome peroxidase, 156t, 281, 282
Cytochromes,
  introduction, 46–53
  non-autoxidizable, 48
  in methanogenesis, 114, 121
  redox potentials, 5f
  spectra, 48–50
  in sulphate reducers, 126, 130
  see also individual cytochromes

Db. propionicus, propionate metabolism in, 132f

DCCD,
  inhibitor of ATP synthase, 75, 76, 120, 123
  inhibitor of proton-pumping oxidases, 241t, 248

Decarboxylase, sodium-translocating, 104

Deazaflavin, 117
  see F_{420}

Delocalized protonmotive force, 459, 478

Denitrification, 141
  role in nitrogen cycle, 8f

'*Desulfo*bacteria', see sulphate-reducing bacteria, 123–135

Desulfobulbus propionicus growth on propionate, 131, 132f

Desulfotomaculum, 123–125f

Desulfovibrio, 123–125t
  cytochrome c_3, 157t, 159, 160
  formate oxidation in, 161t
  hydrogen cycling in, 129f
  hydrogen oxidation in periplasm, 167
  hydrogenase of, 31, 159
  nitrite reductase in, 53, 163t, 166
  periplasmic proteins in, 155t, 157t, 159, 161t, 166
  respiratory chain of, 60f
  rubredoxin in, 32
  signal peptides in, 159, 162

Desulfuromonas, 124–125t, 159

Desulphurication, 123–128
  role in nitrogen cycle, 8f

Detailed balance, principle of, 443

Diacetyl as fermentation product, 95f

Diethylthiocarbonate, inhibition of lactose permease, 419

Diffusion potential, 460, 462

Dihydrolipoate, reaction with thiosulphate, 217

Dihydroorotate dehydrogenase, 30

Dimethylsulphone, 40

Dimethylsulphoxide reductase, 39, 146, 321f

Dimethylsulphoxide reduction in periplasm, 156t, 160

Dissimilatory nitrate reduction, 141–146

Dissimilatory sulphate reduction, 123–135

Dithionate, 208, 210, 220

Dithionite, 210

Dinitrophenol (DNP), 400, *see also* uncouplers

Double inhibitor titrations, in study of pmf, 478–480

EDTA, in preparation of periplasm, 154, 387

Elasticity coefficients, 455

Electrochemical gradient of protons, 11–15
  in methanogenesis, 120
  see protonmotive force

Electrochemical gradient of sodium, 104, 122

Electrochemical gradient of solute, 23, 402t, 414

Electrochemical potential, 11, 12t, 436–439,

Electrochemical potential of ions in measurement of pmf, 467

Electrochemical potential of protons, 439, *see* thermodynamics, protonmotive force

Electrochromism as measure of membrane potential, 336–338, 470, 476
  ATP synthase, 362–364
  cytochrome bc_1 complex, 357–360
  proton currents and bacterial growth rate, 368–370
  reaction centre, 334–341, 349–351

Electrogenic processes in photosynthesis,
  in cytochrome bc_1 complex, 351–361
  in reaction centre, 324–351
  summary, 350f

Electrogenic proton pumping, 473

Electron transfer in and out of photosynthetic reaction centre, 332

Electron transport
  in *Acinetobacter calcoaceticus* during glucose and NADH oxidation, 171, 311, 312f
  in ammonia oxidation, 196f, 200–202

Electron transport (*cont.*)
  in chemolithotrophs (general), 189
  in methanogens, 116–121
  in *M. methylotrophus*, 308, 309f
  in *Rhodobacter*, 321f, 322f, 335f, 355f
  in *Thiobacillus ferrooxidans*, 221f
    see also cyclic electron transport
    electrogenic reaction, quino-
    proteins, respiratory chains,
Electroporation, 464–466
Embden-Meyerhof-Parnas pathway, 90
Emperor's new clothes, 480
Energetics of enzymes (transition state
    theory), 447f
'Energized' proton, 454
Energy coupling,
  criteria for assignment of type in
    transport systems, 380
  in ion-linked transport systems, 398–
    403
  in methanogens, 116–123, *see* metha-
    nogenic bacteria,
  and non-equilibrium thermodyna-
    mics, 439–441
  in periplasmic transport systems,
    162–169, 389–396
  by way of quinoproteins, 293–313
    see also chemiosmotic mechanisms,
    respiratory chains, proton-
    motive force
Energy coupling membrane, 11, 26, 169
Energy-depenent binding of probe ions,
    468
Energy metabolism,
  classification of bacteria by, 3
Energy flow in *E. coli* cells, 390f
Energy utilization and efficiency in
    chemolithotrophs, 194
Enteric bacteria,
  mixed acid fermentation in, 101
  PEP-dependent transport in, 381–386
    see also Escherichia coli and
    Enterobacter
Enterobacter, anaerobic citrate degra-
    dation in, 104
Enthalpy-entropy compensations, 449
Entner-Doudoroff pathway for sugar
    fermentation, 90, 91f
'Energy-rich' compounds, in substrate-
    level phosphorylation reactions, 88t

EPR (electron paramagnetic resonance)
    studies
  of *a*-type cytochromes, 236–238, 243–
    245
  of *o*-type oxidases, 252, 254, 256
  of *d*-type oxidases, 266, 269–271
  of cytochrome cd_1, 280
Ergodic hypothesis, 443, 444
Erwinia, ethanol fermentation in, 90
Escherichia coli,
  anaerobic growth on glycerol and
    lactate, 138
  ATP synthase of, 73–75
  catabolite repression in, 385
  cell turgor regulation, 420–424
  cytochrome a_1-like protein, 265, 274,
    275t
  cytochrome *b*, 254t, 260
  cytochrome b_{562} (periplasmic), 157t,
    162
  cytochrome *c* (absence), 50
  cytochrome *d*, 261–263t, 266–273
  cytochrome *o* of, 171, 251t, 252–260,
    262, 311, 312, 464
  DMSO reductase in, 39
  energy flow in, 390f
  formate dehydrogenase of, 36, 101
  formate hydrogen lyase, 101
  fumarate reductase of, 34, 138
  galactoside carrier, 413f
  genetics, *see* individual proteins
  glucose dehydrogenase (quino-
    protein), 171, 297t, 305, 311
  haemoprotein *b*-590, 266, 276f
  hydroperoxidase, 274, 281, 282t, 283
  lactate transport in, 105
  lactose permease, 406–420
  mixed acid fermentation, 101
  nitrate reductase of, 38, 142–144
  nitrite reduction in, 156t
  periplasmic proteins in, 156t, 157t,
    172
  periplasmic transport in, 387–389,
    394
  potassium transport in, 387, 394,
    422–424
  PTS in, 381
  quinones in, 44
  proline uptake by vesicles, 410f
  proton translocation in, 68f, 258, 259f

500   *Index*

Escherichia coli (cont.)
  reaction with CO, 270f
  respiratory chains of, 60f
  spectra of cytochromes in membranes, 49f
  succinate dehydrogenase of, 34
  superoxide dismutase of, 283
  transhydrogenase of, 28
  TMAO reductase in, 39
Ethanol as substrate for sulphate-reducers, 125t, 129
Ethanol fermentations, 90–92, 101
Exiguobacterium, pH regulation in, 425

F_{420}, 117, 118f, 120, 121f
  in carbon assimilation, 115
F_{430}, 117, 118f, 119
Facilitated diffusion, 378, 379f, 383, 386, 396, 404
FAD, structure, 29f
FCCP, mode of action, 19, 64, 400, 472–474
Fermentation, 89–112
  and ATPase, 20f, 21
  definition, 87
  general comments on importance in nature, 111
  and pmf, 104–105,
    see also pmf
  see also individual fermentations – acetate, butyrate, ethanol, lactate, mixed acid, propionate, succinate
Ferredoxin, 32, 133
  in acetyl-CoA pathway, 108
  in butyrate, butanol, acetone fermentation, 96–98
  in *Clostridium,* 96
  in pyruvate oxidation, 96
  in sulphate reduction, 126, 134
Field-induced potential, 466
Flash photolysis, *see* photolysis
Flavin adenine dinucleotide, *see* FAD
Flavin mononucleotide, *see* FMN
Flavobacterium, polyethylene glycol dehydrogenase, 297t
Flavodoxin, 30, 108, 126
Flavoproteins, redox potentials, 5f

Flavoproteins, introduction, 28–40
Flavoprotein dehydrogenases, 28–38
Flux control coefficients, 456
FMN, structure, 29f
FMN, absorption spectrum, 29f
Folate derivatives in acetyl-CoA pathway, 107f
Formaldehyde, substrate for methanogenesis, 120–123
Formate dehydrogenase, 37
  in acetyl-CoA pathway, 107f, 108
  and F_{420}, 117
Formate oxidation
  and fumarate reduction, 139–141f
  in periplasm, 161t
Formate production, in mixed acid fermentation, 101
Formate-hydrogen lyase, and formate dehydrogenase, 37, 101
Formyl-methanofuran, 117–119
Formyl-tetrahydrofolate synthetase, in SLP reactions, 88t
Formyl-tetrahydromethanopterin, 117–119
Free energy, *see* thermodynamics
Fumarate,
  as sole substrate for anaerobic growth, 138f
  as terminal electron acceptor in anaerobic respiration, 69f, 136–141
Fumarate reductase, 34
  in anaerobic respiration, 138–141
  in propionate and succinate fermentation, 102, 103f
  in sulphide oxidation, 141f, 212f, 216
Fumarate reduction,
  in periplasm, 156t
  and proton translocation, 18

Galactose transport, 383
Galactoside transport, *see* lactose transport
Gallionella, growth on iron salts, 223
Genetics,
  of ATP synthase, 73
  of *a*-type oxidases, 249
  of *d*-type oxidases, 272

Genetics (*cont.*)
  of formate dehydrogenase, 38
  of fumarate reductase, 35
  of hydrogenase, 133
  of hydroperoxidases, 281
  of nitrate reductase, 39
  of o-type oxidases, 260
  of photosynthetic cytochrome bc_1, 351
  of photosynthetic cytochrome c_2, 332
  of potassium transport, 422–424
  of succinate dehydrogenase, 34
  of transport systems,
    group translocation 383–386
    ion-linked systems, 418–420
    periplasmic systems, 387–389, 392–395
Giant mitochrondria, 463
Gibbs free energy, 12t, 434, *see also* thermodynamics
Gluconate transport, 24f
Gluconobacter,
  alcohol dehydrogenase of, 301
  glucose dehydrogenase of, 297t, 305–307
  glycerol dehydrogenase of, 297t
  see also Acetobacter
Glucose oxidation in periplasm, 161t, 305, 312
Glucose dehydrogenase (quinoprotein), 297t, 305–307
  affinity for glucose, 41
  electron transport and energy transduction from, 171, 306, 311, 312t
Glucose phosphate transport, 24f, 397
Glucose transport, 381–386, 390f
Glutathione in regulation of potassium transport and turgor, 423
Glutamate transport, 23, 312
Glyceraldehyde phosphate dehydrogenase, in SLP reactions, 88t
Glycerol, anaerobic growth on, 138, 144
Glycerol dehydrogenase (quinoprotein), 297t, 307
Glycerol phosphate,
  oxidation by electron transport to fumarate, 140f, 141f
  transport, 24f

Glycerol phosphate dehydrogenase, 30, 140f
Glycolysis in fermentation, 90, 91f
Glycine synthase pathway, 109
Gram-positive bacteria,
  cytochrome c in, 176–178
  equivalent of periplasmic proteins?, 176–179
Gramicidin, 400, 401, 425
Green bacteria, electron transfer systems in, 61f
Group translocation, 381–386
  molecular basis of transport, 384
  organization of the PTS, 382
  the PTS and catabolite repression, 384
  the reactions of the PTS, 382
Growth rate *vs.* growth efficiency, 72, 441
Growth yields,
  and ATP/2e quotients (introduction) 71
  during fumarate reduction, 139
  during oxidation of inorganic sulphur compounds, 213, 218
  of sulphate-reducers, 129

Haem,
  in cytochrome a_1, 278
  in cytochrome a_1-like haemoproteins, 275t
  in cytochrome a, aa_3, caa_3 and c_1aa_3, 47f, 50, 236f, 237, 241t, 242–244
  in cytochrome b, 47f
  in cytochrome c, 50
  in cytochrome c_{554}, 201
  in cytochrome co, 50, 251t–252
  in cytochrome d, 47f, 261–263t
  in cytochrome o, 50, 250–252t
  in cytochrome P-460, 199
  in hydroperoxidases, 281, 282t
Haem a, 47f, 236f, 275t
Haem b, 47f. 250, 261, 263t, 282t
Haem c, 47f, 281, 282t
Haem d, 47f, 261, 263t, 280
Haem, covalent attachment to periplasmic cytochromes, 162
  in hydroperoxidases (including catalase), 282t

Haem (*cont.*)
    structures, 47f
Haemoglobin, analogous proteins in
    bacteria, 279
Haemophilus, cytochromes, 256, 271
Haemoprotein *b*-590, 266, 274, 276f
Haemoproteins able to react with
    oxygen, 278–283
    See also cytochromes, cytochrome
        oxidases
Halobacterium,
    pmf and pH, 399f
    'Cytochrome a_1', 275t
Heterolactic fermentation pathway, 93f
Hill diagrams and the proton pump,
    456–459
HIPIP proteins, 32
Histidine transport system, 387, 388f,
    391–394
Homeostasis,
    of pH, 417, 424–426
    of potassium content, 422–424
    properties of regulated transport
        systems, 421f
    and transport systems, 420–426
Homoacetogens, 106–109
Homology between ATP-binding pro-
    teins and transport proteins,
    388
HPr (heat stable protein involved in
    transport), 381–386
Hydrazine, and hydroxylamine oxi-
    dation, 197
Hydrogen oxidation by chemolitho-
    trophs, 186
Hydrogen utilization by sulphate-
    reducers, 125t, 128–133
Hydrogen cycling in sulphate-reducing
    bacteria, 129–133
Hydrogen metabolism in sulphate-
    reducing bacteria, 128–133
Hydrogen oxidation,
    in periplasm, 155t, 161t
    and proton translocation, 69f
Hydrogen-producing acetogens, 109–
    111
Hydrogenase, 31
    and F_{420}, 117, 120
    and hydrogen production in fermen-
        tation pathways, 97–100

Hydrogenase (*cont.*)
    signal peptide of, 159
    in sulphate-reducers, 128–133
    in *Rhodobacter*, 321f
Hydrogenobacter, novel quinones in, 43
Hydropathy profile of lactose permease,
    418f
Hydroperoxidase, cytochrome a_1-like
    protein in, 274t
Hydroperoxidases, 275t, 280–283
Hydroxybenzyl alcohol oxidation in
    periplasm, 155t
Hydroxylamine, as intermediate in
    ammonia oxidation, 196–203
Hydroxylamine dehydrogenase, 171,
    199, 200
Hydroxylamine oxidation to nitrite,
    171, 199–202
Hydroxylamine oxidation in periplasm
    (introduction), 155t
Hydroxylamine oxidoreductase 171,
    199, 196f
    see hydroxylamine dehydrogenase
Hyphomicrobium, periplasmic proteins
    in, 155t, *see also* methanol and
    methylamine oxidation

Inducer exclusion, 386
Inorganic elements and chemolitho-
    trophs, 186–188
Inorganic nitrogen compounds,
    pK values, 195f
    redox potentials, 192f, 195f
Inorganic sulphur compounds,
    oxidation of, 208–219
    pK values, 210f
    redox potentials, 192f, 211f
    see also sulphur
Interspecies hydrogen transfer, 100,
    106, 110, 112, 129–131
Iodoacetate, inhibition of transport,
    405
Ion distribution method for measuring
    pmf, 466–470, 475, 480
Ion-linked transport systems, 396–420
    classes, 396
    energy coupling, 398–403
    experimental systems, 403–406

Ion-linked transport systems (*cont.*)
fundamental tenets, 406–410
thermodynamic considerations, 414–417
molecular mechanisms, 417–420
protonmotive force and kinetics, 410–414
see also individual substrates
Ionophores,
and ion-linked solute transport, 398–405, 425
mode of action, 19, 398–403
Iron, redox potentials, 221–225f
Iron oxidation,
at low pH, 220–223
at neutral pH, 223–225
in periplasm, 69f, 155t, 164, 221–223
and proton translocation, 18, 69f, 164
Iron-sulphur flavoproteins, 32
Iron-sulphur proteins, 30
prosthetic group, 30f
in reaction centres, 325–330, 334
in photosynthetic cytochrome bc_1, 351–356
redox potentials, 5f
Isethionylacetamide, reagent for periplasmic proteins, 154
Isopropanol, fermentation product, 98

Kell, *see* bluff(?)
Ketothiolase, in SLP reactions, 88t
Kinetics, of ion-linked transport systems, 410–414

Lactate,
anaerobic growth of *E. coli* on, 138
substrate for fermentation to propionate, 103f
substrate for nitrate-reducing bacteria, 144
substrate for sulphate-reducing bacteria, 125t, 128–131
Lactate dehydrogenase, 30, 103
Lactate as fermentation product, *Bifidum* pathway, 94f

Lactate (*cont.*)
citrate fermentation, 95f
heterolactic fermentation, 93f
homolactic fermentation, 93
mixed acid fermentation, 101
Lactate transport, 23f, 105, 406
Lactic acid bacteria, 92, 95, 96
Lactobacillus, lactate fermentation, 92
Lactose permease, 397, 406–410, 418–420
driving proton uptake, 407f
effect of membrane potential and pH gradient on, 413, 414f
permease structure, 418f
regulation, 385
and thermodynamic equilibrium, 413f
Lactose transport, 24, 312, 406–420
Langmuir-Blodgett film with reaction centre, 346
Laser, as example of energy-transducing system, 451f–455
Leaks, in chemiosmotic coupling, 458, 473–477, 479
Leaks, in ion-linked transport, 415
Leuconostoc cremoria, fermentation of citrate by, 95f
Light bulb as an energy-transducing system, 451f
Light harvesting complex, 11f, 53, 63, 322
Linear free energy relationships (LFER), 449
Lipid composition and uncouplers, 401
Lipopolysaccharide mutants, 403
Lyase reactions, in fermentation, 88t
Lysine transport, 398, 409

Macroscopic and microscopic descriptions of bioenergetic systems, 422–446
Magnesium, 330, *see* bacteriochlorophyll structure
Malate dehydrogenase, 30
in sulphate-reducers, 134
Maltose transport system, 388f
Malo-lactate fermentation, 96
Manganese oxidation at neutral pH, 223–225

Manganese, redox potentials, 225f
Manganese in superoxide dismutase, 283
Manganese protein, in photosynthesis, 61f
Mannitol transport, 382–384
Mass action effects on kinetics of lactose transport, 411
Melibiose transport, 24f, 380, 419
Membranes (artificial), in study of photosynthesis, 342–347
Membrane,
    diodic nature, 368
    occupancy of sites on, 169
    periplasmic, 153f, 169
    structure in photosynthetic bacteria, 26, 323
Membrane, energy coupling (structure and function), 11, 26
Membrane proteins, 169
    catalysing reactions in periplasm, 161t, 163t
Membrane-associated energy conservation (introduction), 2–77
Membrane potential, 12t, 436–439
    and bacterial growth, 367–370
    dependence on pH, 399f
    effect on galactoside carrier, 413f
    effect on lactose transport, 414f
    generation of artificial potential, 408f
    generation in reaction centres, 332–351
        in artificial membranes, 342–347
        in two steps, 338–342
    and ion-linked transport systems, 398–403, 406–417
    measurement
        by electrochromism, 336–339, 470, 476
        with electrodes, 342–349, 462–466
        by ion distribution, 466–470
        by spectroscopic methods, 470
    problems with magnitude, 461–482
    produced by proton translocation? 472–478
    proline uptake driven by, 410f
    threshold value for ATP synthesis and growth, 363, 368–370, 461
    see also protonmotive force

Menaquinone,
    in fumarate reduction, 140f
    spectrum, 43
    structure, 42
    in reaction centre, 326, 328f
Mercaptoethane sulphonate, 117
    see coenzyme M
Mercaptoheptanoylthrionine phosphate, 118, 119f
Mesohaem, 47
    see haem *c*
Metabolic control theory, 455, 479
Metal corrosion due to chemolithotrophs, 188
Metal-ion oxidation at acid pH, 220–223
Methanobacillus omelianskii, 109, 114
Methanobacterium bryantii, 110, 114
Methanobacterium thermoautotrophicum, 114
    carbon assimilation in, 115
    methanogenesis from acetate, 115
    role of sodium in, 121–123
Methanobrevibacterium, acetate requirement, 114
Methanogenium, growth on propan-2-ol, 114
Methanofuran (CDR), 117, 118f, 119
Methanogenic bacteria, 112
    carbon assimilation in, 115
    chemolithotrophic and methylotrophic metabolism in, 114
    cytochromes in, 114, 121
    energy coupling in, 116–123
        during methanogenesis from carbon dioxide, 119
        during methanogenesis from methanol, 120, 121f
        during methanogenesis from formaldehyde and acetate, 120
        during methanogenesis from formate, 121f
    role of sodium in, 121–123
    in microbial consortia, 135
    unusual coenzymes in, 117–119
    see methanogenesis
Methanogenesis,
    from acetate, 115, 120–123
    from carbon dioxide, 116–120
    from methanol, 120–123
    see methanogenic bacteria

Methanol, methanogenesis from, 120–123

Methanol dehydrogenase, 294, 297t, 298–301
absorption spectrum of, 299
electron transport and energy transduction from, 307–310
interaction with cytochrome *c*, 175, 299–301
mobility in periplasm, 173–176
NAD-dependent, 301
periplasmic location, 155t, 164, 166, 170, 175, 298, 307
prosthetic group, 294–299

Methanol oxidation,
and proton translocation, 17, 164, 166
see methanol dehydrogenase, quinoproteins

Methanopterin, 117–119

Methanosarcina barkeri, 115
carbon assimilation in, 115
co-culture with sulphate-reducers, 129
cytochromes in, 114
methanogenesis from C_1 compounds and acetate, 115f, 120–122

Methanospirillum, growth on propan-2-ol, 114

Methoxatin, 295
see pyrrolo-quinoline quinone

Methoxyindole-2-carboxylic acid, inhibitor of periplasmic transport systems, 391

Methylamine dehydrogenase, 297t, 302–305
electron transport and energy transduction from, 18, 310–311
interaction with amicyanin and cytochromes, 174–176, 303–305, 310–311
mobility in periplasm, 173–176
periplasmic location, 155t 170f, 173–176, 302
prosthetic group of, 302

Methylamine oxidase, 297

Methyl-coenzyme M, 117–120

Methyl-CoM reductase, 118, 119f, 123

Methyl-tetrahydromethanopterin, 117–119

Methylene-tetrahydromethanopterin, 117–119

Methylmalonyl-CoA:pyruvate transcarboxylase, 103, 131, 132f

Methylobacterium AM1 (*Pseudomonas* AM1)
amicyanin, 174, 303–305
a-type oxidase of, 241t, 245, 308
ATP production in, 310, 311
azurin, 174, 304t
interactions of dehydrogenases, cytochromes and cupredoxins, 174, 299–301, 303–305, 307–311
proton translocation in, 310
mutant lacking cytochrome *c*, 179
see also methanol dehydrogenase

Methylomonas J, amicyanin and methylamine oxidation, 304t

Methylophilus methylotrophus,
cytochrome *co* (cytochrome *o*) of, 49f, 251t, 255, 308–310
interaction of MDH and cytochromes in periplasm, 175, 307–310
periplasmic enzymes in, 155t, 175, 309f
proton translocation in, 68f, 308
regulation of oxidases, 239, 255, 308, 309f
respiratory chain of, 60f, 309f
spectrum of membranes of, 49f
see also methanol dehydrogenase

Methylotrophic metabolism in methanogens, 114

Methylotrophs, 3, 297t
see individual organisms, methanol dehydrogenase, methylamine dehydrogenase, periplasmic electron transport proteins

Methyltetrahydrofolate (pteridine) in acetyl-CoA pathway, 107f

Methyl-tetrahydromethanopterin, 118f
in carbon assimilation, 115

Methyltransferase, 120

Micrococcus,
a-type oxidase in, 241t
pmf and pH, 399f

Microelectrodes, for estimating pmf, 462–466

Microscopic and macroscopic descriptions of bioenergetic systems, 442–446
Microscopic reversibility, 443
Mini-electron transport chains in periplasm, 172–176
Mitochondrial cytochrome oxidase, 236–238
Mixathiazol, effect on photosynthesis, 354, 358, 360
Mixed acid fermentation, 100–102
    see also heterolactic fermentation
Modifier protein of methanol dehydrogenase, 299
Molybdoferredoxins, introduction, 32
Molybdoproteins, 35–40, 214
    redox potentials, 5f
Molybdenum cofactor,
    in thiosulphate oxidation, 219, 220
    see also molybdopterin
Molybdopterin, 35, 36f, 214
Monensin, 401, 403
Monooxygenase,
    for ammonia, 197–199
    for methane, 198
Mycobacterium, cytochrome c from, 177

NADH production by reverse proton flow, 24
    in chemolithotrophs, 191
NADH dehydrogenase, 32
    in chemolithotrophs, 189
    in methylotrophs, 309
    in *Acinetobacter,* 312f
N-ethylmaleimide, inhibitor of lactose permease, 419
Nicotinamide nucleotide transhydrogenase, 27
    see transhydrogenase
Nickel in hydrogenase, 132
Nickel tetrapyrrole, 117–119
    see F_{430}
Nigericin, mode of action of, 19, 400f–403
Nitrate, as ligand to cytochrome d, 265
Nitrate reductase, 38
    in anaerobic respiration, 142–144

Nitrate reductase (*cont.*)
    in *Rhodobacter,* 321f, 354
        see also specific organisms
Nitrate reduction,
    in cytoplasm, 169
    in periplasm, 156t, 169
    and proton translocation, 18, 69f
Nitrate respiration, 141–144
Nitrate transport, 143, 205, 207
Nitric oxide, reaction with cytochromes, 46
Nitric oxide reductase, in anaerobic respiration, 145
Nitrification, 195–207
    pK values and redox potentials for, 195f
    role in the nitrogen cycle, 8f
Nitrite,
    as ligand to cytochrome d, 265
    product of ammonia oxidation, 195
    transport, 205, 207
Nitrite dehydrogenase, *see* nitrite oxidoreductase
Nitrite oxidation, 203–207
    coupling to generation of protonmotive force, 17, 69f, 169, 205, 207
    enzymology, 203–205
    location, 205
Nitrite oxidoreductase, 38, 204, 207
    cytochrome a_1-like protein in, 204, 275t
Nitrite reductase, 52
    in anaerobic respiration, 144
    cytochrome cd_1, 280
    independent of cytochrome bc_1 complex, 172
    in *Nitrosomonas,* 202
    periplasmic location, 158, 162, 170f
    rusticyanin and, 46
Nitrite reduction in periplasm, 156t, 162, 163t
Nitrobacter,
    acid corrosion by, 188
    a-type oxidase in, 190t, 204, 207, 241t, 246, 248
    cytochrome a_1-like protein, 275t
    cytochrome c of, 189
    cytochrome oxidase in, 205–207
    electron transport in, 206–207
    nitrite oxidation, 169, 203–207

Nitrobacter (*cont.*)
    nitrite oxidoreductase in, 38, 204,
        275t
Nitrogen compounds (inorganic), pK
    values of, 195f
Nitrogen compounds (inorganic), redox
    potentials of, 192f, 195f
Nitrogen cycle, 8f
Nitrogen fixation,
    and flavodoxin, 30
    role in nitrogen cycle, 8f
Nitrosomonas, 195
    acid corrosion by, 188
    ammonia oxidation to nitrite, 196–
        203
    a-type oxidase in, 190t, 241t, 278
    cytochrome *c*, 157t, 189
    electron transport in, 196f
    periplasmic proteins in, 155t, 157t
Nitrous oxide reductase,
    in *Rhodobacter,* 156t, 321f, 354
Nitrous oxide reduction, in periplasm,
    156t, 166, 168, 170f
Nocardia, methanol dehydrogenase of,
    301
Non-cyclic electron transfer in photo-
    synthesis, 10, 11f, 61f, 62
Non-equilibrium thermodynamics
    (NET), 439–441
Non-ohmic conductance, 412–414f, 458
Non-ohmic leakage, 458, 473–475
Non-thermally activated processes,
    450–455
Novices, *see* tyros

O$_2$-cycle, 18
Oligomycin, inhibitor of ATP synthase,
    75
    see also venturicidin
Oligopeptide transport system, 388f
Organism 4025,
    amicyanin, 157t, 174, 175, 303–305
    azurin, 174, 304t
    cytochromes *c*$_H$ and *c*$_L$ in, 174
    periplasmic proteins in, 155t, 157t,
        174–176, 303–305
    similarity to *Methylophilus,* 303
Ornithine transcarbamoylase, in SLP
    reactions, 88t

Osmoregulation, 422
    see also homeostasis, turgor, potas-
        sium accumulation
Osmotic shock, in study of periplasmic
    proteins, 154, 387
    see also periplasmic transport systems
OSS transport systems, *see* periplasmic
    transport systems
Oxidases, *see* cytochrome oxidases and
    individual cytochromes
Oxides of nitrogen in anaerobic
    respiration, 136, 141–146
2-Oxoglutarate dehydrogenase, in SLP
    reactions, 88t
Oxygen,
    reaction with *a*-type cytochromes,
        237, 246–248, 274
    reaction with *d*-type cytochromes,
        268–272, 280
    reaction with *o*-type cytochromes,
        256–258
    reaction with other haemoproteins,
        279–283
    reduction products, 235
Oxygenated intermediate of cytochrome
    d, 271
Oxygen-pulse techniques, 64–70, 472–
    478

P,324, *see* bacteriochlorophyll special
    pair
P-460, prosthetic group of hydroxyl-
    amine dehydrogenase, 199
Passive transport systems, 378–379
Paracoccus denitrificans,
    amicyanin, 175, 303, 304t
    a-type oxidase of, 240, 241t, 242, 244,
        245, 248, 249, 308
    azurin, 175, 304t
    'Cytochrome *a*$_1$', 276
    cytochrome *c*, 157t, 167, 175, 176
    methanol dehydrogenase and cyto-
        chrome *c,* 176
    methanol oxidation, 167, 298–301,
        308
    methylamine dehydrogenase, 175,
        303
    NADH dehydrogenase of, 33

Paracoccus denitrificans (*cont.*)
  nitrate reductase of, 38, 142–144
  nitric oxide reduction, 145
  nitrite reductase of, 52, 145, 158, 167
  nitrous oxide reduction, 166, 167, 169
  periplasmic proteins of, 155t, 156t, 157t, 166, 303–305
  protonmotive force measurement, 472
  proton translocation in, 68f, 166, 308, 472
  regulation of oxidases, 239
  respiratory chains of, 60f
  sphaeroplast formation, 154
  transhydrogenase of, 28
Passive transport systems, 378, 379t
Patch clamping for measurement of membrane potential, 464
Pediococcus, lactate fermentation in, 93
PEP-dependent transport system (phosphotransferase system or PTS), 379, 381–386
  and catabolite repression, 384–386
  molecular basis, 384
  reactions, 382
  see also group translocation
Pentose oxidation by GDH, 306
Periplasm, 26, 151–160
  criteria used in determining location of proteins in, 152–160
  definition, 152
  equivalent in Gram-positive bacteria, 176–179
  localization of toxic products to, 166, 193
  location in relation to cell envelope, 153f
  mobility of electron transport proteins in, 172–176, 300
Periplasmic electron transport reactions, 151–179
  rationale, 162
  reductase reactions, 166–169
  oxidation reactions, 155t, 161t, 164–169
Periplasmic electron transport proteins,
  in ammonia and hydroxylamine oxidation, 196f, 200
  arranged as mini-electron transfer chains, 172–177

Periplasmic electron (*cont.*)
  chemical modification, 154
  cytochrome c_2 of *Rhodospirillaceae*; a doubt, 179
  in chemolithotrophs, 192, 196f, 200
  equivalent functions in Gram-positive bacteria, 177–179
  definition, 152
  in glucose oxidation, 305–307, 310–312f
  in iron oxidation, 221–223
  in methanol oxidation, 175, 298–301, 307–309f
  in methylamine oxidation, 173–175, 302–305, 310
  preparation, 154
  in sulphate-reducers, 129–132
  in sulphur oxidations, 213, 215–217, 219, 220
  (survey), 157t, 160–162
  translocation into periplasm, 158
  rationale, 162
Periplasmic membrane, 153f
  full occupancy of sites on, 169
Periplasmic (solute) transport systems, 379, 386–395
  ATP as energy source (evidence), 391–393
  component's homology with ATP-binding proteins, 392
  discovery, 387
  energy coupling, 389–391
  genetics of, 387–389
  mechanism, 393
  and solute-translocating ATPases, 394
  structure, 387–389
Peroxidase, 96, 280–283
Peroxide, intermediates of oxidase function, 237
pH,
  effect on fermentation products, 98
  effect on redox potentials, 211f
  homeostasis, 23, 417, 424–426
pH gradient,
  in acidophiles and alkalophiles, 399f
  effect on galactoside carrier, 413f
  in fermenters, 105
  generation of artificial gradients, 408f

pH gradient (*cont.*)
    measurement, 462–470
    TMG uptake in response to artificial
        gradient, 409f
    see also protonmotive force, pro-
        ton gradient
Phase space, 443, 453
Phenanthroline, 346
Phenomenological descriptions of
    bioenergetic systems, 430
Phenomenological non-equilibrium
    thermodynamics, 439–441
Pheophytin, 11f, *see also* bacterio-
    phaeophytin
Phosphate transport, 24, 388f
Phosphoenolpyruvate-dependent trans-
    port system, see PEP-dependent
    transport system
Phosphoglycerate kinase, in SLP
    reactions, 88t
Phosphoketolase,
    in heterolactic fermentation, 93
    in SLP reactions, 88t
Phosphonium cation in measurement of
    membrane potential, 338, 469
Phosphorylation and bulk phase proton
    movement (car analogy), 477f
Phosphorylation driven by artificially-
    induced pmf, 460f, 461
Phosphorylation potential, 14, 20,
    and ATP synthesis, 22, 432
    and solute gradient, 395
Phosphorylation rate and extent and
    pmf, 471
Phosphorylation rate and inhibitor
    concentration, 478f
Phosphorylation rate and respiration
    rate, 458
Phosphotransacetylase, in SLP
    reactions, 88t
Phosphotransferase system of solute
    transport, *see* PEP-dependent
    transport system
Phosphite oxidation, 187
Photobacterium, cytochrome *d* of, 261–
    263t, 265
Photochemical action spectrum,
    of *a*-type oxidases, 247
    of cytochrome a_1 oxidases, 277t
    of thermophile PS3, 247f

Photochemistry, in bacterial photo-
    synthesis, 328–330, 333, 335f
Photodissociation spectrum, 252, 269
    of cytochrome *o*, 252, 253f, 256, 257t
    of cytochrome *d*, 264
Photolithotrophs (photoautotroph), 3t,
    9, 320
Photolysis of CO-liganded oxidases, *see*
    photodissociation
Photoorganotrophs (photohetero-
    trophs), 3t, 9, 320
Photophosphorylation, general, 322f
Photosynthesis, bacterial, 9–11, 318–
    370
    ATP synthesis in, 361–367
    overview in *Rhodobacter* and
        Rhodopseudomonas, 320–324
Photosynthetic cytochrome bc_1
    complex, 335f, 351–361
    measurement of electrogenic activity
        in, 357–360
    mechanism of electron transfer, 352
    protolytic reactions in, 360
    protonmotive Q-cycle, 353–357
    redox poise and pulse technique in,
        353
    structure, 351–353
Photosynthetic electron transfer systems
    (introduction),
    components, 27–56
    examples (introduction), 61f
    introduction, 9–11, 11f, 57–70
    proton translocation by (introduc-
        tion), 67–70
    sequential organization (introduc-
        tion), 57–62
    spatial organization (introduction),
        62
    see also photosynthetic cytochrome
        bc_1 complex, photosynthetic
        reaction centre
Photosynthetic reaction centre, 324–351
    composition of, 325–328t
    electrochromism as tool, 334–342
    electrogenic processes, 328–330, 332–
        351
    electron transfer in and out of it, 332
    energies of electron transfer through,
        335f
    isolation and structure, 56f, 325–328

Photosynthetic reaction centre (*cont.*)
  kinetics of component reactions, 335f
  membrane potentials in artificial
    membranes, 342–349
  orientation of prosthetic groups of,
    328f, 330
  primary photochemistry, 328–330,
    333, 335f
  proton translocation by, 349
  two-electron gate of, 333f
Photosystem I (introduction), 10, 11f
Photosystem II (introduction), 10, 11f,
  56
Phycobiliprotein, 56
pK values,
  for inorganic nitrogen compounds,
    195f
  for inorganic sulphur compounds,
    210
Plastocyanin, 303
  in photosynthesis, 61f
  see cupredoxins
Plastoquinone,
  spectrum, 43
  structure, 42
Polyethylene glycol dehydrogenase,
  (quinoprotein), 297t
Polysulphides, 211
Polythionates, 212
Poly-vinyl alcohol, PQQ-requiring
  dehydrogenase, 296
Porins, 379
Post-translational modification of
  cytochrome *c*, 176
Potassium, accumulation and pH
  homeostasis, 420–424
Potassium, and proline uptake in
  vesicles, 410f
Potassium ATPase, 394, 395
Potassium ionophores in study of trans-
  port, 398, 400–403, 408, 410
Potassium transport, 23, 24f, 394, 395,
  405, 422–424
Potassium transport in *E. coli*, 422–424
PQQ, *see* pyrrolo-quinoline quinone
Primary amine dehydrogenase, 302
  see methylamine dehydrogenase
Primary photochemistry, 328–330, 333,
  335f
Primary transport systems, 379

Proline transport, 23, 406
  energized by membrane potential,
    410f
Propan-2-ol, substrate for methanogens, 114
Propionic acid bacteria, 102, 104, 142
Propionate,
  substrate for acetogens, 110
  substrate for sulphate-reducers, 125t,
    131, 132f
Propionate fermentation, 102–104
Propionate formation and degradation
  in *Db. propionicus*, 132f
Prosthetic groups of photosynthetic
  reaction centre, orientation,
  328f
Protein translocation into periplasm,
  158
Proteus,
  cytochrome a_1, 277t
  fumarate metabolism in, 138f
Protohaem, 47f, 50
  see haem *b*
Proton channels, 15
Proton electrochemical gradient, 12t,
  398
  see protonmotive force
Proton gradient and ion-linked transport, 398–403, 406–417, 423–
  425
Proton leaks, *see* leaks
Proton slips, *see* slips
Proton pump, 168, 432f, 481f
  and Hill diagrams, 456–459
  see protonmotive force, proton trans-
    location
Proton symport, reaction cycle for, 411f
Proton translocation, 64–66, 472–478
  in acetyl-CoA pathway, 109
  in ammonia and hydroxylamine
    oxidation, 202
  and ATP synthase, 75
  and butyryl-CoA oxidation, 111
  driven by lactose transport, 407f
  measurement, 64, 65f, 472–478
  in methylotrophs, 308–310, 311
  in nitrate reduction, 144
  by oxidases, 241t, 248, 258, 272, 308
  in photosynthetic cytochrome bc_1
    complex, 357–361

Proton translocation (*cont.*)
  by respiratory and photosynthetic
    electron transfer chains (intro-
    duction), 16f, 64–70
  and respiratory chain composition,
    66t, 67f
  in thiosulphate oxidation, 220
  see also chemiosmotic mechanism,
    proton gradient, protonmotive
    force
Protonmotive current and bacterial
  growth, 367–370
Protonmotive energy-transducing
  sytems, Chapter 9, 429–482
Protonmotive force, 11–27
  and ATP synthase, 21, 76, 77f, 364
  and ATP synthesis in vesicles, 461–
    462
  delocalized, 459, 478–482
  and electrochemical gradient of
    solute, 402t
  generation, 15–19
    in acetyl-CoA pathway, 109
    in ammonia and hydroxylamine
      oxidation, 202
    by cytochrome oxidases, f, 171,
      248, 258, 272
    in chemolithotrophs, (introduction)
    in glucose oxidation, 171
    in fermentation, 102, 104–5, 111
    in fumarate reduction, 139
    in iron oxidation, 222
    in methanogens, 116–123
    in methylotrophs, 308–311
    in nitrate reduction, 144
    in nitrite oxidation, 205–207
    in periplasmic oxidations and
      reductions, 164–169
    in photosynthesis, 322–325, 328–
      361
    in sulphate-reducers, 128
    in sulphur oxidations, 216, 217,
      220
  and ion-linked transport systems, 22,
    398, 399, 402t, 404–417
  manipulation of, 19
  measurement, 462–470, 475
  and periplasmic transport systems,
    389–391
  pH dependence of components, 399f

Protonmotive force (*cont.*)
  and rate of respiration, 458
  and reversed electron transfer, 24,
    111
  and surface potential, 438f
  threshold value for ATP synthesis,
    363, 368–370, 461, 470, 471
  and transport kinetics, 410–414
  utilization of (introduction), 20
  see also proton translocation
Protonmotive force as intermediate in
  phosphorylation(?), 364–366,
  459–482
  co-reconstituted systems, 480–482
  direct estimation using microelec-
    trodes, 462
  double inhibitor titrations, 478–480
  ion and acid/base distribution
    methods, 466–470, 475
  phosphorylation driven by
    artificially-applied pmf, 461
  relationships between pmf and rates
    and extents of phosphorylation,
    471
  respiration-driven proton transloca-
    tion, 472–478
  spectroscopic methods for deter-
    mining membrane potential,
    334–342, 470
Protonmotive Q-cycle,
  introduction, 16-19, 65, 68f, 70, 194
  in photosynthesis, 353–357
Protonophores, 64
  see ionophores, uncouplers
PS3, see thermophile PS3
Pseudoazurin, 303, 304t
Pseudomonas,
  azurin, 157t, 281
  cytochrome a_1, 277t
  cytochrome cd_1, 280
  cytochrome *o* of, 251t, 258
  cytochrome peroxidase, 282t
  hydroperoxidase, 156t, 281, 282t
  manganese oxidation, 224
  nitrite and nitrous acid reductases of,
    144, 156t, 158, 280
  periplasmic glucose oxidation, 161t
  periplasmic proteins in, 155t, 157t
  polyethylene glycol dehydrogenase,
    297t

Pseudomonas (*cont.*)
    quinoprotein glucose dehydrogenase
        of, 297t, 305–307, 312
    superoxide dismutase in, 283
    transhydrogenase of, 28
Pseudomonas aeruginosa, quinoprotein
        alcohol dehydrogenase of, 297t,
        301
Pseudomonas AM1, *see Methylo-
        bacterium* AM1
Pseudomonas carboxydovorans,
    CO oxidoreductase of, 36
    cytochrome a_1, 278
    respiratory chain of, 37
Pseudomonas testosteroni,
    PQQ growth requirement, 296
    quinohaemoprotein alcohol dehydro-
        genase of, 297t, 301
Pseudomonas TP-1, extraction of PQQ
        from, 295
PTS *see* PEP-dependent transport
        system
Purple bacteria,
    electron transfer systems in (intro-
        duction), 61f
    *see also Rhodospirillaceae, Rhodo-
        bacter, Rhodopseudomonas,
        Rhodospirillum*
Pyridine haemochrome, of haemo-
        protein *b*-590, 276f
Pyrrolo-quinoline quinone (PQQ), 294–
        296
    fluorescence, 295
    natural occurrence and activity as
        growth factor, 296–298
    prosthetic group of glucose dehydro-
        genase, 297t, 306
    prosthetic group of methanol dehyd-
        rogenase, 294–296, 297t
    prosthetic group of methylamine
        dehydrogenase, 297t, 307
    spectrum, 41f, 295
    structure, 41, 295
    as vitamin (growth factor), 294, 296
Pyruvate decarboxylase in fermen-
        tation, 90, 91f, 96
Pyruvate dehydrogenase,
    in ethanol fermentation, 88–92
    in mixed acid fermentations, 101
    in SLP reactions, 88t

Pyruvate dismutation, 96
Pyruvate:ferredoxin oxidoreductase, 96,
        97f
Pyruvate-formate lyase, in SLP
        reactions, 88t, 101
Pyruvate kinase, in SLP reactions, 88t

Q_A, 56, 61f, 70f, 329, 333–335, 339–342,
        344–349, 355f
Q_B, 56, 61f, 70f, 333–335, 339–342,
        344–349, 355f, 356
Q_c, 70f, 354–357, 359–361
Q_i, 68f, 70f, 355f
Q_o, 68f, 70f, 355f
Q_z, 70f, 354–357, 359–361
Q-cycle, *see* protonmotive Q-cycle
Quinones, 42–44
    amount in reaction centres, 326t
    in *Chlorobium*, 43, 61f
    redox potentials, 5f, 44
    spectra, 43f
    see also menaquinone, plastoquinone,
        protonmotive Q-cycle, pyrrolo-
        quinoline quinone, ubiquinone,
        Q, quinoprotein
Quinohaemoprotein alcohol dehydro-
        genase, 301
Quinoproteins, 293–313,
    dehydrogenases, 297t, 208–307
        see individual enzymes—alcohol
            dehydrogenases, methanol
            dehydrogenase, methylamine
            dehydrogenase, glucose dehyd-
            rogenase
    interactions with electron transport
        chains and energy transduction,
        307–312
    introduction, 40–42, 293
    prosthetic group of, 41f, 295f
        see pyrrolo-quinoline quinone,
        294–298
    reconstitution with PQQ, 296, 299,
        306
    redox potentials, 5f, 295f
    why have them?, 312

Reaction centre chlorophyll, 55t
  see also bacteriochlorophyll
Reaction centre (photosynthetic), *see*
      photosynthetic reaction centre
Reconstituted systems, in study of pmf,
      480–482
Redox arm, 15–18, 65–70
  see proton translocation
Redox cycle, 15–18
  see protonmotive Q-cycle
Redox loop, 15–18, 65–70
  see proton translocation
Redox slip, 432f, *see* slip
Redox poise and flash technique, 339,
      349, 353
Redox potential,
  in bacterial redox systems, 5f
  basic concepts, 3
  of a-type oxidases, 246t
  of b-type cytochromes, 254t
  effect of pH on, 211f
  of d-type oxidase of *E. Coli,* 268t
  in chemolithotrophic metabolism,
      192f
  of components in methanogenesis
      from methanol, 121f
  of compounds supporting growth of
      W. succinogenes, 141f
  as inadequate guide to bioenergetics,
      168
  of inorganic nitrogen compounds,
      195f
  of inorganic sulphur compounds,
      211f
  of iron and manganese, 221–225f
  of pyrrolo-quinoline quinone, 295
  of reaction centre components, 353t
  of rusticyanin, 222
    see also individual compounds
Regulation, of pH and turgor, 420–426
Regulation of transport systems, 421f
Regulation, *see* metabolic control
      theory
Respiration, 4
  role in nitrogen and sulphur cycles,
      8f
Respiratory chains, 57–72
  ATP/2e quotients and growth yields
      (introduction), 71
  branched (introduction), 60f

Respiratory chains (*cont.*)
  examples of different types, 7f, 58f,
      60f
  proton translocation by (introduc-
      tion), 64–70
  redox components, 27–56
  regulation (introduction) 57–62
  sequential organization (introduc-
      tion), 57–62
  spatial organization (introduction),
      62, 63f
    see electron transport,
Reverse electron transport, 24, 25f
  in ammonia oxidation, 202, 207
  and ATP synthase, 24
  in chemolithotrophs, 25f, 191
  in hydrogen production from
      butyryl-CoA, 111f
  in photosynthetic bacteria, 61f, 62
  in sulphate-reducers, 131, 134
'Reverse' sulphite reductase, 212f, 216
Rhizobium,
  ATP-binding protein, 392
  cytochrome a_1-like protein, 275t, 279
  PQQ requirement for GDH produc-
      tion, 298
Rhodanese, 212f, 217
Rhodobacter,
  a-type oxidase of, 239–241t
  bacteriochlorophylls of, 55t
  cytochrome bc_1 of, 352f
  cytochrome c_2, doubt over accepted
      role, 179
  cytochromes c, 157t
  cytochrome o in, 251t
  electron transport pathways of, 61f
  energetics of electron transfer, 335f
  mutant lacking cytochrome c_2, 179
  overview of photosynthesis in, 320–
      324
  photosynthetic cytochrome bc_1
      complex, 351–361
  photosynthetic reaction centre of,
      324–351
  nitrate reductase of, 144, 156t, 160,
      171, 321f
  nitric oxide reductase of, 145
  nitrite reductase of, 145, 156t
  nitrous oxide reductase of, 145f, 171,
      321f

Rhodobacter (*cont.*)
  periplasmic proteins in, 156t, 157t
  protonmotive Q-cycle, 70f
  trimethylamine N-oxide reductase in,
      40, 156t, 160, 171, 321f
  variety of metabolism in, 62, 320
Rhodopseudomonas,
  bacteriochlorophyll of, 55t
  cytochrome *c* (modified), 176
  cytochrome *o* of, 251t, 255, 260
  overview of photosynthesis in, 320–
      324
  photosynthetic reaction centre of,
      56f, 322–330, 334, 338, 342,
      346, 348–350
Rhodopseudomonas capsulata, 320, *see*
      Rhodobacter capsulatus
Rhodopseudomonas sphaeroides, 320, *see*
      Rhodobacter sphaeroides
Rhodospirillum, 320
  ATP synthase, 361
  bacteriochlorophyll of, 55t
  reaction centre, 342, 345
  nitrate, nitrite, nitrous oxide reduc-
      tases of, 144, 145
  transhydrogenase of, 28
Ribose, substrate for fermentation, 94
Ribose transport system, 388f
Rieske Fe/S centre, redox potential in
      Rhodobacter, 351, 353t
RPG effect, 119f
Rubredoxin, 32, 108, 126
Ruminococcus, 100, 110
Rusticyanin, 46, 221–223
  periplasmic location, 157t

Saccharomyces, ethanol fermentation, 90
Salmonella,
  group translocation systems in, 383–
      385, 387
  PEP transport, 404
  periplasmic transport systems, 387–
      389
Sarcina, ethanol fermentation in, 90
Secondary solute transport, 22–24, 379
  and pmf, 22, 24f
      see also ion-linked transport
          systems

Selenium,
  in formate dehydrogenase, 37, 108
  in glycine synthase, 109
  in hydrogenase, 132
Semiquinones,
  of flavins, 29f
  of PQQ, 41f, 295f
  of quinones, 43
Shigella, cytochrome *d* of, 261
Signal peptides, 158, 160
Sirohaem, 127
Slips, in chemiosmotic coupling, 458,
      473–475
Slips, in ion-linked transport, 415
SLP reactions, *see* substrate level phos-
      phorylation reactions
'Snuffing' of microbes, 445
Sodium ions,
  role in methanogenesis, 120–123
      see also ionophores
Sodium-linked symports, 397, 405
Sodium-motive force, 23, 122
Sodium-proton antiport, 397, 407
  in alkalophiles, 397, 425
  in methanogens, 122
Solute translocating ATPase, 394
Sodium-translocating ATP synthase,
      33, 426
Sodium-translocating decarboxylase,
      104
Sodium-translocating NADH dehydro-
      genase, 33
Sodium transport, 23, 104, 122
Solute-motive force, 23, 414
Solvent production, 98
Special pair chlorophyll, 329
      see also bacteriochlorophyll
Spectra,
  of cytochromes, 48–50, 233
  of cytochrome a_1, 49f, 277t
  of cytochrome aa_3, 49f, 241t, 243f
  of cytochrome *b*, 254t
  of cytochrome *co*, 49f
  of cytochrome *d*, 49f, 263t
  of cytochrome *o*, 49f, 241t, 253f, 257f
  of haemoprotein *b*-590, 276f
  of flash-photolysed *E. coli*, 270f
  of pyridine haemochrome, 276f
      see individual compounds, *see* photo-
          chemical action spectrum

Spectroscopic methods for estimating membrane potential, 470
 see also electrochromism
Sphaeroplast preparation, 154
Sporolactobacillus, lactate fermentation in, 93
Standard chemical potential, 434, 437f
Stannous salts, oxidation, 186, 220
Staphylococcus,
 cytochrome *o* in, 249
 group translocation in, 384
 lysine uniport, 398
Statistical thermodynamics, 442–446
Stosszahlansatz, 444
Streptococcus,
 fermentation of citrate by, 95f, 96
 lactate fermentation in, 93, 95f, 96
 lactate transport in, 105, 406
 pmf and pH, 399f
 regulation of sugar utilization, 383
 uptake of TMG by, 408, 409f
Substrate level phosphorylation reactions, 88t, 214, 390
Substrate-level phosphorylation during fermentation, 88t
Succinate,
 as electron donor in photosynthesis, 11f, 61f
Succinate dehydrogenase, 33, 140
 in sulphate-reducers, 134
Succinate fermentation, 102–104
Succinate transport, 24f
Succinate thiokinase, in SLP reactions, 88t
Sulfolobus, cytochrome *o* in, 250
Sulphate,
 as electron donor in photosynthesis, 61f
 pK, 210
Sulphate-reducing bacteria, 123–135
 characteristics of, 123–126
 components of redox systems, 126–128
 hydrogen metabolism in, 128–133
 in microbial consortia, 135
 nutrition and carbon flux in, 133
 protonmotive force in, 128
 rubredoxin in, 32
 sulphite dismutation in, 135
Sulphate transport, 128

Sulphide,
 electron donor in photosynthesis, 11f
 chemistry, 210
Sulphide oxidation, 212f, 216
 coupled to fumarate reductase, 141f, 212f, 216
 in periplasm, 155t
 by reversal of sulphite reductase, 216
 to sulphur, 216
Sulphidogens (sulphate-reducing bacteria), 123–135
 in anaerobic consortia, 135
Sulphite chemistry, 208–211
Sulphite dismutation in sulphate-reducing bacteria, 135
'Sulphite oxidase', 212f, 214, 220
Sulphite oxidation, 214–215
 direct oxidation to sulphate, 214
 periplasmic location, 193, 213, 215
 and proton translocation, 17, 69f
Sulphite reductase, 127
 role in sulphide oxidation, 212f, 216
Sulphite reduction,
 and proton translocation, 69f
 see APS reductase
Sulphite transport, 128
Sulphur,
 chemical reaction with sulphite to give thiosulphate, 208, 211, 212
 elemental, 208, 210–212
Sulphur compounds (inorganic),
 pk values, 210f
 redox potentials, 210, 211f
 transport, 213
Sulphur cycle, 8f, 123, 208
Sulphur oxidation, 212f, 215
'Sulphur oxygenase', 212f, 215
Sulphur-reducing archaebacteria, 124, 133
Sulphur reduction
 and formate oxidation, 141f
 by sulphate reducers, 124, 133
Sulphurication, 8f, 123, 208–220
Sulphur transport, 213
Sulphur oxidation, 208–220
 chemistry, 208–211
 microbiology, 208, 209t
 molar growth yields, 213
 pathways, 212
 transport of inorganic sulphur, 213

Sulphur oxidation (*cont.*)
  see also sulphide oxidation, sulphite
      oxidation, thiosulphate
      oxidation
Superoxide dismutases, 96
Surface potential and the pmf, 438
Symport,
  in ion-linked transport systems, 22–
      24, 396, 397, 402t, 406–420
  of protons with lactose, 406–420
  of protons with TMG, 409f
Syntrophism, 110, 114
Syntrophobacter, 110, 114
Syntrophomonas,
  and interspecies hydrogen transfer,
      110
  reverse electron transport in, 111f

TCA cycle
  in nitrate-reducers, 144
  in sulphate-reducers, 133, 134
Tetrahydromethanopterin, 117–119
Tetracycline transport, 415
Tetrathionate, 211f, 219
Thermally-activated reactions, 448
Thermodynamics,
  basic concepts, 3, 12t–15, 433–438
  chemical potentials, 12t, 433–436
  electrical potentials, 12t
  and ion-linked transport systems,
      414–417
  phenomenological non-equilibrium,
      439–441
  statistical, 442–446
Thermophile PS3,
  cytochrome aa_3, 241t
  cytochrome caa_3, 241t, 242–244,
      243f, 246t, 247f, 259f
  cytochrome o of, 251t
  photochemical action spectrum, 247f
Thermus,
  cytochrome aa_3 of, 237, 241, 242,
      245, 246t, 248, 259f
  cytochrome o of, 251t
Thiobacilli, 208–225
  major species (nutritional require-
      ments), 209t

Thiobacillus A2, *see Thiobacillus*
      versutus
Thiobacillus denitrificans, cytochrome d
      in, 265
Thiobacillus ferrooxidans, 209t, 210–225
  acid corrosion by, 188
  acidophilic growth on iron salts, 220–
      223
  cytochrome a_1, 51, 190t, 221f, 275t,
      277t, 278
  electron transport in, 221f
  oxidation of stannous salts, 186
  periplasmic proteins in, 155t, 157t,
      221f
  physiological properties, 209t
  pmf and pH, 399f
  rusticyanin in, 46, 157t
Thiobacillus novellus, 209t
  a-type oxidase of, 241t
Thiobacillus versutus, 209t
  cytochrome aa_3 in, 190t, 246t
  periplasmic proteins in, 155t
Thiocyanate for dissipation of mem-
      brane potential, 406, 472–476
Thiosulphate,
  chemistry, 208–211
  electron donor in photosynthesis, 11f,
      61f
Thiosulphate-combining enzyme, 212f,
      219
Thiosulphate oxidation, 212f, 217–220
  cleavage by rhodanese, 212f, 217
  and proton translocation, 17
  single enzyme oxidation to sulphate,
      219
  role of thiosulphate-combining
      enzyme, 212f, 219
  in periplasm, 155t, 219
Thiosulphate-oxidizing enzyme, 212f,
      219
Thiosulphate reductase, 217
Thiosulphate transport, 128
Thermus, periplasmic cytochrome c,
      157t
Threshold value of membrane poten-
      tial, ATP synthesis and growth,
      363, 368–370, 470
Threshold value of pmf for ATP
      synthesis, 461, 470
TMG transport, 409f, 415

Transcarboxylase, 103, 131, 132f
Transhydrogenase, 27, 66t, 67f, 365, 368
Transition state theory, 447–450
Transport systems, 377–386
  classification, 378–380
  criteria for energy coupling, 380
  see group translocation, periplasmic transport systems, ion-linked transport systems, transport systems and homeostasis, *see also individual substrates*
Transport systems and homeostasis, 420–426
  control of potassium accumulation, 422–424
  control of internal pH, 417, 424–426
Trichlorocarbanilide (ionophore), 400, 403
Trimethylamine N-oxide reduction in periplasm, 156t, 160
Trimethylamine N-oxide reductase, 39, 145, 160
  in anaerobic respiration, 145
  in *Rhodobacter*, 146, 160, 321f, 354
Turgor regulation and potassium transport, 395, 422–424
Tyros, *see* novices

Ubiquinone,
  in photosynthesis, 322f, 326t, 328f, 353–361
  see also Q_A, Q_B, Q_c and Q_z
  spectrum, 43
  structure, 42
unc mutants, 389
Uncouplers, 19, 64, 398–400, 474
  and ammonia oxidation, 203
  and nitrite oxidation, 205
  in study of chemolithotrophs, 203
  in study of photosynthesis, 362, 365
  in study of solute transport, 380, 425
Uniport, 24f, 396–398, 402t
Unusual coenzymes in methanogenesis, 117–119
Uranous salts, oxidation, 220

Valinomycin, 19f, 65f, 400–403, 408
  in methanogens, 117
  in study of proton translocation, 406, 408, 472
  in study of reaction centres, 342
  in study of photosynthetic cytochrome bc_1 complex, 357, 362
  see also ionophores
Variable coupling ion-linked transport, 416
Venturicidin, 364
Vesicles for study of photosynthetic reactions, *see* chromatophores
Vesicles for study of pmf, 459–468
Vesicles for study of transport, 403–405, 408–410
Vibrio alginolyticus,
  NADH dehydrogenase of, 33
  sodium-based energy transduction, 426
Vitreoscilla, 'cytochrome o', 250, 256, 258, 278

Wire-like structure for electron transport, 173
Wolinella succinogenes,
  co-culture with *Ruminococcus*, 100
  formate dehydrogenase of, 37
  formate oxidation, 139–141f, 161t
  fumarate reductase of, 34, 138–141, 216
  hydrogen oxidation, 161t
  hydrogenase of, 31
  periplasmic reactions in, 161t
  redox potentials of growth substrates, 141f
  sulphide oxidation, 141f, 216
  sulphur reduction, 141f

Xylose transport, 380, 419

Y_{ATP}, 72

Zymomonas, ethanol fermentation in, 90